STABLE ADAPTIVE SYSTEMS

KUMPATI S. NARENDRA
Center for Systems Science
Yale University

ANURADHA M. ANNASWAMY
Department of Mechanical Engineering
Massachusetts Institute of Technology

DOVER PUBLICATIONS, INC.
Mineola, New York

Bibliographical Note

This Dover edition, first published in 2005, is an unabridged republication of the work originally published in 1989 by Prentice-Hall, Inc., Englewood Cliffs, New Jersey, as part of the Information and System Sciences series. An errata list (pages 495–496) has been added to the present edition.

Library of Congress Cataloging-in-Publication Data

Narendra, Kumpati S.
Stable adaptive systems / Kumpati S. Narendra, Anuradha M. Annaswamy.
p. cm.
Originally published: Englewood Cliffs, N.J. : Prentice Hall, c1989.
Includes bibliographical references and index.
ISBN 0-486-44226-8 (pbk.)
1. Adaptive control systems. I. Annaswamy, Anuradha M., 1956– II. Title.

TJ217.N37 2005
629.8'36—dc22

2004065741

Manufactured in the United States of America
Dover Publications, Inc., 31 East 2nd Street, Mineola, N.Y. 11501

To the memory of my mentor, Philippe LeCorbeiller,
and
to my graduate students

K.S.N.

To my grandfather, Pandit Narayan Jayaram Sharma

A.M.A.

Contents

6 *Persistent Excitation* 238

7 *Error Models* 273

8 *Robust Adaptive Control* 291

Preface

This is an exciting time to be working in the field of adaptive control. Research in recent years has led to the emergence of a wide spectrum of problems and the field is sufficiently mature to attract theoreticians looking for an area in which nonlinear systems arise naturally. In addition, the adaptive algorithms being studied complement current computing technology resulting in a powerful approach with great potential impact on the world of applications. Sophisticated, yet practical, adaptive controllers are now feasible. For these reasons, it is not surprising that adaptive control has found a large following in all segments of the control community as a whole.

Three decades after the term *adaptation* was introduced into the engineering literature, it is now generally realized that adaptive systems are special classes of nonlinear systems and hence capable of exhibiting behavior not encountered in linear systems. The difficulty experienced in coming up with a universally acceptable definition of adaptive systems, as well as in generating techniques for their analysis and synthesis, may be traced ultimately to this fact. It is well known that design techniques for dynamical systems are closely related to their stability properties. Since necessary and sufficient conditions for the stability of linear systems have been developed over the past century, it is not surprising that well known design methods have been established for such systems. In contrast to this, general methods for the analysis and synthesis of nonlinear systems do not exist since conditions for their stability can be established only on a system by system basis. To design tractable synthesis procedures, adaptive systems are structured in such a fashion that their behavior asymptotically approaches that of linear systems. The central theme of this book is stability of adaptive systems, and it is based on the conviction that adaptive systems can be designed with confidence only when their global stability properties are well understood.

In 1980, the stability problem of an idealized version of an adaptive system was resolved and it has come to represent a landmark in the development of adaptive systems. Following this, in recent years, a multiplicity of ideas have been generated in the field. The flood of new information is so great that the beginner tends to be overwhelmed by the numerous techniques and perspectives adopted. The time appears to be appropriate to attempt a unified presentation of results that are currently well known and to establish the close connections that exist between seemingly independent developments in the field.

The entire book is written in a self-contained fashion and is meant to serve as a text book on adaptive systems at the senior undergraduate or first-year graduate level. A knowledge of linear algebra and differential equations as well as an acquaintance with basic concepts in linear systems theory is assumed. The book can be used for an intensive one-semester course, or a two-semester course with emphasis on recent research in the second semester. The problems included at the end of the chapters should be considered as an integral part of the book. Following the approach used by the authors while teaching this course at Yale, the problems are divided into three categories I, II, and III. Problems in Part I are relatively easy and are meant primarily to test the student's knowledge of mathematical prerequisites and systems concepts. Part II contains problems that can be solved by the application of results derived in the chapters. Problems in Part III are substantially more difficult and occasionally include open questions in adaptive control for which solutions are not currently available.

The book is arranged in such a manner that Chapters 1-7 are accessible to the beginner while Chapters 8-11 are meant for the more advanced, research-oriented student. A fairly extensive first chapter sets the tone for the entire book. Besides introducing basic concepts, it also attempts to trace the evolution of the field to its present state from early approaches that were popular in the 1960s. In particular, an effort has been made to delineate clearly the basis for many of the assumptions that are made in the following chapters which are generally scattered in the technical literature. The authors believe that the importance of these ideas for a broad understanding of the field justifies the unusual length of the introduction. Chapter 2 is devoted to a discussion of results in stability theory with emphasis on those results which are directly relevant to the study of adaptive systems. While Chapters 3-5 deal with the stability properties of adaptive observers and controllers, Chapter 6 introduces the important concept of persistent excitation. In Chapter 7 it is shown that all the systems discussed in Chapters 3-6 can be analyzed in a unified fashion using error models. Chapters 8-10 deal with areas where there has been intense research activity in the last eight years and the final chapter contains five detailed case studies of systems where adaptive control has proved successful.

Developments in the field of adaptive control have proceeded in a parallel fashion in both discrete and continuous systems, and results in one area usually have counterparts in the other. Many of the problems formulated using discrete or continuous time models can also be studied in the presence of stochastic disturbances. This book deals with continuous time finite dimensional deterministic systems. We believe that a thorough understanding of this class will provide the essential foundation for establishing analogous results in both the discrete time and stochastic cases.

It is our privilege to thank a long list of friends and colleagues who have helped us in the preparation of the book in different ways. We are especially grateful to Petar Kokotovic for his careful scrutiny of the first seven chapters and numerous valuable suggestions. We would also like to thank Peter Dorato, Petros Ioannou, Robert Kosut, Gerhard Kreisselmeier, Steve Morse, and R.P. Singh for their comments on portions of the manuscript and to Job van Amerongen, Dan Koditschek, Heinz Unbehauen, and Eric Ydstie for critically evaluating parts of Chapter 11. Manuel Duarte and Jovan Boskovic took part in numerous technical discussions and Jovan's help in collecting the material for Chapter 11 is particularly appreciated. Both of them worked very hard toward proofreading the final manuscript. We are most grateful to them for their commitment to the completion of the book. Finally, we would like to thank Eileen Bernadette Moran for encouraging us to do this project with Prentice Hall, Tim Bozik, engineering editor, for his enthusiasm and help, and Sophie Papanikolaou, production editor, for her remarkable efficiency.

The first author would like to thank his doctoral students in the area of adaptive control during the period 1970-88. Many of the ideas in this book first appeared in papers that they coauthored with him. The time spent with them was instructive, productive, and thoroughly enjoyable.

The second author gratefully acknowledges the IBM Corporation for awarding a postdoctoral fellowship during her stay at Yale from 1985-87. This enabled her to devote her time fully to the book. She would also like to thank Rose Schulz for her hospitality during the entire time the book was in preparation.

Finally, without the patience and encouragement of Barbara Narendra and Mandayam Srinivasan, this book would not have been undertaken, nor completed.

Kumpati S. Narendra

Anuradha M. Annaswamy

1

Introduction

1.1 INTRODUCTION

Questions of control in general and automatic control in particular, are assuming major importance in modern society even as social and technological systems are becoming increasingly complex and highly interconnected. By "control of a process" we mean, qualitatively, the ability to direct, alter, or improve its behavior, and a control system is one in which some quantities of interest are maintained more or less accurately around a prescribed value. Control becomes truly automatic when systems are made to be self-regulating. This is brought about by the simple concept of *feedback* which is one of the fundamental ideas of engineering. The essence of the concept consists of the triad: measurement, comparison, and correction. By measuring the quantity of interest, comparing it with the desired value, and using the error to correct the process, the familiar chain of cause and effect in the process is converted into a closed loop of interdependent events. This closed sequence of information transmission, referred to as feedback, underlies the entire technology of automatic control based on self-regulation. Although the existence of self-regulating mechanisms in living organisms has been recognized for a long time and the deliberate construction of self-regulating systems, such as float regulators and water clocks can be traced back to antiquity [30], the abstract concept of the closed causal loop is a distinct achievement of the twentieth century.

Up to the beginning of the twentieth century, automatic control remained a specialty of mechanical engineering. This was followed by a period when electrical regulators and controllers became quite common. About the 1940s, electrical, mechanical, and chemical engineers were designing automatic control devices in their respective fields using very similar methods arrived at by different routes and disguised under completely different terminologies. Although at first no connection between these developments was recognized, it gradually became clear that the concepts had a common basis, and at the end of World War II, a theory that was mathematically elegant and universal in its scope came into being. In 1948, Wiener named this newly founded discipline *Cybernetics* [43]. In the last forty years, feedback control has evolved from an art into a scientific discipline which cuts across boundaries extending from design, development, and production on one hand, to mathematics on the other. In fact, even about the early 1960s, Bellman [6] felt that, having spent its fledgling years in the shade of the engineering world, control theory had emerged as a mathematical discipline that could exist independent of its applications.

The history of automatic control has witnessed a constant striving toward increased speed and accuracy. World War II, with its need for fast and accurate military systems, imparted a large impetus to the growth of the field. Frequency response methods were developed based on the efforts of Black, Nyquist, and Bode in the design of electronic feedback amplifiers. Using these methods, which are now classified under the rubric of classical control, it was possible to carry out both analysis and synthesis of closed loop systems in a systematic fashion based on open-loop frequency responses. In the course of time, these methods formed the foundations of feedback control theory and became ideally suited for the design of linear time-invariant systems. In the 1950s and 1960s, with developments in space technology, the feedback control problem grew more complex. Stringent requirements of accuracy, weight, and cost of space applications spurred the growth of new techniques for the design of optimal control systems. Models with more than one dependent variable were common occurrences and ushered in the era of multivariable control. Finally, the inevitable presence of noise in both input and output variables called for statistical solutions for the estimation and control problems, and the field witnessed a merging of control techniques with those well established in communications theory. The solution of the linear quadratic gaussian (LQG) regulator problem using the separation principle and the development of the Kalman filter became landmarks in the field in the 1960s.

Even as greater efforts were made in the direction of precise control, linear models were often found to be no longer valid and more accurate descriptions of the processes were necessary. Simple models of the process had to be replaced by more complex ones and uncertainties regarding inputs, parameter values, and structure of the system increasingly entered the picture. Their importance in the design of fast and accurate controllers shifted attention in automatic control to new areas such as adaptive, self-optimizing, and self-organizing systems.

1.2 SYSTEMS THEORY

A *system* may be broadly defined as an aggregation of objects united by some form of interaction or interdependence. When one or more aspects of the system change with time, it is generally referred to as a *dynamical* system. The first step in the analysis of any system is to establish the quantities of interest and how they are interrelated.

The principal concern of systems theory is in the behavior of systems deduced from the properties of subsystems or elements of which they are composed and their interaction. Influences that originate outside the system and act on it so they are not directly affected by what happens in the system are called *inputs*. The quantities of interest that are affected by the action of these external influences are called *outputs* of the system. As a mathematical concept, a dynamical system can be considered a structure that receives an input $u(t)$ at each time instant t where t belongs to a time set $\mathcal{T}$ and emits an output $y(t)$. The values of the input are assumed to belong to some set $\mathcal{U}$ while those of the output belong to a set $\mathcal{Y}$. In most cases the output $y(t)$ depends not only on $u(t)$ but also on the past history of the inputs and hence that of the system. The concept of the *state* was introduced to predict the future behavior of the system based on the input from an initial time t_0.

The dynamical systems we will be concerned with in this book are described by ordinary vector differential equations of the form

$$\begin{aligned} \frac{dx(t)}{dt} \triangleq \dot{x}(t) &= f(x(t), u(t), \theta, t) \qquad t \in \mathbb{R}^+ \\ y(t) &= h(x(t), \theta, t) \end{aligned} \tag{1.1}$$

where

$$\begin{aligned} x(t) &\triangleq [x_1(t), \ldots, x_n(t)]^T \in \mathbb{R}^n, \quad \theta \triangleq [\theta_1, \ldots, \theta_r]^T \in \mathbb{R}^r, \\ u(t) &\triangleq [u_1(t), \ldots, u_p(t)]^T \in \mathbb{R}^p, \quad y(t) \triangleq [y_1(t), \cdots, y_m(t)]^T \in \mathbb{R}^m, \end{aligned}$$

f and h are mappings defined as $f : \mathbb{R}^n \times \mathbb{R}^p \times \mathbb{R}^r \times \mathbb{R}^+ \to \mathbb{R}^n$ and $h : \mathbb{R}^n \times \mathbb{R}^r \times \mathbb{R}^+ \to \mathbb{R}^m$. The vector u is the input to the dynamical system and contains both elements that are under the control of the designer and those that are not. The former are referred to as *control inputs*. The vector $x(t)$ denotes the *state* of the system at time t and its elements $x_i(t)(i = 1, 2, \ldots, n)$ are called *state-variables*. The state $x(t)$ at time t is determined by the state $x(t_0)$ at any time $t_0 < t$ and the input u defined over the interval $[t_0, t)$. The output $y(t)$ as defined by Eq. (1.1) is determined by the time t as well as the state of the system $x(t)$ at time t.

The state and output equations in Eq. (1.1), defining a given process, may be considered as an abstract summary of the data obtained by subjecting the process to different inputs and observing the corresponding outputs. Equation (1.1) is generally referred to as the mathematical model of the process. Once such a model is available, the emphasis shifts to the determination of a control function u which achieves the desired behavior of the process. Many of the major developments in control theory during the past two decades are related to this problem.

When the mappings f and h are linear, the system is said to be linear and may be represented in the form[1]

$$\begin{aligned} \dot{x} &= A(\theta, t)x + B(\theta, t)u \\ y &= H(\theta, t)x \end{aligned} \tag{1.2}$$

[1]The functional dependence of variables on t is sometimes suppressed for simplicity of notation.

where $A(\theta,t), B(\theta,t)$, and $H(\theta,t)$ are $(n \times n), (n \times p)$ and $(m \times n)$ matrices of bounded, piecewise-continuous, time functions. In the case of systems whose characteristics do not vary with time, $A(\theta), B(\theta)$, and $H(\theta)$ are constant matrices and the equations may be expressed in the form

$$\begin{aligned} \dot{x} &= A(\theta)x + B(\theta)u \\ y &= H(\theta)x. \end{aligned} \tag{1.3}$$

The constant vector θ in Eqs. (1.1)-(1.3) is referred to as the *parameter vector*. According to Webster's dictionary, a parameter is a constant in any experiment but whose value may vary from experiment to experiment. The elements of θ include both parameters that are associated with the process to be controlled and those that can be chosen at the discretion of the designer. The former are generally called system parameters, while the latter are referred to as control parameters. By an abuse of the language, when θ varies with time, it is also referred to as a time-varying parameter. Both system and control parameters can be either constants or time-varying parameters.

In the type of problems we will be concerned with, a dynamical system is specified in which some uncertainty exists and control action has to be taken to achieve satisfactory performance. The importance of classifying the variables into inputs, parameters, and state-variables becomes immediately evident when definitions related to adaptation or concepts derived from it are discussed. For example, some of the parameters of the given system may not be constant as originally assumed in describing the system but may vary with time. If θ_i, the ith element of the vector θ is such a parameter, it is no longer possible to decide whether θ_i is an input or a time-varying parameter. Analytical expediency may determine how θ_i is categorized. For example, if the system is described by Eq. (1.2) or (1.3), it is more convenient to consider θ_i as a time-varying parameter so that powerful linear techniques can be applied to determine the effects of u on y. We will more often be interested in situations in which the parameter θ_i is adjusted continuously as a function of some of the observed variables of the system. For example, if the time derivative of θ_i is adjusted on the basis of observed outputs as

$$\dot{\theta}_i = h_{m+1}(y_1, y_2, \ldots, y_m, t) \stackrel{\triangle}{=} f_{n+1}(x_1, x_2, \ldots, x_n, t)$$

then θ_i is no longer a parameter but a state variable of the system so that properties such as stability, controllability, observability, and optimality have to be considered in an expanded state space of dimension $n+1$. Note that, in such a case, even if the original equations describing the system are linear and have the form of Eq. (1.2) or (1.3), the modified equations are nonlinear.

By control of a system such as in Eq. (1.1), we mean the attempts made to keep its relevant outputs y_i within prescribed limits. Control is generally effected by either changing the control inputs for a specified structure given by f and the values of the parameter θ or by adjusting the control parameters for specified inputs u. Systems theory deals with the analysis and synthesis of such systems. The best developed aspect of this theory treats linear systems of the form described in Eqs. (1.2) or (1.3). Although well established techniques in the time domain apply to the general class of linear time-varying systems (LTV) [Eq. (1.2)], the powerful frequency domain methods of classical control theory apply to linear time-invariant systems (LTI) [Eq. (1.3)]. In both cases, given a mathematical model of a fixed part of the system (referred to hereafter as the *plant*), the designer can, using linear systems theory, design linear feedforward and feedback controllers so that the outputs of the plant behave in some desired fashion. While similar controllers can also be designed for nonlinear systems, no coherent theory of control design exists at present for any reasonable class of nonlinear systems.

The theory of optimal control developed in the late 1950s enables the desired behavior of systems to be quantified by defining a suitable performance index which incorporates all the desired qualitative features. A typical performance index is given by

$$J = F[x(T), T] + \int_0^T L[x(t), u(t), t]\, dt,$$

where $L(.,.)$ is a convex function of the state and control variables and F is a convex function of the state of the system at the terminal time T. The maximum principle of Pontryagin and his coworkers and the functional equation of Bellman represent landmarks in this theory using which the control input that optimizes J can be determined theoretically subject to various constraints. The development of an optimal regulator for a linear system based on such a theory has become a standard design tool. In such a problem, a quadratic performance index of the form

$$J = x^T(T)Fx(T) + \int_0^T [x^T(t)Q(t)x(t) + u^T(t)R(t)u(t)]\, dt \tag{1.4}$$

where $R(t)$ is a positive-definite matrix and $Q(t)$ and F are positive-semidefinite matrices, is minimized by the proper choice of the control input u.

If measurement and control uncertainties are present in the form of noise signals, J is no longer deterministic, but the same mathematical tools can still be used to optimize its expected value. However, from a practical standpoint, the approach has proved effective mainly in the case of linear feedback systems with quadratic performance indices in the state and control variables and with additive Gaussian noise. The optimization of nonlinear systems is usually carried out in practice by successive linearization and optimization of a quadratic performance index at every stage.

In all the cases discussed above, the mathematical description of the system as well as the characteristics of the noise are assumed to be completely known. This is expressed in terms of relevant transfer functions, system matrices, and statistics of the noise.

1.3 ADAPTIVE SYSTEMS

In contrast to the type of control described in Section 1.2, adaptive control refers to the control of partially known systems. While controlling a real process, the designer rarely knows its parameters accurately. The characteristics of the process can change with time due to a variety of factors. There may also be unforeseen changes in the statistics of the external inputs and disturbances that may be considered as changes in the environment in which the system operates. The tools of conventional control theory, even when used efficiently in the design of controllers for such systems, may be inadequate to achieve satisfactory performance in the entire range over which the characteristics of the system may vary. The term *adaptation* is defined in biology as "an advantageous conformation of an organism to changes in its environment." In 1957, inspired by this definition, Drenick and Shahbender [14] introduced the term *adaptive system* in control theory to represent control systems that monitor their own performance and adjust their parameters in the direction of better performance.

Discussing the paper above, McCausland pointed out the difficulty in distinguishing between a conventional feedback system and an adaptive system as described in [14], since the former may also be considered to monitor its performance and adjust something in the system to improve it. He also noted that the biological use of the term lacks precision, since many systems described to be adaptive are not essentially different from conventional control systems. As anticipated, the years following witnessed considerable debate on these issues. By monitoring different system characteristics and taking different control actions, a large number of systems were designed. This resulted in a profusion of definitions, each containing some property that its proponent considered peculiar to adaptive systems. These definitions were collected and presented in the survey papers of Aseltine *et al.* [2] and Stromer [38]. We discuss some of them briefly in this section.

In [2] an attempt was made to categorize adaptive systems into the following five classes depending on the manner in which the adaptive behavior was achieved: (i) Passive Adaptation, (ii) Input-Signal Adaptation, (iii) System-Variable Adaptation, (iv) System-Characteristic Adaptation, and (v) Extremum Adaptation. Although these classes are not commonly used at the present time, they are nevertheless of historical interest. We include them here not only to acquaint the reader with the developments in the early days of adaptive systems, but also because successive generations of workers in adaptive control as well as related fields have come up with remarkably similar ideas.

The first class contains those systems in which the clever design of a time-invariant controller results in satisfactory performance over wide variations in environment. Typical examples of such systems are described in [25] and [27]. Consider a linear time-invariant system with a transfer function G. Let the input to the system be r and let the output contain additive disturbance d as shown in Fig. 1.1(a)-(c). The conditional feedback system of Lang and Ham in [25] permits the design requirements on input-output and disturbance-output characteristics to be considered independently by the proper choice of the controller structure. It is well known that such a separation between input and disturbance can be achieved [40] using either a single feedback loop [Fig. 1.1(a)] or multiple feedback loops [Fig. 1.1(b)]. Although all the three feedback loops achieve an identical objective, the configuration suggested in [25] [Fig. 1.1(c)] may, in many cases, be simpler to realize. Another example of passive adaptation, also suggested in the early days of adaptive control, is given in [27] where a nonlinear rate feedback is used to improve the damping characteristics of a second-order servomechanism (Fig. 1.2). The overall system in this case can be described by a nonlinear scalar differential equation

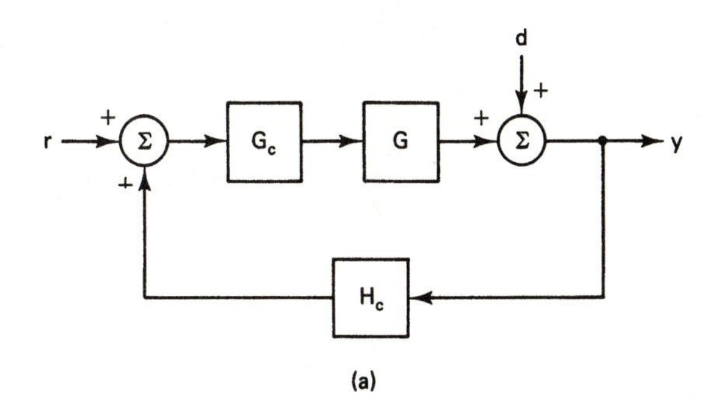

(a)

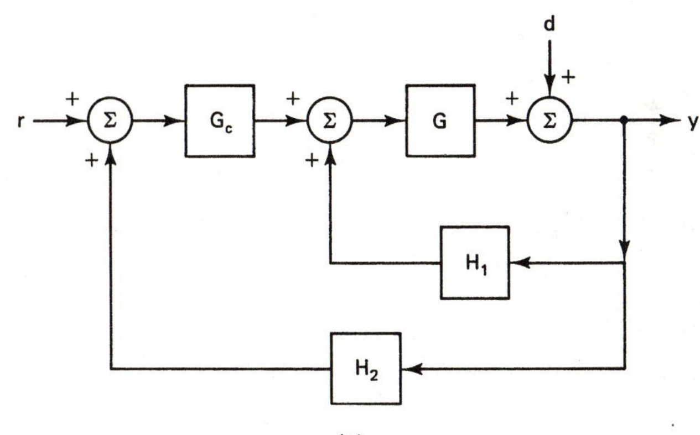

(b)

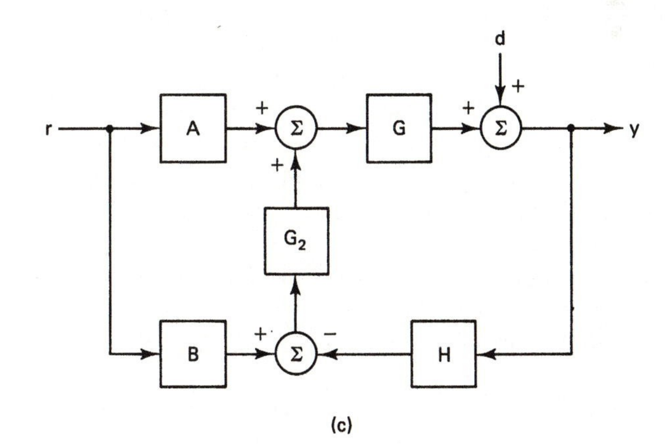

(c)

Figure 1.1 Conditional feedback system. Fig. 1.1(c) ©1955 IEEE.

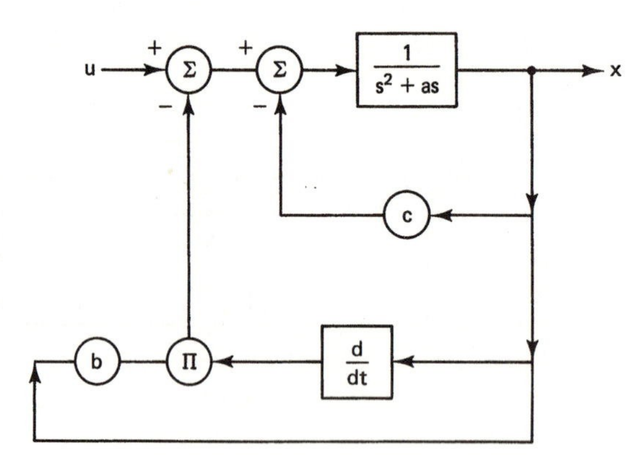

Figure 1.2 A nonlinear feedback system.

of the form

$$\ddot{x} + (a + bx)\dot{x} + cx = u.$$

Although the system considered in [25] is linear and that in [27] is nonlinear, note that the aim in both cases is to determine a controller of a fixed structure and known parameters to improve performance. Innumerable examples of such systems exist in industry. According to present usage, such controllers would be described as robust rather than adaptive.

Class (ii) includes systems that adjust their parameters in accordance with input signal characteristics. The assumption here is that the performance of the overall system depends on some characteristic of the signal and that the optimal parameter values can be determined once this characteristic is known. Automatic gain control systems can be considered to be members of this class. In these systems, the amplitude of the input may vary significantly and a gain parameter is adjusted to maintain the average output amplitude constant. Other features of the input such as autocorrelation function, spectral density, and so on also can be used to classify the input signal. Assuming that optimal parameter values have been computed off-line for each class of input signals, they can be chosen on-line by referring to a look-up table. The latter method is generally referred to as *gain scheduling* in control theory. When the number of classes of inputs to be recognized is reasonably small, such a procedure may prove effective. Gain scheduling has been employed successfully in a variety of applications including process control and the control of aircraft systems, and may be considered as a simple form of adaptive control.

Classes (iii) and (iv) contain systems in which the control input is adjusted based on measurements on the system while it is in operation. In Class (iii), the adjustment is based on system variables such as system outputs, errors or their derivatives, which can be measured. Class (iv) consists of those systems in which the adjustment of the control parameters or the control input depends on system characteristics, as for example, its impulse response or frequency response. For example, from the measured impulse response of a given plant, in [1] the damping ratio is computed and used to generate a feedback control signal.

Adaptive systems that attempt to seek the extremum of a performance criterion are members of Class (v). Although an argument can be made that Class (v) is a proper subset of Class (iii), it is included as a separate class because during the 1960s, these systems received more attention than the others. Furthermore, as shown in Section 1.4.1, they are quite robust and can be generally designed with very little prior information regarding the plant to be controlled. Hence, they find wide application even today in different areas where standard design methods are not applicable. In view of their importance, we deal with them in more detail in the section following.

Implicit in the classification above is the presence of unforeseen and uncertain changes in the system and its environment. At the time these papers were written, the state vector representation of a dynamical system, as given by Eqs. (1.1)-(1.3), was not well established in the control literature in the United States. Hence, it is not surprising that adaptation is described in them only in terms of system inputs, outputs, and characteristics such as impulse and frequency responses. In present day terminology, such classifications would be expressed in terms of the state of the system, its parameters, or the performance index. With the exception of Class (v), the others were not generally accepted as distinct adaptive classes by the control community. This may be attributed partly to the fact that although systems belonging to these categories revealed the ingenuity of the designer in dealing with some aspect of the control problem, they were nevertheless limited in their scope and did not lead to general approaches for the design of adaptive systems.

1.3.1 What Is an Adaptive System?

More than two decades after the survey papers [2], [38], and [20] were written, the definition of an adaptive system continues to be multifaceted and cannot be compressed into a simple statement without loss of vital content. We shall not attempt an exhaustive discussion of the many definitions proposed over the years but will merely examine four representative samples culled from the literature to provide the reader with a better appreciation of the difficulties encountered and the diversity of viewpoints.

Definition 1.1.1 An adaptive system is a system which is provided with a means of continuously monitoring its own performance in relation to a given figure of merit or optimal condition and a means of modifying its own parameters by a closed-loop action so as to approach this optimum [17].

Definition 1.1.2 An adaptive control system is defined as a feedback control system intelligent enough to adjust its characteristics in a changing environment so as to operate in an optimum manner according to some specified criterion [29].

During the 1960s, extremum seeking controllers such as those described in Class (v) in Section 1.3 were the subjects of extensive research in the adaptive control area. Definitions 1.1.1 and 1.1.2 are typical of a number of definitions of adaptive systems proposed at this time. In both cases, the figure of merit is the important output variable, and the system adjusts itself to optimize it. It is not obvious immediately from Definition 1.1.2 why feedback is germane to an adaptive system.

Definition 1.2 In 1959, Bellman and Kalaba introduced the term *adaptive* in the context of multistage decision processes in the absence of complete information [7]. They defined adaptive control processes as belonging to the last of a series of three stages in the evolution of control processes. When the process to be controlled is fully specified and the controller has complete information concerning the behavior of the inputs, they referred to the process as a *deterministic control process.* In Eq. (1.1) this corresponds to the case where u is a control vector and the designer has complete knowledge of the functions $f_i(\cdot)$ as well as the parameter vector θ. When unknown factors are present in the process which appear mathematically as random variables with known distribution functions, the process is referred to as a *stochastic control process.* This corresponds to Eq. (1.1) where some of the inputs are random processes or when some of the parameters are unknown with known distributions. The third stage is when even less information about the process is available. As pointed out by Bellman and Kalaba, such situations are ubiquitous in problems in engineering, economics, biology, and operations research. In all these cases, the entire set of admissible decisions, the effect of the decisions on the process, or the duration of the process itself may not be known. In such situations, the controller has to learn to improve its performance through the observation of the outputs of the process. As the process unfolds, additional information becomes available and improved decisions become possible. In [7], this is defined as an *adaptive control process.*

The definition, as given above, probably has the widest degree of acceptance among control theorists today. The fact that it is not as specific as the earlier definitions partly accounts for this.

A general definition of adaptation, which is fairly precise and yet flexible, was given by Zadeh [46]. According to Zadeh, the difficulty in defining the notion of adaptivity is due to a lack of a clear differentiation between the external manifestations of adaptive behavior and the internal mechanism by which it is achieved. In [46], Zadeh concentrated on the former and expressed adaptation in mathematical terms: Let the performance of a system A be denoted by P and let $\mathcal{W}$ denote the set of acceptable performances. Let $\{S_\gamma\}$ denote a family of time functions, indexed by γ, to which the system A is subjected. If the resultant performance is denoted by P_γ, then the adaptive behavior of A is defined in [46] as follows.

Definition 1.3 The system A is adaptive with respect to $\{S_\gamma\}$ and $\mathcal{W}$ if it performs acceptably well with every source in the family $\{S_\gamma\}$, $\gamma \in \Gamma$, that is, $P_\gamma \in \mathcal{W}$. More compactly, A is adaptive with respect to Γ and $\mathcal{W}$ if it maps Γ into $\mathcal{W}$.

According to the definition above, all systems are adaptive with respect to some Γ and $\mathcal{W}$. An open loop system can be considered to be adaptive if Γ consists of a single element and the tolerance (denoted by the set $\mathcal{W}$) is sufficiently large. It is well known that if a linear time-invariant feedback system is asymptotically stable for some nominal value γ_0 of a parameter γ, then it is also asymptotically stable in some neighborhood of γ_0. This, in a sense, answers the questions raised in the late 1950s as to whether a feedback system is also an adaptive system. The term *adaptive* used in Definition 1.3 is thus seen to resemble quite closely the term *robust* as used in the present control literature [47]. The question then is to determine whether all adaptive controllers would be *more* adaptive than a conventional robust controller. Mathematically this could be posed as follows: If A is an adaptive controller and A^1 is a robust controller and the two are adaptive with respect to Γ and $\mathcal{W}$ and Γ^1 and $\mathcal{W}$ respectively, is Γ^1 a proper subset of Γ? Since Definition 1.3 is not concerned with the internal mechanism by which the control is realized, the answer to such a question is beyond its scope.

Definition 1.4 In 1963 Truxal [39] defined an adaptive system as one that is designed from an adaptive viewpoint. While, at first glance, this may seem to have been said tongue in cheek, closer examination reveals that it captures much of the difficulty encountered in defining adaptive systems. A designer who sets out to control a system in the presence of a certain class of uncertainty may produce a controller that achieves its objective. As far as the designer is concerned, he or she has successfully designed an adaptive system. However, an observer unfamiliar with the status of the process prior to the design of the controller merely observes a complex feedback system that performs satisfactorily; the person, in turn, may attempt to control the system adaptively to take into account a different class of uncertainty. Hence, at any stage, adaptation, like beauty, is only in the eye of the designer.

The definitions discussed thus far reveal the variety of personal visions of adaptation encountered in the field. The seeming divergence between these different viewpoints may be partly explained by the current status of control theory. As mentioned earlier, although linear systems theory is well established, no corresponding general theory is available for the analysis and synthesis of nonlinear systems, which usually have to be carried out on a system-by-system basis. The frontiers of nonlinear systems theory will therefore recede only as analysis and synthesis methods for specific classes of nonlinear systems are developed. Establishing design procedures for such classes invariably opens the doors to unchartered territory.

This fact has an important bearing on our earlier discussions, particularly if one bears in mind that every adaptive system, no matter how complex, is merely a feedback system involving estimation and control. The complexity of adaptation increases as a parameter in an existing system is adjusted to cope with a new uncertainty and hence becomes a state variable at the next level. This agrees with Zadeh's definition, by which any system can be defined to be adaptive by the proper choice of the uncertainty and the acceptable level of performance; successively more sophisticated adaptive procedures are called for as the set $\{S_\gamma\}$ is enlarged or the set $\mathcal{W}$ of acceptable performance is restricted. As for Truxal's definition, it merely affirms the fact that adaptation is an evolving process and at every stage only the designer is aware of the transition of a given system from an existing state to an adaptive state without being concerned with the prior evolution of the system. Bellman and Kalaba's definition may be interpreted to imply that situations for which design procedures exist belong to either deterministic or stochastic processes, while adaptive processes correspond to the unexplored territory ahead. The boundary between adaptive and nonadaptive processes has to be advancing continuously according to this viewpoint, even as the adaptive control problems of today become the established paradigms for the design of nonlinear systems tomorrow.

In this book, attention is focused on a special class of nonlinear systems that we define as being adaptive. Such systems arise while attempting to control linear plants with unknown parameters. The overall control system is nonlinear, but its behavior asymptotically approaches that of a linear time-invariant system. In the chapters following, a general methodology is developed for the systematic design of such adaptive controllers and for the analysis of their performance under different perturbations.

1.3.2 Parameter Adaptive and Structurally Adaptive Systems

If a dynamical system is described by Eq. (1.1), uncertainty may arise due to a lack of knowledge of some of the system or noise parameters that are elements of a parameter vector or some of the functions f_i. In both cases, other elements of the parameter vector or the functions f_i, may be varied appropriately to achieve the desired control. The former are said to be parameter adaptive, while the latter are said to be structurally adaptive. Switching systems, where structural changes can take place due to modifications in the interconnections between subsystems, belong to the latter class. While parameter adaptive systems can also result in structural changes, as for example when certain parameters assume zero values, we shall ignore such special cases and be concerned entirely with parametric adaptation.

Assuming that the vector is comprised of two sub-vectors p and θ, where the elements of p are either unknown and constant or vary with time in an independent fashion, and the elements of θ represent the parameters that are under the control of the designer, the differential equations (1.1)-(1.3) can be rewritten as

$$\begin{aligned} \dot{x} &= f(x, u, p, \theta, t) \\ y &= h(x, p, \theta, t) \end{aligned} \tag{1.5a}$$

$$\begin{aligned} \dot{x} &= A(p, \theta, t)x + B(p, \theta, t)u \\ y &= H(p, \theta, t)x \end{aligned} \tag{1.5b}$$

$$\begin{aligned} \dot{x} &= A(p, \theta)x + B(p, \theta)u \\ y &= H(p, \theta)x \end{aligned} \tag{1.5c}$$

Parametric adaptation then corresponds to the adjustment of the control parameter vector θ to compensate for the unknown parameter p. It may therefore be stated entirely in terms of the two sets of parameters $\{p\}$ and $\{\theta\}$. If $\{p\}$ is constant and unknown, the problem can be considered to be one of adaptive regulation; if the elements of $\{p\}$ are time-varying, the elements of $\{\theta\}$ may need to vary with time as well.

The parameter adaptation problem stated above is still too general for developing systematic analysis and synthesis procedures. The class of systems must be restricted substantially for this purpose. Such a class is given by Eq. (1.5b) where the differential equations governing the plant are linear and the parameters are unknown but constant.

Assuming that a constant parameter vector θ^* exists such that the system described by the equations

$$\begin{aligned} \dot{x} &= A(p,\theta^*,t)x + B(p,\theta^*,t)u \\ y &= H(p,\theta^*,t)x \end{aligned} \tag{1.6}$$

behaves in the desired fashion, the aim of the adaptive procedure is to adjust the parameter vector θ using the measured signals of the system so that $\lim_{t\to\infty}\theta(t) = \theta^*$. In other words, the controller adjusts its characteristics so that the same behavior is realized even when the plant parameters are unknown. Since θ is adjusted dynamically as a function of the states of the system, it is no longer a parameter but a state variable, which makes the overall system nonlinear. The equations governing the system can now be written as follows:

$$\begin{aligned} \dot{x} &= A(p,\theta,t)x + B(p,\theta,t)u \\ \dot{\theta} &= g(x,\theta,t). \end{aligned} \tag{1.7}$$

As time t tends to infinity, it is desired that the behavior of the nonlinear system in Eq. (1.7) approach that of the linear system in Eq. (1.6). As mentioned in the previous section, it is this special class of nonlinear systems that we shall refer to as *adaptive systems*.

1.4 DIRECT AND INDIRECT CONTROL

A living organism, as the prototype of an adaptive system, has to cope with changes in its environment for its survival, growth, and development. Familiarity with the environment results in better understanding, which in turn enables the organism to predict changes that are vital for its survival. Since understanding and controlling the world are two distinct activities, predictive ability does not always imply the ability to control the environment. Although the latter generally reflects the power of the organism to exploit its environment, it may also be desirable at times for the purposes of understanding it. Hence, a close connection exists between the understanding or identification of an environment and its control [26].

The implications of the ideas above in the context of partially known goal-oriented automatic control systems were first discussed by Feldbaum [18] in 1965. Assuming that a controller A is used to control a given plant or process B, the problem can be classified as deterministic, stochastic, or adaptive as discussed earlier, depending on the prior information available to A regarding B. For example, if, as in the deterministic case, the characteristics of the process B, the information about the disturbance acting on the object, knowledge of the state of the system as well as the performance index are all available, the optimal control action can be determined using well known results in control theory. On the other hand, if the characteristics of the process B or the disturbances acting on it are unknown, they can be estimated by making observations on the system. However, if what is required is the control of a partially known system as in the adaptive case, neither of the procedures above is adequate. Identification of the process characteristics alone does not result in satisfactory control while attempts to control the unknown plant without identification may result in poor response. Hence, according to Feldbaum, the controller A in an automatic control system with incomplete information regarding the plant B must simultaneously solve two problems that are closely related but different in character. Feldbaum referred to this as *dual control.* First, on the basis of the information collected, the controller must determine the characteristics and state of the plant B. Second, on the basis of this acquired knowledge, it has to determine what actions are necessary for successful control. The first problem may be considered one of estimation or identification while the second is one of control.

Two philosophically different approaches exist for the solution of the adaptive control problem discussed earlier. In the first approach, referred to as *indirect control*, the plant parameters are estimated on-line and the control parameters are adjusted based on these estimates. Such a procedure has also been referred to as *explicit identification* in the literature [3]. In contrast to this, in what is referred to as *direct control*, no effort is made to identify the plant parameters but the control parameters are directly adjusted to improve a performance index. This is also referred to as *implicit identification.* In conformity with the ideas expressed by Feldbaum, we note that in both cases efforts have to be made to probe the system to determine its behavior even as control action is being taken based on the most recent information available. The input to the process is therefore used simultaneously for both identification and control purposes. However, not every estimation scheme followed by a suitable control action will result in optimal or even stable behavior of the overall system. Hence, the estimation and control procedures have to be blended carefully to achieve the desired objective. The adaptive control schemes described in the chapters following can be considered special cases where successful dual control has been realized.

In Sections 1.4.1 and 1.4.2, we deal with parameter perturbation and sensitivity methods that are examples of the direct and indirect method respectively. These methods were investigated extensively in the 1960s and represented at that time the two principal approaches to adaptive control. We provide a somewhat more-than-cursory treatment of the two methods in this introductory chapter, since many of the adaptive concepts as well as plant parametrizations suggested later have their origins in these two methods. The reader who is not interested in these historical developments may proceed directly to Section 1.5 with no loss of continuity.

1.4.1 Parameter Perturbation Method

Extremum adaptation mentioned in Section 1.3 was perhaps the most popular among the various adaptive methods investigated in the early 1960s. It had considerable appeal to researchers due to its simplicity, applicability to nonlinear plants, and the fact that it did not require explicit identification of plant parameters. For several years it was investigated extensively and its principal features were studied exhaustively. Starting with the work of Draper and Li [13], who suggested the scheme for optimizing the performance of an internal combustion engine, the method collected a large following, and at present a rather extensive literature exists on this subject. Although its scope is somewhat tangential to that of this book, we nevertheless include some details concerning the method since many of the questions concerning the performance and limitations of adaptive systems encountered at present have counterparts in the analyses of the parameter perturbation method carried out over two decades ago. The interested reader is referred to [12,17,31] for further details.

The parameter perturbation method is a direct-control method and involves perturbation, correlation, and adjustment. Consider, for example, a plant in a laboratory excited by appropriate inputs, having an artificial environment, and provided with a means of continuously measuring a performance index. Further assume that knobs can be twiddled or switches operated to affect this performance index. The question that is posed is how one, asked to adjust the knobs and switches to optimize the performance function, should proceed and what kind of problems the person would face. The most direct and, perhaps, simplest procedure to follow would be to adjust the controls and see the effect on the performance index. If the performance improves, one would continue to alter the controls in the same direction; if it worsens, the controls would have to be changed in the opposite direction. In principle this is what is attempted in the parameter perturbation technique. However, in a practical problem, the number of parameters to be adjusted, the presence of output noise, the fact that parameters of the plant vary with time and that nonlinearities may be present in a plant, and so on, would complicate the problem considerably. Whether or not the method would be considered practically feasible would depend on the stability of the overall system as well as the speed with which it would adapt itself, as the plant parameters varied with time. If the procedure is deemed to be practically feasible, it is certainly simple in concept and easy to implement in terms of hardware.

Detailed analyses of extremal adaptation have been carried out by many authors. The analysis in [23] for example, reveals that sophisticated perturbation methods involving differential equations with multiple time scales would be needed for a precise analysis of the behavior of such systems. However, such approaches are not directly relevant to the contents of this book. We shall, instead, merely bring out the salient qualitative features of the approach as well as its limitations by confining our attention to a few simple problems.

Example 1.1

Consider a no-memory plant in which the input is $\theta(t)$ and the output is a performance function $F(\theta(t))$. $F(\theta)$ is a function of θ as shown in Fig. 1.3. Let θ_{opt} correspond to the value of θ for which $F(\theta)$ has a minimum. Assuming that the designer can only choose the value of θ and observe the corresponding value of $F(\theta)$ at every instant, the objective is to determine a procedure for adjusting $\theta(t)$ so that it converges to the optimal value θ_{opt} of θ.

Let the parameter θ be varied sinusoidally around a nominal value θ_0 so that $\theta(t) = \theta_0 + \epsilon\delta_\theta(t)$, where $\delta_\theta(t) = \sin \omega_p t$. ϵ is the amplitude and ω_p is the frequency of the perturbation. The output $F(\theta)$ will then oscillate around the nominal value $F(\theta_0)$ as $F(\theta_0) + \delta_F(t)$. If $\theta_0 + \epsilon < \theta_{\text{opt}}$, it is obvious that $\theta(t)$ and $F(\theta(t))$ are out of phase while if $\theta_0 - \epsilon > \theta_{\text{opt}}$, they are in phase. Hence, by correlating the perturbation and output as

$$\int_t^{t+2\pi/\omega_p} \delta_\theta(t)F(\theta)\,dt \;\triangleq\; R(\theta)$$

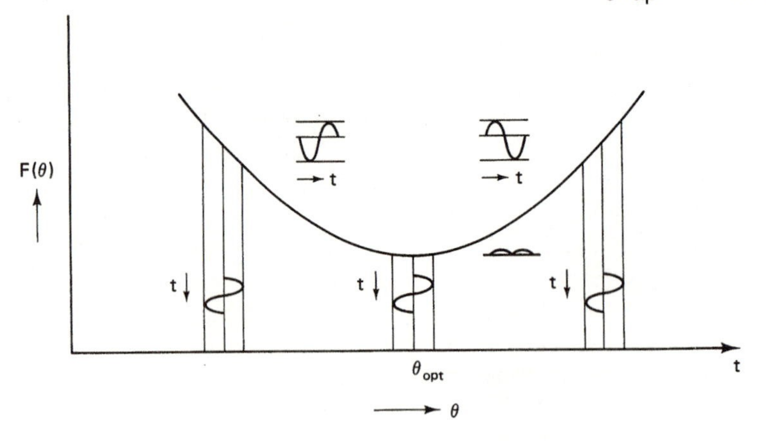

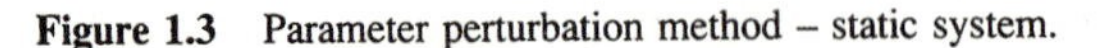
Figure 1.3 Parameter perturbation method – static system.

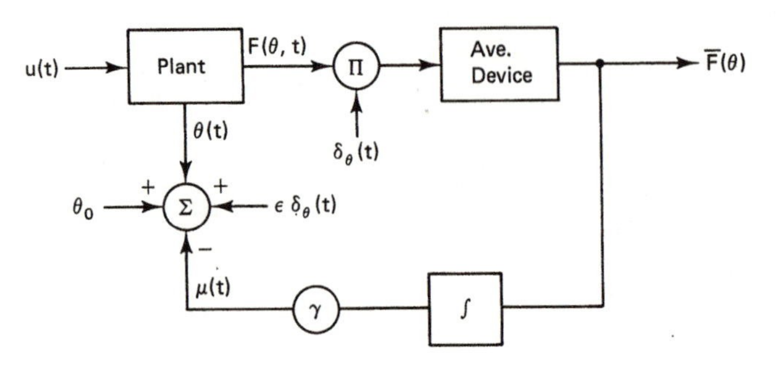

Figure 1.4 Parameter perturbation method – dynamic system.

the side of θ_{opt} on which θ_0 lies can be determined. This in turn indicates whether θ is to be increased or decreased from its nominal value θ_0.

If for every value of θ, the gradient $\nabla_\theta F(\theta)$ with respect to θ is known, it follows directly that if the parameter $\theta(t)$ is adjusted as

$$\dot{\theta}(t) = -\gamma \nabla_\theta F(\theta(t)) \qquad \gamma > 0$$

or alternately

$$\dot{\theta}(t) = -\gamma sgn\left[\nabla_\theta F(\theta(t))\right], \tag{1.8}$$

where $sgn(x) = +1$ if $x \geq 0$ and -1 otherwise, then $\lim_{t \to \infty} \theta(t) = \theta_{\text{opt}}$. Since $\nabla_\theta F(\theta)$ is not known, Eq. (1.8) cannot be implemented in practice. However, the method outlined in Example 1.1 represents an approximation of the above since $R(\theta)$ yields the sign of the gradient and hence can be used for adjusting the parameter.

Example 1.2

Figure 1.4 shows a simple dynamical plant containing a single parameter θ. The input to the plant is u and the only relevant output of the plant is the instantaneous performance index $F(\theta, t)$. It is assumed that using an averaging device, an averaged performance index $\overline{F}(\theta)$ which is independent of time can be obtained. The function $\overline{F}(\theta)$ has the same form as $F(\theta)$ given in Example 1.1. Once again the objective is to determine the optimal value θ_{opt} of θ so that the performance index is minimized.

Note that the problem of optimization of a dynamical system has been reduced to the minimization of the function $\overline{F}(\theta)$ with respect to the parameter θ and that quantities such as the input u are no longer directly relevant. In this equivalent system, the parameter θ can be considered as the input and $\overline{F}(\theta)$ as the output as in Example 1.1.

Assuming once again that $\theta(t)$ is perturbed around a nominal value θ_0 as $\theta(t) = \theta_0 + \epsilon\delta_\theta(t)$, the change in the performance index may be approximated by

$$\overline{F}[t; \theta_0 + \epsilon\delta_\theta(t)] \approx \overline{F}[t; \theta_0] + \epsilon\, \delta_\theta(t)\nabla_\theta \overline{F}\ |_{\theta=\theta_0}$$

if $\overline{F}$ is a smooth function of the parameter θ. Correlating $\delta_\theta(t)$ and $\overline{F}[t; \theta_0 + \epsilon\delta_\theta(t)]$, we obtain

$$\overline{\delta_\theta(t)\overline{F}[t; \theta(t)]} \approx \overline{\delta_\theta(t)\overline{F}[t; \theta_0]} + \overline{\epsilon\delta_\theta^2(t)\nabla_\theta \overline{F}}\ |_{\theta=\theta_0} \tag{1.9}$$

where the overbar denotes an average value over an interval of time T. Assuming that $\delta_\theta(t)$ is independent of the input $u(t)$ and has an average value zero, the first term can be neglected. The second term in Eq. (1.9) yields a quantity which is approximately proportional to the gradient of $\overline{F}$ with respect to θ at the operating point θ_0. This quantity is used for updating the parameter θ. At every instant t, the parameter $\theta(t)$ is composed of the nominal value θ_0, the perturbation signal $\epsilon\delta_\theta(t)$, and the correction term $\mu(t)$, which is based on an estimate of the sign of the gradient $\nabla_\theta \overline{F}(\theta)$. $\mu(t)$ is obtained by integrating the output of the averaging device and using a feedback gain γ as shown in Fig. 1.4.

The model above clearly separates the various aspects of the adaptive problem. As described here, the four parameters of interest are the amplitude ϵ of the perturbing signal, its frequency ω_p, the averaging time T, and the gain γ in the feedback path, which can be considered as the step-size of the correction term. A few comments regarding the choice of the values of these parameters are worthwhile [23]:

(i) Too small a value of ϵ makes the determination of the gradient difficult while too large a value may overlook the optimum value.

(ii) A very high frequency of perturbation ω_p may have a negligible effect on the output while a low value of ω_p requires a large averaging time.

(iii) A small value of T may result in a noisy value of the gradient while a large value of T implies slow adaptation.

(iv) A large step size γ may result in hunting or even instability while a small value of γ would result in very slow convergence.

As mentioned previously, the compromises indicated in (i)-(iv) above make their appearances in all adaptive schemes in one form or another and are closely related to the ideas expressed by Feldbaum.

The detailed analysis of even a simple second-order system with a single control parameter reveals that many assumptions have to be made regarding the drift frequency of the control parameter which has to be tracked, the frequency of the correction term in the adaptive parameters, the frequency of the perturbing signal, the bandwidth of the plant and the bandwidth of the closed-loop system. This becomes even more complex if more than one parameter is to be adjusted. Despite these drawbacks, the procedure is intuitively appealing and easy to implement, and consequently finds frequent application in a variety of situations where very little is known about the detailed mathematical model of the plant and only a few parameters can be adjusted. Although such analyses provide valuable insights into the nature of the adaptive process, they are nevertheless beyond the scope of this book.

1.4.2 Sensitivity Method

An alternate approach to the control of systems when uncertainty is present is through the use of sensitivity models. This method gained great popularity in the 1960s and has wide applicability at present in industrial design. A set of forty-five papers, including three surveys containing several hundred references, representative of the state of the art in 1973, was collected in a single volume by Cruz [11], and the reader is referred to it for further information regarding this subject. In this section we briefly indicate how sensitivity methods find application in the design of adaptive control systems.

Since uncertainty in a system can arise in a variety of forms, the corresponding sensitivity questions can also be posed in different ways. Parametric uncertainty may arise due to tolerances within which components are manufactured and the combined effect of variations in parameter values may affect overall system behavior. In adaptive systems where parameters are adjusted iteratively on-line, it would be useful to know how the system performance is improved. The effect of changes in parameters on the eigenvalues of the overall time-invariant system, the states of the system at any instant of time (for example, the terminal time), or the entire trajectory of the system, can be of interest in different design problems. In such cases the partial derivative of the quantity of interest with respect to the parameter that is perturbed has to be computed. Such partial derivatives are called sensitivity functions and assuming they can be determined on-line, the control parameters can be adjusted for optimal behavior using standard hill-climbing methods. Although the use of dynamic models was originally suggested by Byhovskiy [9] in the late 1940s and later extended by Kokotovic [22], Wilkie and Perkins [44], Meissinger [32], and others, in the following examples we present the results obtained independently by Narendra and McBride [34] in the early 1960s and later generalized in [35].

Example 1.3

Consider the linear time-invariant differential equation

$$\ddot{e} + \theta_2 \dot{e} + \theta_1 e = u \qquad e(t_0) = 0 \tag{1.10}$$

Assume that the input u is specified but that the values of the constant parameters θ_1 and θ_2 are to be determined to keep the output $e(t)$ as small as possible. Quantitatively this can be represented as the minimization of a performance function $J(\theta_1, \theta_2)$ with respect to θ_1 and θ_2 where

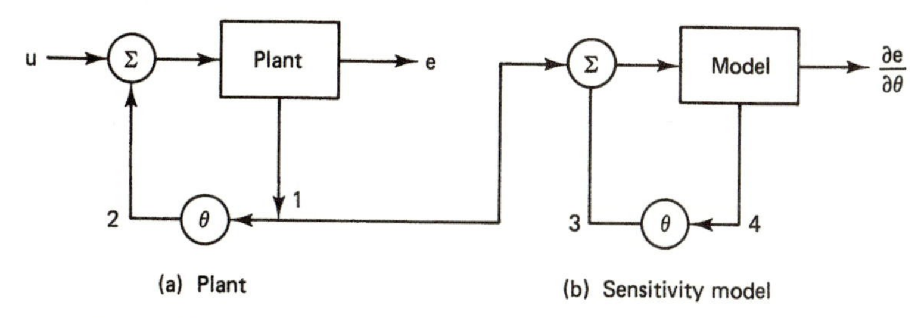

Figure 1.5 Generation of the partial derivative using a sensitivity model.

$$J(\theta_1,\theta_2) = \frac{1}{T}\int_0^T e^2(t)dt.$$

If a gradient approach is to be used to determine the optimal values of θ_1 and θ_2, the partial derivatives $\partial e(t)/\partial\theta_1$ and $\partial e(t)/\partial\theta_2$ are needed. To compute them, we consider the differential equations obtained by taking partial derivatives of the two sides of Eq. (1.10).

$$\begin{aligned}\frac{\partial\ddot{e}(t)}{\partial\theta_1}+\theta_2\frac{\partial\dot{e}(t)}{\partial\theta_1}+\theta_1\frac{\partial e(t)}{\partial\theta_1} &= -e(t)\\ \frac{\partial\ddot{e}(t)}{\partial\theta_2}+\theta_2\frac{\partial\dot{e}(t)}{\partial\theta_2}+\theta_1\frac{\partial e(t)}{\partial\theta_2} &= -\dot{e}(t).\end{aligned}$$

Denoting $\partial e/\partial\theta_1 \triangleq y_1$ and $\partial e/\partial\theta_2 \triangleq y_2$, we have

$$\begin{aligned}\ddot{y}_1+\theta_2\dot{y}_1+\theta_1 y_1 &= -e\\ \ddot{y}_2+\theta_2\dot{y}_2+\theta_1 y_2 &= -\dot{e}\end{aligned}$$

so that y_1 and y_2 can be generated using models identical to the system in Eq. (1.10) but with inputs $-e$ and $-\dot{e}$ respectively. Such models are referred to as sensitivity models.

Since $\partial e/\partial\theta_1$ and $\partial e/\partial\theta_2$ can be obtained using sensitivity models, the gradient of $J(\theta_1,\theta_2)$ in the parameter space can be obtained as

$$\frac{\partial J}{\partial\theta_i} = \frac{2}{T}\left\{\int_0^T e(t)y_i(t)dt\right\} \qquad i=1,2.$$

Hence, by successively adjusting $\theta_i(t)$ over intervals of length T by the algorithms

$$\theta_i(t+T) = \theta_i(t) - \gamma\int_0^T e(t)y_i(t)dt \qquad i=1,2$$

or alternately by adjusting it continuously as $\dot{\theta}_i(t) = -\gamma e(t)y_i(t)$, the performance can be improved. When $\theta_i(t)$ converges to some constant value θ_i^*, the optimum parameters are achieved.

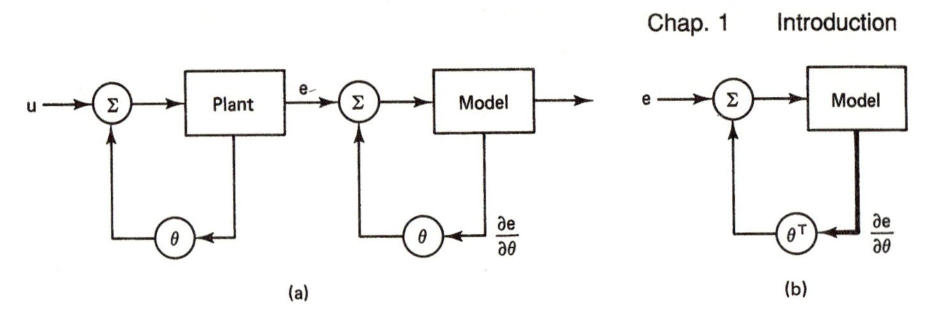

Figure 1.6 Generation of the gradient using a sensitivity model.

Example 1.4 The General Case:

Let a system such as that shown in Fig. 1.5 have u as the input, e an output of interest, and θ an adjustable feedback parameter. It was shown in [35] that if the input signal to the parameter θ in the system denoted by the point 1 is fed into the output of the gain in the sensitivity model whose input (corresponding to u) is identically zero, the output of the model is $\partial e/\partial\theta$. The proof of this is based on the definition of a partial derivative and the linear time-invariant nature of the systems involved. Assuming that the partial derivatives with respect to n parameters $\theta_1, \theta_2, \ldots, \theta_n$ are needed, the same procedure can be repeated using n-sensitivity models of the system. However, it was also shown independently in [22,35] that in special cases where the outputs of all the gains θ_i end in the same summing point, it may be possible to generate all the partial derivatives of y using a single sensitivity model. This is based once again on relatively simple concepts related to linear time-invariant systems.

Since the signal $\partial e/\partial\theta$ can be considered to be obtained as $\partial e/\partial\theta = T_1T_2u$, where T_1 is the transfer function from the point 3 to the output and T_2 is the transfer function from the input of the system to the point 1, and since $T_1T_2 = T_2T_1$, $\partial e/\partial\theta$ also can be generated as the input to the gain parameter θ by making the output e as the input to the model as shown in Fig. 1.6a. This is readily generalized to the many parameter case as shown in Fig. 1.6b since $\partial e/\partial\theta_i, (i = 1, 2, \ldots, n)$ would be the signal generated in the sensitivity model at the input of θ_i. As in the simpler case described earlier, $\partial e/\partial\theta_i$ can be used to generate the gradient of the performance index to adjust the parameters θ_i of the system.

In the discussions thus far, it has been assumed that an exact replica of the overall system can be constructed as a sensitivity model to yield the required sensitivity functions. However, in adaptive systems, which can generally be assumed to be made up of a plant with unknown parameters and a controller whose structure as well as parameters are known, it is no longer a straightforward process to determine the sensitivity model. In such cases, using the input and output of the plant, the parameters of an identification model can be adjusted using the sensitivity approach so that the output of the identification model approximates that of the plant. The sensitivity model for control can then be approximated by the identification model. This makes the sensitivity approach an example of indirect control.

Although simulation results of adaptive systems based on this approach indicate that the convergence of the controller is rapid, very little can be said about the theoretical stability of the overall system. In fact, the problem of assuring stability for arbitrary initial

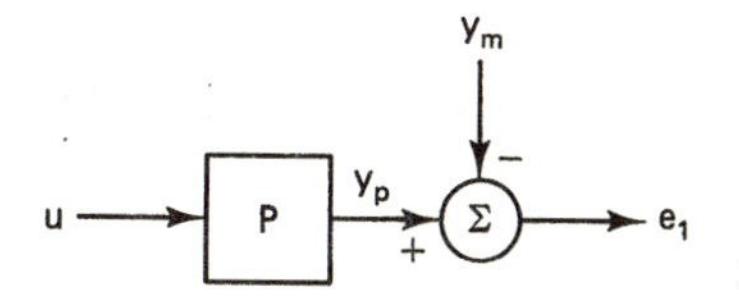

Figure 1.7 The control objective.

conditions encountered in both of the approaches treated in this section led to a search for more systematic procedures for synthesizing stable adaptive controllers and, in turn, resulted in the stability methods discussed in this book.

1.5 MODEL REFERENCE ADAPTIVE SYSTEMS AND SELF-TUNING REGULATORS

As mentioned in Section 1.2, the aim of control is to keep the relevant outputs of a given plant within prescribed limits. If the input and output of a plant P are as shown in Fig. 1.7, the aim of control may be quantitatively stated as the determination of the input u to keep the error $e_1 = y_p - y_m$ between the plant output y_p and a desired output y_m within prescribed values. If y_m is a constant, the problem is one of *regulation* around this value (also known as an operating point or set point). When y_m is a function of time, the problem is referred to as *tracking*. When the characteristics of the plant P are completely known, the former involves the determination of a controller to stabilize the feedback loop around the set point. In the latter case, a suitable controller structure may be employed and control parameters determined so as to minimize a performance index based on the error e_1. As described earlier, powerful analytical techniques based on the optimization of quadratic performance indices of the form of Eq. (1.4) are available when the differential equations describing the behavior of the plant are linear and are known a priori. When the characteristics of the plant are unknown, both regulation and tracking can be viewed as adaptive control problems. Our interest will be in determining suitable controllers for these two cases, when it is known a priori that the plant is linear but contains unknown parameters. *Model reference adaptive systems* (MRAS) and *self-tuning regulators* (STR) are two classes of systems that achieve this objective.

As discussed in Section 1.4, the problem above can be attempted using either an indirect or direct approach. In the indirect approach, the unknown plant parameters are estimated using a model of the plant, before a control input is chosen. In the direct approach, an appropriate controller structure is selected and the parameters of the controller are directly adjusted to reduce some measure of the error e_1. While dealing with the tracking problem, it becomes necessary in both cases to specify the desired output y_m in a suitable form for mathematical tractability. This is generally accomplished by the use of a reference model. Thus, an indirect approach calls for an explicit model of the plant as well as a reference model, while the direct approach requires only the latter. Adaptive systems that make explicit use of such models for identification or control purposes are called MRAS [24]. In view of the important role played by both the identification and reference models in the proper formulation of the adaptive control problem, we provide some additional comments concerning their choice.

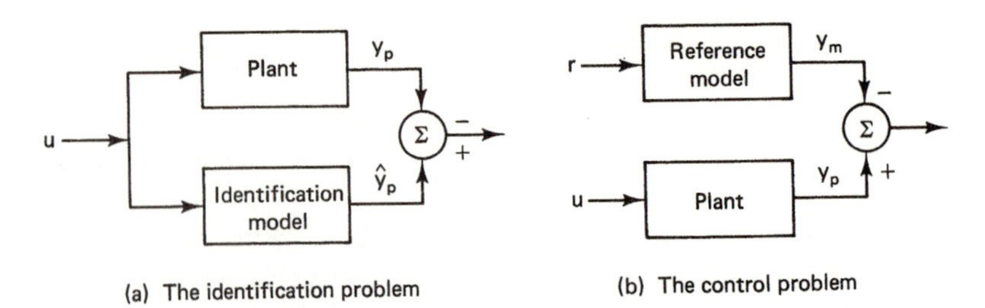

Figure 1.8 Identification and control problems.

1.5.1 Identification Model

The importance of mathematical models in every aspect of the physical, biological, and social sciences is well known. Starting with a phenomenological model structure that characterizes the cause and effect links of the observed phenomenon in these areas, the parameters of the model are tuned so that the behavior of the model approximates the observed behavior. Alternately, a general mathematical model such as a differential equation or a difference equation can be used to represent the input-output behavior of the given process, and the parameters of the model can be determined to minimize the error between the process and model outputs in some sense. The model obtained by the latter procedure is termed an *identification model.* This model is preferred in adaptive systems since models of the first type are either not available or too complex for control purposes. Sometimes the second procedure may be used to approximate and reduce the models derived using the first approach to analytically manageable forms. Based on this, the identification problem may be succinctly stated as follows [Fig. 1.8(a)]:

Problem 1.1 The input and output of a linear time-invariant plant are u and y_p respectively, where u is a uniformly bounded function of time. The plant is assumed to be stable with a known parametrization but the values of the parameters are assumed to be unknown. From all available data, the objective is to construct a suitable model, which when subjected to the same input $u(t)$ as the plant, produces an output $\widehat{y}_p(t)$, which tends to $y_p(t)$ asymptotically.

As stated above, the output is assumed to be free of noise. When additive observation noise is present, some statistical criterion has to be used to determine the model that best fits the observed data. System identification, based on such an approach, is a well developed area of systems theory. Numerous identification procedures and their convergence properties, under different assumptions regarding inputs and disturbances, are currently well known [28]. In this book, however, we do not discuss the realization of an identification model when the noise is stochastic. Instead, we confine our attention to the noise-free case, as well as situations in which the noise is uniformly bounded and deterministic.

1.5.2 Reference Model

The use of a reference model for control purposes is a recent development which had its beginnings in aircraft control. The implicit assumption here is that the designer is sufficiently familiar with the system under consideration; by suitably choosing the structure and parameters of the reference model, the desired response can be specified in terms of the model outputs. Although such a model can be either linear or nonlinear, practical considerations as well as analytical tractability usually limit the models to the class of linear time-invariant systems.

In the design of aircraft systems, the desired dynamics of the aircraft in terms of transient behavior, decoupling of modes, and handling qualities, have been represented by a reference model described by the homogeneous differential equation

$$\dot{y}_m = A_m y_m \tag{1.11}$$

where the constant matrix $A_m \in \mathbb{R}^{m \times m}$ is suitably chosen. The same model may also be satisfactory for many other situations as well. Assuming that a plant can be described adequately by an nth order differential equation with $m(m << n)$ outputs as

$$\begin{aligned} \dot{x}_p &= A_p x_p + B_p u \\ y_p &= C_p x_p \end{aligned} \tag{1.12}$$

where A_p, B_p, and C_p are $(n \times n), (n \times p)$, and $(m \times n)$ constant matrices respectively, the reference model is chosen to have the form of Eq. (1.11) so that the output y_p follows y_m asymptotically in some sense. Two methods that have been studied extensively for such model-following are called explicit and implicit model-following methods, and are based on different performance indices that depend on the error $y_p - y_m$ between the outputs of the plant and the model. In the first case the performance index has the form

$$I_e = \int_0^\infty \left[(y_p - y_m)^T Q_e (y_p - y_m) + u^T R u \right] dt$$

where the subscript e refers to explicit model-following, and the matrices Q_e and R are positive-definite. Using quadratic optimization theory it can be shown [8] that the minimization of I_e with respect to u results in a control input of the form

$$u(t) = K_m y_m(t) + K_p x_p(t)$$

so that the control configuration is as shown in Fig. 1.9(a). In implicit model-following, the model is incorporated in the performance index I_i (the subscript i referring to the implicit model) as

$$I_i = \int_0^\infty \left\{ \left[\dot{y}_p - A_m y_p \right]^T Q_i \left[\dot{y}_p - A_m y_p \right] + u^T R u \right\} dt \qquad Q_i > 0.$$

In this case, the optimal input has the form $u = K_p x_p$. The reference model and the plant, together with the controller, can now be represented as shown in Fig. 1.9(b). The reference model forms a prefilter in Fig. 1.9(a) and occurs as a parallel model in Fig. 1.9(b). Since, in both cases, the control parameters have to be computed on the basis of

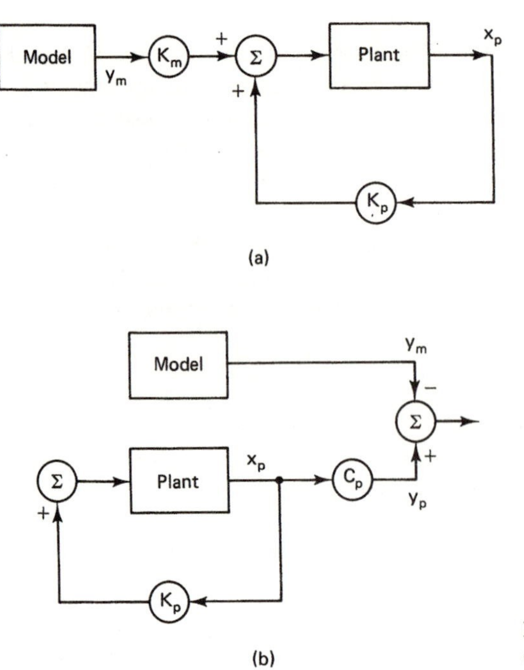

Figure 1.9 (a) Explicit model-following (b) Implicit model-following.

the performance criteria as well as the specified plant parameter values, the procedures cannot be used in adaptive situations. However, even in such situations they provide a rationale for using similar structures for generating the control input to the plant. For the most part the parallel structure shown in Fig. 1.9(b) is used in the adaptive systems discussed in the chapters following, although the series structure in Fig. 1.9(a) is also examined in Chapter 9.

1.5.3 Reference Models with Inputs

In the preceding section it was assumed that the desired output y_m is the solution of a homogeneous differential equation. A more general problem of model-following occurs when the reference model has the form

$$\dot{x}_m = A_m x_m + B_m r; \qquad y_m = C_m x_m$$

where A_m is a stable $n \times n$ matrix with constant elements, B_m and C_m are constant matrices of appropriate dimensions, and r is a piecewise-continuous uniformly bounded function. Given a plant defined by Eq. (1.12), the objective is to determine the input u to the plant so that the output $y_p(t)$ is close in some sense to the output $y_m(t)$ of the model as $t \to \infty$. In this case the reference input r, along with the model specified by the triple $\{A_m, B_m, C_m\}$, determines the reference output y_m and hence the performance expected of the plant. The introduction of the reference input significantly increases the class of desired outputs y_m as compared with that defined by the model in Eq. (1.11).

Perfect model-following is said to occur if $|y_p(t) - y_m(t)| \equiv 0$ for $t \geq t_0$. Asymptotic model-following is said to occur if $\lim_{t\to\infty} |y_p(t) - y_m(t)| = 0$. For perfect model-following it is clear that the differential equations governing $y_p(t)$ and $y_m(t)$ as well as the initial conditions $y_p(t_0)$ and $y_m(t_0)$ must be identical. When $C_m = C_p = I$, so that both model and plant state vectors are accessible, perfect model-following implies choosing an input $u(t)$ for all $t \geq t_0$ so that

$$\begin{aligned} \dot{x}_p(t) &= A_p x_p(t) + B_p u(t), \qquad x_p(t_0) = x_m(t_0), \text{ and} \\ \dot{x}_m(t) &= A_m x_m(t) + B_m r(t) \end{aligned}$$

have identical solutions. This imposes restrictive conditions on the matrices A_p, B_p, A_m, and B_m which have been derived by Erzberger [16] who uses an input of the form

$$u(t) = K_p x_p(t) + K_m x_m(t) + K_r r(t). \tag{1.13}$$

Of greater relevance in the adaptive context is the asymptotic model-following problem. In such a case, using an input u of the form given in Eq. (1.13), the aim is to determine the conditions under which the transfer matrix from r to y_p is identical to that from r to y_m. Solutions to this problem have been obtained by several researchers [33], [41], and [45]. The importance of these results in model reference adaptive control (MRAC) is discussed in the following section.

1.5.4 Model Reference Adaptive Control

The MRAC problem can be qualitatively stated as follows: Let a linear time-invariant plant P with an input-output pair $\{u(\cdot), y_p(\cdot)\}$ be given. Let a stable LTI reference model M be specified by its input-output pair $\{r(\cdot), y_m(\cdot)\}$ where $r : \mathbb{R}^+ \to \mathbb{R}$ is a bounded piecewise-continuous function. The aim is to determine the control input $u(t)$ for all $t \geq t_0$ so that

$$\lim_{t\to\infty} |y_p(t) - y_m(t)| = 0.$$

It soon becomes evident that considerable prior information regarding the plant P is needed to have a well posed problem and that the choice of the control input as well as the reference model depends critically on this information. We shall first briefly discuss some of these issues before presenting a more precise version of the MRAC problem.

The Input Set $\mathcal{U}$. Let the input u to the plant belong to some set $\mathcal{U}$. From a practical standpoint, the input u must be generated using a controller C that has access to all the signals that can be measured. A natural requirement is that u is uniformly bounded. Since noise is invariably present in all systems, we further impose the condition that the controller C that generates u be differentiator-free.

The Class of Reference Models $\mathcal{M}$. From the earlier discussion concerning reference models, it is clear that if the output of the plant is to asymptotically follow the output of the model, the class of models $\mathcal{M}$ has to be constrained in some sense. Obviously, $\mathcal{M}$ depends on the prior information concerning the class $\mathcal{P}$ of linear time-invariant systems to which P belongs. For example, if the reference input r is a square wave and the model M has a unity transfer function, it is clear that the output of the plant (which has a strictly proper transfer function) cannot follow $y_m(t)$ asymptotically with a uniformly bounded input u generated by a differentiator-free controller. To determine $\mathcal{M}$, we draw upon the results in linear systems theory when the plant characteristics are known. The results in [33], [41], and [45] are concerned precisely with this problem and are based on the invariants in the representation of the plant. Hence, given some prior information concerning the plant, the class $\mathcal{M}$ of reference models can be chosen.

The Controller C. From the prior information concerning $\mathcal{P}$ and $\mathcal{M}$ the structure of the controller C that generates u is determined. Theoretically, C can be nonlinear; but since our aim, as stated in Section 1.3.2, is to extend linear design concepts to adaptive control, we shall assume that C is parametrized by a vector $\theta : \mathbb{R}^+ \to \mathbb{R}^m$ so that for all constant values of $\theta \in \mathbb{R}^m$, C is linear and time-invariant. For every $P \in \mathcal{P}$ and $M \in \mathcal{M}$, there exists a $\theta^* \in \mathbb{R}^m$ such that for $\theta \equiv \theta^*$, the transfer function of the plant together with the controller matches the transfer function of the model M. This is the *algebraic part* of the adaptive control problem.

The aim of adaptation is to generate the control input u such that $\lim_{t\to\infty} |y_p(t) - y_m(t)| = 0$. Since $u(t)$ is determined by the manner in which the control parameter $\theta(t)$ is adjusted, the problem can equivalently be stated in terms of $\theta(t)$. The rule by which $\theta(t)$ is adjusted is referred to as the *adaptive law*. The determination of the adaptive law that assures the stability of the overall system constitutes the *analytic part* of the adaptive control problem.

With the preamble above, the adaptive control problem addressed in this book can be stated below [Fig. 1.8(b)]:

Problem 1.2 The input and output of a linear time-invariant plant P with unknown parameters are $u(\cdot)$ and $y_p(\cdot)$ respectively.

(i) Determine the class $\mathcal{M}$ of stable LTI reference models such that if $M \in \mathcal{M}$ has an input-output pair $\{r(\cdot), y_m(\cdot)\}$, a uniformly bounded input $u(\cdot)$ to the plant P, generated by a differentiator-free controller, exists which assures

$$\lim_{t\to\infty} |y_p(t) - y_m(t)| = 0. \tag{1.14}$$

(ii) Determine a differentiator-free controller $C(\theta)$ parametrized by a vector $\theta : \mathbb{R}^+ \to \mathbb{R}^m$, which generates u, such that a constant value $\theta \equiv \theta^* \in \mathbb{R}^m$ exists for which the transfer function of the plant together with the controller is equal to the transfer function of M.

(iii) Determine a rule for adjusting $\theta(t)$ so that Eq. (1.14) is satisfied.

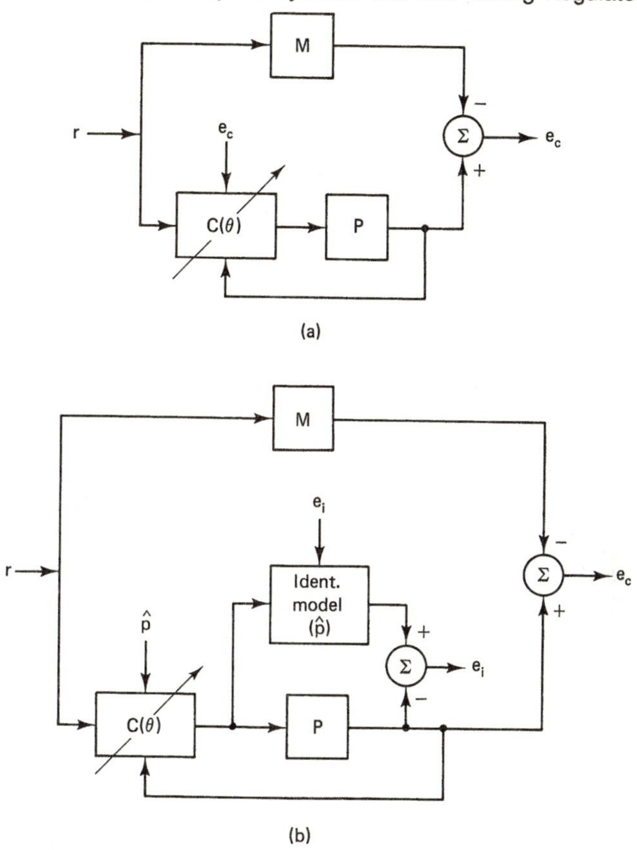

Figure 1.10 (a) Direct adaptive control (b) Indirect adaptive control.

For the solution of (i) and (ii), prior information regarding the plant is essential. The determination of $\mathcal{M}$ in (i) and $C(\theta)$ in (ii) represents the algebraic part, and (iii) constitutes the analytic part of the adaptive control problem. The marriage of algebraic and analytical methods is the essence of adaptation.

In the statement of the adaptive control problem, it is assumed that the plant is linear time-invariant. In practice, the need for adaptation arises primarily because the parameters of the plant vary with time. In addition, the plant may be modeled only partially or be subjected to bounded disturbances. In such cases, it may not be possible to satisfy the condition in Eq. (1.14), and the objective instead is to determine a rule for adjusting $\theta(t)$ so that $|y_p(t) - y_m(t)|$ is bounded. This problem is dealt with in detail in Chapter 8.

As mentioned earlier, adaptive control can be carried out using the direct or indirect control approach. The structure of the adaptive systems using these two approaches is shown in Fig. 1.10(a) and 1.10(b) respectively. As shown in the figure, the structure of the controllers are identical in (a) and (b), but differ only in the nature of the adjustment of the controller $C(\theta)$. Information concerning the output error e_c is directly used to adjust

$C(\theta)$ in (a), while the identification error e_i is used to adjust the identification parameter $\widehat{p}$, which in turn is used to adjust $C(\theta)$ in (b). In both cases, while the identifier and the controller can be parametrized in numerous ways so that a solution to the algebraic part exists, only some of these parametrizations will lead to stable analytical solutions that assure the boundedness or the convergence of the parameter vector to its desired value. Recently, efforts have been made to use both direct and indirect methods to realize their combined advantages [15]. The choice of the reference model, identifier, and controller structure for different classes of problems, and the generation of a stable adaptive law in each case, are the subjects of the chapters following.

1.5.5 Self-Tuning Regulators

The MRAS described above arose in investigations concerned with the optimal control of deterministic servomechanisms. In contrast to this, self-tuning regulators evolved during the study of stochastic regulator problems. The principal approach of the STR is indirect control where the parameters of the plant are estimated prior to determining the control parameters. Originally proposed by Kalman [21], STRs consist of a parameter estimator, a linear controller, and a block which determines the control parameters from the estimated parameters. Depending on the estimation and control schemes, various possibilities exist for designing such a regulator. Over the years, STRs have been investigated in great detail [3,4,5,37] and enjoy a wide popularity due to their flexibility and ease of understanding and implementation.

Self-tuning regulators were originally developed for the stochastic minimum variance control problem. Since the design approach is flexible, many extensions have subsequently been made. Stochastic approximation, least-squares, extended and generalized least-squares, extended Kalman filtering, as well as instrumental variable method, and the maximum likelihood method can be used for identification, and all of these have been incorporated in the design of the STR. At every time instant the corresponding control can be computed using phase and amplitude margins, pole placement, PID control, or LQG design methods.

Explicit STRs consist of an explicit estimation of the process to be controlled followed by a tuning of the regulator parameters; implicit STRs are based on an implicit estimation of the process and a direct update of the regulator parameters. Hence the two methods are identical to the indirect and direct methods mentioned earlier in the context of MRAC. Although an extensive literature exists on the subject of STRs, most of it is concerned with plants operating in discrete time. Since this book deals only with continuous time systems, we will not have occasion to deal with much of this literature. However, in many cases, a direct correspondence exists between MRAS and STRs [24].

1.6 STABLE ADAPTIVE SYSTEMS

In almost all the examples of adaptive systems that were discussed earlier, the control parameters were adjusted based on the measurements of some system outputs. As mentioned in Section 1.2, this makes the modified system nonlinear. If external inputs are also present, the system could also be nonautonomous. Hence, the analysis or synthesis of adaptive systems invariably involves the study of system theoretic properties, such as stability, controllability, observability, and optimality of nonlinear and nonautonomous systems. In most of the approaches used during the 1960s for the study of adaptive systems, the emphasis was on their optimality. Using heuristic arguments based on the behavior of linear time-invariant systems, analytical procedures were developed to adjust control parameters that would optimize a chosen performance function. The parameter perturbation method as well as the sensitivity method described in the previous section are typical examples of such approaches. In both cases, the adjustment of the parameters is based on the assumption that the system is almost time-invariant. Once the design was completed, the stability of the overall nonlinear nonautonomous system was analyzed and conditions for local stability were derived. In some cases, no stability analyses were given but extensive computer simulations were used to justify the use of the methods developed.

Since stability is vital to the satisfactory operation (see Chapter 2) of any system and adaptive systems are invariably nonlinear, one of the major difficulties encountered in designing adaptive systems lies in the stability analysis of the corresponding nonlinear differential equations. In 1963, Grayson [19] suggested that a reversal of the procedure adopted earlier would be conceptually more efficient. He argued that adaptive systems should be designed in such a manner that they are stable for all values of a parameter γ belonging to a set S_γ. The optimization of the system can then be carried out by choosing the optimum value γ_{opt} of γ, where $\gamma_{\text{opt}} \in S_\gamma$. For example, let the differential equations governing an adaptive system when the control parameters are constant be given by

$$\dot{x} = f(x, p, \theta, t) \tag{1.15a}$$

where x is the state of the system, p is a fixed but unknown parameter vector, and θ is the adjustable control parameter vector. The argument t in Eq. (1.15a) is due to external inputs. Since the choice of θ is at the discretion of the designer, we assume that a rule given by Eq. (1.15b) exists that makes the combined nonlinear system [Eqs. (1.15a,b)] stable for all values of γ in a set S_γ:

$$\dot{\theta} = g(x, \theta, \gamma, t) \tag{1.15b}$$

Once stability is assured, the optimization of the system merely involves the search for γ_{opt} in S_γ. This is very similar to the procedure adopted in the design of linear time-invariant systems where the optimal controller parameters are chosen under the constraint that the poles of the overall system should lie in the open left-half of the complex plane.

The methodology used throughout this book for the analysis and synthesis of adaptive systems is based on the strong conviction that the stability of an adaptive system should be assured before any attempts are made to achieve optimal response. This implies that the form of the adaptive law should be chosen first so that the overall nonlinear system has desirable stability properties. In view of the importance attached to stability in this approach, Chapter 2 is devoted entirely to the collection and presentation of different results in stability theory that are essential for the study of adaptive systems. Since the direct method of Lyapunov is one of the general methods known for the stability analysis of dynamical systems, many of the results included pertain to this method.

In the early days of adaptive system design, when Lyapunov's direct method was the principal tool used, errors were made in the application of the theory to adaptive systems. When the adaptive law in Eq. (1.15b) is used, the m-dimensional parameter vector θ becomes part of the $(n+m)$ dimensional state vector of the modified nonlinear system in Eq. (1.15a) so that the stability analysis has to be carried out in the modified state space $\mathbb{R}^{n+m}$ rather than in $\mathbb{R}^n$. Further, since the system being analyzed is nonlinear, questions concerning the existence and uniqueness of the solutions of the corresponding differential equations also arise. These were frequently ignored in earlier analyses. Also, the stability properties of the solutions of nonlinear time-varying differential equations with external inputs or disturbances were not well understood. These and similar questions resulting from the nonlinear nature of the systems under consideration are addressed throughout this book.

1.6.1 Error Models

The first step in the stability approach to adaptive system design, as emphasized earlier, is the choice of the adaptive law for adjusting the control parameters to assure stability. Let θ^* be a constant unknown vector such that the output of the adaptive system follows the output of the reference model exactly when $\theta(t) \equiv \theta^*$. If the state error vector $e(t)$ and the parameter error vector $\phi(t)$ are defined as

$$e(t) \triangleq x(t) - x^*(t), \qquad \phi(t) \triangleq \theta(t) - \theta^*$$

where $x^*(t)$ is the desired state trajectory, we are interested in $e(t)$ tending to zero as $t \to \infty$, in the absence of external disturbances. In many cases, it is also desirable to assure that $\lim_{t\to\infty} \phi(t) = 0$. This approach was first suggested by Narendra *et al.* [36] in 1971. In terms of the error vector (e, ϕ), the problem may be stated as follows:

Given that the evolution of the error vector $e(t)$ is determined by the vector differential equation

$$\dot{e}(t) = f_1(e(t), \phi(t), t) \tag{1.16a}$$

determine the adaptive law

$$\dot{\phi}(t) = f_2(e(t), t) \tag{1.16b}$$

so that the origin of Eqs. (1.16a,b) is (i) stable with $\lim_{t\to\infty} e(t) = 0$, (ii) asymptotically stable.

Focusing attention directly on the error vectors, rather than on the actual response of the plant or the reference model, is the major change proposed in this approach. As shown throughout the book, this procedure enables the designer to concentrate on the essential features of the problem without getting bogged down in a morass of mathematical details.

Several points are worth noting in the problem discussed above. The error differential equations are used merely for analysis purposes and to ensure that $e(t)$ and $\phi(t)$ behave in an acceptable fashion. Once the analysis is complete, the adaptive law is implemented using the fact that $\phi(t) = \theta(t) - \theta^*$, so that the adaptive law has the form

$$\dot{\theta}(t) = f_2(e(t), t).$$

Chapter 7 is devoted to the study of the stability properties of error models that are used in earlier chapters for the design of adaptive observers and controllers in the absence of external disturbances. The modifications in the adaptive laws proposed in subsequent chapters when different types of perturbations are present, are based on the insights provided by these error models.

The error equations (1.16a,b) represent a set of nonlinear and nonautonomous differential equations. While Eq. (1.16a) can be readily derived using the information provided regarding the process to be controlled and the structure of the controller, the adaptive law given by Eq. (1.16b) has to be chosen by the designer. At first glance it might appear that the designer has considerable freedom to achieve the objective, that is, the stability of the equilibrium state, by suitably choosing $f_2(.,.)$. However it should be noted that θ^* and, hence, $\phi(t)$ are unknown so that the adaptive law cannot explicitly include $\phi(t)$. This places a substantial constraint on the choice of the adaptive law and hence makes the problem theoretically considerably more interesting.

1.7 APPLICATIONS

The demands of a rapidly growing technology for faster and more accurate controllers have had a strong influence on the progress of automatic control theory ever since its birth five decades ago. Developments in computer technology have also had an equally great impact on the thinking of control theorists. Since the 1950s, when the computer emerged as a powerful computational tool, major changes in computer technology have almost immediately been followed by significant changes in control system design. In the first stage of this interaction, the advent of the computer freed the systems theorist from the philosophy that systems problems had to be solved in closed form before practically implementable solutions could be obtained. In the second stage, during the middle 1960s, special-purpose computer programs based on powerful results in optimal control theory were applied to a wide spectrum of control problems including attitude control, interplanetary guidance, chemical reactors, and high speed trains. In the current third stage, when revolutionary progress is being made in microprocessor technology, the speed and size of the computers coupled with their low cost are making it feasible to implement adaptive control algorithms in situations that were inconceivable a decade ago. In fact, we are rapidly approaching a stage where the scope of the application will be limited only by the available theory and the imagination of the designer.

In recent years both MRAS and STR have been applied to numerous industrial problems. Feasibility studies have also been carried out in such widely different areas as autopilots for aircrafts and ships, cement mills, paper machines, and power systems, indicating the broad applicability of the methods developed. Although the general reason for requiring adaptive control in all applications is to cope with uncertainty in the system, the specific reason can vary from one area to another or even for different applications within the same area. For example, process control systems are described by very complex nonlinear equations but have the advantage of having long time constants so that indirect control methods can be effectively used. In aircraft systems, however, where flight regimes are constantly being extended, adaptive control may be needed to cope with rapid changes in dynamics that make gain scheduling infeasible. Sensors for measuring the relevant output variables may be unavailable in some processes as, for example, a cement mixer, or they may be available but economically unattractive as in aircraft systems.

While the examples above indicate that the use of adaptive control may be motivated by diverse considerations, it should also be noted that similar problems occurring in different industries may also call for adaptive solutions. In the chemical industry, for example, hundreds of parameters have to be tuned continuously to keep processes operating at or near their optimal values. In the automotive industry, the tuning or testing of thousands of engines coming off the production line may call for similar automatic tuning procedures. In these situations, adaptive control may result in only a slight increase in efficiency, which in turn may correspond to substantial economic savings. These and similar questions related to the reasons for using adaptive methods in different contexts are treated in Chapter 11 through case studies.

The last few years have witnessed numerous feasibility studies of adaptive control methods both in universities and industrial research laboratories. Increasingly, we are beginning to hear that such initial efforts are proving successful. The appearance of commercial products generally signifies a state of maturity of a field. The self-tuning regulators introduced by ASEA of Sweden and the Foxboro Company and Leeds and Northrup in the United States indicate that adaptive control has come of age. We can anticipate accelerated growth of adaptive systems in the coming years. To cope with increased uncertainty in extremely complex systems, even as different features of the adaptation such as perturbation levels, variation of gains, averaging times, performance indices, and priorities change, multilevel adaptation will be required to achieve adaptation at reasonable rates. Combining the concepts of knowledge-based expert systems of artificial intelligence at the highest levels where the adaptive process is on a slow time scale, with adaptive methods such as those described in this book developed by control theorists at the lower fast time-scale levels, multilevel systems can be synthesized and expected to achieve new capabilities.

1.8 SUMMARY

In this chapter, we have attempted to acquaint the reader with some of the developments in control theory and practice that led to the emergence of adaptive control theory in recent years as a viable area of research within the general field of mathematical systems theory. The evolution of linear systems theory, which is currently well understood, is first outlined in Section 1.1 and some of the mathematical concepts used in the analysis and synthesis of feedback systems are reviewed in Section 1.2.

The presence of uncertainties in the structure or parameters of a dynamical system precludes the direct application of well known linear principles for its control. Qualitatively, one can say that the information needed to improve the performance of such a system has to be collected even as the system is in operation. The term adaptation, borrowed from biology, was introduced into the control literature to describe such a class of systems. In Section 1.3, early attempts to characterize an adaptive system are discussed, and reveal that a universal definition of an adaptive system is quite elusive. Four definitions proposed in the 1960s by well known workers in the field, also examined in Section 1.3, address different aspects of adaptation and serve to emphasize the diversity of viewpoints.

Most of the systems that are currently considered as being adaptive can also be viewed as nonlinear feedback systems. The principal difficulty in developing systematic procedures for their design stems from the fact that methods for dealing with nonlinear systems are not available. Hence, for mathematical tractability, we confine our attention in this book, as stated in Section 1.3.2, to the analysis of a class of nonlinear systems. Under ideal conditions, these systems converge asymptotically to desired linear time-invariant systems. We refer to this class of systems in the following chapters as being adaptive.

Direct and indirect control, which are two distinct approaches to adaptive control, are discussed in Section 1.4. The parameter perturbation method as an example of direct control and the sensitivity method as an example of indirect control are examined. Although the basis for parameter adjustment using the two methods is quite clear, global stability of the resultant nonlinear systems cannot be demonstrated. This fact led to a search for alternate methods for designing stable adaptive controllers. However, the importance of these early methods lies in the fact that many of the ideas presented in the following chapters can be traced back to them.

Model reference adaptive control and self-tuning regulation are the two well known methods that are currently in vogue for the adaptive control of dynamical systems. While direct or indirect control can be used in the two cases, historically MRAC has been associated with direct control and STR with indirect control. In both cases, either an identification model and/or a reference model is needed and these are discussed in Section 1.5. Although identification models have been in use for a long time and are extensively developed in the literature, control models are of more recent origin. Reference models for control purposes, when the plant to be controlled is linear and time-invariant with known parameters, are examined in this section and provide a rationale for the choice of similar models in adaptive situations. Following this, the adaptive identification problem and the adaptive control problem are stated at the end of Section 1.5.

The fundamental requirement of any dynamical system is its stability. The earlier adaptive methods described suffer from the serious drawback that they do not assure the stability of the overall system for arbitrary initial conditions. In contrast to these methods, the approach developed in this book results in systematic methods for designing stable adaptive systems. Such systems, in turn, can be optimized by the adjustment of the adaptive gains under the control of the designer. This is stated in Section 1.6. Error models introduced in Section 1.6 are found to be particularly convenient for the choice of stable adaptive laws as well as for the analysis and synthesis of the overall adaptive system. In fact, most of the stability analyses carried out in the book are based on such error models.

Section 1.7 concludes with a brief statement of the present status of the applications of adaptive systems theory and provides an introduction to Chapter 11 where five applications are treated in detail.

1.9 SCOPE OF THE BOOK

The principal aim of this book is to present, in a self-contained fashion, a general approach based on Lyapunov's direct method for the control of deterministic continuous time linear systems with unknown parameters. These parameters can be either constant or they can vary with time. In both cases, the aim is to adjust the parameters of a controller so that the overall system has bounded solutions. Since such an adjustment is based on the observed outputs, the overall system is invariably nonlinear. Hence, the book deals with the analysis and synthesis of a class of systems that are intentionally made nonlinear. As mentioned in Section 1.1, when these parameters are completely known, linear systems theory can be used to design controllers so that the performance of the closed-loop system is improved or optimized according to some criterion. When the parameters are unknown, our aim will be to determine control methods that will result in the performance of the overall system comparable to that obtained above.

The chapters of the book are arranged to address the basic questions related to the problem above. In Chapter 3, some simple adaptive situations are introduced and the manner in which the stability approach finds application in them is discussed. These, in turn, provide the basis for more general problems encountered in Chapters 4 and 5. In Chapter 4, the design of adaptive observers that estimate the state and parameters of unknown systems using input-output data is considered. In Chapter 5, the fundamental problem of adaptive control is formulated and solved. The conditions sufficient to guarantee stable solutions are stated using a direct approach. It is shown that, by a proper parametrization of the plant, identical results can also be obtained using an indirect approach. The solution to this problem, which is referred to as the *ideal case*, provides a benchmark for the comparison of different adaptive schemes and their modifications treated in the chapters that follow. Chapter 6 deals with the important concept of persistent excitation, which has a wide application in adaptive systems. Following this, the relation between persistent excitation and the convergence of the adjusted parameters to their true values in the ideal case, is discussed. Chapter 7 introduces the concept of error models and describes the central role they play in the analysis of all the adaptive systems described in the book. It is shown that the convergence of the various adaptive procedures described in Chapters 3-6 can be studied in a unified fashion using such error models.

Chapters 3-7 deal with the developments in the field from the late 1960s to the end of the 1970s when the first complete solution to the adaptive control problem was given. The concepts and techniques developed during this period are now well established and are included in the chapters above. In our opinion, the repertoire of any aspiring adaptive control engineer should include the methods described here.

Research in the adaptive control field has proceeded in three distinct directions in the 1980s. The first, concerned with the effect of different types of perturbations on the stability and performance of adaptive systems, has become increasingly important and is currently an active area of research. The contents of Chapter 8 describe some of the basic difficulties encountered in the analysis of such systems and efforts that have proved successful in coping with them. In view of the present level of interest and activity in these problems, it is only to be expected that many of the specific results presented in this chapter will be superseded by newer developments.

A second direction of research deals with some of the basic theoretical issues in the control of uncertain systems. The main thrust of this effort is in determining the least restrictive assumptions under which stable adaptation can be achieved in the presence of plant uncertainty. Some of the results using this approach are included in Chapter 9 to acquaint the reader with this different viewpoint. The methods used in this chapter are distinctly different from those used in previous chapters. They are still in the early stages of development and consequently in a state of flux. While many of the results cannot be applied directly to practical problems, they are nevertheless of considerable importance and are bound to have a significant impact on the thinking in the field in the future.

Chapter 10 deals with multivariable adaptive systems, which represents the third direction in which research proceeded in the 1980s. It is shown that essentially all the results derived for single-input single-output systems have their counterparts in multiple-input multiple-output systems as well.

Finally, in Chapter 11, five case studies of applications of adaptive systems theory to problems in industry are discussed.

PROBLEMS

Part I

1. In the following systems identify the inputs, disturbances, parameters, state variables, and outputs. In (a) and (b), make reasonable assumptions to derive a mathematical model of the system.

(a) A string of N cars travelling along a single lane highway.

(b) An electrical network consisting of resistances, inductances, and capacitances.

(c) The dynamics of an aircraft are represented by the differential equations

$$\begin{aligned}
\dot{x}_1 &= \frac{f_1(x_3, M)}{m} \cos x_5 + \frac{f_2(x_3, M, x_5)}{m} - g \sin x_2 \\
\dot{x}_2 &= \frac{f_3(x_3, M, x_5)}{m\dot{x}_1} + \frac{f_1(x_3, M)}{mx_1} \sin x_5 - g\frac{\cos x_2}{x_1} \\
\dot{x}_3 &= x_1 \sin x_2 \\
\dot{x}_4 &= x_1 \cos x_2
\end{aligned}$$

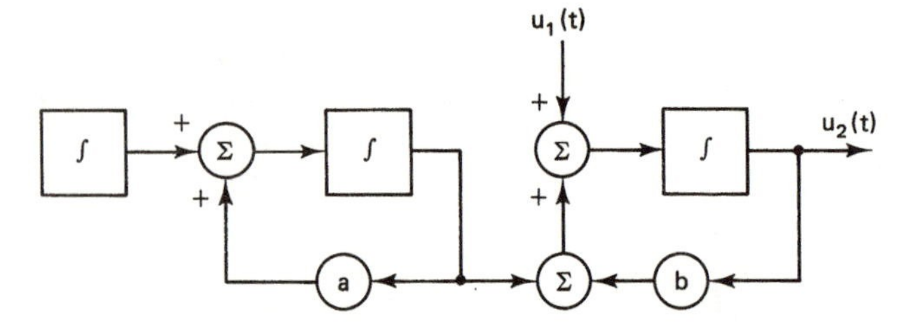

Figure 1.11 Representation of a dynamical system.

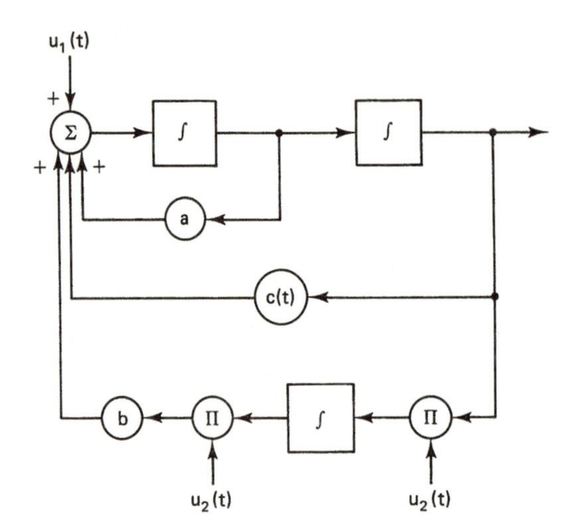

Figure 1.12 A dynamical system with multipliers.

where M = mach number, m = mass, g = gravity and x_1, x_2, x_3, x_4, and x_5 are, respectively, the velocity, its angle with the horizontal, the altitude, the horizontal distance, and angle of attack. $f_1(x_3, M)$, $f_2(x_3, M, x_5)$ and $f_3(x_3, M, x_5)$ are stored functions of their arguments. It is assumed that the mach number M varies with time.

2. **(a)** A block diagram representation of a dynamical system is shown in Fig. 1.11 and contains integrators, summers, and constant gains. Determine the set of differential equations describing the system. Is the system linear? What are the inputs, state variables, and outputs of the system?

   (b) The dynamical system shown in Fig. 1.12 contains multipliers (denoted by the symbol Π) in addition to the elements contained in Problem 2(a). Is this system linear? Determine a set of differential equations describing the system. Indicate the signals you would consider as inputs and those you would describe as time-varying parameters. Give the reasons for your choice.

Part II

3. **(a)** Define direct control and indirect control in adaptive systems. Give a typical example for each system.

   (b) (i) A person drives a car on an icy road. (ii) A person maneuvers a boat in rough weather.

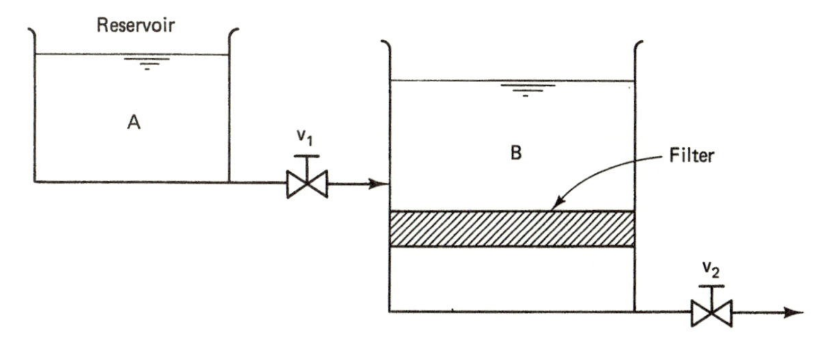

Figure 1.13 Water filtration system.

According to you, are these adaptive systems? If so, are they examples of direct control or indirect control?

4. In a dynamical system only the input and the output can be observed and a single parameter can be adjusted. The object is to adjust the parameter so that the output is minimized in some sense.

   (a) Briefly describe how the parameter perturbation method can be used to determine the optimal value of k. Why is this a direct approach? Explain why it is difficult to prove the stability of the overall system in this case.

   (b) Can a sensitivity method be used for the problem above? If so, what is the prior information that will be needed to use such an approach? Why is this an indirect method?

5. Some simple feedback systems are described below. In each case, discuss whether or not you would consider the system to be adaptive. Give reasons for your answer. If you decide that it is not adaptive, indicate the modifications in the problem that would require an adaptive solution. Once again, give reasons why you consider the solution to be adaptive.

   (a) A water filtration and purification system is shown in Fig. 1.13. Water from a reservoir A is admitted through a valve v_1 into a tank B where it is filtered. The tank B is connected through a valve v_2 to chemical and biochemical purifiers downstream. The objective is to control the flow through the system by adjusting v_1 and v_2 so that the level of the water in B is regulated around a fixed value.

   When the resistance in B is a constant, fixed openings of v_1 and v_2 are found to be adequate. However, when the resistance of the filtering tank changes due to the deposition of impurities, the valves v_1 and v_2 have to be adjusted continuously.

   (b) A turbo generator shown in Fig. 1.14, consisting of an air pressure turbine that drives a synchronous generator, is used to keep the output voltage and frequency at constant values. The actuating valve of the turbine has a nonlinear characteristic, and the relationship between the field current and the voltage shows a strong nonlinear behavior. Denoting the valve position and the field current by u_1 and u_2 respectively, and the output frequency f and voltage v by y_1 and y_2 respectively, a two-input two-output dynamical system can be described. When the load R_L varies, the turbo generator works in different operating points resulting in significant variations in

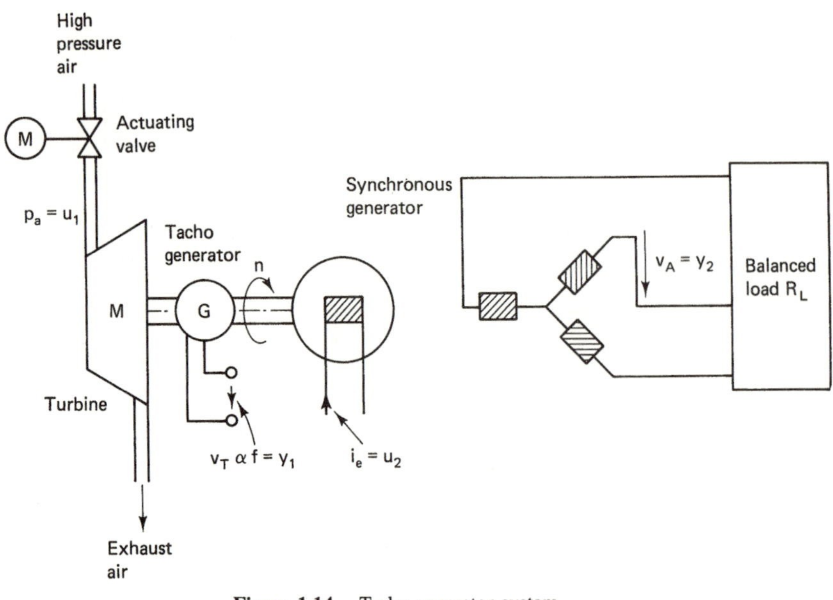

Figure 1.14 Turbo generator system.

frequency and voltage. Hence, the valve position u_1 and field current u_2 have to be adjusted continuously [42].

(c) The object of control in paper machines is to adjust the control variables u_1 and u_2 in such a manner as to minimize the deviations of the basis weight y_1 and moisture y_2 of the manufactured paper from predetermined values. The variables u_1 and u_2 correspond to the thick stock valve and the steam pressure valve in the drier section respectively (Fig. 1.15). u_1 and u_2 must be adjusted when the quality of the paper changes due to moisture deviations, pressure changes in the headbox, or due to varying conditions in the drying section [10].

(d) A simple model of an aircraft system can be described by the differential equation

$$\ddot{x} + 2[\zeta(t) + k(t)]\dot{x} + x = r(t)$$

where $r(t)$ is a reference input. The desired response of the system is achieved when the damping factor $\zeta(t) + k(t)$ is a constant and equals .7. It is known that $\zeta(t)$ varies with time due to changes in air density. The step response of the system is determined periodically (by using an additive step input) and the damping parameter is estimated. This in turn is used to adjust the velocity feedback gain k so that the estimate of the damping factor of the overall system is .7. Under what conditions do you expect the system above to perform satisfactorily?

Part III

6. When a direct approach is used to optimize the parameters of the controller in a dynamical system, explicit identification of the unknown plant parameters is not used in the control process. When adaptation is complete, can the controller parameter values be used to estimate the plant parameters? Describe a simple example to indicate when this is possible.

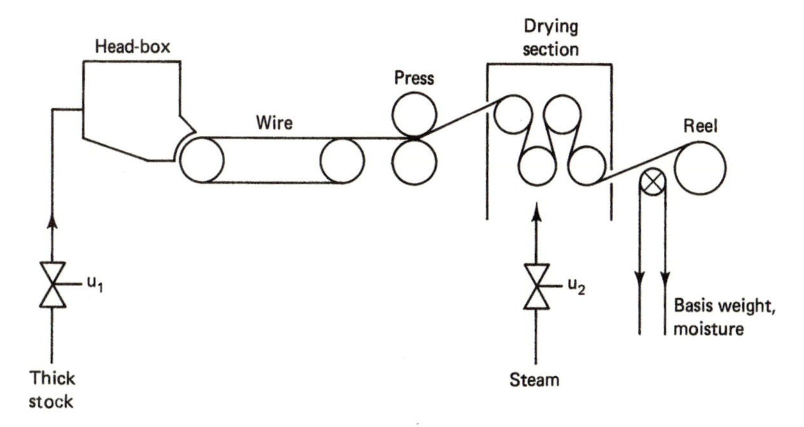

Figure 1.15 Control of a paper machine. Courtesy Pergamon Press.

7. A single-loop feedback system has an unstable linear time-invariant block in the forward path and a single adjustable parameter $k(t)$ in the feedback path. It is further known that there exists a constant value k^* of k with $k_1 \leq k^* \leq k_2$ for which the feedback system is asymptotically stable. Starting from some arbitrary initial condition k_0 in the given interval $[k_1, k_2]$, how would you adjust $k(t)$ so that it converges to a stable value of the parameter k? If you are convinced that your method will work, can you formalize the procedure to synthesize an adaptive controller that will stabilize the system?

REFERENCES

1. Anderson, G.W. "A self-adjusting system for optimum dynamic performance." *IRE National Conventional Record* Pt. 4:182–190, 1958.
2. Aseltine, J.A., Mancini, A.R., and Sarture, C.W. "A survey of adaptive control systems." *IRE Transactions on Automatic Control* 3:102–108, Dec. 1958.
3. Åström, K.J. "Self-tuning regulators - Design principles and applications." In *Applications of Adaptive Control*, edited by K.S. Narendra and R.V. Monopoli, pp. 1–68, New York: Academic Press, 1980.
4. Åström, K.J., Borisson, U., Ljung, L., and Wittenmark, B. "Theory and applications of self-tuning regulators." *Automatica* 13:457–476, 1977.
5. Åström, K.J., and Wittenmark, B. "On self-tuning regulators." *Automatica* 9:185–199, 1973.
6. Bellman, R. *Adaptive control processes - A guided tour*. Princeton:Princeton University Press, 1961.
7. Bellman, R., and Kalaba, R. "On adaptive control processes." *IRE Transactions on Automatic Control* 4:1–9, Nov. 1959.
8. Bryson, Jr., A.E., and Ho, Y.C. *Applied optimal control*. Waltham, MA:Blaisdell Publishing Company, 1969.
9. Bykhovskiy, M.L. "Sensitivity and dynamic accuracy of control systems." *Engineering Cybernetics*, pp. 121–134, Nov.–Dec. 1964.

10. Cegrell, T., and Hedqvist, T. "Successful adaptive control of paper machines." *Automatica* 11:53–59, 1975.

11. Cruz, Jr, J.B. *System sensitivity analysis*. Stroudsburg, PA:Dowden, Hutchinson & Ross, Inc., 1973.

12. Douce, J.L., Ng, K.C., and Gupta, M.M. "Dynamics of the parameter perturbation process." In *Proceedings of the IEE*, vol. 113, no. 6, pp. 1077–1083, June 1966.

13. Draper, C.S., and Li, Y.T. "Principles of optimalizing control systems and an application to the internal combustion engine." *ASME Publication*, Sept. 1951.

14. Drenick, R.F., and Shahbender, R.A. "Adaptive servomechanisms." *AIEE Transactions* 76:286–292, Nov. 1957.

15. Duarte, M.A., and Narendra, K.S. "Combined direct and indirect approach to adaptive control." Technical Report No. 8711, Center for Systems Science, Yale University, New Haven, CT, September 1987.

16. Erzberger, H. "Analysis and design of model following control systems by state space techniques." *Proceedings of the JACC*, Ann Arbor, Michigan, 1968.

17. Eveleigh, V.W. *Adaptive control and optimization techniques*. New York:McGraw-Hill Book Company, 1967.

18. Fel'dbaum, A.A. *Optimal control systems*. New York:Academic Press, 1965.

19. Grayson, L.P. "Design via Lyapunov's second method." *Proceedings of the 4th JACC*, Minneapolis, Minnesota, 1963.

20. Jacobs, O.L.R. "A review of self-adjusting systems in automatic control." *Journal of Electron. Control* 10:311–322, 1961.

21. Kalman, R.E. "Design of a self optimizing control system." *Transactions of the ASME* 80:468–478, 1958.

22. Kokotovic, P.V. "Method of sensitivity points in the investigation and optimization of linear control systems." *Automation and Remote Control* 25:1512–1518, 1964.

23. Kronauer, R.E. and Drew, P.G. "Design of the adaptive feedback loop in parameter-perturbation adaptive controls." *Proceedings of the IFAC Symposium on the Theory of Self-Adaptive Control Systems*, Teddington, England, September 1965.

24. Landau, I.D. *Adaptive control*. New York:Marcel Dekker, Inc., 1979.

25. Lang, G. and Ham, L.M. "Conditional feedback systems - A new approach to feedback control." *AIEE Transactions* 74:152–161, July 1955.

26. Laszlo, E. *The relevance of general systems theory*. New York:George Braziller, Inc., 1972.

27. Lewis, J.B. "The use of nonlinear feedback to improve the transient response of a servomechanism." *AIEE Transactions* 71:449–453, Jan. 1953.

28. Ljung, L. and Söderstrom, T. *Theory and practice of recursive identification*. Cambridge, MA:M.I.T. Press, 1985.

29. Margolis, M. and Leondes, C.T. "A parameter tracking servo for adaptive control systems." *IRE Transactions on Automatic Control* 4:100–111, Nov. 1959.

30. Mayr, O. *The origins of feedback control*. Cambridge, MA:M.I.T. Press, 1970 (Originally published by R. Oldenbourg Verlag, Munich, 1969).

31. McGrath, R.J., Rajaraman, V., and Rideout, V.C. "A parameter perturbation adaptive control system." *IRE Transactions on Automatic Control* 6:154–161, May 1961.

32. Meissinger, H.F. "Parameter influence coefficients and weighting functions applied to perturbation analysis of dynamical systems." In *Proc. III ASICA Conf.*, Optija, 1961.
33. Morse, A.S. "Structure and design of linear model following systems." *IEEE Transactions on Automatic Control* 18:346–354, 1973.
34. Narendra, K.S., and McBride, Jr., L.E. "Multiparameter self-optimizing systems using correlation techniques." *IRE Transactions on Automatic Control* 9:31–38, Jan. 1964.
35. Narendra, K.S., Thathachar, M.A.L., and Baker, T. "Adaptive control using correlation techniques." *Proceedings of the Allerton Conference*, Urbana-Champaign, Illinois, 1965.
36. Narendra, K.S., Tripathi, S.S., Luders, G., and Kudva, P. *"Adaptive control using Lyapunov's direct method"*. Technical Report No. CT-43, Becton Center, Yale University, New Haven, CT, Oct. 1971.
37. Peterka, V. "Adaptive digital regulation of noisy systems." In *Proceedings of the IFAC Symposium on Identification and Process Parameter Estimation*, Prague, 1970.
38. Stromer, P.R. "Adaptive or self-optimizing control systems – A bibliography." *IRE Transactions on Automatic Control* 4:65–68, May 1959.
39. Truxal, J.G. "Adaptive control." In *Proceedings of the 2nd IFAC*, Basle, Switzerland, 1963.
40. Truxal, J.G. *Automatic feedback control system synthesis*. New York:McGraw-Hill Book Company, 1955.
41. Wang, S.H., and Desoer, C.A. "The exact model matching of linear multivariable systems." *IEEE Transactions on Automatic Control* 17:347–349, 1972.
42. Wiemer, P., and Unbehauen, H. "Decentralized adaptive control of a turbogenerator." In *Proceedings of the Fifth Yale Workshop on Applications of Adaptive Systems Theory*, New Haven, CT, 1987.
43. Wiener, N. *Cybernetics*. Cambridge, MA:M.I.T. Press, 1948.
44. Wilkie, D.F., and Perkins, W.R. "Essential parameters in sensitivity analysis." *Automatica* 5:191–197, 1969.
45. Wolovich, W.A. "The application of state feedback invariants to exact model matching." *Proceedings of th 5th Annual Princeton Conference Information Science and Systems*, 1971.
46. Zadeh, L.A. "On the definition of adaptivity. In *Proceedings of the IEEE* 51:569–570, March 1963.
47. Zames, G., and Francis, B.A. "Feedback, minimax sensitivity, and optimal robustness." *IEEE Transactions on Automatic Control* 28:585–601, 1983.

2

Stability Theory

2.1 INTRODUCTION

The first step in the quantitative analysis of any dynamical system is the formulation of a set of mathematical equations that describe its behavior. However, even in the most rigorous of such formulations, the effect of all unknown disturbing forces cannot be taken into account. The concept of stability, which first arose in mechanics, is concerned with the investigation of the effect of such forces. In some cases, the effect of disturbances may be insignificant while in others it may result in the behavior of the system deviating considerably from that when no disturbances are present. Qualitatively, behavior of the first kind is considered to be stable and that of the second kind, unstable. Since disturbing forces are invariably present in all physical systems, the study of stability problems is of great importance for both theoretical and practical reasons.

The concept of stability has been investigated extensively in the past century and a rich literature currently exists. Readers interested in a more detailed treatment of the subject than is contained in this brief chapter are referred to the books by Hahn [15,16], Coppel [11], Krasovskii [20], and LaSalle and Lefschetz [26]. Since the systems that are studied, the disturbances that can occur, and the phenomena of interest, may take a wide variety of forms, different definitions of stability are found to be appropriate depending on the context. The stability theory of Lyapunov [28], and input-output stability theory based on functional analysis techniques, are the two general approaches that are most widely used to address stability problems that arise in control systems. Lyapunov stability considers stability as an internal property of the system, and deals with the effect of momentary perturbations resulting in changes in initial conditions. Input-output stability theory, as the name suggests, considers the effect of external inputs on the system. Among the many other notions of stability prevalent in the literature, two kinds that deserve special mention here include structural stability and stability in the presence of persistent disturbances (total stability). Structural stability is concerned with the effect of parameter changes on the behavior of the system, while the term persistent disturbance has been used in the Russian literature while referring to both external as well as state-dependent disturbances.

Control theory deals with the analysis and synthesis of dynamical systems in which one or more variables are kept within prescribed limits. The effectiveness of any control system is judged mainly by the accuracy as well as the speed with which the variables are controlled. But before efforts are made to satisfy specific performance criteria, or to optimize the system in any manner, it is imperative that its stability be assured. Hence, one of the primary goals in the analysis of control systems is a clear understanding of their stability properties. As mentioned in Chapter 1, for centuries the existence of feedback was recognized in many natural phenomena and used to build some ingenious practical systems. However, it was only after the pioneering works of Black, Nyquist, and Bode, concerning the stability of linear feedback systems, were well established that the concept could be exploited extensively in a systematic fashion. It is not surprising that the best developed part of control theory at present deals with systems defined by linear time-invariant operators. The stability properties of such systems are well understood and a vast body of frequency domain and time-domain techniques exists for their analysis and design, which are directly or indirectly related to their stability properties.

In the preceding chapter, the need for using adaptive control for improving the performance of a system was discussed, when uncertainty regarding its parameters exists. In parameter adaptive systems, the control parameters are adjusted using all available data, even as the system is in operation. This renders the overall system nonlinear and nonautonomous. The feedback structure of the adaptive system also introduces the possibility of instability. In the early days of adaptive control, the emphasis in design was on the optimization of a performance criterion. The stability of the overall system was analyzed after a suitable adaptive law was chosen. Since, in general, the stability problem of nonlinear systems is a formidable one, these efforts for the most part resulted in stability results of a local nature. In contrast to this procedure, the stability approach started in the 1960s aims to ensure the global stability of the overall adaptive system from the outset by the proper choice of an adaptive law. Optimization of its performance is subsequently carried out within a stability framework. This spirit pervades all aspects of this book; it is based on the premise that it is only when the stability properties of adaptive systems are clearly understood that their design will become a routine matter, even as linear feedback system design is today.

In this chapter, important results in stability theory, which are directly relevant to our study of adaptive systems in the chapters following, are discussed. Although some of these results are general stability theorems that can be found in standard textbooks on the subject, others are of more recent origin developed in the specific context of adaptive systems. All of them are collected and presented here in a coherent fashion and we shall call on them, when necessary, throughout the remainder of the book. Due to space limitations, only a few of the theorems are stated and proved in detail; the proofs of others are either sketched briefly for the sake of completeness or omitted entirely. Appropriate references are provided for readers interested in further details.

In Section 2.2, the notation used throughout the book is introduced. In Section 2.3, we recall briefly well known results in linear time-invariant systems. Definitions of Lyapunov stability, related theorems, and their implications to linear and nonlinear systems are discussed in Section 2.4. Strict positive-real functions and the Kalman-Yakubovich lemma, which play key roles in the design of adaptive systems, are stated in Sections 2.5 and 2.6 respectively. In Section 2.7, the concept of input-output stability and related theorems are presented. In Section 2.8, results peculiar to the stability of parameter adaptive systems are discussed. Finally, other stability concepts such as total stability and bounded-input bounded-output stability are defined and related theorems presented in Section 2.9.

2.2 NOTATION

$\mathbb{R}^n$ denotes the n-dimensional Eulidean space. $\mathbb{R}^1$ is also denoted as $\mathbb{R}$. $\mathbb{R}^+$ is the set of nonnegative real numbers. $\mathbb{C}$ denotes the set of all complex numbers. $\mathbb{C}^-$ refers to the open left half of the complex plane, and $\mathbb{C}^+ = \mathbb{C} - \mathbb{C}^-$. The symbol s is used to denote both the complex variable as well as the differential operator d/dt.

$\mathbb{R}^{n \times n}$ is the set of all real $(n \times n)$ matrices. I denotes the identity matrix. The dimension of I is not explicitly stated in the following chapters but can be inferred from the context. All vectors are column vectors and the superscript ^{T} denotes the transpose. The symbol $\triangleq$ defines the variable before it. If $x = [x_1, x_2, \ldots, x_n]^T$, its norm, denoted by $\|x\|$, is the Euclidean norm

$$\|x\|^2 \triangleq x_1^2 + \cdots + x_n^2.$$

The norm of an $n \times n$ matrix A is defined as

$$\|A\| \triangleq \sup_{\|x\|=1} \{\|Ax\|, \ x \in \mathbb{R}^n\}.$$

If a symmetric matrix A is positive-definite, then we write $A > 0$. If A_1 and A_2 are symmetric positive-definite matrices, we write $A_1 > A_2$ if $A_1 - A_2 > 0$. Some additional notational details pertaining to multivariable systems can be found in Appendix E.

In the study of adaptive systems, we restrict our attention to dynamical systems that are governed by ordinary differential equations containing a finite number of parameters. If $x_1, x_2, \ldots, x_n$ represent the n coordinates in a Euclidean n-space $\mathbb{R}^n$, and t the time, the behavior of a typical finite-dimensional dynamical system is described by the differential equations

$$\dot{x}_i(t) = f_i(x_1(t), x_2(t), \ldots, x_n(t), t) \qquad i = 1, 2, \ldots, n$$

or equivalently by the vector differential equation

$$\dot{x} = f(x, t) \qquad f(0, t) = 0, \ \forall\, t \geq t_0 \tag{2.1}$$

where x and f are column vectors with components x_i and $f_i (i = 1, 2, \ldots, n)$ respectively. The variables x_i are referred to as state variables and x is referred to as the state vector. The notation $x(t)$ indicates that the components x_i of x are functions of t. All time-varying functions are defined for all $t \geq t_0$, where $t_0 \geq 0$. We shall refer to the $(n+1)$ dimensional space of the quantities $x_1, \ldots, x_n, t$ as the motion space M. If these are continuous, then the point $(x(t), t)$ moves along a segment of a curve in motion space as t varies over an interval $[t_1, t_2]$. The projection of the motion on the state space is called a state curve, trajectory of the motion, or the solution of Eq. (2.1). If the initial time is t_0, and the initial value of $x(t)$ is $x(t_0) = x_0$, we shall denote the solution of Eq. (2.1) as $x(t; x_0, t_0)$, where $x(t_0; x_0, t_0) = x_0$. In cases where the initial condition is evident from the context, the solution is simply denoted as $x(t)$.

From the existence theorem of Caratheodory ([10], page 43), it is known that if for all (x, t) belonging to a domain $\mathcal{B}$ in M, (i) $f(x, t)$ is continuous in x for all fixed t, (ii) measurable in t for all fixed x, and (iii) $\|f(x, t)\| \leq \mu(t)$ where $\mu(t)$ is integrable over $|t - t_0| \leq \alpha$, then for some $\alpha_1 > 0$, there exists a solution $x(t; x_0, t_0)$ for $|t - t_0| \leq \alpha_1$. We shall denote the class of functions that satisfy conditions (i)-(iii) by $\mathcal{F}$. Under the same conditions (i)-(iii), the solution can be continued up to the boundary of the domain $\mathcal{B}$. Throughout the book, our interest will be in solutions of differential equations that exist for all $t \geq t_0$ (see Comment 2.1).

A constant solution $x(t; x_0, t_0) \equiv x_0$ is an equilibrium state or a singular point of Eq. (2.1). If x_0 is the only constant solution in the neighborhood of x_0, it is called an isolated equilibrium state. If the right-hand side of Eq. (2.1) does not depend on t, the equation is called autonomous; if $f(x, t+T) = f(x, t)$ for some constant T, the equation is said to be periodic. If $f(x, t)$ is linear in x, Eq. (2.1) is said to be a linear equation and the corresponding dynamical system that it represents is referred to as a linear system.

2.3 LINEAR SYSTEMS

The stability of linear systems is a well studied topic and can be considered the precursor to almost all results in the stability of nonlinear systems. In this section, we summarize some well known results pertaining to linear systems.

Consider a linear system described by the differential equation

$$\dot{x}(t) = A(t)x(t) \qquad x(t_0) = x_0 \tag{2.2}$$

where $x : \mathbb{R}^+ \to \mathbb{R}^n$ and $A : \mathbb{R}^+ \to \mathbb{R}^{n \times n}$. The elements of $A(t)$ are assumed to be bounded and piecewise-continuous for all $t \in \mathbb{R}^+$. Then Eq. (2.2) has a unique solution that can be expressed as

$$x(t) = \Phi(t, t_0)x_0$$

for any initial condition x_0, where $\Phi(t, t_0)$ is the transition matrix of Eq. (2.2) satisfying the differential equation

$$\dot{\Phi}(t, t_0) = A(t)\Phi(t, t_0) \qquad \Phi(t_0, t_0) = I.$$

Each column of $\Phi(t, t_0)$ represents a linearly independent solution of the differential equation (2.2) and any solution of the latter can be expressed as a linear combination of the columns of $\Phi(t, t_0)$.

The behavior of the solutions of Eq. (2.2) can be summarized as follows:

Theorem 2.1 (i) All solutions of Eq. (2.2) are bounded if, and only if, $\|\Phi(t, t_0)\| \leq k$, where k is a positive constant.
(ii) All solutions of Eq. (2.2) tend to zero as $t \to \infty$ if, and only if, $\|\Phi(t, t_0)\| \to 0$ as $t \to \infty$.

For general time-varying systems, the determination of the analytic form of the transition matrix is a difficult task and, hence, conditions of Theorem 2.1 are hard to verify. For the special case when the system is time-invariant, the analytic conditions of Theorem 2.1 can be translated into algebraic conditions.

2.3.1 Linear Time-Invariant Systems

When the system is linear and time-invariant, described by the differential equation

$$\dot{x}(t) = Ax(t) \tag{2.3}$$

where A is a constant matrix, the transition matrix $\Phi(t, t_0)$ is given by

$$\Phi(t, t_0) = \exp\{A(t - t_0)\}$$

where $\exp(At)$ is the matrix defined by

$$\exp(At) \triangleq I + At + \frac{A^2t^2}{2!} + \cdots + \frac{A^n t^n}{n!} + \cdots$$

and converges for all A and all $t \in \mathbb{R}$. Using the similarity transformation $A = T\Lambda T^{-1}$, where Λ is the Jordan canonical form of A, the transition matrix and, hence, the conditions of Theorem 2.1 can be evaluated using the identity

$$\exp(At) = T \exp(\Lambda t) T^{-1}.$$

Example 2.1

If

$$A = \begin{bmatrix} 0 & 1 \\ 0 & -3 \end{bmatrix} \quad \text{then} \quad \Lambda = \begin{bmatrix} 0 & 0 \\ 0 & -3 \end{bmatrix},$$

$$T = \begin{bmatrix} 1 & 1 \\ 0 & -3 \end{bmatrix} \quad \text{and} \quad \exp(At) = \begin{bmatrix} 1 & \frac{1}{3}(1 - e^{-3t}) \\ 0 & e^{-3t} \end{bmatrix}.$$

Then by Theorem 2.1, the solutions of $\dot{x} = Ax$ are bounded. Since T is a bounded matrix, the information concerning the boundedness or unboundedness of the solutions of Eq. (2.3) is contained in the term $\exp[\Lambda t]$, or equivalently, in the matrix Λ. Hence, in this case the analytic conditions of Theorem 2.1 can be expressed algebraically in terms of the eigenvalues of Λ and, hence, those of A. These are given in Theorem 2.2 and the reader is referred to [9] for a proof.

Theorem 2.2 [9]

1. All solutions of Eq. (2.3) are bounded if, and only if, all the eigenvalues of A have nonpositive real parts and those with zero real parts are simple zeros of the minimal polynomial of A[1].
2. All solutions of Eq. (2.3) tend to zero as $t \to \infty$ if, and only if, all the eigenvalues of A have negative real parts.

We shall refer to matrices that satisfy condition 1 as stable matrices and those that satisfy condition 2 as asymptotically stable matrices.

The eigenvalues of A can be determined as the roots of the characteristic equation

$$g(\lambda) \stackrel{\triangle}{=} det[\lambda I - A] = 0.$$

If the roots of $g(\lambda)$ lie in $\mathbb{C}^-$, $g(\lambda)$ is referred to as a Hurwitz polynomial and the solutions of Eq. (2.3) decay to zero. If the roots satisfy condition (1) in Theorem 2.2, $g(\lambda)$ is referred to as a stable polynomial. These conditions can be checked by using the Routh-Hurwitz criterion, obtained about a hundred years ago. This shows the importance of the results of Routh and Hurwitz in the stability theory of linear time-invariant systems. Since, in general, the transition matrix $\Phi(t, t_0)$ of a linear time-varying system cannot be expressed directly in terms of the matrix $A(t)$, similar algebraic conditions are not available for time-varying systems.

2.3.2 Almost Time-Invariant Systems

In the class of adaptive systems we will be discussing in the following chapters, conditions under which the behavior of a nonlinear system tends asymptotically to that of a linear time-invariant system are of interest. The well known results concerned with the stability of almost time-invariant systems are found to be relevant in this context.

A linear dynamical system is said to be almost time-invariant if it can be described by the linear time-varying differential equation

$$\dot{x} = (A + B(t))\, x \qquad \forall\, t \geq t_0 \tag{2.4}$$

where A is a constant matrix and $B(t)$ is small, in some sense, as $t \to \infty$. Two particularly important cases are those where $\|B(t)\| \to 0$, or $\int_{t_0}^{\infty} \|B(t)\| dt < \infty$. Given the properties of the solutions of the LTI differential equation

$$\dot{z} = Az \tag{2.5}$$

the question that arises is the manner in which the solutions of Eq. (2.4) deviate from those of Eq. (2.5) for the same initial conditions. For a detailed study of such equations the reader is referred to the books by Bellman [6] and Cesari [8]. If the matrix A in Eq. (2.5) is stable, but not asymptotically stable, very little can be concluded about the solutions of Eq. (2.4) from the fact that $\lim_{t \to \infty} B(t) = 0$. This can be seen from the following example.

---

[1] The *minimal polynomial* of a matrix A is the monic polynomial $\psi(\lambda)$ of least degree, such that $\psi(A) = 0$.

Example 2.2

If in Eq. (2.4)

$$A = \begin{bmatrix} 0 & 1 \\ -1 & 0 \end{bmatrix} \quad \text{and} \quad B(t) = \begin{bmatrix} 0 & 0 \\ 0 & -2/t \end{bmatrix}, \qquad \forall\, t \geq t_0$$

then the fundamental solutions of Eq. (2.4) are $(1/t)\sin t$ and $(1/t)\cos t$ and, hence, are bounded for all $t > 0$, and tend to zero as $t \to \infty$. If

$$A = \begin{bmatrix} 0 & 1 \\ -1 & 0 \end{bmatrix} \quad \text{and} \quad B(t) = \begin{bmatrix} 0 & 0 \\ 0 & 2/t \end{bmatrix},$$

then the fundamental solutions are of the form $\sin t - t\ \cos t$ and $\cos t + t\ \sin t$, and, hence, grow in an unbounded fashion as $t \to \infty$.

Stronger conditions on $B(t)$ are needed to assure the boundedness of the solutions of Eq. (2.4) and are contained in Theorem 2.3. The following important lemma, generally referred to as the Gronwall-Bellman lemma, is used to derive stability results for equations of the form in Eq. (2.4), and is stated below [6].

Lemma 2.1 If $u, v \geq 0$, if c_1 is a positive constant and if

$$u \leq c_1 + \int_0^t uv dt$$

then

$$u \leq c_1 \exp\left(\int_0^t v dt\right).$$

Theorem 2.3 deals with the case where A is a stable matrix. In the problems discussed later in this book, the matrix A is generally asymptotically stable. Theorem 2.4 deals with the solutions of Eq. (2.4) under less restrictive conditions on $B(t)$. In particular, two specific results when (i) $\|B(t)\| < b_0$ for all $t \geq t_0$, and (ii) $\|B(\cdot)\| \in \mathcal{L}^1$, which were given in [6] are presented. Lemma 2.2, which deals with the case $\|B(\cdot)\| \in \mathcal{L}^2$, was derived more recently in the context of adaptive systems.

Theorem 2.3 [6] If all solutions of Eq. (2.5) are bounded for $t \geq t_0$, the same is true of the solutions of Eq. (2.4) if $\int_{t_0}^{\infty} \|B(t)\| dt < \infty$.

Proof. Every solution of Eq. (2.4) is of the form

$$x(t) = \exp\{A(t-t_0)\}\, x(t_0) + \int_{t_0}^t \exp\{A(t-\tau)\}\, B(\tau)x(\tau)\, d\tau \qquad \forall\, t \geq t_0.$$

Let

$$c_1 = \max\left\{ \sup_{t\geq t_0} \left(\|\exp[A(t-t_0)]\, x(t_0)\|\right), \sup_{t\geq t_0} \left(\|\exp[A(t-t_0)]\,\|\right) \right\}.$$

Since the solutions of Eq. (2.5) are bounded, c_1 is a positive constant. Hence,

$$\|x(t)\| \leq c_1 + c_1 \int_{t_0}^{t} \|B(\tau)\| \, \|x(\tau)\| \, d\tau.$$

Applying Lemma 2.1, we can derive that

$$\|x(t)\| \leq c_1 \exp\left(c_1 \int_{t_0}^{t} \|B(\tau)\| \, d\tau \right) \leq c_2$$

where c_2 is a positive constant.

Theorem 2.4 If all the solutions of Eq. (2.5) approach zero as $t \to \infty$, and constants b_0 and b_1 exist, where b_0 depends on A, such that (i) $\|B(t)\| \leq b_0$ for $t \geq t_0$, or (ii) $\int_{t_0}^{\infty} \|B(t)\| dt \leq b_1$, then the same holds for the solutions of Eq. (2.4).

Proof. Since A is asymptotically stable, there exists positive constants c_3 and c_4 such that

$$\|x(t)\| \leq c_3 \exp\{-c_4(t-t_0)\} + c_3 \exp\{-c_4 t\} \int_{t_0}^{t} \exp\{c_4 \tau\} \|B(\tau)\| \|x(\tau)\| d\tau \quad \forall\, t \geq t_0$$

where

$$c_3 \exp\{c_4(t-t_0)\} \geq \max\left\{ \sup_{t \geq t_0} \left(\|\exp[A(t-t_0)]\, x(t_0)\|\right), \sup_{t \geq t_0} \left(\|\exp[A(t-t_0)]\|\right) \right\}.$$

Applying Lemma 2.1, we have

$$\|x(t)\| \exp\{c_4(t-t_0)\} \leq c_3 \exp\left(c_3 \int_{t_0}^{t} \|B(\tau)\| \, d\tau \right). \tag{2.6}$$

If assumption (i) in Theorem 2.4 is satisfied, then

$$\|x(t)\| \leq c_3 \exp\{-(c_4 - b_0 c_3)(t-t_0)\}$$

and, hence, the solution tends to zero for $b_0 < c_4/c_3$. If assumption (ii) is satisfied, then

$$\|x(t)\| \leq c_5 \exp\{-c_4(t-t_0)\}$$

where c_5 is a positive constant and, hence, the solution tends to zero as $t \to \infty$.

The following lemma is an extension of Theorem 2.4 and is used frequently in the stability analysis of adaptive systems.

Lemma 2.2 Let A be an asymptotically stable matrix and $\int_{t_0}^{\infty} \|B(t)\|^2 dt \leq b_2^2 < \infty$. Then the origin of Eq. (2.7) is exponentially stable, where

$$\dot{x} = [A + B(t)]\, x \qquad \forall\, t \geq t_0. \tag{2.7}$$

Proof. Proceeding as in the proof of Theorem 2.4, we have from the inequality in Eq. (2.6)

$$\|x(t)\| \leq c_3 \exp\{-c_4(t-t_0)\} \exp\left(c_3 \int_{t_0}^{t} \|B(\tau)\| d\tau\right).$$

From the Cauchy-Schwarz inequality, it follows that

$$\int_{t_0}^{t} \|B(\tau)\| d\tau \;\leq\; \sqrt{t-t_0}\left[\int_{t_0}^{t} \|B(\tau)\|^2 d\tau\right]^{1/2} \;\leq\; b_2\sqrt{t-t_0}.$$

Therefore,

$$\begin{aligned} \|x(t)\| &\leq c_3 \exp\left\{-c_4(t-t_0) + c_3 b_2\sqrt{t-t_0}\right\} \qquad \forall\, t \geq t_0 \\ &\leq c_3 \exp\left\{-c_5(t-t_0)\right\} \qquad \forall t \geq t_1 > t_0 \end{aligned}$$

for some positive constant c_5 and finite time t_1.

Corollary 2.1 Let $x : \mathbb{R}^+ \to \mathbb{R}^+$ be a nonnegative function and

$$\dot{x} \;\leq\; [-a + b(t)]\, x \qquad a > 0. \tag{2.8}$$

If (i) $\lim_{t\to\infty} b(t) = 0$, (ii) $b(\cdot) \in \mathcal{L}^1$, or (iii) $b(\cdot) \in \mathcal{L}^2$, then

$$\lim_{t\to\infty} x(t) \;=\; 0.$$

Proof: From Eq. (2.8), it follows that

$$x(t) \;\leq\; \exp\left\{\int_{t_0}^{t} [-a + b(\tau)]\, d\tau\right\} x(t_0).$$

Hence, the results follow from Theorem 2.4 and Lemma 2.2.

2.4 LYAPUNOV STABILITY

In this book, our primary interest is in the stability of adaptive systems represented by nonlinear vector differential equations of the form in Eq. (2.1). In such cases, we first have to assure the existence of solutions for all initial conditions (x_0, t_0) where $x_0 \in \mathbb{R}^n$ and $t_0 \in \mathbb{R}^+$. The origin of the state space $x = 0$ is an equilibrium state of Eq. (2.1) and we will be mainly concerned with its stability properties.

Among the many definitions that have been proposed for the concept of stability, the one formulated by Lyapunov is particularly suited to our discussions and is presented below.

2.4.1 Definitions

Let a system be described by the nonlinear differential equation

$$\dot{x} \;=\; f(x,t) \qquad f(0,t) = 0, \;\; \forall\, t \geq t_0 \tag{2.1}$$

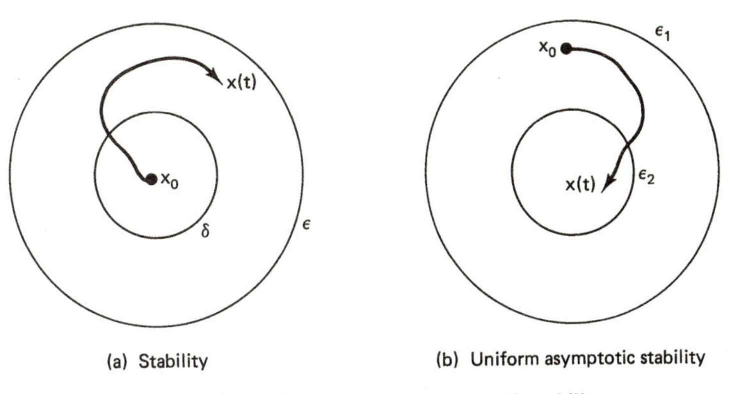

Figure 2.1 Stability and uniform asymptotic stability.

where $x(t_0) = x_0$ and $f : \mathbb{R}^+ \rightarrow \mathbb{R}^n$ is such that a solution $x(t; x_0, t_0)$ exists for all $t \geq t_0$ (see Comment 2.1). Since $f(0, t) \equiv 0$, this implies that the origin is an equilibrium state. The following definitions pertain to some of the basic notions in Lyapunov stability of such an equilibrium state [15,37].

Definition 2.1 The equilibrium state $x = 0$ of Eq. (2.1) is said to be stable if for every $\epsilon > 0$ and $t_0 \geq 0$, there exists a $\delta(\epsilon, t_0) > 0$ such that $\|x_0\| < \delta$ implies that $\|x(t; x_0, t_0)\| < \epsilon \ \forall \, t \geq t_0$ [Fig. 2.1(a)].

Essentially, Definition 2.1 states that small disturbances result in small deviations from the equilibrium state or more precisely, that we can keep the trajectory close to the origin by starting sufficiently close to it.

Definition 2.2 The equilibrium state $x = 0$ of Eq. (2.1) is said to be attractive if for some $\rho > 0$ and every η and $t_0 > 0$, there exists a number $T(\eta, x_0, t_0)$ such that $\|x_0\| < \rho$ implies that $\|x(t; x_0, t_0)\| < \eta$ for all $t \geq t_0 + T$.

Attractivity implies that all trajectories starting in a neighborhood of the origin eventually approach the origin. If the initial vector x_0 has a norm less than ρ, then for all time t greater than $t_0 + T$, the trajectory is less than a distance η away from the origin. One may be tempted to conclude that attractive systems should also be stable. But attractivity and stability were shown to be independent concepts. The reader is referred to [15] for an interesting example by Vinograd in which it is shown that an autonomous system can be attractive but unstable.

Definition 2.3 The equilibrium state $x = 0$ of the differential equation (2.1) is said to be asymptotically stable if it is both stable and attractive.

In this case the trajectory tends to the origin as $t \rightarrow \infty$ and at the same time remains near the origin if it starts sufficiently close to it.

The stability properties of many systems do not depend on the initial time t_0. Autonomous and some periodic systems are typical examples. This motivates the definitions of uniform stability and uniform asymptotic stability.

Definition 2.4 The equilibrium state $x = 0$ of the differential equation (2.1) is said to be uniformly stable if in Definition 2.1, δ is independent of the initial time t_0.

For example, the origin of $\dot{x} = -x$ is uniformly stable, but the origin of $\dot{x} = -x - x(1 - x^2t^3)/t$ is not [15].

Definition 2.5 The equilibrium state $x = 0$ of Eq. (2.1) is uniformly asymptotically stable (u.a.s.) if it is uniformly stable and for some $\epsilon_1 > 0$ and every $\epsilon_2 > 0$, there exists $T(\epsilon_1, \epsilon_2) > 0$ such that if $\|x_0\| < \epsilon_1$, then $\|x(t; x_0, t_0)\| < \epsilon_2$ for all $t \geq t_0 + T$ [Fig. 2.1(b)].

The importance of uniform asymptotic stability in the analysis of dynamical systems will become clear in the following sections.

Definition 2.6 The equilibrium state $x = 0$ of Eq. (2.1) is exponentially stable if there exist constants $a > 0$ and $b > 0$ such that $\|x(t; x_0, t_0)\| \leq a \exp\{-b(t - t_0)\}\|x_0\|, \forall\, t \geq t_0$, for all t_0 and in a certain neighborhood $\mathcal{B}$ of the origin.

All the definitions above imply certain properties of the solutions of differential equations in a neighborhood of the equilibrium state. While stability implies that the solution lies near the equilibrium state, asymptotic stability implies that the solution tends to the equilibrium as $t \to \infty$ and uniform asymptotic stability implies that the convergence of the solution is independent of the initial time. If $\lim_{\epsilon \to \infty} \delta(\epsilon) = \infty$ in the definitions above, then the stability is said to hold *in the large* or the equilibrium state is globally stable[2]. If Definition 2.5 holds for every $\epsilon_1 > 0$, $x = 0$ in Eq. (2.1) is uniformly asymptotically stable in the large (u.a.s.l.).

Exponential stability of a system always implies uniform asymptotic stability. The converse is not true in general. A simple example is the origin of the differential equation $\dot{x} = -x^3$, which is u.a.s. but not exponentially stable. In the case of linear systems, however, uniform asymptotic stability is equivalent to exponential stability. Also, all stability properties hold in the large for linear systems. If the system is autonomous, all stability properties are uniform. If $\mathcal{B} = \mathbb{R}^n$ in Definition 2.6, then the equilibrium state is globally exponentially stable.

Stability of Motion. Although all the definitions above concern the behavior of the equilibrium state of a free dynamical system, the question of stability of a fixed motion is also of interest. However, as suggested by Lyapunov, such a stability problem can be transformed into an equivalent problem of stability of the equilibrium state, where the fixed motion of interest is the equilibrium point under investigation. This is described briefly below:

Let $x^*(t; x_0^*, t_0)$ be the solution of the equation

$$\dot{x} = g(x, t) \qquad x(t_0) = x_0^* \tag{2.9}$$

[2] We shall use the terms global stability and stability in the large interchangeably in this book.

in whose stability we are interested. We shall refer to $x^*(t; x_0^*, t_0)$ as the nominal solution. To determine its stability, we perturb the initial condition to $x(t_0) = x_0^* + \delta x_0$ and observe the deviation of the corresponding solution $x(t; x_0, t_0)$ from the nominal. Since $x(t)$ and $x^*(t)$ are solutions of the same differential equation, we have

$$\begin{aligned} \dot{x}^* &= g(x^*, t) \qquad x^*(t_0) = x_0^* \\ \dot{x} &= g(x, t) \qquad x(t_0) = x_0 = x_0^* + \delta x_0. \end{aligned}$$

If

$$e(t) \triangleq x(t) - x^*(t)$$

then $e(t)$ satisfies the differential equation

$$\dot{e} \triangleq f(e, t) = g(x^* + e, t) - g(x^*, t) \qquad e(t_0) = \delta x_0. \tag{2.10}$$

Since $f(0, t) \equiv 0$, the solution $e \equiv 0$ corresponds to the equilibrium state of the modified differential equation (2.10). This implies that the stability problem associated with the nominal solution $x^*(t; x_0^*, t_0)$ of Eq. (2.9) can be transformed into an equivalent problem of stability of the equilibrium state of Eq. (2.10).

2.4.2 Lyapunov's Direct Method

The direct method of Lyapunov enables one to determine whether or not the equilibrium state of a dynamical system of the form in Eq. (2.1) is stable without explicitly determining the solution of Eq. (2.1). The method has proved effective in the stability analysis of nonlinear differential equations whose solutions are generally very difficult to determine. It involves finding a suitable scalar function $V(x, t)$ and examining its time derivative $\dot{V}(x, t)$ along the trajectories of the system. The reasoning behind this method is that in a purely dissipative system, the energy stored in the system is always positive whose time derivative is nonpositive. In textbooks on stability theory, the Lyapunov theorems for stability are stated separately for autonomous and nonautonomous systems and restrictions are placed on the functions $V(x, t)$ and $\dot{V}(x, t)$ to ensure the different kinds of stability defined in Section 2.4.1. Since the discussions of all these theorems are beyond the scope of this chapter, we shall state below the theorem for the general case of nonlinear nonautonomous differential equations as given by Kalman and Bertram, and provide a brief outline of the proof in [19].

Theorem 2.5 The equilibrium state of Eq. (2.1) is uniformly asymptotically stable in the large if a scalar function $V(x, t)$ with continuous first partial derivatives with respect to x and t exists such that $V(0, t) = 0$ and if the following conditions are satisfied:

(i) $V(x, t)$ is *positive-definite*, that is, there exists a continuous nondecreasing scalar function α such that $\alpha(0) = 0$ and $V(x, t) \geq \alpha(\|x\|) > 0$ for all t and all $x \neq 0$;

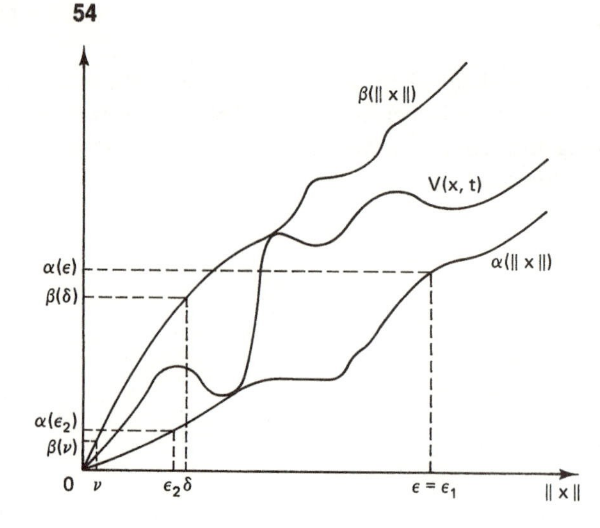

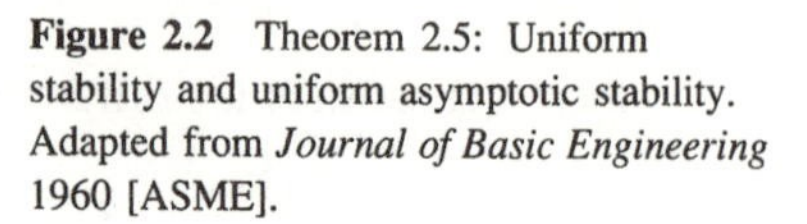
Figure 2.2 Theorem 2.5: Uniform stability and uniform asymptotic stability. Adapted from *Journal of Basic Engineering* 1960 [ASME].

(ii) $V(x,t)$ is *decrescent*, that is, there exists a continuous nondecreasing scalar function β such that $\beta(0) = 0$ and $\beta(\|x\|) \geq V(x,t)$ for all t;

(iii) $\dot{V}(x,t)$ is *negative-definite*, that is,

$$\dot{V}(x,t) = \frac{\partial V}{\partial t} + (\nabla V)^T f(x,t) \leq -\gamma(\|x\|) < 0$$

where γ is a continuous nondecreasing scalar function with $\gamma(0) = 0$, ∇V is the gradient of V with respect to x, and the time derivative of V is evaluated along the trajectory of the differential equation (2.1).

(iv) $V(x,t)$ is *radially unbounded*, that is, $\alpha(\|x\|) \to \infty$ with $\|x\| \to \infty$.

Proof. An outline of the proof is given below in two stages:

(1) Uniform Stability. From (i) and (ii), we have

$$\alpha(\|x\|) \leq V(x,t) \leq \beta(\|x\|).$$

For any $\epsilon > 0$, there exists a $\delta > 0$ such that $\beta(\delta) < \alpha(\epsilon)$. Let the initial condition x_0 be chosen such that $\|x_0\| < \delta$. Then

$$\begin{aligned}
\alpha(\epsilon) > \beta(\delta) &\geq V(x_0, t_0) \\
&\geq V(x(t; x_0, t_0), t) \quad \text{since } \dot{V}(x,t) \leq 0 \text{ by condition (iii)} \\
&\geq \alpha(\|x(t; x_0, t_0)\|).
\end{aligned}$$

Since $\alpha(\cdot)$ is nondecreasing, this implies that (Fig. 2.2)

$$\|x(t; x_0, t_0)\| < \epsilon \qquad \forall t \geq t_0$$

and, hence, the origin of Eq. (2.1) is uniformly stable.

(2) Uniform Asymptotic Stability. To establish uniform asymptotic stability, we need to show that given any ϵ_2, there exists $T(\epsilon_1, \epsilon_2) > 0$ such that

$$\|x_0\| < \epsilon_1 \Rightarrow \|x(t)\| < \epsilon_2 \qquad \forall\ t \geq t_0 + T.$$

Let δ and ϵ_1 be such that $\beta(\delta) < \alpha(\epsilon_1)$ and, hence, by (1), if $\|x_0\| < \delta$, then $\|x(t; x_0, t_0)\| < \epsilon_1$ for all $t \geq t_0$. If we choose $\epsilon_2 < \delta$ and $\nu > 0$ such that $\beta(\nu) < \alpha(\epsilon_2)$, then if $\|x(t; x_0, t_0)\| < \nu$ for some time $t = t_0 + T$, then $\|x(t; x_0, t_0)\| < \epsilon_2$ for all $t \geq t_0 + T$. The existence of such a T is shown by contradiction. Let $\|x(t)\|$ lie in the compact interval $[\nu, \epsilon_1]$ for all $t \geq t_0$. The continuous function $\gamma(\|x\|)$ in condition (iii) of Theorem 2.5 assumes a minimum value $\gamma_0 > 0$ in this compact interval. Defining $T = \beta(\delta)/\gamma_0$, then there exists a time $t_2 = t_0 + T$ at which

$$\begin{aligned} 0 \ < \ \alpha(\nu) \ &\leq \ V(x(t_2; x_0, t_0), t_2) \\ &\leq \ V(x_0, t_0) - T \cdot \gamma_0 \\ &\leq \ \beta(\delta) - \beta(\delta) = 0, \end{aligned}$$

which is a contradiction. Hence, the origin of Eq. (2.1) is u.a.s. Since by condition (iv), $\alpha(\cdot)$ is radially unbounded, an ϵ can be found such that $\beta(\delta) < \alpha(\epsilon)$ for any δ. In addition, δ can be made arbitrarily large. Hence, $x = 0$ in Eq. (2.1) is uniformly asymptotically stable in the large.

If $\dot{V}(x,t) \leq 0$ and condition (i) is satisfied, the origin of Eq. (2.1) is stable; if in addition condition (ii) is satisfied, uniform stability follows and $V(x,t)$ is referred to as a *Lyapunov function*.

Comment 2.1 In the definitions of stability in Section 2.4.1 as well as in this section, we have tacitly assumed that the solutions of the differential equation for a given initial condition can be extended for all $t \geq t_0$. A sufficient condition for this is that $f(x,t)$ satisfies a global uniform Lipschitz condition, that is, a constant k exists such that

$$\|f(x_1, t) - f(x_2, t)\| \ \leq \ k\|x_1 - x_2\|$$

for all (x_1, t), (x_2, t) in M [19]. However, in most of the problems of interest to us, this cannot be verified. Instead, we assume that $f \in \mathcal{F}$ so that a solution exists in some interval $[t_0 - \alpha_1, t_0 + \alpha_1]$ around t_0 and use the existence of a Lyapunov function to assure that $x(t; x_0, t_0)$ belongs to a compact set. This enables the solution to be extended for all $t \geq t_0$.

When the function f in Eq. (2.1) does not explicitly depend on time t, the system is referred to as an autonomous system. The theorem of Lyapunov can be stated in the following simplified form for such systems [23].

Theorem 2.6 Let an autonomous system be described by the equation

$$\dot{x} \ = \ f(x) \tag{2.11}$$

where $f(0) = 0$. Uniform asymptotic stability in the large of the equilibrium state of Eq. (2.11) is assured by the existence of a scalar function $V(x)$ with continuous partial derivatives with respect to x, such that $V(0) = 0$, and (i) $V(x) > 0$ for all $x \neq 0$, (ii) $\dot{V}(x) < 0$ for all $x \neq 0$, and (iii)$V(x) \to \infty$ with $\|x\| \to \infty$.

The proof of this theorem follows directly from that of Theorem 2.5.

Example 2.3

Consider the differential equations

$$\begin{aligned} \dot{x}_1 &= -x_1 + x_1 x_2 \\ \dot{x}_2 &= -x_1^2 \end{aligned} \qquad (a) \qquad \begin{aligned} \dot{x}_1 &= -x_1 + x_1 x_2 \\ \dot{x}_2 &= -x_2 - x_1^2. \end{aligned} \qquad (b)$$

If a scalar function $V(x_1, x_2)$ is chosen as

$$V(x_1, x_2) = \frac{1}{2}\left(x_1^2 + x_2^2\right)$$

the time derivatives of V along (a) and (b) are given respectively, by

$$\dot{V}\big|_{(a)} = x_1\dot{x}_1 + x_2\dot{x}_2\big|_{(a)} = -x_1^2 \leq 0, \qquad \dot{V}\big|_{(b)} = x_1\dot{x}_1 + x_2\dot{x}_2\big|_{(b)} = -x_1^2 - x_2^2 < 0.$$

Hence, from Theorem 2.6, we can only conclude that the origin of (a) is uniformly stable in the large, while the origin of (b) is uniformly asymptotically stable in the large.

Despite its generality, there is no explicit prescription for finding Lyapunov functions satisfying the conditions of Theorem 2.5 except in the case of LTI systems. If the origin of Eq. (2.1) is u.a.s., an important theorem of Massera [30] assures us of the existence of a Lyapunov function. In Section 2.4.4, we discuss theorems that assure the existence of Lyapunov functions for linear systems. These, in turn, can be modified suitably to apply to specific classes of nonlinear systems.

2.4.3 $V(x, t) > 0, \;\; \dot{V}(x, t) \leq 0$

Condition (iii) in Theorem 2.5, requiring that $\dot{V}(x, t)$ evaluated along any trajectory be negative-definite, is a particularly stringent one to satisfy. In practice, it is quite often considerably simpler to construct a Lyapunov function $V(x, t)$ with a negative semidefinite rather than a negative-definite time derivative. More relevant to our purposes is the fact that condition (iii) can *never* be satisfied in the entire state space for adaptive systems. This is because the parameter error vector, which is part of the state of the system, is unknown and cannot appear in the expression for $\dot{V}(x, t)$. Hence, the best one can hope for is to make $\dot{V}(x, t)$ negative-semidefinite. One of the fundamental problems in the stability analysis of adaptive systems consequently is the determination of the asymptotic behavior of the solutions of the differential equations governing the system from a knowledge of the Lyapunov function $V(x, t)$ whose time derivative is negative-semidefinite.

The problem above has been discussed at great length in the stability literature, and for certain classes of systems LaSalle has demonstrated that uniform asymptotic stability can be concluded. We first consider autonomous nonlinear systems before proceeding to nonautonomous systems, which are of greater interest in adaptive control.

(i) Autonomous Systems.

Theorem 2.7 Let Ω be a compact set with the property that every solution of Eq. (2.11) which begins in Ω remains for all future time in Ω. Suppose there is a scalar function $V(x)$ that has continuous first partials in Ω and is such that $\dot{V}(x) \leq 0$ in Ω. Let $E = \{x | x \in \Omega, \dot{V}(x) = 0\}$ and let M be the largest invariant set in E. Then every solution starting in Ω approaches M as $t \to \infty$.

Proof. Since $V(x)$ is continuous on the compact set Ω, it is bounded from below on Ω. Since $\dot{V}(x) \leq 0$ in Ω, it is a nonincreasing function of t and, hence, $V(x(t))$ has a finite limit c as $t \to \infty$. Further, it is known that the positive limiting set Γ^+ of $x(t)$ is nonempty, compact, and invariant. Since Ω is a closed set, Γ^+ lies in Ω. Further, since V is continuous on Ω, $V(x) \equiv c$ on Γ^+. Since Γ^+ is an invariant set, $\dot{V}(x) = 0$ on Γ^+. Thus Γ^+ is in M and $x(t) \to M$ as $t \to \infty$.

Comment 2.2 In the proof above, the negative semidefiniteness of $\dot{V}(x)$ is used to prove that $V(x(t))$ tends to a limit c. The continuity of $V(x)$ is used to show that $V(x) \equiv c$ on Γ^+ and the invariance of Γ^+ is used to show that $\dot{V}(x) = 0$ on Γ^+. Hence the solution approaches the set E and, more specifically, the largest invariant set M in E.

Comment 2.3 The objective of stability theory here is to provide information on the location of the positive limit sets Γ^+ of solutions and to do this in terms of conditions that can be verified in practice. In Theorem 2.7, it is clear that solutions tend to the set $\Gamma^+ \subset E$. Further, since Γ^+ is known to be invariant, we look for the largest invariant set M in E. Example 2.4 indicates how M can be readily found in specific cases.

Comment 2.4 Autonomous differential equations arise in adaptive systems when the reference input as well as disturbances on the system are constants. In such problems, the construction of a Lyapunov function $V(x)$ will itself guarantee the existence of the set Ω. When $V(x)$ is radially unbounded, the set Ω is defined by $V(x_0) = c$ for any initial condition x_0.

Example 2.4

(a) Consider the differential equations

$$\begin{aligned} \dot{x}_1 &= -x_1 + x_1 x_2 \\ \dot{x}_2 &= -x_1^2. \end{aligned} \tag{2.12}$$

A quadratic function $V = (x_1^2 + x_2^2)/2$ yields the time derivative along the solutions of Eq. (2.12) as $\dot{V} = -x_1^2 \leq 0$. Hence, the origin of Eq. (2.12) is uniformly stable. The set E defined as in Theorem 2.7 is given by $E = \{x | \, x_1 = 0\}$. Since $(x_1(t) = 0, x_2(t) = x_{20})$ is a solution of Eq. (2.12) for any initial condition $(0, x_{20})$, it follows that any solution that starts in E lies in E for all $t \geq t_0$. Hence, the largest invariant set M in E is E itself.

(b) Consider the equations

$$\begin{aligned} \dot{x}_1 &= -x_1 + x_2(x_1 + c) \\ \dot{x}_2 &= -x_1(x_1 + c) \end{aligned} \tag{2.13}$$

where c is a nonzero constant. If $V = (x_1^2 + x_2^2)/2$, we have as in case (a), $\dot{V} = -x_1^2 \leq 0$ and uniform stability of the origin of Eq. (2.13) follows. The set E is the same as in case (a). However, since $\dot{x}_1 = x_2 c$ on E, solutions starting in E leave E for all $x_2 \neq 0$. Hence, the origin is the largest invariant set in E and by Theorem 2.7, all solutions tend to the origin as $t \to \infty$.

(ii) Periodic Systems. The results of Theorem 2.7 were subsequently extended to nonautonomous but periodic systems by LaSalle [22]. A periodic system is defined as one described by the differential equation

$$\begin{aligned} \dot{x} &= f(x,t) \qquad f(0,t) = 0 \; \forall\, t \geq t_0 \\ f(x,t) &= f(x,t+T) \quad \forall t \text{ and finite } T. \end{aligned} \tag{2.14}$$

It is assumed that f is such that the solutions are continuous functions of initial conditions. In such a case, the following theorem may be stated.

Theorem 2.8 Let $V(x,t)$ be a uniformly positive-definite function with continuous first partials for all (x,t), radially unbounded, and $V(x,t) = V(x,t+T)$. Let its time derivative be $\dot{V}(x,t)$ which is uniformly negative-semidefinite. Define $E = \{(y_0,t_0) | \; \dot{V}(y_0,t_0) = 0\}$ and let M be the union over all (y_0,t_0) in E of all trajectories $x(t;y_0,t_0)$ with the property that $(x(t;y_0,t_0),t_0)$ lies in E for all t. Then all solutions approach M as $t \to \infty$.

The proof of this theorem is beyond the scope of this book and the interested reader is referred to [22]. Qualitatively, the proof can be stated as follows: Let $x(t;x_0,t_0)$ be a bounded solution with initial condition (x_0,t_0). Since $V(x(t;x_0,t_0),t)$ is nonincreasing and bounded below, it has a limit c as $t \to \infty$. This in turn is used to show that on all trajectories on the limit set $L^+(x_0,t_0)$ of the solution, $\dot{V} = 0$ and, hence, $L^+(x_0,t_0) \subset M$. Hence, $x(t;x_0,t_0) \to M$ as $t \to \infty$.

(iii) Nonautonomous Systems. In the elegant proof above, use is made of the periodicity of the vector $f(x,t)$ to conclude that the limiting set is nonempty, compact, and invariant. However, most of the adaptive systems we will have occasion to discuss in the following chapters do not enjoy this property of periodicity. Hence, our interest is in determining, with the aid of Lyapunov functions, the set to which the solution of a general nonautonomous system described by the differential equation

$$\dot{x} = f(x,t) \qquad f(0,t) = 0 \;\; \forall\, t \geq t_0 \tag{2.1}$$

converge.

Assumption A. Let f be continuous in x, measurable in t, bounded in t for bounded x, so that the usual theorems on existence and maximal extension of solutions are satisfied [10].

The following theorem, based on a theorem of LaSalle [25], characterizes the limiting behavior of the solutions of Eq. (2.1).

Theorem 2.9 Let V be a Lyapunov function for Eq. (2.1) satisfying conditions (i), (ii), and (iii) of Theorem 2.5 and $f(.,.)$ satisfy assumption A. Let the time derivative $\dot{V}$ along any solution of Eq. (2.1) satisfy the inequality

$$\dot{V}(x,t) \quad \leq \quad -W(x) \leq 0 \qquad \forall\, t \geq t_0,\ x \in \mathbb{R}^n$$

where $W(x)$ is a continuous function of x. Then all solutions approach E as $t \to \infty$, where the set E is defined as $E = \{x |\ W(x) = 0\}$.

In the theorem above, $f(x,t)$ was assumed to be bounded for all $t \geq t_0$ for any bounded x. Extensions of the result to the case when this is not satisfied have been suggested by LaSalle [24] and Artstein [5]. The function f then satisfies the following assumption:

Assumption B. For every compact set K in $\mathbb{R}^n$, there is a nondecreasing function $\alpha_K : [t_0, \infty) \to [t_0, \infty)$, continuous at t_0, with $\alpha_K(t_0) = 0$ and such that, whenever $u : [a,b] \to K$ is continuous, the integral $\int_a^b f(u(\tau),\tau)d\tau$ is well defined, and $f(.,.)$ satisfies the inequality

$$\left\| \int_a^b f(u(\tau),\tau)d\tau \right\| \ \leq \ \alpha_K(b-a).$$

Example 2.5

Consider the following set of nonlinear nonautonomous differential equations.

$$\begin{aligned} \dot{x}_1 &= -x_1 + x_2(x_1 + c(t)) \\ \dot{x}_2 &= -x_1(x_1 + c(t)) \end{aligned} \tag{2.15}$$

where $c : \mathbb{R}^+ \to \mathbb{R}$ is a bounded time-varying function. As in Example 2.4, if $V = (x_1^2 + x_2^2)/2$, then $\dot{V} = -x_1^2 \leq 0$ and uniform stability of the origin of Eq. (2.15) follows and the set $E = \{x | x_1 = 0\}$.

(i) When $c(t) \equiv 0$, the system is autonomous and Theorem 2.7 applies. Hence, all solutions tend to E (see Example 2.4).

(ii) When $c(t)$ is periodic, Theorem 2.8 can be applied and the solutions tend to the origin that is the largest invariant set in E.

(iii) When $c(t)$ is a general time-varying function such that either Assumption A or Assumption B is satisfied in Eq. (2.15), then all solutions tend to E.

Comment 2.5 Many of the results presented in Chapter 6 concerning the uniform asymptotic stability in the large of adaptive systems deal with general nonautonomous systems. From the properties of $f(x,t)$, it is shown that the solutions tend to the origin in E. In this sense, these can be considered to be efforts aimed at generalizing LaSalle's results for autonomous and periodic cases.

2.4.4 LTI Systems and Lyapunov Stability

The stability of LTI systems can be determined using Lyapunov's direct method, without explicit knowledge of the solutions, and is stated in the following theorem.

Theorem 2.10 The equilibrium state $x = 0$ of the linear time-invariant system

$$\dot{x} = Ax \tag{2.3}$$

is asymptotically stable if, and only if, given any symmetric positive-definite matrix Q, there exists a symmetric positive-definite matrix P, which is the unique solution of the set of $n(n+1)/2$ linear equations

$$A^T P + PA = -Q. \tag{2.16}$$

Therefore, $V(x) = x^T Px$ is a Lyapunov function for Eq. (2.3).

Proof. Sufficiency follows directly by choosing a matrix $P = P^T > 0$ satisfying Eq. (2.16). If $V(x) = x^T Px > 0$, the time derivative of V along the solutions of Eq. (2.3) is given by

$$\dot{V}(x) = x^T[A^T P + PA]x = -x^T Qx < 0.$$

Hence, $V(x)$ is a Lyapunov function and satisfies the conditions of Theorem 2.6 so that the asymptotic stability of $x = 0$ of Eq. (2.3) follows.

To prove necessity, assume that $x = 0$ in Eq. (2.3) is asymptotically stable. Then a matrix P defined as

$$P \triangleq \int_0^\infty \exp(A^T t) Q \exp(At)\, dt \tag{2.17}$$

exists, is symmetric and positive-definite. To show that P as defined by Eq. (2.17) is the unique solution of Eq. (2.16), let $\overline{P}$ be any other solution of Eq. (2.16). Then

$$\begin{aligned} P &= -\int_0^\infty \exp(A^T t)(A^T\overline{P} + \overline{P}A)\exp(At)dt \\ &= -\int_0^\infty \frac{d}{dt}\left[\exp(A^T t)\overline{P}\exp(At)\right] dt \qquad \text{since } \exp(At) \text{ commutes with } A\ \forall t \\ &= -\exp(A^T t)\overline{P}\exp(At)\Big|_0^\infty = \overline{P} \qquad \text{since } x = 0 \text{ in Eq. (2.3) is a.s.} \end{aligned}$$

Eq. (2.16) is referred to as the Lyapunov equation.

Example 2.6

Let a system be described by

$$\dot{x} = Ax \qquad A = \begin{bmatrix} 0 & 1 \\ -6 & -5 \end{bmatrix}. \tag{2.18}$$

If we choose $Q = I$, where $I \in \mathbb{R}^{2\times 2}$ is an Identity matrix, and P as

$$P = \begin{bmatrix} p_1 & p_2 \\ p_2 & p_3 \end{bmatrix},$$

then $A^T P + PA = -I$ leads to the solution

$$p_1 = \frac{67}{60}, \qquad p_2 = \frac{5}{60}, \quad \text{and } p_3 = \frac{7}{60}.$$

P is positive-definite, and the origin of Eq. (2.18) is asymptotically stable.

Corollary 2.2 If $C \in \mathbb{R}^{m\times n}$ and (C, A) is observable, the origin of Eq. (2.3) is asymptotically stable if, and only if, there exists a symmetric positive-definite matrix P which is the unique solution of the equation $A^T P + PA = -C^T C$.

Proof. Let $V = x^T Px$ where P is a symmetric positive-definite matrix such that $A^T P + PA = -C^T C$. $\dot{V}$ can be evaluated along the solutions of Eq. (2.3) as $\dot{V} = -\|Cx\|^2 \leq 0$. Uniform asymptotic stability follows from Theorem 2.7 as shown below. The set E is defined by

$$E \triangleq \{x | \, Cx = 0\}.$$

On an invariant set in E, $\dot{V}(x(t)) \equiv 0$ or,

$$C\left[x(t),\ \dot{x}(t),\ \ldots,\frac{d^{n-1}x(t)}{dt^{n-1}}\right] = C\left[I, A, \ldots, A^{n-1}\right] x(t) \equiv 0.$$

Since (C, A) is observable and, hence, $C[I, A, \ldots, A^{n-1}]$ is of rank n, it follows that the largest invariant set in E is the origin.

Necessity follows directly by substituting $Q = C^T C$ in Eq. (2.17). Since (C, A) is observable, P is positive definite.

As mentioned in Section 2.2, the Routh-Hurwitz conditions are both necessary and sufficient for the asymptotic stability of Eq. (2.3). It therefore follows that the Routh-Hurwitz criterion as well as the condition given in Theorem 2.10 are equivalent. This was established by Parks [38].

2.4.5 Linear Time-Varying Systems

Many of the techniques for determining the conditions for the stability of an LTI system are not applicable when the linear system is time-varying. The following example deals with the simplest class of linear time-varying systems.

Example 2.7

Consider the scalar time-varying differential equation

$$\dot{x}(t) = -\beta(t)x(t). \tag{2.19}$$

The solution of the differential equation above is given by

$$x(t) = \exp\left[-\int_{t_0}^{t}\beta(\tau)d\tau\right]x_0 = \exp[-b(t,t_0)]x_0.$$

It then follows directly that the origin in Eq. (2.19) is

(i) stable if $|b(t,t_0)| \le M$, for some constant M, $\forall\, t \ge t_0$,

(ii) asymptotically stable if $b(t,t_0) \to \infty$ as $t \to \infty$, and

(iii) uniformly asymptotically stable if $b(t,t_0) \ge c_1(t-t_0)+c_2$ for some constant $c_1 > 0$ and c_2 for all $t \ge t_0$.

Example 2.7 shows that the integral of the time-varying parameter $\beta(t)$ rather than its value at an instant t determines the stability properties of Eq. (2.19).

For a vector time-varying differential equation,

$$\dot{x}(t) = A(t)x(t) \qquad x : \mathbb{R}^{+} \to \mathbb{R}^{n} \tag{2.2}$$

the stability properties above can be stated by replacing $\exp[-b(t,t_0)]$ in (i)-(iii) by the transition matrix $\Phi(t,t_0)$ of Eq. (2.2). Since, in general, the transition matrix of a linear time-varying differential equation is difficult to determine, the conditions above are hard to verify. Theorem 2.11 shows that a quadratic Lyapunov function exists if, and only if, the equilibrium state of Eq. (2.2) is uniformly asymptotically stable.

Theorem 2.11 [19] Let a time-varying system be described as in Eq. (2.2) where $A(t)$ has uniformly bounded elements for all time t. A necessary and sufficient condition for uniform asymptotic stability of Eq. (2.2) is that for any positive-definite matrix $Q(t)$ which is bounded, a scalar function

$$V(x,t) \triangleq \int_{t}^{\infty} x^T(t)\Phi^T(\tau,t)Q(\tau)\Phi(\tau,t)x(t)d\tau$$

exists and is a Lyapunov function.

We shall refer to the time-varying matrix $A(t)$ with bounded elements for which the origin of Eq. (2.2) is u.a.s. as an exponentially stable matrix.

Proof. While the reader is referred to [19] for a proof, the following comments may prove helpful. If $\Phi(t,\tau)$ is the transition matrix of Eq. (2.2), then $\lim_{t\to\infty}\Phi(t,\tau) = 0$ uniformly in τ. Expressing $V(x,t) = x^T(t)P(t)x(t)$, it follows that $P(t)$ is defined by

$$P(t) \triangleq \int_{t}^{\infty} \Phi^T(\tau,t)Q(\tau)\Phi(\tau,t)d\tau.$$

Although $P(t)$ is defined in terms of the transition matrix, it can be shown that it also satisfies the matrix differential equation

$$\dot{P}(t) = -Q(t) - A^T(t)P(t) - P(t)A(t). \tag{2.20}$$

In contrast to the case of LTI systems where a Lyapunov function $V = x^T Px$ can be determined by solving the algebraic equation (2.16), in LTV systems the differential equation (2.20) has to be solved to obtain the quadratic function $x^T P(t)x$. It is therefore not surprising that Lyapunov functions are much harder to determine in the case of LTV systems.

The following theorem presents a set of weaker conditions that are necessary and sufficient for uniform asymptotic stability [20].

Theorem 2.12 A necessary and sufficient condition for the uniform asymptotic stability of the origin in Eq. (2.2) is that a matrix $P(t)$ exists such that (i) $V = x^T P(t)x$ is positive-definite, (ii) $\dot{V} = x^T[A^T P + PA + \dot{P}]x \leq k(t)V$ where $\lim_{T\to\infty} \int_{t_0}^{T} k(\tau)d\tau = -\infty$ uniformly with respect to t_0.

In Theorem 2.12, the derivative $\dot{V}$ is not negative-semidefinite and, hence, $V(x,t)$ may tend to zero along a trajectory in a nonmonotonic fashion.

Comment 2.6 In Theorem 2.12, if $k(t) = -\alpha + \beta(t)$ where $\alpha > 0$ and $\beta \in \mathcal{L}^1$ or $\mathcal{L}^2$, it follows that $\lim_{t\to\infty} V(t) = 0$.

2.5 POSITIVE REAL FUNCTIONS

The concept of positive realness plays a central role in stability theory in general and in many of the stability proofs of adaptive systems in the chapters following. The definition of a positive real (PR) function of a complex variable s arose in the context of circuit theory. It was shown by Brune [7] that the driving point impedance of a passive network (consisting of only inductances, resistances, and capacitances) is rational and positive real. If the network is dissipative, due to the presence of resistors, the driving point impedance is strictly positive real (SPR). Alternately, a PR (SPR) rational function can be realized as the driving point impedance of a passive (dissipative) network.

The properties of PR functions have been studied extensively in network theory and we confine our attention in what follows to rational positive real functions. The following equivalent definitions of PR functions are now well accepted in the literature.

Definition 2.6.1 [37] A rational function $H(s)$ of the complex variable $s = \sigma + j\omega$ is PR if (i) $H(s)$ is real for real s, (ii) Re$[H(s)] \geq 0$ for all Re$[s] > 0$.

Definition 2.6.2 [3] A rational function $H(s)$ is PR if (i) $H(s)$ is real for real s, (ii) $H(s)$ is analytic in Re$[s] > 0$ and the poles on the imaginary axis are simple and such that the associated residue is nonnegative, and (iii) for any real value of ω for which $j\omega$ is not a pole of $H(j\omega)$, Re$[H(j\omega)] \geq 0$.

Strictly positive real functions, however, have not received the same attention [37], [41] as PR functions, and many definitions have been given in the literature of an SPR function $H(s)$ both in terms of the complex variable s as well as the frequency response $H(j\omega)$. In the following section, the Kalman-Yakubovich lemma is discussed, which is a vital link in relating SPR functions to the existence of a Lyapunov function and hence the stability of a corresponding dynamical system. This underscores the importance of SPR in most of the stability problems dealt with in this book. In view of this, we shall start with the definition of an SPR function based on [37] and [41].

Definition 2.7 A rational function $H(s)$ is strictly positive real if $H(s-\epsilon)$ is PR for some $\epsilon > 0$.

According to Definition 2.7, a function $H(s)$ is SPR if an associated function $H(s-\epsilon)$ is PR. In the literature, other definitions of SPR have been given directly in terms of $H(s)$ and the behavior of the real part of $H(j\omega)$:

(a) If $H(s)$ is analytic for $\text{Re}[s] \geq 0$ and $\text{Re}[H(j\omega)] > 0 \; \forall \omega \in (-\infty, \infty)$.

(b) If $H(s)$ is analytic for $\text{Re}[s] \geq 0$ and $\text{Re}[H(j\omega)] \geq \delta > 0 \; \forall \omega \in (-\infty, \infty)$.

If $H(s) = 1/(s+\alpha)$ with $\alpha > 0$, it follows that $H(s)$ is SPR according to Definition 2.7. However, it satisfies (a) but not (b). Hence, condition (b) is sufficient but not necessary. Alternately, if $H(s) = (s+\alpha+\beta)/[(s+\alpha)(s+\beta)]$, $\alpha, \beta > 0$, then $\text{Re}[H(j\omega)] > 0$ but no $\epsilon > 0$ exists such that $H(s-\epsilon)$ is PR. Hence, according to Definition 2.7, (a) is necessary, but not sufficient.

It was shown by Taylor [41] that if $H(s)$ is SPR and strictly proper, then $\text{Re}[H(j\omega)]$ can go to zero as $|\omega| \to \infty$ no faster than ω^{-2}. In [17], necessary and sufficient conditions for SPR in terms of the analyticity of $H(s)$ and the rate at which $\text{Re}[H(j\omega)]$ tends to zero is given.

Definition 2.8 [17] A rational function $H(s)$ is SPR if, and only if,

1. $H(s)$ is analytic in $\text{Re}[s] \geq 0$,
2. $\text{Re}[H(j\omega)] > 0 \quad \forall\, \omega \in (-\infty, \infty)$, and
3. **(a)** $\lim_{\omega^2 \to \infty} \omega^2 \text{Re}[H(j\omega)] > 0$ when $n^* = 1$, and
   (b) $\lim_{|\omega| \to \infty} \frac{H(j\omega)}{j\omega} > 0$ when $n^* = -1$,

where n^* is the relative degree of $H(s)$ and is defined as the number of poles of $H(s)$ - the number of zeros of $H(s)$.

In Definition 2.8, when $H(s)$ is proper but not strictly proper, conditions (1) and (2) are necessary and sufficient for $H(s)$ to be SPR [17].

From the definitions of rational PR and SPR functions above, it is clear that if $H(s)$ is PR, its phase shift for all frequencies lies in the interval $[-\pi/2, \pi/2]$. Hence, n^* can only be either 0 or 1, if $H(s)$ is the transfer function of a dynamical system that is causal.

For many other interesting properties of PR functions and their applications to different problems in systems theory, the reader is referred to [14]. We list below a few of the properties that are frequently used in the chapters following.

(i) If $H(s)$ is PR(SPR), then $1/H(s)$ is also PR(SPR).

(ii) If $H_1(s)$ and $H_2(s)$ are PR (SPR), then $\alpha H_1(s)+\beta H_2(s)$ is PR (SPR) for $\alpha, \beta > 0$.

(iii) If $H_1(s)$ and $H_2(s)$, the feedforward and feedback transfer functions in a negative feedback system are PR (SPR), then $H_{ov}(s)$, the overall transfer function is also PR (SPR).

2.5.1 Positive Real Transfer Matrices

The concept of a rational positive real function can be generalized to rational positive real transfer matrices. Such matrices find application in this book in the stability analysis of multivariable adaptive systems in Chapter 10.

Definition 2.9 [2] An $n \times n$ matrix $Z(s)$ of functions of a complex variable s is called PR if

1. $Z(s)$ has elements that are analytic for $\text{Re}[s] > 0$,
2. $Z^*(s) = Z(s^*)$, $\quad$ $\text{Re}[s] > 0$, and
3. $Z^T(s^*) + Z(s)$ is positive-semidefinite for $\text{Re}[s] > 0$

where the asterisk * denotes complex conjugate transpose.

As in the case of the scalar functions, SPR matrices can be defined as follows:

Definition 2.10 An $n \times n$ matrix $Z(s)$ is SPR if $Z(s - \epsilon)$ is PR for some $\epsilon > 0$.

2.6 KALMAN-YAKUBOVICH LEMMA

The concept of positive realness found significant application in system theoretic investigations in the 1960s when Popov developed the stability criterion for a feedback system with a single memoryless nonlinearity. The elegant criterion of Popov is stated in terms of the frequency response of the linear part of the system. He proved that if a Lyapunov function, which contains a quadratic form in the state variables and an integral of the nonlinear function, exists, the conditions of his theorem would be satisfied. Further work by Yakubovich [42], Kalman [18], Meyer [31] and Lefschetz [27] showed that the frequency condition of Popov was also sufficient for the existence of the Lyapunov function. A lemma that played a crucial role in establishing the relationship between the frequency domain conditions of the LTI part of the system and the existence of a Lyapunov function is due to Kalman and Yakubovich.

As mentioned earlier, our interest is in the stability of a class of nonlinear systems using Lyapunov's method. The choice of the Lyapunov function is simplified substantially when the transfer function of the relevant linear time-invariant system is strictly positive real. In this context the Kalman-Yakubovich lemma finds application in adaptive control theory. Although many forms of this lemma exist, we present below only two versions due to Lefschetz and Meyer. The Lefschetz version of the lemma replaces a negative-semidefinite matrix in earlier versions of the lemma by a negative-definite matrix.

Lemma 2.3 **[Lefschetz-Kalman-Yakubovich (LKY) lemma].** Given a scalar $\gamma \geq 0$, a vector h, an asymptotically stable matrix A, a vector b such that (A, b) is controllable, and a positive-definite matrix L, there exist a scalar $\epsilon > 0$, a vector q and a symmetric positive-definite matrix P satisfying

$$\begin{aligned} A^T P + PA &= -qq^T - \epsilon L \\ Pb - h &= \sqrt{\gamma} q \end{aligned} \tag{2.21}$$

if, and only if,

$$H(s) \triangleq \frac{1}{2}\gamma + h^T (sI - A)^{-1} b \tag{2.22}$$

is SPR.

By the lemma above, the strict positive realness of $H(s)$ is necessary and sufficient for the existence of a matrix P and a vector q that simultaneously satisfy the matrix and vector equations in Eq. (2.21). The proof of the lemma is contained in Appendix A.

As stated above, the dynamical system represented by the transfer function $H(s)$ is asymptotically stable and controllable. The relaxation of controllability in the K-Y lemma was given by Meyer [31]. Although several versions of the lemma are given in [31], the following extension of Lemma 2.3 is most relevant for our purpose.

Lemma 2.4 **[Meyer-Kalman-Yakubovich (MKY) lemma].** Given a scalar $\gamma \geq 0$, vectors b and h, an asymptotically stable matrix A, and a symmetric positive-definite matrix L, if

$$\mathrm{Re}\,[H(i\omega)] \triangleq \mathrm{Re}\left[\frac{\gamma}{2} + h^T (i\omega I - A)^{-1} b\right] > 0 \;\; \forall \text{ real } \omega,$$

then there exist a scalar $\epsilon > 0$, a vector q and a symmetric positive-definite matrix P such that

$$\begin{aligned} A^T P + PA &= -qq^T - \epsilon L \\ Pb - h &= \sqrt{\gamma} q. \end{aligned}$$

For the proof of the lemma above, the reader is referred to Appendix A. The following comments pertain to the applications of the Kalman-Yakubovich lemma in adaptive control problems.

Comment 2.7 Using the differential versions of the K-Y lemma, a positive-definite matrix P can be determined satisfying a matrix and vector equation simultaneously when a transfer function is SPR. Although this matrix is used in the Lyapunov function to prove stability, it occurs explicitly in the adaptive laws only in some of the cases. In others, the existence of P is assured by the K-Y lemma.

Comment 2.8 Due to the very nature of the problems that arise in adaptive systems, the constant γ in Eq. (2.22) is zero in all the cases that we shall consider. Hence, the strict positive realness of $h^T(sI - A)^{-1}b$ assures the existence of a matrix $P = P^T > 0$, which satisfies

$$A^T P + PA = -Q; \qquad Pb = h$$

where $Q = Q^T > 0$.

Comment 2.9 In many of the adaptive systems discussed, the relevant transfer function is obtained by stable pole-zero cancellations. Hence, although the system is asymptotically stable, it may not be controllable or observable. This accounts for the need for Meyer's lemma in proving their stability.

In Chapter 10, where the adaptive control of multivariable systems is treated, matrix versions of the two lemmas above are needed. In [37], this is dealt with in great detail. For our discussion, we need only the following lemma due to Anderson [2].

Lemma 2.5 Let $Z(s)$ be a matrix of rational functions such that $Z(\infty) = 0$ and $Z(s)$ has poles only in $\mathrm{Re}[s] < -\mu$. Let (H, A, B) be a minimal realization of $Z(s)$. Then, $Z(s)$ is SPR if and only if there exist a symmetric positive-definite matrix P and a matrix L such that

$$\begin{aligned} A^T P + PA &= -LL^T - 2\mu P = -Q \\ PB &= H. \end{aligned}$$

2.7 INPUT-OUTPUT STABILITY

Another general approach to stability is based on the techniques of functional analysis, pioneered by Popov, and subsequently developed by Sandberg and Zames. Using such techniques, a number of results have been obtained over the past two decades concerning the input-output properties of nonlinear feedback systems. In fact, most of the results derived in this book, using Lyapunov theory, may also be derived using this method. For a comprehensive treatment of the subject, the reader is referred to [13]. The advantage of the functional analytic approach is that distributed and lumped systems, discrete and continuous time systems, and single-input and multi-input systems can be treated in a unified fashion. While we do not use this approach to analyze any of the problems in the chapters following, we shall nevertheless have occasion to call on specific results that have been derived using it. Further, for some of the discussions regarding robust adaptive systems, some knowledge of input-output stability is found to be valuable. Hence, in this section we outline some of the principal concepts of the approach, state without proof a few of the significant results we shall require from time to time, and indicate how these are relevant to problems that arise in adaptive systems.

If the given system is described by an operator W that maps an input space $\mathcal{U}$ into an output space $\mathcal{Y}$, the concept of stability is based on the properties of $\mathcal{U}$ and $\mathcal{Y}$. If a property $\mathcal{L}$ of the input is invariant under the transformation W, the system is said to be $\mathcal{L}$-stable. This is stated more precisely below.

Definition 2.11 For any fixed $p \in [1, \infty)$, $f : \mathbb{R}^+ \to \mathbb{R}$ is said to belong to $\mathcal{L}^p$ if, and only if, f is locally integrable and

$$||f||_p \triangleq \left(\int_0^\infty |f(t)|^p \, dt \right)^{1/p} < \infty.$$

When $p = \infty$, $f \in \mathcal{L}^\infty$ if, and only if,

$$||f||_\infty \triangleq \sup_{t \geq 0} |f(t)| < \infty.$$

With the definition above, the system represented by the operator W is said to be $\mathcal{L}^p$-stable if $u \in \mathcal{L}^p$ is mapped into $y \in \mathcal{L}^p$. When $p = \infty$, $\mathcal{L}^p$-stability is also referred to as bounded-input bounded-output (BIBO) stability. For ease of exposition, the statements "$f \in \mathcal{L}^\infty$" and "f is uniformly bounded" are used synonymously throughout the book.

For linear systems, many stability results can be stated in a concise fashion.

Theorem 2.13 [13] Let a linear system be described by the differential equations

$$\begin{aligned} \dot{x}(t) &= Ax(t) + bu(t) \\ y(t) &= h^T x(t) \end{aligned}$$

where $A \in \mathbb{R}^{n \times n}$ is an asymptotically stable matrix, $b, h \in \mathbb{R}^n$, and $H(s) = h^T(sI - A)^{-1}b$ is the transfer function from u to y. The following results then hold:

(i) If the input u belongs to $\mathcal{L}^\infty$, then the output y belongs to $\mathcal{L}^\infty$, $\dot{y} \in \mathcal{L}^\infty$, and y is uniformly continuous. If, in addition, $\lim_{t \to \infty} u(t) = u_0$, a constant in $\mathbb{R}$, then $y(t) \to y_0 = H(0)u_0$, a constant in $\mathbb{R}$.

(ii) If u belongs to $\mathcal{L}^1$, then y belongs to $\mathcal{L}^1 \cap \mathcal{L}^\infty$, $\dot{y} \in \mathcal{L}^1$, y is absolutely continuous, and $\lim_{t \to \infty} y(t) = 0$.

(iii) If u belongs to $\mathcal{L}^2$, then y belongs to $\mathcal{L}^2 \cap \mathcal{L}^\infty$, $\dot{y} \in \mathcal{L}^2$, y is continuous, and $\lim_{t \to \infty} y(t) = 0$.

(iv) If u belongs to $\mathcal{L}^p$, then y belongs to $\mathcal{L}^p$ and $\dot{y} \in \mathcal{L}^p, [1 < p < \infty]$.

For a proof of Theorem 2.13, the reader is referred to [13].

Systems that arise frequently in adaptive systems are often of the form shown in Fig. 2.3. Many of the results derived using input-output analysis also pertain to such

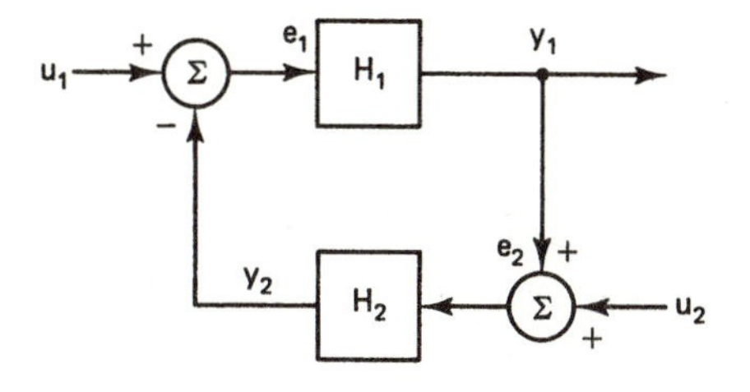

Figure 2.3 A canonical feedback system. Adapted from Desoer and Vidyasagar, Academic Press, 1975.

systems. Let H_1 and H_2 be two operators that act on inputs e_1 and e_2, respectively, to produce outputs y_1 and y_2. u_1 and u_2 are two external inputs into the system and the equations describing the system are given by

$$\begin{aligned} y_1 &= H_1 e_1 = H_1(u_1 - y_2) \\ y_2 &= H_2 e_2 = H_2(u_2 + y_1). \end{aligned} \tag{2.23}$$

The problem then is to determine conditions on H_1, H_2, so that if u_1, u_2 belong to some class $\mathcal{L}^p$, then e_1, e_2, y_1, and y_2 belong to the same class.

Central to the proofs for demonstrating the $\mathcal{L}^p$-stability of various systems of the type in Eq. (2.23) is the concept of an extended space, $\mathcal{L}_e$, associated with a normed space $\mathcal{L}$. Let $\mathcal{L}$ be a normed linear space defined by

$$\mathcal{L} \stackrel{\triangle}{=} \{f : \mathbb{R}^+ \to \mathbb{R}^n |\ \|f\| < \infty\}$$

where $\|\cdot\|$ corresponds to any of the norms introduced earlier. If

$$f_T(t) = \begin{cases} f(t), & t \le T \\ 0 & t > T \end{cases}$$

where $T \in \mathbb{R}^+$, associated with $\mathcal{L}$ is the extended space $\mathcal{L}_e$ defined by

$$\mathcal{L}_e \stackrel{\triangle}{=} \{f : \mathbb{R}^+ \to \mathbb{R}^n |\ \|f_T\| < \infty,\ \forall\, T \in \mathbb{R}^+\}.$$

In all cases, it is assumed that H_1 and H_2 map $\mathcal{L}_e$ into $\mathcal{L}_e$. A general theorem that gives sufficient conditions under which a bounded input provides a bounded output is the small gain theorem and can be stated as follows:

Theorem 2.14 Let $H_1, H_2 : \mathcal{L}_e \to \mathcal{L}_e$, $e_1, e_2 \in \mathcal{L}_e$ and be defined as in Eq. (2.23). Suppose H_1 and H_2 satisfy the inequality

$$\begin{aligned} \|(H_1 e_1)_T\| &\le \gamma_1 \|e_{1T}\| + \beta_1 \\ \|(H_2 e_2)_T\| &\le \gamma_2 \|e_{2T}\| + \beta_2 \end{aligned}$$

$\forall\, T \in \mathbb{R}^+$ and

$$\gamma_1 \gamma_2 < 1$$

where $\gamma_1, \gamma_2 \in \mathbb{R}^+$, $\beta_1, \beta_2 \in \mathbb{R}$, then

$$\|e_{1T}\| \leq (1 - \gamma_1\gamma_2)^{-1} \left(\|u_{1T}\| + \gamma_2\|u_{2T}\| + \beta_2 + \gamma_2\beta_1 \right) \tag{2.24}$$

$$\|e_{2T}\| \leq (1 - \gamma_1\gamma_2)^{-1} \left(\|u_{2T}\| + \gamma_1\|u_{1T}\| + \beta_1 + \gamma_1\beta_2 \right) \tag{2.25}$$

$\forall\, T \in \mathbb{R}^+$. If, in addition, u_1 and u_2 have finite norms, then e_1, e_2, y_1, and y_2 have finite norms, and the norms of e_1 and e_2 are bounded by the right-hand sides of Eqs. (2.24) and (2.25), provided all subscripts T are dropped.

According to the theorem above, if the product $\gamma_1\gamma_2$ of the gains of H_1 and H_2 is smaller than 1, then provided a solution exists, a bounded pair (u_1, u_2) produces a bounded output pair (y_1, y_2). The small gain theorem holds for continuous time as well as discrete time, and single-input single-output as well as multi-input multi-output systems.

A second concept, which is systematically used in input-output stability, is that of passivity. This concept is closely related to that of positive realness discussed in Section 2.5, and is identical to it under the conditions mentioned below. The passivity theorem presented here along with the small gain theorem presented earlier is used to derive stability results for many nonlinear feedback systems in [13].

The concept of passivity can be introduced in terms of a single port network whose voltage and current are v and i respectively. If $E(t_0)$ denotes the energy stored in the single port at time t_0, then the system is said to be passive if

$$E(t_0) + \int_{t_0}^{t} v(\tau)i(\tau)\, d\tau \geq 0 \qquad \forall\, v, i,\ \forall\, t \geq t_0.$$

Generalizing this to n-port networks and assuming that the stored energy at $t = -\infty$ is zero, the n-port network is said to be passive if, and only if,

$$\text{Re}\left[\int_{-\infty}^{t} v^T(\tau)i(\tau)\, d\tau \right] \geq 0 \qquad \forall\, v, i,\ \forall\, t \in \mathbb{R}.$$

To give a formal definition of passivity, the inner product of two vectors in $\mathbb{R}^n$ is needed. Let

$$< x|y > = \int_0^{\infty} x^T(t)y(t)\, dt.$$

Then the space $\mathcal{H}$ is defined by

$$\mathcal{H} \triangleq \left\{ x \in \mathbb{R}^n |\ \|x\|^2 = \langle x|x \rangle < \infty \right\}$$

and the extended space $\mathcal{H}_e$ is defined as

$$\mathcal{H}_e \triangleq \left\{ x \in \mathbb{R}^n |\ \|x_T\|^2 = \langle x_T|x_T \rangle < \infty\ \forall\, T \in \mathbb{R}^+ \right\}.$$

We now state the definition of passive and strictly passive systems.

Definition 2.12 $H : \mathcal{H}_e \to \mathcal{H}_e$ is said to be passive if there exist $\beta \in \mathbb{R}$ such that

$$< Hx|x >_T \geq \beta \qquad \forall\, T \in \mathbb{R}^+,\ \forall\, x \in \mathcal{H}_e.$$

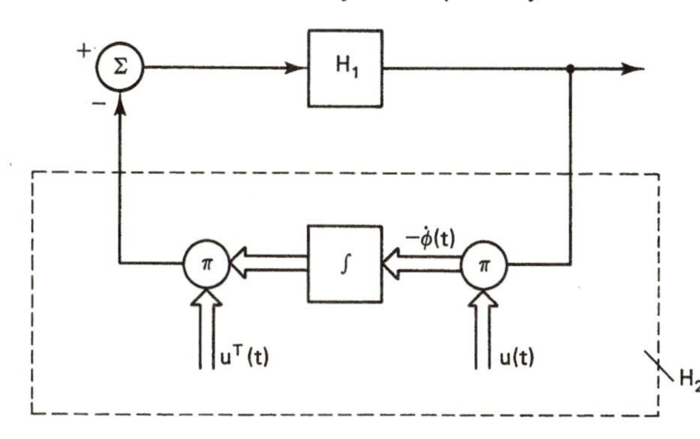

Figure 2.4 Passivity in adaptive systems.

Definition 2.13 $H : \mathcal{H}_e \to \mathcal{H}_e$ is said to be strictly passive if there exist $\delta \in \mathbb{R}^+$ and $\beta \in \mathbb{R}$ such that

$$< Hx|x >_T \geq \delta \|x_T\|^2 + \beta \qquad \forall T \in \mathbb{R}^+, \forall x \in \mathcal{H}_e.$$

When H is a linear time-invariant operator, if H is passive (strictly passive), then H is PR (SPR).

The passivity theorem can be stated as follows:

Theorem 2.15 If in Eq. (2.23), $u_2(t) \equiv 0$, H_1 is passive, H_2 is strictly passive, and $u_1 \in \mathcal{L}^2$, then $y_1 \in \mathcal{L}^2$.

Proof. The reader is referred to [13] for a proof.

In many of the stability problems discussed in later chapters, the error equations describing the system can be expressed in the form given in Fig. 2.4. H_1 in Fig. 2.4 is a linear time-invariant operator that corresponds to an unknown plant together with a fixed controller, while the feedback system H_2 corresponds to the mechanism generating the parameter error ϕ and the corresponding control input. Hence, Theorem 2.15 can be applied directly to such systems to prove $\mathcal{L}^2$-stability if H_1 is strictly passive and H_2 is passive.

2.8 STABILITY OF ADAPTIVE SYSTEMS

As mentioned earlier, many of the stability properties of adaptive systems discussed in this book can be derived using well known theorems in stability theory collected thus far in this chapter. Some differential equations, however, arise frequently for which standard theorems are not directly applicable. In view of their importance in our investigations, they are presented in a concise manner in this section. Detailed proof of the theorems are provided wherever such proofs are needed to follow the stability arguments in later chapters.

2.8.1 Two Differential Equations That Arise in Adaptive Systems

Although an adaptive system can generally be described only by nonlinear time-varying differential equations, its stability analysis is carried out by considering subsystems that are linear. Two linear differential equations with time-varying coefficents are found to arise frequently in this context. The reasons for this become evident in the chapters following and particularly in Chapters 6 and 7 where these two equations are prominently featured. We discuss the properties of the solutions of the two equations in some detail since they are used extensively throughout this book.

(i) The first differential equation has the form

$$\dot{x}(t) = -u(t)u^T(t)x(t) \tag{2.26}$$

where $u : \mathbb{R}^+ \to \mathbb{R}^{n\times m}$ is a piecewise continuous and bounded function. If

$$Q(t) \triangleq u(t)u^T(t),$$

then $Q(t)$ is a positive-semidefinite matrix. If a quadratic function $V(x) = x^T x/2$, its time derivative along the solutions of Eq. (2.26) is given by $\dot{V}(x) = -x^T Q(t)x$. Hence, the equilibrium state $x = 0$ of Eq. (2.26) is uniformly stable. It does not follow by any of the Theorems 2.5-2.8 discussed in Section 2.4 that the equilibrium state is uniformly asymptotically stable, since $Q(t)$ is time-varying and positive-semidefinite. The conditions on $Q(t)$ [and, hence, $u(t)$] under which the latter holds are of interest.

This problem has been studied in depth by many authors [1,33,40] in the past. Although [1] and [40] derive sufficient conditions for the uniform asymptotic stability of the origin, it is shown in [33] that the conditions are also necessary. We follow the development in [33] here. The principal theorem of interest to us can be stated as follows.

Theorem 2.16 The following conditions (a)-(d) are equivalent and assure the uniform asymptotic stability of the equilibrium state $x = 0$ of the differential equation (2.26):

(a) Positive constants t_0, T_0, and ϵ_1 exist such that for all unit vectors $\mathrm{w} \in \mathbb{R}^n$,

$$\int_t^{t+T_0} \mathrm{w}^T u(\tau)u^T(\tau)\mathrm{w}\,d\tau \geq \epsilon_1 \qquad \forall\, t \geq t_0. \tag{2.27a}$$

(b) There exist positive constants t_0, T_0, and ϵ_2 such that for all unit vectors $\mathrm{w} \in \mathbb{R}^n$,

$$\int_t^{t+T_0} \|u(\tau)u^T(\tau)\mathrm{w}\|\,d\tau \geq \epsilon_2 \qquad \forall\, t \geq t_0. \tag{2.27b}$$

(c) There exist positive constants t_0, T_0, and ϵ_3 such that for all unit vectors $\mathrm{w} \in \mathbb{R}^n$,

$$\frac{1}{T_0}\int_t^{t+T_0} \|u^T(\tau)\mathrm{w}\|\,d\tau \geq \epsilon_3 \qquad \forall\, t \geq t_0. \tag{2.27c}$$

(d) There exist positive constants t_0, T_0, and ϵ_4 such that

$$\lambda_i \left[\int_t^{t+T_0} u(\tau)u^T(\tau)d\tau \right] \geq \epsilon_4, \ i = 1, 2, \ldots, n, \ \forall\, t \geq t_0 \tag{2.27d}$$

where $\lambda_i[A]$ denotes the ith eigenvalue of the matrix A.

In proving the theorem it is first shown that conditions (a)-(d) are equivalent, that is, any one of them implies the others. Next it is shown that (c) assures the uniform asymptotic stability of $x = 0$ of Eq. (2.26).

Theorem 2.16 states that conditions (a)-(d) are sufficient for $x = 0$ to be u.a.s. This is adequate for all the discussions of stability of adaptive systems given in this book. However, the uniform asymptotic stability of $x = 0$ of Eq. (2.26) also assures that conditions (a)-(d) are satisfied, that is, they are necessary. For a proof of the latter, we refer to the reader to the source paper [33].

Proof.

(i) Let $u_{\max}$ be a constant such that $\|u(t)\| \leq u_{\max}$. If w is any unit vector, it follows that

$$\mathrm{w}^T u(t)u^T(t)\mathrm{w} \leq \|u(t)u^T(t)\mathrm{w}\| \leq \|u^T(t)\mathrm{w}\|u_{\max}.$$

Hence, (a)$\Rightarrow$(b), and (b)$\Rightarrow$(c). Further,

$$\int_t^{t+T_0} \mathrm{w}^T u(\tau)u^T(\tau)\mathrm{w}d\tau \geq \frac{1}{T_0}\left[\int_t^{t+T_0} \|u^T(\tau)\mathrm{w}\|d\tau\right]^2$$

by the Cauchy-Schwarz inequality so that (c)$\Rightarrow$(a). The smallest eigenvalue of the matrix

$$\overline{Q} \triangleq \int_t^{t+T_0} Q(\tau)d\tau$$

is equal to

$$\inf_{\|\mathrm{w}\|=1} \mathrm{w}^T\overline{Q}\mathrm{w} = \inf_{\|\mathrm{w}\|=1} \int_t^{t+T_0} \mathrm{w}^T Q(\tau)\mathrm{w}d\tau.$$

Hence, (d)$\Leftrightarrow$(a).

(ii) Let $t_1 \geq t_0$. If $V(x) = x^T x/2$, the time derivative of V along the solutions of Eq. (2.26) is $\dot{V} = -(u^T x)^2$. Integrating $\dot{V}$ over an interval $[t_1, t_1 + T_0]$, we have

$$-\int_{t_1}^{t_1+T_0} \dot{V}(x,t)dt = \int_{t_1}^{t_1+T_0} \{u^T(\tau)x(\tau)\}^2 d\tau \geq \frac{1}{T_0}\left(\int_{t_1}^{t_1+T_0} \|u^T(\tau)x(\tau)\|d\tau\right)^2$$

or equivalently

$$\int_{t_1}^{t_1+T_0} \|u^T(\tau)x(\tau)\|d\tau \leq \sqrt{T_0}\,\{V(t_1) - V(t_1+T_0)\}^{1/2}. \tag{2.28}$$

The left-hand side can also be expressed as

$$\int_{t_1}^{t_1+T_0} \|u^T(\tau)x(\tau)\|d\tau \;\geq\; \int_{t_1}^{t_1+T_0} \|u^T(\tau)x(t_1)\|d\tau - \int_{t_1}^{t_1+T_0} \|u^T(\tau)\left[x(t_1)-x(\tau)\right]\|d\tau. \tag{2.29}$$

From Eq. (2.27c) in condition (c), it follows that

$$\int_{t_1}^{t_1+T_0} \|u^T(\tau)x(t_1)\|d\tau \;\geq\; \|x(t_1)\|T_0\epsilon_3.$$

The second term on the right-hand side of Eq. (2.29) is given by

$$\begin{aligned}\int_{t_1}^{t_1+T_0} \|u^T(\tau)\left[x(t_1)-x(\tau)\right]\|d\tau &\leq u_{\max}T_0 \sup_{\tau\in[t_1,t_1+T_0]} \|x(t_1)-x(\tau)\| \\ &\leq u_{\max}T_0 \int_{t_1}^{t_1+T_0} \|\dot{x}(\tau)\|d\tau\end{aligned}$$

since the distance between two points $x(t_1)$ and $x(\tau)$, $\tau \in [t_1, t_1+T_0]$, is less than the arc length $\int_{t_1}^{t_1+T_0} \|\dot{x}(\tau)\|d\tau$. Hence, evaluating $\dot{x}$ from the differential equation (2.26), Eq. (2.29) becomes

$$\int_{t_1}^{t_1+T_0} \|u^T(\tau)x(\tau)\|d\tau \;\geq\; \|x(t_1)\|T_0\epsilon_3 - u_{\max}^2 T_0 \int_{t_1}^{t_1+T_0} \|u^T(\tau)x(\tau)\|d\tau.$$

Therefore,

$$\int_{t_1}^{t_1+T_0} \|u^T(\tau)x(\tau)\|d\tau \;\geq\; \frac{\|x(t_1)\|T_0\epsilon_3}{1+u_{\max}^2 T_0}. \tag{2.30}$$

From Eqs. (2.28) and (2.30), we have

$$V(t_1+T_0) \;\leq\; (1-\gamma)\,V(t_1) \qquad \text{where } \gamma = 1 - \frac{2T_0\epsilon_3^2}{(1+u_{\max}^2 T_0)^2}.$$

Since $\epsilon_3 \leq u_{\max}$, $\gamma \in (0,1)$ and, hence, the origin in Eq. (2.26) is uniformly asymptotically stable.

Comment 2.10 If a linear time-varying system is described by

$$\dot{x} = -\Gamma u(t)u^T(t)x \qquad x, u : \mathbb{R}^+ \to \mathbb{R}^n,\ \Gamma = \Gamma^T > 0 \tag{2.31}$$

then the uniform asymptotic stability of $x = 0$ in Eq. (2.31) can be proved using similar arguments to that in Theorem 2.16.

(ii) The second differential equation has the form

$$\begin{aligned}\dot{x}_1(t) &= A(t)x_1(t) + u^T(t)x_2(t) \\ \dot{x}_2(t) &= -u(t)x_1(t)\end{aligned} \tag{2.32}$$

where $x_1 : [0, \infty) \to \mathbb{R}^m$, $x_2 : [0, \infty) \to \mathbb{R}^n$, $A(t)$ and $u(t)$ are, respectively, $n \times n$ and $n \times m$ matrices of bounded piecewise-continuous functions, and $A(t) + A^T(t)$ is uniformly negative-definite, that is, $A(t) + A^T(t) \leq -Q < 0$. Let $x = [x_1^T, x_2^T]^T$. Our aim is to determine necessary and sufficient conditions on u such that the equilibrium state $x = 0$ of Eq. (2.32) is u.a.s.

That the equilibrium state $x = 0$ of Eq. (2.32) is uniformly stable can be shown directly by considering a function $V(x) = x^T x$ whose time derivative $\dot{V}(x, t)$ along the solutions of Eq. (2.32) is given by

$$\dot{V}(x,t) = x_1^T(A(t) + A^T(t))x_1 \leq -x_1^T Q x_1 \leq 0.$$

Theorem 2.17 due to Morgan and Narendra [32] gives a complete characterization of the stability of Eq. (2.32).

Theorem 2.17 $x = 0$ in Eq. (2.32) is uniformly asymptotically stable if, and only if, positive constants T_0, δ_0, and ϵ_0 exist with a $t_2 \in [t, t+T_0]$ such that for any unit vector w$\in \mathbb{R}^n$,

$$\left\| \frac{1}{T_0} \int_{t_2}^{t_2+\delta_0} u^T(\tau) \mathrm{w} d\tau \right\| \geq \epsilon_0 \qquad \forall\, t \geq t_0. \tag{2.33}$$

When $\|\dot{u}(t)\|$ is uniformly bounded, the condition in Eq. (2.33) in Theorem 2.17 can be relaxed as given in Corollary 2.3.

Corollary 2.3 If $u(t)$ is smooth, $\dot{u}(t)$ is uniformly bounded, and $u(t)$ satisfies the condition

$$\frac{1}{T_0} \int_t^{t+T_0} \left\| u^T(\tau) \mathrm{w} \right\| d\tau \geq \epsilon_0 \qquad \forall\, t \geq t_0 \tag{2.34}$$

for $t_0 \in \mathbb{R}^+$, and positive constants T_0 and ϵ_0, and all unit vectors w$\in \mathbb{R}^n$, then the solution $x = 0$ of Eq. (2.32) is u.a.s.

The following corollary shows that the condition in Eq. (2.34) is also necessary for the uniform asymptotic stability of $x = 0$ of Eq. (2.32).

Corollary 2.4 If the equilibrium state $x = 0$ of Eq. (2.32) is u.a.s., then constants t_0, T_0, and ϵ_0 exist such that Eq. (2.34) holds for all unit vectors w$\in \mathbb{R}^n$.

Corollary 2.4 follows directly from Theorem 2.17 since, if the condition in Eq. (2.33) is satisfied, then the condition in Eq. (2.34) is also satisfied. Corollary 2.3 also follows from Theorem 2.17 because, when $\dot{u}(t)$ is bounded, the integral condition in Eq. (2.34) implies the integral condition (2.33) in the theorem.

That Corollary 2.3 will not hold when the boundedness of $\dot{u}(t)$ is not assumed, is evident from the following example from [32].

Example 2.8

Consider the system

$$\begin{aligned} \dot{x}_1 &= -x_1 + u(t)x_2 \\ \dot{x}_2 &= -u(t)x_1 \end{aligned}$$

where

$$u(t) = \begin{cases} +1 & t \in [t_n, t_n + \frac{1}{2(n+1)}) \\ -1 & t \in [t_n + \frac{1}{2(n+1)}, t_{n+1}) \end{cases} \qquad \text{where } t_n = \sum_{i=1}^{n} \frac{1}{i} \qquad n = 1, 2, \ldots$$

While

$$\frac{1}{T}\int_t^{t+T} |u(\tau)| d\tau = 1 \qquad \text{for some finite } T,$$

u does not satisfy the condition of the theorem and, hence, does not assure u.a.s.

The proof of Theorem 2.17 is quite involved and the reader is referred to the paper [32] for further details. As in the case of the differential equation (2.26), we deal with only the sufficiency part of Theorem 2.17. Comments (i)-(iii) given below provide a qualitative outline of the proof, and are followed by highlights of the proof.

(i) If $\|x(t_0)\| \leq \epsilon_1$, by the Lyapunov function $V(x) = x^T x$, it follows that the solution $x(t)$ satisfies the inequality $\|x(t)\| \leq \epsilon_1$ for all $t \geq t_0$.

(ii) Since $\dot{V}(x) = -x_1^T Q x_1$, $x_1 \in \mathcal{L}^2$. Since $\dot{x}_1$ is also bounded, $\lim_{t\to\infty} x_1(t) = 0$ (refer to Corollary 2.9).

(iii) Let $\|x(t)\| > \epsilon_2$ for some $\epsilon_2 < \epsilon_1$ for all $t \geq t_0$. If $\|x_1(t_1)\|$ is small at time $t = t_1$, this implies that $\|x_2(t_1)\|$ is large. If $u(t)$ satisfies the condition in Eq. (2.33) and $\|x_2(t)\| > \delta$ for $t \in [t_1, t_1 + T]$, then $\|x_1(t_2)\| \geq \epsilon$, for some ϵ and for $t_2 \in [t_1, t_1 + T]$. This contradicts (ii) and, hence, $\|x(t)\| < \epsilon_2$ at time $t = t_3 > t_0$, leading to uniform asymptotic stability of $x = 0$ in Eq. (2.32).

In the proof that follows, it is shown that if u satisfies the condition in Eq. (2.33), $x_1(t)$ has to assume a large value at some instant in every interval $[t, t + T_0]$. Since $\dot{V} \leq -x_1^T Q x_1$, this implies that V decreases over every interval of length T_0 which assures uniform asymptotic stability.

Proof. Let $\|x_1(t)\| < c\|x(t)\|$ $\forall t \in [t_0, t_0 + T_0]$, where $c \in (0, 1)$. Since u satisfies the condition in Eq. (2.33), there exists a $t_2 \in [t_0, t_0 + T_0]$ so that

$$\int_{t_2}^{t_2+\delta_0} \left\|u^T(\tau)x_2(\tau)\right\| d\tau \geq \epsilon_0 T_0 \|x_2(t_2)\| - u_{\max}^2 \delta_0^2 \sup_{\tau\in[t_2, t_2+\delta_0]} \|x_1(\tau)\|.$$

Integrating the first equation over the interval $[t_2, t_2 + \delta_0]$ in Eq. (2.32) it follows that

$$\|x_1(t_2 + \delta_0)\| \geq \left\{\epsilon_0 T_0 \sqrt{1 - c^2} - c\left(u_{\max}^2 \delta_0^2 + A_{\max}\delta_0 + 1\right)\right\} \|x(t_2 + \delta_0)\| \tag{2.35}$$

since $\|x_2(t_2)\| \geq \sqrt{1 - c^2}\|x(t_2)\|$ and $\|x(t_2)\| \geq \|x(t_2 + \delta_0)\|$. If

$$c^2 = \frac{T_0^2 \epsilon_0^2}{T_0^2 \epsilon_0^2 + (1 + b)^2}, \qquad b = u_{\max}^2 \delta_0^2 + A_{\max}\delta_0 + 1,$$

we have from Eq. (2.35)

$$\|x_1(t_2+\delta_0)\| \geq c\|x(t_2+\delta_0)\| \tag{2.36}$$

which is a contradiction of the assumption regarding $\|x_1(t)\|$.

Integrating $\dot{V}(x)$ over an interval $[t_1, t_1+T] \subset [t_0, t_0+T_0]$, we obtain that

$$\begin{aligned} V(t_1) - V(t_1+T) &\geq \lambda_{Q-\min} \int_{t_1}^{t_1+T} \|x_1(\tau)\|^2 d\tau \\ &\geq \frac{\lambda_{Q-\min}}{T} \left(\int_{t_1}^{t_1+T} \|x_1(\tau)\| d\tau \right)^2 \end{aligned}$$

where $\lambda_{Q-\min}$ is the minimum eigenvalue of Q. Further, by integrating the first equation in Eq. (2.32) over $[t_1, t_1+T]$

$$\int_{t_1}^{t_1+T} \|x_1(\tau)\| d\tau \geq T\|x_1(t_1)\| - T^2 d\sqrt{V(t_1)}$$

where $d = A_{\max} + u_{\max}$, $\|A(t)\| \leq A_{\max}$, and $\|u(t)\| \leq u_{\max}$. Choosing $t_1 = t_2 + \delta_0$, from Eq. (2.36), it follows that

$$V(t_1+T) \leq (1-\gamma)V(t_1) \text{ where } \gamma = \lambda_{Q-\min} T(c-Td)^2.$$

Choosing $T = \min(t_0 + T_0 - t_1, c/d)$, we have

$$V(t_0+T_0) \leq V(t_1+T) \leq (1-\gamma)V(t_1) \leq (1-\gamma)V(t_0).$$

Since $\lambda_{Q-\min}/d \leq 2$, $c \in (0,1)$, and $Td(c-Td)^2 \leq 4c^3/27$, it follows that $\gamma \in (0, 8c^3/27)$ and hence $x = 0$ in Eq. (2.32) is u.a.s.

Comment 2.11 In many identification and control problems, the underlying differential equations have somewhat different forms than those described in Eq. (2.32). They are given by

$$\begin{aligned} \dot{x}_1(t) &= Ax_1(t) + u^T(t)x_2(t) \\ \dot{x}_2(t) &= -u(t)Px_1(t) \end{aligned} \tag{2.37}$$

where A is asymptotically stable, and P is a symmetric positive-definite matrix satisfying the Lyapunov equation $A^TP + PA = -Q$, $Q = Q^T > 0$. The uniform asymptotic stability of the origin of Eq. (2.37) can however be proved along similar lines to that given above, when u satisfies the condition in Eq. (2.33) [32].

2.8.2 Growth Rates of Signals in Dynamical Systems

In many of the problems discussed in Chapters 5-9, the global boundedness of signals in a nonlinear system has to be demonstrated. However, Lyapunov's direct method cannot be used, since a Lyapunov function for the overall system cannot be determined. The best one can hope for is to assure the boundedness of the parameters. In such cases, the stability analysis is based on the characterization and ordering of the unbounded signals (see Appendix B) in the adaptive loop. Since many of the subsystems are linear, this generally depends on order relations between their inputs and outputs. In this section, we discuss situations when the inputs and outputs of linear systems grow at the same rate as well as those in which they grow at different rates, using the concepts described in Appendix B.

The following three classes of functions feature prominently in our discussions [36]:

1. *Linear time-invariant systems:* For the differential equation $\dot{x} = Ax$ where $x(t) \in \mathbb{R}^n$ and A is a constant matrix, let $\psi_i(t)$ and $\psi_j(t)$ be any two distinct solutions. Then, $\sup_{\tau \leq t} \|\psi_i(t)\| \sim \sup_{\tau \leq t} \|\psi_j(t)\|$, $\sup_{\tau \leq t} \|\psi_i(t)\| = o[\sup_{\tau \leq t} \|\psi_j(t)\|]$, or $\sup_{\tau \leq t} \|\psi_j(t)\| = o[\sup_{\tau \leq t} \|\psi_i(t)\|]$, $i \neq j$.

2. *Homogeneous linear systems:* If $\dot{x} = A(t)x$ where the elements of $A(t)$ are bounded, then all the components of $x(t)$ can grow at most exponentially. Alternately, there exists a $\gamma > 0$ such that $\lim_{t\to\infty} x(t)\exp(-\gamma t) = 0$.

3. *Nonhomogeneous linear systems:* If the system is described by the equation $\dot{x} = A(t)x + b(t)r$ where the elements of the matrix $A(t)$ and vector $b(t)$ are uniformly bounded, we have

$$\left|\frac{d}{dt}\|x\|\right| \leq \|\dot{x}\| \leq M_1\|x\| + M_2 \qquad M_1, M_2 \in \mathbb{R}^+ \tag{2.38}$$

Then, there exist constants $c_1 > 0$, c_2 and $T > 0$ such that

$$c_1\|x(t)\| - c_2 < \|x(t+T)\| < \frac{1}{c_1}\|x(t)\| + c_2. \tag{2.39}$$

The set of all functions x that satisfy conditions in Eq. (2.38) is denoted by $\overline{\mathcal{E}}$.

Note that $\|x(t)\|$ as defined by the inequality in Eq. (2.38) can grow at most exponentially as defined in (2) above. However, signals which grow at most exponentially need not belong to $\overline{\mathcal{E}}$ as seen in the following example.

Example 2.9

If

$$\dot{x} = Ax, \qquad x(0) = [1,\ 0]^T$$

where

$$A = \begin{bmatrix} 1 & -1 \\ 1 & 1 \end{bmatrix},$$

then $x(t) = [e^t \cos t, e^t \sin t]^T$. Hence, $\|x\| \in \overline{\mathcal{E}}$ but x_1 and $x_2 \notin \overline{\mathcal{E}}$.

Motivated by the example above, we define below a larger class of signals than that given by Eq. (2.38):

$$\mathcal{E} \stackrel{\triangle}{=} \{x | \, \|\dot{x}(t)\| \leq M_1 \sup_{\tau \leq t} \|x(\tau)\| + M_2, \;\; M_1, M_2 \in \mathbb{R}^+\}$$

It follows that $\overline{\mathcal{E}} \subset \mathcal{E}$. Also, if $\dot{x} = Ax$, where $x : \mathbb{R}^+ \to \mathbb{R}^n$, then every component of x can be shown to belong to $\mathcal{E}$.

Signals growing at the same rate. The concepts and definitions developed in Appendix B can be used directly in linear dynamical systems to establish the fact that some of the signals grow at the same rate. This is particularly easy in the case of feedback systems that contain linear subsystems. The following lemma forms the starting point for many of the results derived in this section.

Lemma 2.6 Let a linear system be described by

$$\begin{aligned} \dot{x}(t) &= A(t)x(t) + b(t)u(t) \qquad\qquad x(t_0) = x_0 \\ y(t) &= c^T(t)x(t) \end{aligned} \tag{2.40}$$

where $u, y : \mathbb{R}^+ \to \mathbb{R}$, $x : \mathbb{R}^+ \to \mathbb{R}^n$, and $u \in \mathcal{PC}_{[0,\infty)}$. Let the elements of $A(t), b(t)$, and $c(t)$ be uniformly bounded, and $A(t)$ be an exponentially stable matrix. Then

$$\|x_s(t)\| = O[u_s(t)] \qquad \text{and} \qquad |y(t)| = O[u_s(t)].$$

Proof. The solution of Eq. (2.40) can be expressed as

$$x(t) = \Phi(t, t_0)x_0 + \int_{t_0}^{t} \Phi(t, \tau)b(\tau)u(\tau)d\tau,$$

where $\Phi(t, \tau)$ is the transition matrix. If $\|\Phi(t, t_0)x_0\| \leq c_2$ and $\lim_{t\to\infty} \int_{t_0}^{t} \|\Phi(t, \tau)b(\tau)\|d\tau \leq c_1$ where c_1 and c_2 are finite constants, we have

$$\|x(t)\| \leq c_1 \sup_{\tau \leq t} |u(\tau)| + c_2.$$

Since $|y(t)| \leq \|c(t)\| \, \|x(t)\|$, we have $|y(t)| = O[u_s(t)]$.

We note that the result of Lemma 2.6 follows from the exponential stability of the unforced system and controllability need not be assumed.

Corollary 2.5 Let $W(s) = p(s)/q(s)$ be a transfer function with $p(s)$ and $q(s)$ n^{th} degree Hurwitz polynomials. If $W(s)u = y$, then u and y grow at the same rate.

Proof. $W(s)$ and $W^{-1}(s)$ can be expressed as $W(s) = c_1 + W_1(s)$ and $W^{-1}(s) = c_2 + W_2(s)$ where $W_1(s)$ and $W_2(s)$ are asymptotically stable strictly proper transfer functions[3] and c_1 and c_2 are constants. Hence, $y = W(s)u = c_1 u + W_1(s)u = O[u_s(t)]$; similarly, $u = O[y_s(t)]$ and therefore u and y grow at the same rate.

[3] A transfer function $W(s)$ is said to be stable if its poles are in the closed left-half of the complex plane and the poles on the imaginary axis are simple, and asymptotically stable if its poles are in $\mathbb{C}^-$.

Lemma 2.7 Let a linear time-varying system be described by

$$\begin{aligned} \dot{x}_1(t) &= A_1(t)x_1(t) + b_1(t)y_2(t) + r_1(t) \qquad y_1(t) = c_1^T(t)x_1(t) \\ \dot{x}_2(t) &= A_2(t)x_2(t) + b_2(t)y_1(t) + r_2(t) \qquad y_2(t) = c_2^T(t)x_2(t) \end{aligned} \tag{2.41}$$

where $A_1(t)$ and $A_2(t)$ are exponentially stable matrices, $b_1(t), b_2(t), c_1(t)$, and $c_2(t)$ are uniformly bounded vectors for all $t \in \mathbb{R}^+$ and r_1 and r_2 are piecewise continuous uniformly bounded inputs. Then the states $\|x_1(t)\|$ and $\|x_2(t)\|$ and the outputs $y_1(t)$ and $y_2(t)$ grow at the same rate.

Proof. From Lemma 2.6, $y_1(t) = O[\|x_1(t)\|] = O[y_{2s}(t)]$ and $y_2(t) = O[\|x_2(t)\|] = O[y_{1s}(t)]$. Hence, y_1, y_2, $\|x_1\|$, and $\|x_2\|$ grow at the same rate.

Corollary 2.6 A special case of the system in Eq. (2.41) can be described by

$$\dot{x}(t) = Ax(t) + b[\phi^T(t)x(t) + r(t)] \tag{2.42}$$

where $x, \phi : \mathbb{R}^+ \to \mathbb{R}^n$, A is asymptotically stable, and $\|\phi(t)\|$ and $r(t)$ are uniformly bounded. Then $|\phi^T x|$ and $\|x\|$ grow at the same rate.

Lemmas 2.6 and 2.7 deal with systems whose outputs y are related to the corresponding inputs u by $y(t) = O[u_s(t)]$. To complete the mathematical preliminaries we consider situations that yield $u_s(t) = O[y_s(t)]$ so that sufficient conditions for input and output of a dynamical system to grow at the same rate can be established. This is given by Lemma 2.8 below.

Lemma 2.8 If a linear system is described by

$$\dot{x} = Ax + bu \tag{2.43}$$

where $A \in \mathbb{R}^{n \times n}$, (A, b) is controllable and $u \in \mathcal{E}$, then

$$u_s(t) = O[\|x(t)\|_s]. \tag{2.44}$$

Lemma 2.8 plays a central role in the proofs of stability given in Chapters 5 and 8. According to the lemma, if $u \in \mathcal{E}$, then the input cannot grow faster than the state of the system, if the system is controllable. It is well known that if the matrix A is asymptotically stable, a bounded input u results in a bounded output. If A is unstable, $\|x(t)\|_s$ can grow in an unbounded fashion so that Eq. (2.44) is verified. The principal result of the lemma is that the same holds for the class of inputs $\mathcal{E}$. The importance of the lemma arises from the fact that all the relevant signals used in the proofs in Chapters 5 and 8 belong to the set $\mathcal{E}$.

We first show that if the lemma is true for a first order system, it is also true for an nth-order system for $n > 1$. Following this, the lemma is proved for $n = 1$. While the proof of the lemma given in [36] treats directly the general case, the proof as given here brings out the importance of zeros in $\mathbb{C}^-$ in stability arguments based on growth rates of signals.

Proof of Lemma 2.8. If $x = Tz$, where T is a nonsingular transformation which yields the Jordan canonical representation

$$\dot{z} = Jz + du \tag{2.45}$$

and Eq. (2.44) holds for the system in Eq. (2.45), then it also holds for the system in Eq. (2.43), since $\|z\| \leq c\|x\|$, where c is a positive constant. Hence it is sufficient to prove the lemma when the pair (A, b) is in Jordan form. The following arguments indicate that if the lemma is true for a first order system, it is also true for the system in Eq. (2.43).

(i) Let $y_1 = W_1(s)u$, where $W_1(s) = 1/(s-\alpha)$, $\alpha \in \mathbb{R}$. Let $u_s(t) = O[y_{1s}(t)]$. Since $\dot{y}_1 = \alpha y_1 + u$, and $u \in \mathcal{E}$, it follows that $y_1 \in \mathcal{E}$.

(ii) Let (i) hold and $y_2 = W_2(s)u$, where $W_2(s) = 1/q(s)$, and $q(s)$ is an nth degree polynomial with real roots. By the successive application of (i) which holds for a first order system, it follows that $u_s(t) = O[y_{2s}(t)]$ and $y_2 \in \mathcal{E}$.

(iii) Let (i) hold and $y_3 = W_3(s)u$, where $W_3(s) = 1/q(s)$, and $q(s)$ is an arbitrary n^{th} degree polynomial. y_3 can also be expressed as

$$y_3 = \frac{p(s)}{q(s)} \cdot \frac{1}{p(s)} u = \frac{p(s)}{q(s)} u_1,$$

where $p(s)$ is an nth degree Hurwitz polynomial with only real roots and $u_1 = [1/p(s)]u$. Since $u \in \mathcal{E}$, we have from (ii), $u_s(t) = O[u_{1s}(t)]$. Since $p(s)/q(s)$ is of relative degree zero and $p(s)$ is Hurwitz, from Corollary 2.5, it follows that $u_{1s}(t) = O[y_{3s}(t)]$. Hence, $u_s(t) = O[y_{3s}(t)]$.

(iv) From (ii) and (iii), it follows that

$$u_s(t) = O[\|z(t)\|_s]$$

where z is defined as in Eq. (2.45).

Therefore, the following proof that $u_s(t) = O[y_s(t)]$, where $y = [1/(s-\alpha)]u$, completes the proof of the lemma.

Let $\Omega_u = \{t \mid |u(t)| = \sup_{\tau \leq t} |u(\tau)|\}$. Let $t_1 \in \Omega_u$ where

$$|y(t_1)| < M|u(t_1)| \qquad \text{for some constant } M.$$

If no such t_1 exists, it implies that $u_s(t) = O[y_s(t)]$ and the lemma is true. Positive constants c_0, c_1 and T exist with

$$|\exp\{\alpha T\}| \leq c_0, \qquad \left|\int_0^T \exp\{\alpha\tau\}\, d\tau\right| \geq c_1.$$

The output $y(t)$ at time $t_1 + T$ can be expressed as

$$y(t_1 + T) = \exp(\alpha T)y(t_1) + \int_{t_1}^{t_1+T} \exp(\alpha(t_1 + T - \tau))u(\tau)d\tau.$$

Let $[t_1, t_1 + T] \subset \Omega_u$. Then

$$\begin{aligned} |y(t_1+T)| &\geq c_1|u(t_1)| - c_0|y(t_1)| \\ &\geq (c_1 - c_0 M)\,|u(t_1)|. \end{aligned}$$

Further since $u \in \mathcal{E}$, a constant $c_2 \in (0,1)$ exists with $|u(t_1)| \geq c_2|u(t_1+T)|$. Hence,

$$|y(t)| \;\geq\; (M+\delta)|u(t)| \qquad \text{where } M = \frac{c_1c_2 - \delta}{c_0c_2 + 1} > 0 \tag{2.46}$$

at time $t = t_1 + T$ for some $\delta > 0$. If the inequality in Eq. (2.46) holds for all $t \geq t_1 + T$, then Eq. (2.44) follows. If it is not satisfied, then there exists an unbounded monotonic sequence $\{t_i\}$ such that $|y(t_i)| = M|u(t_i)|$. This in turn implies that at time $t = t_i + T$,

$$|y(t_i+T)| \;\geq\; (M+\delta)|u(t_i+T)|.$$

Since $\sup_{\tau\leq t}|y(\tau)| \geq |y(t)|$ we have

$$|u(t)| \;\leq\; \frac{1}{c_2M}\sup_{\tau\leq t}|y(\tau)| \tag{2.47}$$

for all $t \in \Omega_u \cap [t_1, \infty)$. When $t \notin \Omega_u$, the inequality in Eq. (2.47) is strengthened further. Hence, it follows that

$$u_s(t) \leq \frac{1}{c_2M}\sup_{\tau\leq t}|y(\tau)| \qquad \forall t \geq t_1 + T.$$

Comment 2.12 If $y = W(s)u$, where $W(s) = p(s)/q(s)$, $p(s)$ is a Hurwitz polynomial of degree m, and $q(s)$ is an arbitrary polynomial of degree $n \geq m$, using the same arguments as in (iii), Corollary 2.7 can be proved. If however $p(s)$ is not Hurwitz, it cannot be concluded that $u(t)$ grows at a slower rate than $y(t)$. For example, if $p(s) = s - \alpha$, $q(s)$ is Hurwitz, and $u(t) = \exp(\alpha t), \alpha > 0$, then $\lim_{t\to\infty} y(t) = 0$.

If A is asymptotically stable in Eq. (2.43), by lemmas 2.6 and 2.8 u and $\|x\|$ grow at the same rate. Corollaries 2.7 and 2.8 are used often to draw conclusions regarding growth rates of signals in adaptive systems.

Corollary 2.7 (i) If $y(t) = W(s)u(t)$, where the zeros of $W(s)$ lie in $\mathbb{C}^-$, and $u \in \mathcal{E}$, then

$$u_s(t) = O[y_s(t)].$$

(ii) If in (i), the condition that $u \in \mathcal{E}$ is replaced by

$$|\dot{u}(t)| = O\left[\sup_{\tau\leq t}\|z(\tau)\|\right]$$

where $z^T \stackrel{\Delta}{=} [u, y]$, then once again $u_s(t) = O[y_s(t)]$.

The proof of Corollary 2.7 follows along the same lines as that of Lemma 2.8.

Corollary 2.8 In the dynamical system in Eq. (2.42) defined in Corollary 2.6, assume further that $\dot{\phi}(t)$ is uniformly bounded and let $W_i(s)$ denote the transfer function from the input $(\phi^T x + r)$ to $x_i(t)$, the ith element of the state vector x. Then $\sup_{\tau \leq t} |x_i(\tau)| \sim \|x(t)\|_s$ for all i such that $W_i(s)$ is a transfer function with zeros in $\mathbb{C}^-$.

Proof. Since

$$\frac{d}{dt}\left[\phi^T x\right] = \phi^T A x + \phi^T b[\phi^T x + r] + \dot{\phi}^T x$$

and since $\|x(t)\| = O[\sup_{\tau \leq t} |\phi^T(\tau) x(\tau)|]$, we have $\phi^T x \in \mathcal{E}$. Hence, the conditions of Lemma 2.8 are satisfied and $|\phi^T x| = O[\|x(t)\|_s]$. Hence, $\phi^T x$ and $\|x\|$ grow at the same rate. If $W_i(s)$ is a transfer function with zeros in $\mathbb{C}^-$, then it follows from Lemma 2.8 that $\|x\|$ and x_i grow at the same rate.

Comment 2.13 By Corollary 2.6, when $\|\phi(t)\|$ is bounded, $\phi^T x$ and $\|x\|$ grow at the same rate. By imposing the additional constraint that $\|\dot{\phi}\|$ is bounded in Lemma 2.8, it follows that the components x_i's corresponding to transfer functions with zeros in $\mathbb{C}^-$ grow at the same rate as $\|x\|$.

Signals that grow at different rates. In the following paragraphs, we discuss conditions under which the output of a system grows at a slower rate than the input. In Theorem 2.13, we stated well known results in input-output stability theory that provided conditions under which the output $y(t)$ of a linear system tends to zero as $t \to \infty$. Based on these results, the following lemma can be stated which pertains to outputs that grow at slower rates than the inputs.

Lemma 2.9 If $\beta \in \mathcal{L}^1$ or $\mathcal{L}^2, u \in \mathcal{PC}_{[0,\infty)}$, and βu is the input to an asymptotically stable linear time-invariant system whose transfer function is $W(s)$, and y is the corresponding output, then

$$|y(t)| = o[u_s(t)].$$

Proof. Let w be the impulse response of $W(s)$. Then

$$y(t) = \int_{t_0}^{t} w(t-\tau)\beta(\tau)u(\tau)d\tau + y_w(t)$$

where $y_w(t)$ is due to initial conditions and decays exponentially. Therefore

$$\begin{aligned} |y(t)| &\leq \sup_{\tau \leq t} |u(\tau)| \int_{t_0}^{t} |w(t-\tau)\beta(\tau)| d\tau \\ &= o[\sup_{\tau \leq t} |u(\tau)|] \end{aligned}$$

since the integral term tends to zero as $t \to \infty$.

Lemma 2.10 In the system in Eq. (2.42) described in Corollary 2.6, if $\phi \in \mathcal{L}^1$, $\mathcal{L}^2$, $\lim_{t\to\infty} \phi(t) = 0$, or $|\phi^T(t)x(t)| = o[\|x(t)\|_s]$, then $x(t)$ is uniformly bounded.

Proof. This is a direct consequence of Lemma 2.9 since $\|x(t)\|_s = o[\sup_{\tau\leq t}\|x(\tau)\|]$ $+y(t)$ where $y(t)$ is the response due to $r(t)$ and is therefore uniformly bounded.

The following lemma represents one of the main tools used in proving the stability of adaptive systems in Chapters 5 and 8, and is described variously as "the flipping method" or "the swapping lemma."

Lemma 2.11 Let $W(s)$ be a proper transfer function whose poles and zeros are in $\mathbb{C}^-$, $u, \phi : \mathbb{R}^+ \to \mathbb{R}^n$, $y : \mathbb{R}^+ \to \mathbb{R}$ and

$$y(t) = \phi^T(t)W(s)Iu(t) - W(s)\phi^T(t)u(t).$$

If $\dot{\phi} \in \mathcal{L}^2$, then

$$y(t) = o\left[\sup_{\tau\leq t}\|u(\tau)\|\right].$$

Proof. Since $W(s)$ is an asymptotically stable transfer function, the initial conditions can be assumed to be zero with no loss of generality. Defining $\zeta(t) = W(s)u(t)$, the output y can be expressed in terms of ζ as

$$y(t) = \left[\phi(t) - W(s)\phi(t)W^{-1}(s)\right]^T \zeta(t).$$

Let

$$\mathbf{L}(\phi) \triangleq W(s)\phi W^{-1}(s).$$

The following properties of $\mathbf{L}(\phi)$ can be derived:

(i) If $\phi(t)$ is a constant, $\mathbf{L}(\phi) = \phi$.

(ii) If $W(s) = 1/(s+a)$, $\mathbf{L}(\phi) = \phi - \big(1/(s+a)\big)\,\dot{\phi}$.

(iii) If $W(s) = 1/(s^2+as+b)$,

$$\mathbf{L}(\phi) = \phi - \frac{s+a/2}{s^2+as+b}\dot{\phi} - \frac{1}{s^2+as+b}\dot{\phi}(s+a/2).$$

(iv) If

$$W(s) = 1/\Pi_{i=1}^{p}(s+a_i), \qquad \mathbf{L}(\phi) = \phi - \sum_{i=0}^{p-1} A_i^{-1}(s)\dot{\phi}A_{i+1}(s)$$

where

$$A_i(s) = \frac{W^{-1}(s)}{\Pi_{j=1}^{i}(s+a_j)} \qquad i = 1, 2, \ldots, p, \quad A_0(s) = W^{-1}(s).$$

(v) If

$$W(s) = \Pi_{i=1}^{p}(s+b_i)/\Pi_{i=1}^{p}(s+a_i), \quad \mathbf{L}(\phi) = \phi - \sum_{i=1}^{p} B_i(s)\dot{\phi}A_{i+1}(s) + \sum_{i=1}^{p} B_i(s)\dot{\phi}A_i(s)$$

where

$$A_i(s) = \Pi_{j=i}^{p}\left(\frac{s+a_j}{s+b_j}\right), \; i = 1, \ldots, p, \qquad A_{p+1}(s) = 1$$

$$B_i(s) = \frac{\Pi_{j=i+1}^{p}(s+b_j)}{\Pi_{j=i}^{p}(s+a_j)}, \; i = 1, \ldots, p-1, \qquad B_p(s) = \frac{1}{s+a_p}.$$

If $W(s)$ is a proper transfer function whose zeros and poles are real and lie in $\mathbb{C}^-$, using (iv) and (v), we obtain

$$y(t) = \sum C_i(s)\dot{\phi}D_i(s)u(t)$$

where $C_i(s)$ and $D_i(s)$ are asymptotically stable proper transfer functions for all i. Hence,

$$y(t) = o[\sup_{\tau \leq t} \|u(\tau)\|].$$

Using similar arguments, the same result can also be obtained when $W(s)$ has complex poles and zeros.

In almost all the stability problems discussed in the book, it is only possible to show that an output error $e \in \mathcal{L}^2$. The fact that $\dot{e}$ is bounded is then used to draw the desired conclusion that $\lim_{t\to\infty} e(t) = 0$. Lemma 2.12 and Corollary 2.9 consequently feature prominently throughout the book. Lemma 2.12 has been derived by various authors, but the original work was attributed to Barbalat by Popov in his book [39].

Lemma 2.12 If $f : \mathbb{R}^+ \to \mathbb{R}$ is uniformly continuous for $t \geq 0$, and if the limit of the integral

$$\lim_{t\to\infty} \int_0^t |f(\tau)|d\tau$$

exists and is finite, then

$$\lim_{t\to\infty} f(t) = 0.$$

Proof. Let $\lim_{t\to\infty} f(t) \neq 0$. Then there exists an infinite unbounded sequence $\{t_n\}$ and $\epsilon > 0$ such that $|f(t_i)| > \epsilon$. Since f is uniformly continuous,

$$|f(t) - f(t_i)| \leq k|t - t_i| \qquad \forall t, t_i \in \mathbb{R}^+$$

for some constant $k > 0$. Also,

$$|f(t)| \geq \epsilon - |f(t) - f(t_i)|.$$

Integrating the previous equation over an interval $[t_i, t_i + \delta]$, where $\delta > 0$,

$$\int_{t_i}^{t_i+\delta} |f(\tau)| d\tau \quad \geq \quad \epsilon\delta - k\delta^2/2.$$

Choosing $\delta = \epsilon/k$, we have

$$\int_{t_i}^{t_i+\delta} |f(\tau)| d\tau \quad \geq \quad \epsilon\delta/2 \qquad \forall\ t_i.$$

This contradicts the assumption that $\lim_{t\to\infty} \int_0^t |f(\tau)|\, d\tau$ is finite.

Corollary 2.9 If $g \in \mathcal{L}^2 \cap \mathcal{L}^\infty$, and $\dot{g}$ is bounded, then $\lim_{t\to\infty} g(t) = 0$.

Proof. Choose $f(t) = g^2(t)$. Then $f(t)$ satisfies the conditions of Lemma 2.12. Hence, the result follows.

2.8.3 Stability in Terms of a Parameter Vector θ

As described in Chapter 1, in the adaptive identification and control problems treated in this book, a parameter vector θ is adjusted using the measured signals in the system. For constant values of θ, the overall system is linear and time-invariant. In most cases, the adaptive law ensures the boundedness of $\theta(t)$ for all $t \geq t_0$ and, hence, the stability of the system can be analyzed by considering θ as a time-varying parameter. From a knowledge of the differential equations describing the plant as well as the adaptive law used for adjusting θ, it is also generally possible to determine the behavior of $\dot{\theta}(t)$ for all $t \geq t_0$. This takes the form

$$\lim_{t\to\infty} \dot{\theta}(t) = 0, \quad \text{or} \quad \dot{\theta} \in \mathcal{L}^2.$$

Hence, the stability problem of the adaptive system can be stated in terms of the stability of a linear time-varying system depending on a parameter θ when some constraints are imposed on the time derivative $\dot{\theta}$. In this section, we state some stability theorems that parallel those given in Section 2.3.2. The system matrices in the present case are expressed in terms of the parameter vector $\theta(t)$ rather than in terms of time t. The stability theorems are consequently directly applicable to many of the problems discussed in the book.

Let $\theta : \mathbb{R}^+ \to S \subset \mathbb{R}^m$, where S is a compact set with the property that for every constant vector $\overline{\theta} \in S$, the origin of the differential equation

$$\dot{x} \;=\; A(\overline{\theta})x \qquad x(t_0) \;=\; x_0$$

is asymptotically stable. The results in this section pertain to the stability of the equilibrium state of the differential equation

$$\dot{x} = A(\theta(t))x \qquad \theta : \mathbb{R}^+ \to S, \; x(t_0) = x_0, \tag{2.48}$$

under different assumptions on the matrix $A(\theta(t))$.

Lemma 2.13 Let $A(\theta)$ be a continuously differentiable function of the parameter θ and let $A(\overline{\theta})$ be an asymptotically stable matrix for any constant value $\overline{\theta} \in S$. Then a symmetric positive-definite matrix $P(\theta)$ exists which is a continuously differentiable function of θ and satisfies the matrix equation

$$A^T(\theta)P(\theta) + P(\theta)A(\theta) = -Q \qquad Q = Q^T > 0, \; \forall \; \theta \in S. \tag{2.49}$$

Proof. The existence of $P(\theta)$ follows directly from Theorem 2.10 and the fact that $A(\theta)$ is asymptotically stable. The fact that P has continuous partial derivatives with respect to θ follows from the implicit function theorem. Since S is a compact set, it follows that $\partial P/\partial\theta$ is bounded for $\theta \in S$.

Theorem 2.18 If in Eq. (2.48), (i) $\lim_{t\to\infty} \theta(t) = \overline{\theta} \in S$, (ii) $\dot{\theta} \in \mathcal{L}^1$, or (iii) $\dot{\theta} \in \mathcal{L}^2$, then the solutions of Eq. (2.48) tend to zero exponentially.

Proof. (i) is a direct consequence of Theorem 2.4.

To show that the asymptotic behavior of the solutions is similar in cases (ii) and (iii), we use the quadratic function

$$V(x(t)) = x^T(t)P(\theta(t))x(t)$$

as a Lyapunov function candidate. The time derivative of V along the trajectories of Eq. (2.48) are given by

$$\dot{V}(x) = x^T\left[A^T(\theta)P(\theta) + P(\theta)A(\theta)\right]x + x^T\frac{d}{dt}[P(\theta)]x.$$

Choosing $P(\theta)$ as the solution of Eq. (2.49) and using the fact that

$$\sup_{\theta\in S}\frac{\partial P_{ij}}{\partial\theta_k} \le c_2 \qquad \forall i, j = 1,\ldots,n, \; k = 1,\ldots,m$$

for some constant c_2, we obtain

$$\begin{aligned}\dot{V} &\le -x^T Qx + c_2\|\dot{\theta}\|x^T x \\ &\le \left[-c_1 + \|\dot{\theta}\|c_3\right]V\end{aligned}$$

for positive constants c_1 and c_3. By Corollary 2.1,

$$V(x) \to 0 \qquad \text{if } \lim_{t\to\infty} \dot{\theta}(t) = 0, \;\; \dot{\theta} \in \mathcal{L}^1, \text{ or } \in \mathcal{L}^2.$$

In most cases of interest, the stability of the adaptive system has to be concluded on the basis of the properties of $\dot{\theta}$ as well as the output of the system. This is given by the following theorem based on the work of Morse [34]. We denote the pair (C, A) to be *detectable* provided every eigenvalue λ of A, whose eigenvector q is in the kernel of C, lies in $\mathbb{C}^-$.

Theorem 2.19 Let a dynamical system be described by the equation

$$\dot{x} = A(\theta(t))x, \qquad y = C(\theta(t))x. \tag{2.50}$$

where $C(\theta)$ and $A(\theta)$ are continuously differentiable functions of θ. Further let $(C(\overline{\theta}), A(\overline{\theta}))$ be detectable for every constant value of $\overline{\theta} \in S'$ where $S \subset S' \subset \mathbb{R}^m$ and S' is a compact set. If $\theta : \mathbb{R}^+ \to S'$, then

(i) if $\lim_{t\to\infty} \theta(t) = \overline{\theta} \in S'$ and $\lim_{t\to\infty} y(t) = 0$, then $\lim_{t\to\infty} x(t) = 0$.
(ii) If $\dot{\theta} \in \mathcal{L}^2$ and $y \in \mathcal{L}^2$, then $\lim_{t\to\infty} x(t) = 0$.

Proof. Since $(C(\theta), A(\theta))$ is detectable for $\theta \in S'$, there exists a matrix $H(\theta)$, whose elements are continuous with respect to θ, such that

$$\overline{A}(\theta) \triangleq A(\theta) + H(\theta)C(\theta)$$

is an asymptotically stable matrix for $\theta \in S'$. If a vector function z is defined by

$$\dot{z} = \overline{A}(\theta(t))z,$$

and if either (i) or (ii) in Theorem 2.19 is satisfied, then $z(t)$ decays exponentially to zero from Theorem 2.18. Equation (2.50) can be rewritten as

$$\dot{x} = \overline{A}(\theta(t))x - H(\theta(t))y.$$

Since $\overline{A}(\theta(t))$ is an exponentially stable matrix, and S' is a compact set, if $\lim_{t\to\infty} y(t) = 0$, or $y \in \mathcal{L}^2$, it follows that $x(t) \to 0$ as $t \to \infty$ [13].

2.9 OTHER STABILITY CONCEPTS

The primary interest of this book is in the global stability of adaptive systems and the principal approach used to assure this is the direct method of Lyapunov. In many situations, however, Lyapunov's method cannot be used exclusively and one has to call upon other concepts that have been treated extensively in the stability literature. These include BIBO stability which is more appropriate when continuous external disturbances are present, and total stability when the disturbances can be considered to be state-dependent. In this section we describe some of the significant ideas related to these concepts that are directly relevant to our discussions in the chapters following.

Yet another approach, which has been used extensively in the adaptive literature, is based on hyperstability. This notion, which grew out of the concept of absolute stability and was introduced by Popov in 1963, is also found to be suitable for dealing with adaptive control problems. All the results derived in this book using Lyapunov's method can, in theory, also be derived using this approach. Consequently, the two approaches can be considered to be equivalent, and whether one or the other is used depends on one's personal preference. We do not use hyperstability arguments in this book and refer the reader to [39] for a definition and discussion of the concept and to [21] for its application in adaptive systems.

2.9.1 Total Stability

When persistent disturbances are present in a system that are due to both external as well as internal signals, the equations to be analyzed take the following nonlinear form:

$$\begin{aligned} \dot{x}(t) &= f(x,t) \qquad f(0,t)=0 \;\; \forall\, t \geq t_0 \quad (S) \\ \dot{x}(t) &= f(x,t)+g(x,t) \qquad\qquad\qquad (S_p) \end{aligned} \tag{2.51}$$

where (S) represents the unperturbed system with $x=0$ as the equilibrium state and (S_p) is a perturbed system with a perturbation $g(x,t)$ which could be due to nonlinearities, parameter variations, or structural perturbations. We are often interested in deducing the effect of the additional term $g(x,t)$ on the stability or instability of the equilibrium of S_p. In contrast to the results derived in the book, which are mostly global in nature, the results related to this problem are local in character.

We first consider one of the simplest cases where $f(x,t)=Ax$ is a linear function with constant coefficients and $g(x,t)=g(x)$, where the elements $g_i(x)$ of $g(x)$ have, in the neighborhood of the origin, power series expansions in the variables $x_1, x_2, \ldots, x_n$ beginning with terms of at least second order. Then the two equations (S) and (S_p) of Eq. (2.51) can be written as

$$\dot{x} = Ax \qquad (S_\ell) \qquad\qquad \dot{x} = Ax+g(x) \qquad (S_{p\ell})$$

and (S_ℓ) is called the equation of first approximation to $(S_{p\ell})$ and is said to be obtained from $(S_{p\ell})$ by linearization [16]. The equilibrium of (S_ℓ) is known to be asymptotically stable if all the eigenvalues of A have negative real parts and unstable if at least one of the eigenvalues of A has a positive real part. A differential equation (S_ℓ) in which the matrix A satisfies one of the conditions above is said to have *a significant behavior*. The following fundamental theorem on the stability in the first approximation is due to Lyapunov.

Theorem 2.20 [28] If the differential equation of the first approximation (S_ℓ) has significant behavior, then the equilibrium state of the complete differential equation $(S_{p\ell})$ has the same stability behavior.

A more general case is when the equations of first approximation are time-varying so that (S) and (S_p) specialize to

$$\dot{x} = A(t)x \qquad (S_t) \qquad\qquad \dot{x} = A(t)x + g(x,t) \quad (S_{pt})$$

respectively. A number of results concerning the stability behavior of (S_{pt}) based on the stability properties of (S_t) have been derived by various authors. We present below, in a single theorem, several results culled from the excellent book by Coppel [11]. The reader is referred to [11] for the proof of the theorem.

Let $\Phi(t, t_0)$ be the transition matrix of (S_t). Let $g(x, t)$ be continuous for all $t \geq t_0$ and $\|x\| < c$, where c is a positive constant.

Theorem 2.21

(i) If in the differential equation (S_{pt}), $\|\Phi(t, \tau)\| \leq k$ for all $t \geq \tau \geq t_0$, and $k \in \mathbb{R}^+$,

$$\|g(x,t)\| \leq \gamma(t)\|x\|$$

where $\gamma(t) \geq 0$ for all $t \geq t_0$, and $\int_{t_0}^{\infty} \gamma(\tau)d\tau < \infty$, then there exists a positive constant L such that

$$\|x(t)\| \leq L\|x(t_1)\| \qquad \forall\, t \geq t_1, \;\; \forall\, \|x(t_1)\| < L^{-1}c.$$

(ii) If, in addition, $\|\Phi(t, \tau)\| \to 0$ uniformly in τ as $t \to \infty$, then $\|x(t)\| \to 0$.

(iii) If $\int_{t_0}^{t} \|\Phi(t, \tau)\|d\tau \leq k < \infty$ and $\|g(x, t)\| \leq \gamma\|x\|$ for some $\gamma < 1/k$, then the equilibrium state $x = 0$ of (S_{pt}) is asymptotically stable.

(iv) If $\|\Phi(t, \tau)\| \leq k \exp\{-\alpha(t - \tau)\}$ for some $k, \alpha \in \mathbb{R}^+$ and all $t, \tau \geq t_0$ and $\|g(x, t)\| \leq \gamma\|x\|$, where $\gamma < (\alpha/k)$, then

$$\|x(t)\| \leq k \exp\{-\beta(t - \tau)\}\|x(\tau)\| \;\; \forall t \geq \tau \geq t_0 \text{ and } \|x(t_0)\| < c/k,$$

where $\beta = \alpha - \gamma k > 0$.

Comment 2.14 According to Theorem 2.21-(i), if the transition matrix is bounded or the equilibrium state is uniformly stable, all solutions are bounded provided $\|g(x, t)\|$ is smaller than $\gamma(t)\|x\|$. By (ii), $x(t)$ tends to zero if the transition matrix tends to zero uniformly in t. If $\Phi(.,.) \in \mathcal{L}^1$, the linear system as well as the nonlinear system are asymptotically stable. Finally when the equilibrium of (S_t) is exponentially stable, then all the solutions of the nonlinear system also tend to zero if $\|g(x, t)\|$ is bounded above by a linear function of $\|x\|$.

Thus far, we have expressed a nonlinear system in terms of a linearized system together with perturbations $g(x, t)$. At the next stage, we consider the stability behavior of the equilibrium state $x = 0$ of the nonlinear system (S) when it is also the equilibrium state of (S_p). The following theorem by Krasovskii addresses this problem.

Theorem 2.22 [20] If the equilibrium state of (S) is exponentially stable and in a neighborhood of the origin, the term $g(x, t)$ satisfies the inequality

$$\|g(x,t)\| \leq b\|x\| \qquad b > 0$$

then, for a sufficiently small b, the equilibrium state of the differential equation (S_p) has the same stability behavior.

The proof of the theorem above is based on the existence of Lyapunov functions $V(x)$ whose time derivatives are negative-definite for the unperturbed as well as the perturbed systems in a neighborhood of the origin. In most physical situations, the condition $g(0,t) \equiv 0$ required in Theorem 2.22 is not generally met and this gives rise to the concept of total stability defined below.

Definition 2.14 [29] The equilibrium state $x = 0$ of (S) is *totally stable* if for every $\epsilon > 0$, two positive numbers $\delta_1(\epsilon)$ and $\delta_2(\epsilon)$ exist such that $\|x(t_0)\| < \delta_1$ and $\|g(x,t)\| < \delta_2$ imply that every solution $x(t; x_0, t_0)$ of (S_p) satisfies the condition $\|x(t; x_0, t_0)\| < \epsilon$.

The following theorem by Malkin [26,29] indicates the relation between total stability and uniform asymptotic stability in the sense of Lyapunov.

Theorem 2.23 If the equilibrium state of (S) is uniformly asymptotically stable, then it is totally stable.

Proof. Since $x = 0$ of (S) is uniformly asymptotically stable, there exist positive-definite functions $\alpha(\|x\|)$, $\beta(\|x\|)$, and $\gamma(\|x\|)$ such that $\alpha(\|x\|) \leq V(x,t) \leq \beta(\|x\|)$ and $\dot{V}(x,t) \leq -\gamma(\|x\|)$ in a region $D = \{x | \; \|x\| < \epsilon_A\}$, where ϵ_A is a positive constant. Let the partial derivatives $\partial V/\partial x_i$ be bounded in D for $i = 1, \ldots, n$, $t \geq t_0$ so that

$$|\partial V/\partial x_i| \leq c_0 \qquad \forall\, x \in D,\ \forall\, t \geq t_0 \text{ and } c_0 \in \mathbb{R}^+.$$

We then have

$$\dot{V}_{(S_p)} = \dot{V}_{(S)} + \sum_{i=1}^{n} g_i \, \partial V/\partial x_i$$

where $\dot{V}_{(S)}$ and $\dot{V}_{(S_p)}$ are the time derivatives of V evaluated along the solutions of (S) and (S_p) respectively, and g_i is the ith component of g. Given any $\epsilon < \epsilon_A$, there exist δ_1 and δ_2 such that $\alpha(\epsilon) > \beta(\delta_1)$, and

$$\|g(x,t)\| < \delta_2 \qquad \forall x \in D.$$

Let

$$\gamma_0 = \min_{\delta_1 \leq \|x\| \leq \epsilon} (\gamma(\|x\|)).$$

For all solutions $x(t; x_0, t_0)$ with $\|x_0\| < \delta_1$ entering the region $\delta_1 \leq \|x\| \leq \epsilon$ for some time $t = t_1$, we have

$$\dot{V}_{(S_p)} \leq -\gamma_0 + n c_0 \delta_2.$$

Choosing $\delta_2 = k\gamma_0/(nc_0)$, where $0 < k < 1$, $\dot{V}_{(S_p)} < 0$ for all time $t \geq t_1$, so that $\|x(t; x_0, t_0)\| < \epsilon$. Hence, the origin of (S) is totally stable.

The definition above implies that the solutions of (S_p) are small if the initial condition x_0 as well as the perturbations are sufficiently small. Although problems of the type described here have been studied by many authors in the Russian literature, we have referred to only a few theorems here. These suffice for some of the discussions in Chapter 8 where brief comments are made concerning local stability results. The reader is referred to the books by Hahn [16] and Coppel [11] for a detailed treatment of the subject and to [4] for its application in adaptive systems.

2.9.2 BIBO Stability

It is well known that a linear system of the form

$$\dot{x} = A(t)x + B(t)u,$$

where $A(t)$ and $B(t)$ are matrices with bounded piecewise-continuous elements, results in a bounded x with a bounded input u if $A(t)$ is exponentially stable. As mentioned earlier, uniform asymptotic stability and exponential stability are equivalent in linear systems. This fact makes uniform asymptotic stability (rather than asymptotic stability) an important attribute in linear systems.

From Theorem 2.23 in the previous section, it follows that for sufficiently small disturbances and initial conditions, that is, $\|u\| \leq \delta_2$ and $\|x_0\| \leq \delta_1$ where δ_1 and δ_2 are sufficiently small, the solutions of the nonlinear differential equation

$$\dot{x} = f(x,t) + u(t) \qquad x(t_0) = x_0 \tag{2.52}$$

will be uniformly bounded for all $t \geq t_0$. However, in practice, it is of greater interest to determine the nature of the solutions, when some bound on $\|u\|$ is specified. The following example due to Desoer [12] shows conclusively that, even in simple cases, unbounded solutions can result. Hence, uniform asymptotic stability in the large in nonlinear systems is not sufficient to ensure bounded-input bounded-output stability.

Example 2.10

[12] Consider the nonlinear system described by the second-order equation

$$\ddot{x} + f(\dot{x}) + x = u. \tag{2.53}$$

where $f(\cdot)$ is a first and third quadrant nonlinear function defined by

$$|f(y)| \leq 1 \quad -\infty < y < \infty$$

$$yf(y) > 0 \quad \text{for } y \neq 0$$

$$f(0) = 0.$$

When $u(t) \equiv 0$, the equilibrium state of Eq. (2.53) is u.a.s.l. However, with an input $u(t) = A \sin t$, $A > 4/\pi$, the output of the system is unbounded, which can be shown as follows:

The uniform stability in the large of the origin when $u = 0$ can be shown by choosing $V = (x^2+\dot{x}^2)/2$ for which $\dot{V} = -\dot{x}f(\dot{x}) \leq 0$. By LaSalle's theorem, the origin can be shown to be u.a.s.l. since it is the largest invariant set in the set E on which $\dot{V} = 0$. $u - f(\dot{x})$ can be considered to be an external input into a system whose homogeneous part is given by $\ddot{x} + x = 0$. Hence, the solution can be written as

$$\begin{aligned} x(t) &= \frac{A}{2}(\sin t - t \cos t) - \int_0^t \sin(t-\tau)f(\dot{x})\, d\tau \\ &\geq \frac{A}{2}(\sin t - t \cos t) - \int_0^t |\sin(t-\tau)|\, d\tau. \end{aligned}$$

The second term on the right-hand side can be bounded by $(2/\pi)(1+\epsilon)t$ for $t \geq t_0$ for any $\epsilon > 0$ and some t_0. Choosing $t_n = (2n+1)\pi$, we have

$$x(t_n) \geq (2n+1)\pi \left[\frac{A}{2} - \frac{2}{\pi}(1+\epsilon)\right].$$

Hence, if $A > 4/\pi$, we can find an ϵ such that $|x(t_n)| \to \infty$.

The following example, which arises in adaptive systems, is discussed in greater detail in Chapter 8.

Example 2.11

[35] Let a system with a bounded disturbance $\nu(t)$ be described by the equations

$$\begin{aligned} \dot{x}_1 &= -x_1 + x_2(x_1 + c_0) + \nu(t) \\ \dot{x}_2 &= -x_1(x_1 + c_0) \end{aligned} \tag{2.54}$$

where c_0 is a positive constant. The equilibrium state of Eq. (2.54) is u.a.s. when $\nu(t) \equiv 0$. It can further be shown that when $\nu(t) \equiv \nu_0$ where ν_0 is a constant and $0 < \nu_0 < c_0$, the solutions of Eq. (2.54) are uniformly bounded. However, if $\nu(t) \equiv -\nu_0 < -c_0$, the solutions are unbounded for some initial conditions.

The two examples above indicate that BIBO stability is considerably more difficult to establish in nonlinear systems of the form in Eq. (2.52) than in linear systems, where such stability follows if the unforced system is u.a.s.l.(or exponentially stable). However, the analysis of adaptive systems in the presence of disturbances invariably leads to questions related to BIBO stability. Despite extensive efforts, very few results are currently known that can be applied directly to problems of interest to us. In Chapter 8, problems are encountered where the following theorem is used to establish boundedness.

Theorem 2.24 [23] Consider the differential equation

$$\dot{x} = f(x) \qquad x : \mathbb{R}^+ \to \mathbb{R}^n \tag{2.55}$$

Let Ω be a bounded neighborhood of the origin and let Ω^c be its complement. Assume that $V(x)$ is a scalar function with continuous partial derivatives in Ω^c and satisfying 1) $V(x) > 0 \;\; \forall \, x \in \Omega^c$, 2) $\dot{V}(x) \leq 0 \;\; \forall \, x \in \Omega^c$ and 3) $V(x) \to \infty$ as $\|x\| \to \infty$. Then each solution of Eq. (2.55) is bounded for all $t \geq 0$.

2.10 SUMMARY

Some of the stability concepts that arise in adaptive systems are discussed in this chapter. Since the analysis of adaptive systems, which are nonlinear, is generally carried out by studying subsystems that are linear, Section 2.3 deals with well known stability results in linear systems theory. Lyapunov's direct method, which is the general tool used throughout the book, is introduced in Section 2.4. Positive real functions and the Kalman-Yakubovich lemma are discussed in Sections 2.5 and 2.6 respectively. Two versions of the lemma, which are used in proving the stability of adaptive systems, are presented here. The well known concept of input-output stability is treated briefly in Section 2.7. Stability theorems that are used frequently in the analysis of adaptive systems are collected in Section 2.8. In many problems in Chapters 5 and 8, arguments based on relative growth rates of signals within the overall system are used to establish their boundedness. Some key lemmas that pertain to the growth rates of signals in dynamical systems are presented in this section. Two specific time-varying differential equations that find application in adaptive systems are discussed here. A brief introduction to total stability and BIBO stability is given in Section 2.9. Although many of the theorems stated here are not used in the derivation of stability results in the book, they are nevertheless useful for the discussions that follow.

PROBLEMS

Part I

1. Determine conditions on $\alpha(t)$ so that the equilibrium state of the differential equation

$$\dot{x} + \alpha(t)x = 0$$

is (i) stable, (ii) asymptotically stable, (iii) uniformly asymptotically stable.

2. Discuss the asymptotic behavior of the solutions of the system described by

$$\begin{aligned} \dot{x}_1 &= -x_2 - \epsilon\left[1 - x_1^2 - x_2^2\right] x_1 \\ \dot{x}_2 &= x_1 - \epsilon\left[1 - x_1^2 - x_2^2\right] x_2 \end{aligned} \qquad \epsilon > 0$$

for various initial conditions.

3. Is the matrix

$$A = \begin{bmatrix} 1 & 2 \\ -11 & -16 \end{bmatrix}$$

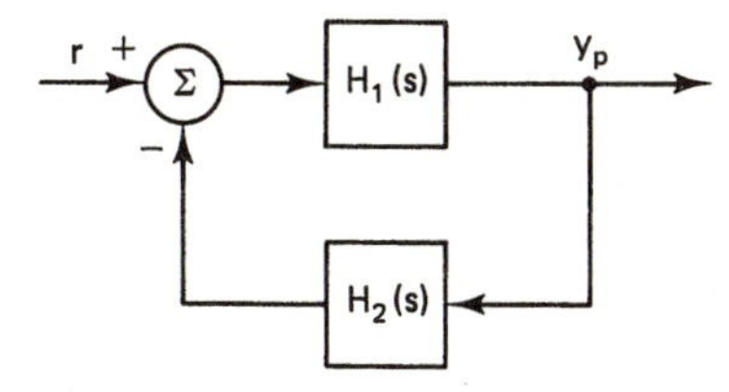

Figure 2.5 Feedback system with positive real subsystems.

stable? Is it asymptotically stable? Show this by solving the Lyapunov equation $A^T P + PA = -I$.

4. Are the following transfer functions PR? Are they SPR?

$$(a)\ \frac{1}{s^2+3s+4} \qquad (b)\ \frac{s+2}{s^2+4s+3} \qquad (c)\ \frac{s+5}{s^2+4s+3}$$

For what values of a is the transfer function $(s+a)/(s^2+2\zeta\omega_n s+\omega_n^2)$ SPR?

5. If $W(s)$ is a strictly proper transfer function that is asymptotically stable with input x and output y, show that

   (a) $y(t)$ is uniformly bounded if $x(t)$ is uniformly bounded,

   (b) $y(t) \to 0$ as $t \to \infty$ if $x(t) \to 0$ as $t \to \infty$,

   (c) $x \in \mathcal{L}^1$ or $\in \mathcal{L}^2 \Rightarrow y(t) \to 0$ as $t \to \infty$.

Part II

6. The second-order differential equation

$$\begin{aligned}\dot{x}_1 &= x_2\\ \dot{x}_2 &= -kx_1 - \alpha x_2\end{aligned}$$

where α and k are constants, is asymptotically stable for all $\alpha > 0$ and $k > 0$. Determine a Lyapunov function $V = x^T P x$, ($x = [x_1, x_2]^T$) that assures the asymptotic stability of the system when $\alpha = 1$ for the maximum range $0 < k < k_1$.

7. In the feedback system shown in Fig. 2.5, $H_1(s)$ and $H_2(s)$ are positive real transfer functions. Show that the overall transfer function from the input r to the output y_p is also positive real.

8. Show that if $e : \mathbb{R}^+ \to \mathbb{R}$, $e \in \mathcal{L}^2$ and $\dot{e} \in \mathcal{L}^\infty$, then $\lim_{t\to\infty} e(t) = 0$.

9. Consider the differential equation

$$\dot{x}(t) = A(t)x(t) \qquad x(t_0) = x_0$$

where $A(t) \in \mathbb{R}^{n\times n}$. If the elements of $A(t)$ are bounded piecewise continuous functions, show that $\|x(t)\|$ can grow at most exponentially, that is,

$$\|x(t)\| \exp(-\alpha t) = 0$$

for some $\alpha > 0$.

10. A feedback system contains a feedforward transfer function $W(s)$ (defined as in Problem 5) and a feedback gain $\beta(t)$ where $\beta \in \mathcal{L}^2$. Show that the output y is uniformly bounded if the input u is uniformly bounded for all $t \geq t_0$. Also, show that $y(t) \to 0$ as $t \to \infty$ if $u(t) \to 0$ as $t \to \infty$.

11. **(a)** In Lemma 2.8, if A is an asymptotically stable matrix, show that $u_s \sim \|x\|_s$.

(b) If $W(s)$ is an asymptotically stable transfer function with all its zeros in $\mathbb{C}^-$, show that the input and the output grow at the same rate.

(c) Give a simple example to show that (b) is violated if at least one zero of $W(s)$ lies in the open right half plane.

Part III

12. **(a)** In Problem 6, determine a Lyapunov function that verifies the asymptotic stability of the system for $\alpha = 1$ and any constant value $k > 0$.

(b) If the gain k is nonlinear and is a continuous function of x_1, it is known that the system

$$\begin{aligned} \dot{x}_1 &= x_2 \\ \dot{x}_2 &= -k(x_1)x_1 - x_2 \end{aligned}$$

is asymptotically stable in the large for all $k(x_1) > 0$. Determine a Lyapunov function to demonstrate this.

REFERENCES

1. Anderson, B. D. O. "Exponential stability of linear equations arising in adaptive identification." *IEEE Transactions on Automatic Control* 22:83–88, 1977.
2. Anderson, B.D.O. "A system theory criterion for positive real matrices." *SIAM Journal of Control* 5:171–182, Feb. 1967.
3. Anderson, B.D.O. *Network analysis and synthesis: A modern systems theory approach.* Englewood Cliffs, NJ:Prentice-Hall, Inc., 1973.
4. Anderson, B.D.O., Bitmead, R., Johnson, C.R., Kokotovic, P.V., Kosut, R.L., Mareels, I., Praly, L., and Riedle, B. *Stability of adaptive systems.* Cambridge, MA:M.I.T Press, 1986.
5. Artstein, Z. Limiting equations and stability of nonautonomous ordinary differential equations. In *The Stability of Dynamical Systems*, Society of Industrial and Applied Mathematics, Philadelphia, PA, 1976.
6. Bellman, R. *Stability of differential equations.* New York:Dover Publications, 1953.
7. Brune, O. "Synthesis of a finite two terminal network whose driving-point impedance is a prescribed function of frequency." *Journal of Mathematics and Physics* 10:191–236, 1931.
8. Cesari, L. *Asymptotic behavior and stability problems in ordinary differential equations.* Berlin: Springer Verlag, 1959.

9. Chen, C.T. *Introduction to linear system theory.* New York, NY:Holt, Rinehart and Winston, Inc., 1970.

10. Coddington, E.A., and Levinson, N. *Theory of ordinary differential equations.* New York:McGraw-Hill Book Company, 1965.

11. Coppel, W.A. *Stability and asymptotic behavior of differential equations.* Boston, MA:D.C. Heath and Company, 1965.

12. Desoer, C.A., Liu, R., and Auth, Jr., L.V. "Linearity vs. nonlinearity and asymptotic stability in the large." *IRE Transactions on Circuit Theory* 12:117–118, March 1965.

13. Desoer, C.A., and Vidyasagar, M. *Feedback systems: Input-output properties.* New York:Academic Press, 1975.

14. Faurre, P., Clerget, M., and Germain, F. *Opérateurs rationnels positifs.* BORDAS, Paris, 1979.

15. Hahn, W. *Stability of motion.* New York:Springer-Verlag, 1967.

16. Hahn, W. *Theory and application of Lyapunov's direct method.* Englewood Cliffs, NJ:Prentice-Hall, Inc., 1963.

17. Ioannou, P.A., and Tao, G. "Frequency domain conditions for strictly positive real functions." *IEEE Transactions on Automatic Control* 32:53–54, 1987.

18. Kalman, R.E. "Lyapunov functions for the problem of Lur'e in automatic control." In *Proceedings of the National Academy of Sciences* 49:201–205, 1963.

19. Kalman, R.E., and Bertram, J.E. "Control systems analysis and design via the 'second method' of Lyapunov." *Journal of Basic Engineering* 82:371–392, June 1960.

20. Krasovskii, N.N. *Stability of motion.* Stanford, CA:Stanford University Press, 1963.

21. Landau, I.D. *Adaptive control.* New York:Marcel Dekker, Inc., 1979.

22. LaSalle, J.P. "Asymptotic stability criteria." In *Proceedings of Symposium on Appl. Math.* 13: 299–307, Hydrodynamic Instability, Providence, R.I., 1962.

23. LaSalle, J.P. "Some extensions of Lyapunov's second method." *IRE Transactions on Circuit Theory* 7:520–527, Dec. 1960.

24. LaSalle, J.P. "Stability of nonautonomous systems." *Nonlinear Analysis: Theory, Methods and Applications* 1:83–91, 1976.

25. LaSalle, J.P. "Stability theory for ordinary differential equations." *Journal of Differential Equations* 4:57–65, 1968.

26. LaSalle, J.P., and Lefschetz, S. *Stability by Lyapunov's direct method.* New York:Academic Press, 1961.

27. Lefschetz, S. *Stability of nonlinear control systems.* New York:Academic Press, 1963.

28. Lyapunov, A.M. "Problème général de la stabilité du mouvement." *Ann. Fac. Sci. Toulouse* 9:203–474, 1907. Reprinted in Ann. Math Studies, No. 17, 1949, Princeton, NJ:Princeton University Press, 1947.

29. Malkin, I.G. "Theory of stability of motion." Technical Report Tr. 3352, U.S. Atomic Energy Commission, English ed. 1958.

30. Massera, J.L. "Contributions to stability theory." *Annals of Mathematics* 64:182–206, 1956.

31. Meyer, K.R. "On the existence of Lyapunov functions for the problem on Lur'e." *SIAM Journal of Control* 3:373–383, Aug. 1965.

32. Morgan, A.P., and Narendra, K.S. "On the stability of nonautonomous differential equations $\dot{x} = [A+B(t)]x$ with skew-symmetric matrix $B(t)$." *SIAM Journal of Control and Optimization* 15:163–176, Jan. 1977.

33. Morgan, A.P., and Narendra, K.S. "On the uniform asymptotic stability of certain linear nonautonomous differential equations." *SIAM Journal on Control and Optimization* 15:5–24, Jan. 1977.

34. Morse, A.S. "High-gain adaptive stabilization." In *Proceedings of the Carl Kranz Course*, Munich, W. Germany, 1987.

35. Narendra, K.S., and Annaswamy, A.M. "Robust adaptive control in the presence of bounded disturbances." *IEEE Transactions on Automatic Control* 31:306–315, April 1986.

36. Narendra, K.S., Annaswamy, A.M., and Singh, R.P. "A general approach to the stability analysis of adaptive systems." *International Journal of Control* 41:193–216, 1985.

37. Narendra, K.S., and Taylor, J.H. *Frequency domain criteria for absolute stability.* New York:Academic Press, 1973.

38. Parks, P.C. "A new proof of the Routh-Hurwitz stability criterion using the second method of Lyapunov." In *Proceedings of the Cambridge Phil. Soc.* 58:694–702, 1962.

39. Popov, V.M. *Hyperstability of control systems.* New York:Springer-Verlag, 1973.

40. Sondhi, M.M., and Mitra, D. "New results on the performance of a well-known class of adaptive filters." *Proceedings of the IEEE* 64:1583–1597, Nov. 1976.

41. Taylor, J. H. "Strictly positive-real functions and the Lefschetz-Kalman-Yakubovich (LKY) lemma." *IEEE Transactions on Circuits and Systems*, 1974.

42. Yakubovich, V.A. "Solution of certain matrix inequalities in the stability theory of nonlinear control systems." *Dokl. Akad. Nauk. USSR*, No. 143, 1962.

3

Simple Adaptive Systems

3.1 INTRODUCTION

The basic principles used throughout this book for the design of identifiers and controllers for plants with unknown parameters, are introduced in this chapter through some simple examples. These examples include plants whose input-state-output relations are described by first-order linear time-invariant differential equations with unknown coefficients. The same ideas are later extended to plants described by vector differential equations when the state variables of the plant are accessible. In both cases we are interested in the identification problem where the plant parameters are to be estimated using input-output data, as well as the control problem where the parameters of a suitably chosen controller are to be adjusted to make the error between the plant output and the output of a reference model tend to zero asymptotically.

Lyapunov's direct method is the principal tool used for the derivation of stable adaptive laws. It is shown in Chapters 3-8 that the method can be used to systematically design a fairly large class of stable adaptive systems. Since in all the cases considered, global stability is assured by demonstrating the existence of a Lyapunov function for the overall system, the choice of such a function plays a crucial role in the selection of a suitable adaptive law. It is well known that although the existence of a Lyapunov function assures the stability of a given system, the selection of such a function is rarely straightforward. This is particularly true of nonlinear systems. The simple examples included in this chapter serve to introduce the reader to the manner in which the Lyapunov function candidate and the adaptive law are chosen when all the state variables of the plant are accessible. The reader's understanding of the basic ideas involved in these simple cases will facilitate an appreciation of their extension to more complex problems described in subsequent chapters.

3.2 ALGEBRAIC SYSTEMS

Many identification and control problems can be formulated in terms of a set of algebraic equations. We first consider such a formulation of the identification problem [6].

3.2.1 Identification

Scalar Case. Consider a system in which the input $u(t)$ and the output $y(t)$ at time t are related by an algebraic equation $y(t) = \theta u(t)$ where θ is an unknown constant. Assuming that $u(t)$ and $y(t)$ can be measured for all values of $t \geq t_0$, it is desired to obtain an estimate $\widehat{\theta}$ of θ. The obvious solution is to observe $\{u(t), y(t)\}$ over an interval $[t_0, t_0 + T]$ so that $u(t_i) \neq 0$ for some $t_i \in [t_0, t_0 + T]$ and choose the estimate $\widehat{\theta}$ as

$$\widehat{\theta} = \frac{y(t_i)}{u(t_i)}. \tag{3.1}$$

This is not normally desirable since $u(t_i)$ may be either zero or assume values arbitrarily close to it so that $\widehat{\theta}$ has to be computed as the ratio of two small numbers. The situation may be further aggravated by the presence of noise which may result in a large spread in the values of $\widehat{\theta}$ corresponding to different values of t_i.

The solution of Eq. (3.1) may be considered as one that minimizes the squared error $\left[y(t_i) - \widehat{y}(t_i)\right]^2$ at time t_i where $\widehat{y}(t_i) = \widehat{\theta} u(t_i)$. Alternately, we can attempt to minimize the integral square error J_0 over an interval $[t_0, t_0 + T]$ so that $\widehat{\theta}$ minimizes

$$J_0 = \int_{t_0}^{t_0+T} \left[y(t) - \widehat{y}(t)\right]^2 dt = \int_{t_0}^{t_0+T} \left[y(t) - \widehat{\theta} u(t)\right]^2 dt.$$

The value of $\widehat{\theta}$ may be computed as

$$\widehat{\theta} = \frac{\int_{t_0}^{t_0+T} u(t)y(t)dt}{\int_{t_0}^{t_0+T} u^2(t)dt} \tag{3.2}$$

if the denominator in Eq. (3.2) is not equal to zero.

When no noise is present, it is evident that the estimate $\widehat{\theta}$ can be obtained in a single step using either Eq. (3.1) or (3.2). Such a procedure, however, does not yield the true value of θ if noise is present, independent of its magnitude. Efforts to improve the estimate must be based on the prior information assumed regarding the noise. Under fairly general conditions, increasing the length T of the observation interval is found to improve the estimate $\widehat{\theta}$. Alternately, the estimate $\widehat{\theta}$ at time nT can be used together with the observation of the input and the output over the interval $[nT, (n+1)T]$, to determine a new estimate $\widehat{\theta}[(n+1)T]$. The asymptotic properties of such schemes have been studied extensively but are not germane to our present discussion.

Another method of obtaining $\widehat{\theta}(t)$ is by solving the differential equation

$$\dot{\widehat{\theta}}(t) = -u^2(t)\widehat{\theta}(t) + u(t)y(t) = -u^2(t)\left[\widehat{\theta}(t) - \theta\right] \tag{3.3}$$

whose equilibrium state is $\widehat{\theta}(t) \equiv \theta$. The right-hand side of Eq. (3.3) can be considered as the negative gradient of the performance index

$$J_1 = \frac{1}{2}\left[y(t) - \widehat{\theta}(t)u(t)\right]^2$$

with respect to $\widehat{\theta}$. By Eq. (3.3), $\widehat{\theta}$ is adjusted along this direction at every instant of time. Since θ is a constant, defining

$$\phi(t) \triangleq \widehat{\theta}(t) - \theta$$

Eq. (3.3) can also be expressed as

$$\dot{\phi}(t) = -u^2(t)\phi(t).$$

If $\int_t^{t+T} u^2(t)dt \geq \epsilon > 0 \quad \forall t > t_0$ and some constant T, then it can be shown that $\lim_{t\to\infty} \phi(t) = 0$ or $\widehat{\theta}(t)$ tends to θ asymptotically with time. It is this procedure that we will adopt in the chapters following. The main idea behind the approach is to convert the solution of an algebraic equation into the solution of a differential equation, by assuring the equilibrium state of the latter to be uniformly asymptotically stable in the large.

Vector Case. We now consider the identification of an n-dimensional vector θ where

$$\theta^T u(t) = y(t)$$

and $u : \mathbb{R}^+ \to \mathbb{R}^n$ and $y : \mathbb{R}^+ \to \mathbb{R}$. The same procedure as in the scalar case can be used here with the exception that computations of matrices and their inverses need to be carried out. In particular, the solution

$$\widehat{\theta} = \left[\int_{t_0}^{t_0+T} u(t)u^T(t)dt\right]^{-1}\left[\int_{t_0}^{t_0+T} u(t)y(t)dt\right]$$

corresponds to Eq. (3.2) in the scalar case. Similarly, the extension of Eq. (3.3) to the vector case is given by

$$\begin{aligned}\dot{\widehat{\theta}}(t) &= -u(t)u^T(t)\widehat{\theta}(t) + u(t)y(t)\\ &= -u(t)u^T(t)\left[\widehat{\theta}(t) - \theta\right]\end{aligned}$$

or alternately

$$\dot{\phi}(t) = -u(t)u^T(t)\phi(t)$$

where

$$\phi(t) \triangleq \widehat{\theta}(t) - \theta.$$

If

$$\int_t^{t+T} u(\tau)u^T(\tau)d\tau \geq \alpha I > 0 \qquad \forall t \geq t_0; \tag{3.4}$$

for positive constants α and T, it can then be shown that $\lim_{t\to\infty} \widehat{\theta}(t) = \theta$. The condition in Eq. (3.4) implies that the matrix $u(\tau)u^T(\tau)$, which has a rank of unity at every instant, should attain full rank when integrated over an interval, that is, the vector $u(\tau)$ should span the entire n-dimensional space as τ varies from t to $t+T$. This property of u, which is referred to as *persistent excitation*, is central to many adaptive systems and is discussed in detail in Chapter 6.

3.3 DYNAMICAL SYSTEMS

In this section we discuss both identification and control of plants that are described by first-order linear time-invariant differential equations with unknown coefficients [5].

3.3.1 The Identification Problem: Scalar Case

Consider the problem of identification of a dynamical plant with a bounded input u and an output x_p, described by a first-order differential equation:

$$\dot{x}_p(t) = a_p x_p(t) + k_p u(t) \tag{3.5}$$

where the parameters a_p and k_p are constant but unknown. The equilibrium state of the unforced plant is assumed to be asymptotically stable so that $a_p < 0$. The problem of identification then reduces to the determination of a_p and k_p from the observed input-output pair $u(t)$ and $x_p(t)$.

To realize this we describe two estimator models that have different structures. Although the stability analyses of the two are almost identical, only the second model can be readily extended to higher order systems as shown in Section 3.4. The input u and output $\widehat{x}_p$ of the two models are related by the differential equations (Fig. 3.1)

$$\dot{\widehat{x}}_p(t) = \widehat{a}_p(t)\widehat{x}_p(t) + \widehat{k}_p(t)u(t) \qquad \text{(Model 1)} \tag{3.6a}$$

and

$$\dot{\widehat{x}}_p(t) = a_m\widehat{x}_p(t) + (\widehat{a}_p(t) - a_m)x_p(t) + \widehat{k}_p(t)u(t) \qquad a_m < 0 \qquad \text{(Model 2).} \tag{3.6b}$$

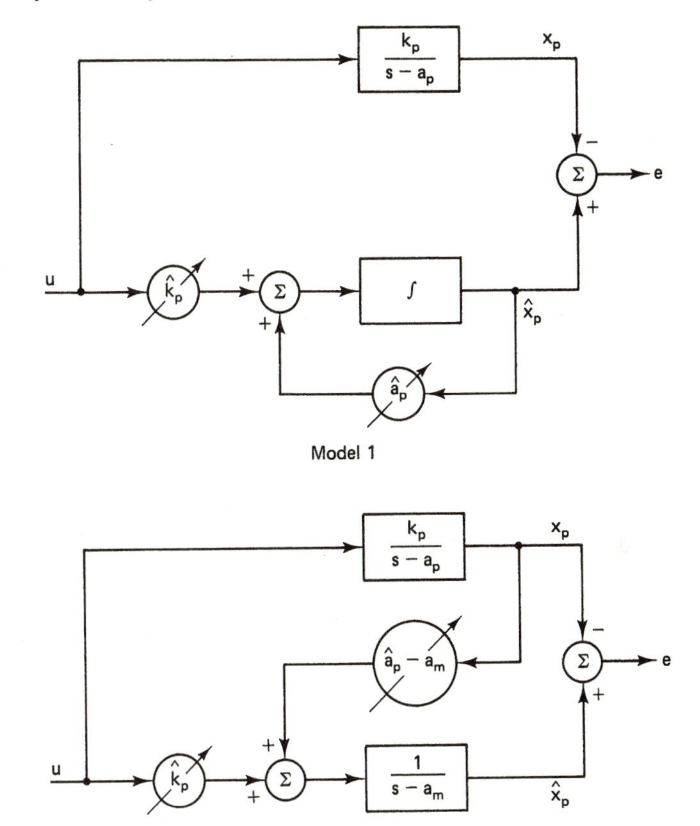

Figure 3.1 The identification problem.

$\widehat{a}_p(t)$ and $\widehat{k}_p(t)$ in the two models represent respectively the estimates of a_p and k_p at time t. The objective is to adjust $\widehat{a}_p(t)$ and $\widehat{k}_p(t)$ in such a manner that

$$\lim_{t\to\infty} \widehat{a}_p(t) = a_p \qquad \text{and} \qquad \lim_{t\to\infty} \widehat{k}_p(t) = k_p$$

even as all the signals in the overall system remain bounded. Further, as stated in Section 1.5, this has to be accomplished without using the time derivatives of the observed signals, since noise is invariably present.

Defining the error between the true output $x_p(t)$ of the plant and the estimated output $\widehat{x}_p(t)$ as $e(t)$, and the parameter errors between $\widehat{a}_p(t)$ and a_p and $\widehat{k}_p(t)$ and k_p as $\phi(t)$ and $\psi(t)$ respectively, we have the error differential equations

$$\dot{e}(t) = a_p e(t) + \phi(t)\widehat{x}_p(t) + \psi(t)u(t) \qquad \text{(Model 1)} \tag{3.7a}$$

$$\dot{e}(t) = a_m e(t) + \phi(t)x_p(t) + \psi(t)u(t) \qquad \text{(Model 2)} \tag{3.7b}$$

with

$$e(t) \stackrel{\Delta}{=} \widehat{x}_p(t) - x_p(t), \quad \phi(t) \stackrel{\Delta}{=} \widehat{a}_p(t) - a_p, \text{ and } \psi(t) \stackrel{\Delta}{=} \widehat{k}_p(t) - k_p.$$

In terms of the error equation (3.7a) or (3.7b), the objective can be stated as the determination of a rule for adjusting $\widehat{a}_p(t)$ and $\widehat{k}_p(t)$ so that the parameter errors $\phi(t)$ and $\psi(t)$ tend to zero as $t \to \infty$. It is clear that if $u(t)$ and $x_p(t)$ are uniformly bounded and $\phi(t)$ and $\psi(t)$ tend to zero, the error $e(t)$ will also tend to zero asymptotically. Hence, the convergence of the output error to zero will necessarily follow if exact parameter estimation is achieved. Since the parameter error is never known explicitly, the success of the identification procedure must be inferred in practice from the behavior of the measured output error.

Adaptive Law. In this section we describe the choice of the adaptive laws only for Model 2. Since the same analysis can also be used for Model 1, we merely comment on the essential differences between the two at the end of this section.

Equation (3.7b) represents the first important step in the mathematical description of the identification process using Model 2. It indicates how the parameter errors affect the output error in the system. Since the identifier is to be differentiator free, we attempt to obtain adaptive identification laws of the form

$$\begin{aligned} \dot{\widehat{a}}_p(t) &= f_1\left[x_p(t), u(t), \widehat{x}_p(t), \widehat{a}_p(t)\right] \\ \dot{\widehat{k}}_p(t) &= f_2\left[x_p(t), u(t), \widehat{x}_p(t), \widehat{k}_p(t)\right] \end{aligned} \tag{3.8}$$

so that the adjustment of the parameters $\widehat{a}_p(t)$ and $\widehat{k}_p(t)$ are based on all the signals that can be directly measured. The functions f_1 and f_2 in Eq. (3.8) have to be chosen in such a manner that Eqs. (3.7b) and (3.8) constitute a third-order system whose equilibrium state is given by

$$\widehat{a}_p = a_p, \quad \widehat{k}_p = k_p, \quad \text{and } e = 0$$

and is globally stable. One such rule is

$$\begin{aligned} \dot{\widehat{a}}_p(t) &= -e(t)x_p(t) \\ \dot{\widehat{k}}_p(t) &= -e(t)u(t). \end{aligned} \tag{3.9}$$

Since a_p and k_p are constants, the adaptive laws given in Eq. (3.9) can also be expressed in terms of the parameter errors $\phi(t)$ and $\psi(t)$ as

$$\begin{aligned} \dot{\phi}(t) &= -e(t)x_p(t) \\ \dot{\psi}(t) &= -e(t)u(t). \end{aligned} \tag{3.10}$$

In this case the equilibrium state $e = \phi = \psi = 0$ can be shown to be globally stable.

The justification for the choice of the adaptive laws [Eq. (3.10)] is based on the Lyapunov function candidate

$$V(e, \phi, \psi) = \frac{1}{2}\left[e^2(t) + \phi^2(t) + \psi^2(t)\right]. \tag{3.11}$$

$V(e, \phi, \psi)$ is a quadratic form in $e(t)$, $\phi(t)$ and $\psi(t)$ and is positive-definite (see Theorem 2.5). Evaluating the time derivative

$$\dot{V}(e, \phi, \psi) \triangleq \frac{dV}{dt}(e, \phi, \psi)$$

along any trajectory of the differential equations (3.7b) and (3.10) yields

$$\dot{V}(e, \phi, \psi) = a_m e^2(t) + \phi(t)e(t)x_p(t) + \psi(t)e(t)u(t) + \phi(t)\dot{\phi}(t) + \psi(t)\dot{\psi}(t). \tag{3.12}$$

$\dot{\phi}(t)$ and $\dot{\psi}(t)$ in Eq. (3.12) must be such as to cancel the terms containing $\phi(t)$ and $\psi(t)$ if $\dot{V}$ is to be negative-semidefinite. This suggests the adaptive law in Eq. (3.10) from which we obtain

$$\dot{V}(e, \phi, \psi) = a_m e^2(t) \leq 0. \tag{3.13}$$

Hence, $V(e, \phi, \psi)$ is a Lyapunov function and the origin is uniformly stable in the large. It follows that $e(t)$, $\phi(t)$ and $\psi(t)$ are bounded for all $t \geq t_0$.

It should be noted that since the parameter errors $\phi(t)$ and $\psi(t)$ at any time t are not known (since a_p and k_p are unknown), the adaptive law cannot be implemented as given in Eq. (3.10). Hence, while the analysis of the adaptive system can be conveniently carried out using the error equations (3.7b) and (3.10), the practical implementation of the adaptive law has to be carried out in terms of the parameters $\hat{a}_p(t)$ and $\hat{k}_p(t)$ as given by Eq. (3.9).

Since

$$-\int_{t_0}^{\infty} \dot{V}\left(e(\tau), \phi(\tau), \psi(\tau)\right)\, d\tau = V(t_0) - V(\infty) < \infty,$$

we have

$$0 \leq \int_{t_0}^{\infty} e^2(\tau)\, d\tau < \infty$$

or $e \in \mathcal{L}^2$. Since $\dot{e}(t)$, as given by Eq. (3.7b), is bounded, it follows from Lemma 2.12 that

$$\lim_{t \to \infty} e(t) = 0.$$

In the early days of adaptive control it was concluded from Eqs. (3.11) and (3.13) that $\lim_{t\to\infty} e(t) = 0$. However, as seen from the analysis above, all that we can conclude is that $e \in \mathcal{L}^2$. Since a function can be square integrable without tending to zero as shown by the following simple example, it is clear that additional conditions are needed for $e(t)$ to tend to zero asymptotically.

Example 3.1

Let $u : \mathbb{R}^+ \to \mathbb{R}$ be defined by

$$u(t) = \begin{cases} 1 & n < t \leq n + \frac{1}{n^2} \\ 0 & n + \frac{1}{n^2} < t \leq n+1 \end{cases}$$

for all positive integers n. Then $u \in \mathcal{L}^2$ but $\lim_{t\to\infty} u(t)$ does not exist.

The problem arises in the example above because the time-derivative of $u(t)$ is unbounded. By Lemma 2.12, if $u(t)$ and $\dot{u}(t)$ are bounded and $u \in \mathcal{L}^2$, then $\lim_{t\to\infty} u(t) = 0$. It is this procedure that has been used above to show that $e(t) \to 0$.

Alternately, as shown in Chapter 2, if Eqs. (3.76) and (3.10) were either autonomous or periodic, their solutions can be shown by using Theorem 2.7 or 2.8, to converge to the largest invariant set in $E = \{(e, \phi, \psi) | e = 0\}$. Since the system described by Eqs. (3.7b) and (3.10) is nonautonomous, only Theorem 2.9 can be applied. Once again, since the conditions of the theorem are satisfied, we can conclude that $\lim_{t\to\infty} e(t) = 0$.

Using a similar analysis with Model 1 and using $V(e, \phi, \psi)$ as in Eq. (3.11) leads to adaptive laws of the form

$$\begin{aligned} \dot{\hat{a}}_p(t) &= -e(t)\hat{x}_p(t) \\ \dot{\hat{k}}_p(t) &= -e(t)u(t) \end{aligned}$$

resulting in

$$\dot{V}(e, \phi, \psi) = a_p e^2(t) \leq 0. \tag{3.14}$$

Hence, the same conclusions as in the previous case can also be drawn here. However, while a_p is unknown in Eq. (3.14), a_m is known in Eq. (3.13). This accounts for the fact that, unlike Model 1, Model 2 can be readily extended to the higher dimensional case discussed in Section 3.4.

The parameter errors $\phi(t)$ and $\psi(t)$ do not necessarily converge to zero, that is, parameters $\hat{a}_p(t)$ and $\hat{k}_p(t)$ need not converge to their desired values a_p and k_p. For example, if $u(t) \equiv 0$ or $\lim_{t\to\infty} u(t) = 0$, the outputs of both plant and estimator can tend to zero with $\phi(t)$ and $\psi(t)$ assuming constant nonzero values. Perfect identification depends on the nature of the input. In Chapter 6, the general concept of persistent excitation of the reference input is discussed in detail and it is shown to be closely related to parameter convergence in adaptive identification and control problems. Whether or not the parameter errors tend to zero, we can conclude from Eq. (3.10) that

$$\lim_{t\to\infty} \dot{\phi}(t) = \lim_{t\to\infty} \dot{\psi}(t) = 0 \tag{3.15}$$

or the parameter errors change more and more slowly as $t \to \infty$. This is characteristic of many of the adaptive schemes discussed in the following chapters. It is worth noting that Eq. (3.15) does not necessarily imply that $\phi(t)$ or $\psi(t)$ tends to a constant value asymptotically.

Adaptive Gains. In the discussion above, the adaptive laws in Eq. (3.10) assured the existence of a Lyapunov function of the form $V = (e^2 + \phi^2 + \psi^2)/2$. If, on the other hand, $V(e, \phi, \psi)$ is chosen to have the form

$$V = \frac{1}{2}\left[e^2 + \frac{1}{\gamma_1}\phi^2 + \frac{1}{\gamma_2}\psi^2\right]$$

where γ_1 and γ_2 are positive constants, the adaptive laws must have the form

$$\begin{aligned}\dot{\phi} &= -\gamma_1 e x_p \\ \dot{\psi} &= -\gamma_2 e u\end{aligned}$$

to make $V(e, \phi, \psi)$ a Lyapunov function. γ_1 and γ_2 are referred to as *adaptive gains.* Since the overall system is globally stable for all positive values of γ_1 and γ_2, the latter can be chosen to optimize the performance of the overall system in some sense. In all the problems discussed in this chapter, we shall merely determine the basic adaptive laws of the form in Eq. (3.10). These in turn can be modified to include adaptive gains.

Simulation 3.1 The identification of a plant with $a_p = -1$ and $k_p = 1$, using the procedure discussed in this section, is shown in Fig. 3.2. The parameter a_m in Eq. (3.6b) is equal to -3 and the simulations are carried out when (i) $u(t) \equiv 0$, (ii) $u(t) \equiv 2$, and (iii) $u(t) = 2\cos t + 3\cos 2t$. The initial conditions are chosen as indicated in Fig. 3.2. In each case, the output error $e(t)$ and the norm of the parameter error $\|\varphi(t)\| = \sqrt{\phi^2(t) + \psi^2(t)}$ are shown as functions of time.

(i) When $u(t) \equiv 0$ it is seen that the output error converges to zero asymptotically while the parameter errors converge to nonzero constant values [Fig. 3.2a].

(ii) When $u(t) \equiv 2$, similar behavior of the errors is observed. The asymptotic value of $\|\varphi(t)\|$ is smaller in this case [Fig. 3.2b].

(iii) When $u(t) = 2\cos t + 3\cos 2t$, $e(t), \phi(t)$ and $\psi(t)$ tend to zero asymptotically [Fig. 3.2c].

Identification of Nonlinear Plants. In the error differential equation (3.7b), the right-hand side is a linear function of the parameter errors ϕ and ψ. Further, these errors are associated with signals x_p and u that can be measured. This fact enabled us to construct stable identification laws of the form in Eq. (3.10). Hence, the same procedure can also be extended to the identification of a special class of nonlinear plants as shown below:

Let the plant to be identified be of the form

$$\dot{x}_p = a_p x_p + \alpha f(x_p) + k_p g(u) \tag{3.16}$$

where a_p, α, and k_p are constant scalar parameters that are unknown. $f(\cdot)$ and $g(\cdot)$ are known smooth nonlinear functions such that $x_p = 0$ is the only equilibrium state of Eq. (3.16) when $g(u) \equiv 0$, and Eq. (3.16) has bounded solutions for a bounded input u. To estimate the unknown parameters, we construct an estimator of the form

$$\dot{\hat{x}}_p = a_m \hat{x}_p + \left(\hat{a}_p(t) - a_m\right) x_p + \hat{\alpha}(t) f(x_p) + \hat{k}_p(t) g(u)$$

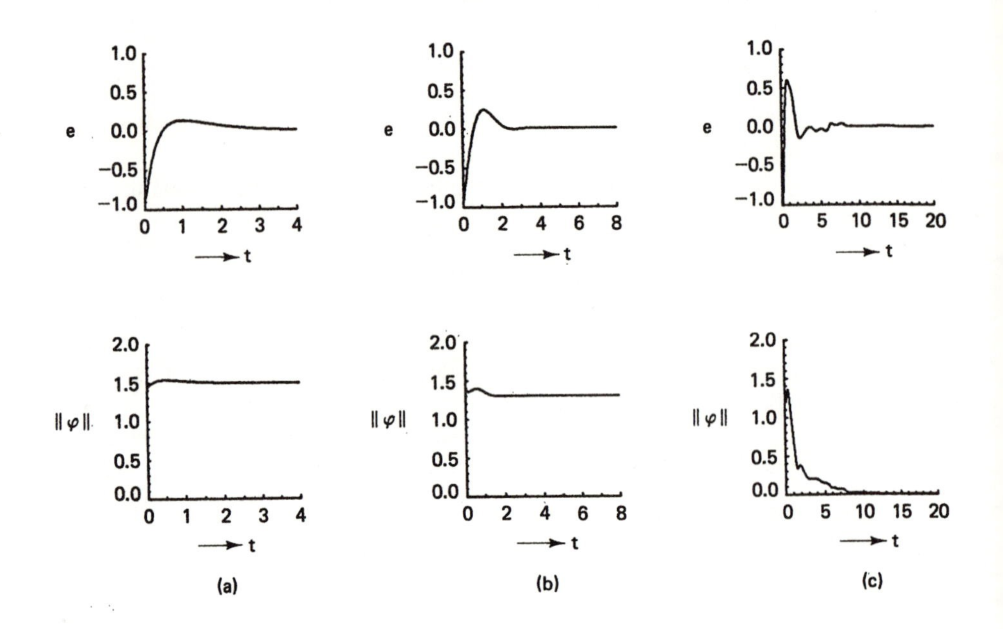

$$e = \hat{x}_p - x_p$$
$$\phi = \hat{a} + 1$$
$$\psi = \hat{k} - 1$$
$$\|\varphi\| = \sqrt{\phi^2 + \psi^2}$$
$$\dot{x}_p = -x_p + u$$
$$\dot{\hat{x}}_p = -3\hat{x}_p + (\hat{a}(t) + 3)\, x_p + \hat{k}(t)u$$
$$\dot{\hat{a}} = -e x_p$$
$$\dot{\hat{k}} = -eu$$
$$x_p(0) = 1$$
$$\hat{x}_p(0) = 0$$
$$\hat{a}(0) = 0$$
$$\hat{k}(0) = 0$$

(a) $u(t) = 0$ (b) $u(t) = 2$ (c) $u(t) = 2\cos t + 3\cos 2t$

Figure 3.2 Simulation 3.1.

so that the error e satisfies the differential equation

$$\dot{e} = a_m e + \left(\hat{a}_p(t) - a_p\right) x_p + (\hat{\alpha}(t) - \alpha) f(x_p) + \left(\hat{k}_p(t) - k_p\right) g(u)$$

and has the same form as Eq. (3.7b). Therefore it follows that adaptive identification laws of the form

$$\begin{aligned} \dot{\hat{a}}_p &= -e x_p \\ \dot{\hat{\alpha}} &= -e\, f(x_p) \\ \dot{\hat{k}}_p &= -e\, g(u) \end{aligned}$$

can be used to ensure the global stability of the overall system with the output error tending to zero asymptotically.

Simulation 3.2 The results obtained from the identification of a nonlinear plant described by the differential equation

$$\dot{x}_p = -x_p - 2x_p^3 + u$$

are shown in Fig. 3.3 with $a_m = -3$, initial conditions as shown in Fig. 3.3 and the same inputs as in Simulation 3.1. The output error is seen to tend to zero for $u(t) \equiv 0$ while the parameter errors do not. For $u(t) = 2 \cos t + 3 \cos 2t$, the output as well as the parameter errors tend to zero. The speed of convergence of the parameter error in Fig. 3.3(c) is much slower than that in Fig. 3.2(c).

3.3.2 The Control Problem: Scalar Case (Direct Control)

The simplest problem of adaptively controlling a dynamical system can be stated in terms of a plant of the form described in Section 3.3.1.

A plant with an input-output pair $\{u(\cdot), x_p(\cdot)\}$ is described by the scalar differential equation

$$\dot{x}_p(t) = a_p(t)x_p(t) + k_p(t)u(t)$$

where $a_p(t)$ and $k_p(t)$ are plant parameters. A reference model is described by the first-order differential equation

$$\dot{x}_m(t) = a_m x_m(t) + k_m r(t) \tag{3.17}$$

where $a_m < 0$, a_m and k_m are known constants, and r is a piecewise-continuous bounded function of time. It is assumed that a_m, k_m and r have been chosen so that $x_m(t)$ represents the output desired of the plant at time t (refer to Section 1.5). The aim is to determine a bounded control input u so that all the signals in the system remain bounded and

$$\lim_{t\to\infty} |x_p(t) - x_m(t)| = 0. \tag{3.18}$$

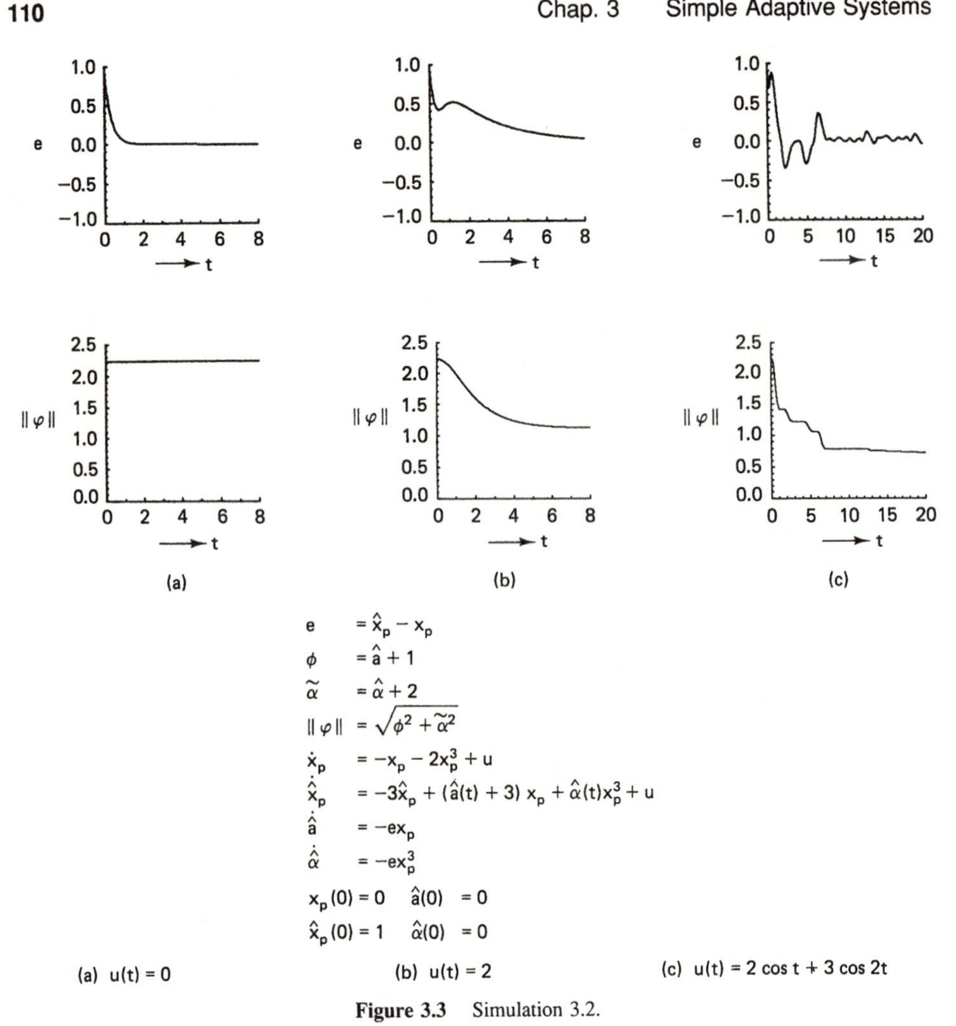

Figure 3.3 Simulation 3.2.

If $a_p(t)$ and $k_p(t)$ can be directly adjusted and their values are known, this can be achieved trivially by choosing

$$a_p(t) \equiv a_m; \qquad k_p(t) \equiv k_m; \text{ and } \qquad u(t) = r(t), \forall\, t \geq t_0.$$

To make the problem more realistic, it is assumed that the initial values of $a_p(t)$ and $k_p(t)$ are unknown but that their time derivatives can be adjusted using the measured signals in the system. When stated in this fashion, the control problem is seen to be identical to the identification problem using Model 1 discussed in Section 3.3.1, with the role of the plant and the model reversed. In this case, the error between the plant and the reference model outputs satisfies the differential equation

$$\dot{e}(t) = a_m e(t) + \left(a_p(t) - a_m\right) x_p(t) + (k_p(t) - k_m)r(t). \tag{3.19}$$

If the adaptive laws [13]

$$\begin{aligned} \dot{a}_p(t) &= -e(t)x_p(t) \\ \dot{k}_p(t) &= -e(t)r(t) \end{aligned} \tag{3.20}$$

are used, Eqs. (3.19) and (3.20) are similar to Eqs. (3.7b) and (3.9). Their global stability can be shown in the same manner as that described in the previous section.

Feedback Control. In practice, the direct adjustment of the plant parameters as described above may not be possible and the plant behavior may have to be altered only by using feedforward and feedback gains. If a_p and k_p are known constants so that the plant is given by

$$\dot{x}_p(t) = a_p x_p(t) + k_p u(t), \tag{3.21}$$

then a control input of the form

$$u(t) = \theta^* x_p(t) + k^* r(t) \tag{3.22}$$

can be chosen where

$$\theta^* \triangleq \frac{a_m - a_p}{k_p} \quad \text{and} \quad k^* \triangleq \frac{k_m}{k_p}.$$

Using the expression of Eq. (3.22) in Eq. (3.21), it is seen that the transfer function of the plant together with the controller will be the same as that of the reference model so that the objective in Eq. (3.18) is realized. We are assured of the existence of such a θ^* and k^*, provided $k_p \neq 0$, that is, when the plant is controllable. When a_p and k_p are unknown, the control input is chosen to have the form (Fig. 3.4)

$$u(t) = \theta(t)x_p(t) + k(t)r(t) \tag{3.23}$$

where $\theta(t)$ and $k(t)$ are the adjustable parameters of the controller. The adaptation should be such that $\theta(t)$ and $k(t)$ evolve to the constant values θ^* and k^* respectively. The plant given by Eq. (3.21) together with the adaptive controller in Eq. (3.23), which we shall refer to henceforth as the *system*, can be described by

$$\dot{x}_p(t) = \left(a_p + k_p\theta(t)\right) x_p(t) + k_p k(t)r(t). \tag{3.24}$$

Defining the output error e and the parameter errors ϕ and ψ as

$$e(t) \triangleq x_p(t) - x_m(t), \; \phi(t) \triangleq \theta(t) - \theta^*, \; \text{and } \psi(t) \triangleq k(t) - k^*$$

we obtain the error equation from Eqs. (3.17) and (3.24) as

$$\dot{e}(t) = a_m e(t) + k_p\phi(t)x_p(t) + k_p\psi(t)r(t). \tag{3.25}$$

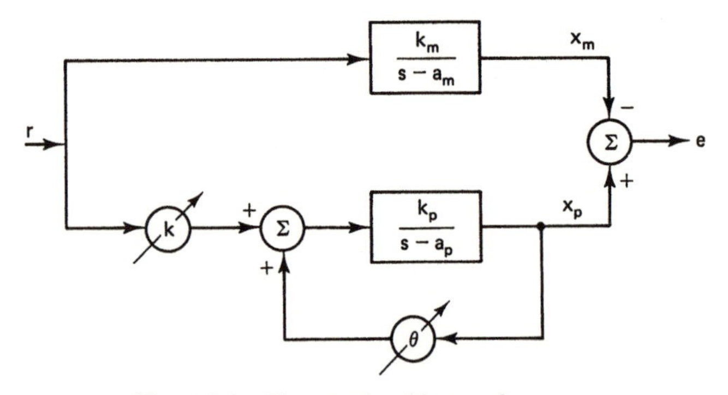

Figure 3.4 The control problem: scalar case.

The problem then is to determine how $\dot{\theta}(t)$ and $\dot{k}(t)$, or alternately, $\dot{\phi}(t)$ and $\dot{\psi}(t)$ are to be adjusted using all available information.

Case (i) k_p known. In this case, the control parameter $k(t)$ can be directly chosen to be equal to k^* so that the control input, from Eq. (3.23), is of the form

$$u(t) = \theta(t)x_p(t) + k^*r(t)$$

and only one parameter $\theta(t)$ has to be adjusted. The error equation can be derived as

$$\dot{e}(t) = a_m e(t) + k_p \phi(t) x_p(t). \tag{3.26}$$

The rule for updating $\theta(t)$ (or $\phi(t)$) is given by

$$\dot{\theta}(t) = \dot{\phi}(t) = -k_p e(t) x_p(t) \tag{3.27}$$

and the justification for the choice of such an adaptive law is provided, as in the identification problem, by demonstrating the existence of a Lyapunov function that assures the global stability of the resultant nonlinear system.

Stability Analysis. Let a candidate for the Lyapunov function be given by

$$V(e,\phi) \triangleq \frac{1}{2}\left[e^2 + \phi^2\right]. \tag{3.28}$$

The time derivative of Eq. (3.28) evaluated along the trajectories of Eqs. (3.26) and (3.27) is

$$\dot{V}(e,\phi) = e\dot{e} + \phi\dot{\phi} = a_m e^2 \leq 0.$$

This implies that the origin in Eqs. (3.26) and (3.27) is uniformly stable in the large. As in the identification problem, we can show that

$$\lim_{t\to\infty} e(t) = 0.$$

However, asymptotic stability of the origin in the (e, ϕ) space cannot be directly concluded, since $\dot{V}$ is only negative-semidefinite. The convergence of $\theta(t)$ to the desired value θ^*, which depends on the persistent excitation of the reference input r, is discussed further in Chapter 6.

Case (ii) k_p unknown. (sign of k_p known) In this case the input has the form of Eq. (3.23). As shown below, the presence of the term k_p in the error equation (3.25) requires a modification of the adaptive laws to maintain stability. In particular the adaptive laws are chosen as

$$\begin{aligned} \dot{\theta}(t) = \dot{\phi}(t) &= -sgn(k_p)e(t)x_p(t) \\ \dot{k}(t) = \dot{\psi}(t) &= -sgn(k_p)e(t)r(t) \end{aligned} \tag{3.29}$$

where $sgn(x)$ is defined as

$$sgn(x) = \begin{cases} +1 & \text{if } x \geq 0 \\ -1 & \text{if } x < 0. \end{cases}$$

The Lyapunov function candidate

$$V(e, \phi, \psi) \triangleq \frac{1}{2}\left[e^2 + |k_p|(\phi^2 + \psi^2)\right] \tag{3.30}$$

has a time derivative $\dot{V}(e, \phi, \psi)$, which when evaluated along the solutions of Eqs. (3.25) and (3.29) leads to

$$\begin{aligned} \dot{V}(e, \phi, \psi) &= e\dot{e} + |k_p|\left[\phi\dot{\phi} + \psi\dot{\psi}\right] \\ &= a_m e^2 + k_p\phi e x_p + k_p\psi e r - |k_p|\left[sgn(k_p)\phi e x_p + sgn(k_p)\psi e r\right] \\ &= a_m e^2 \leq 0. \end{aligned} \tag{3.31}$$

As in case (i), we conclude from Eqs. (3.30) and (3.31) that the equilibrium state of the system described by Eqs. (3.25) and (3.29) is uniformly stable in the large. In both cases, since $\dot{e}$ is bounded and $e \in \mathcal{L}^2$, it can be shown that

$$\lim_{t \to \infty} e(t) = 0.$$

As in the identification problem, the results are unaffected by the introduction of positive adaptive gains γ_1 and γ_2 in the adaptive laws.

Simulation 3.3 The results of controlling an unstable linear time-invariant plant described by $\dot{x}_p = x_p + 2u$, using the methods described in this section, are shown in Fig. 3.5. $a_p = 1$ and $k_p = 2$ used in the simulations are assumed to be unknown. The reference model is described by $\dot{x}_m = -x_m + r$.

(i) When $r(t) \equiv 0$ [Fig. 3.5(a)] and $r(t) \equiv 5$ [Fig. 3.5(b)], the plant is stabilized but the parameter errors do not tend to zero as $t \to \infty$.

(ii) When $r(t) = 2 \cos t + 3 \cos 2t$ [Fig. 3.5(c)] both output and parameter errors tend to zero.

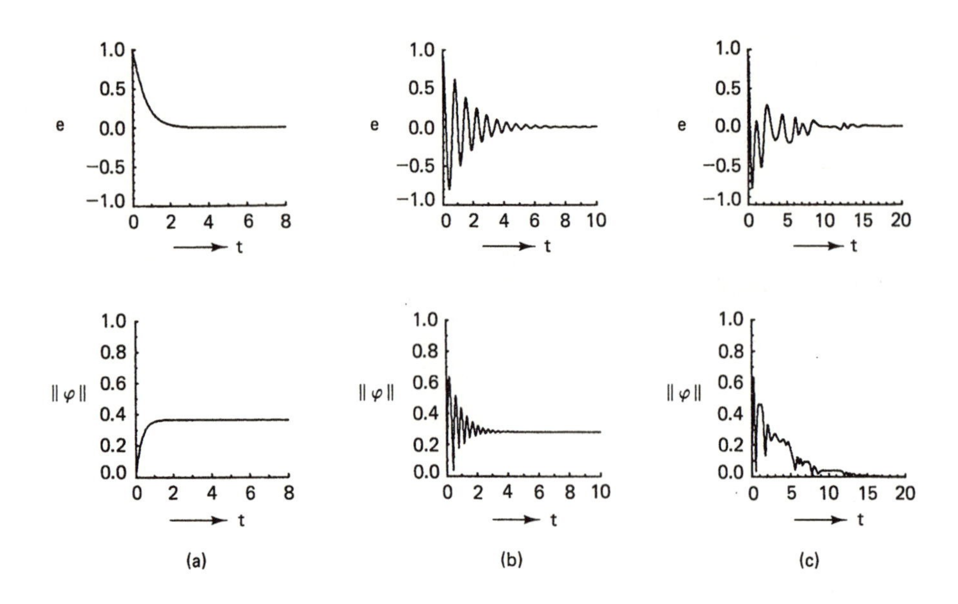

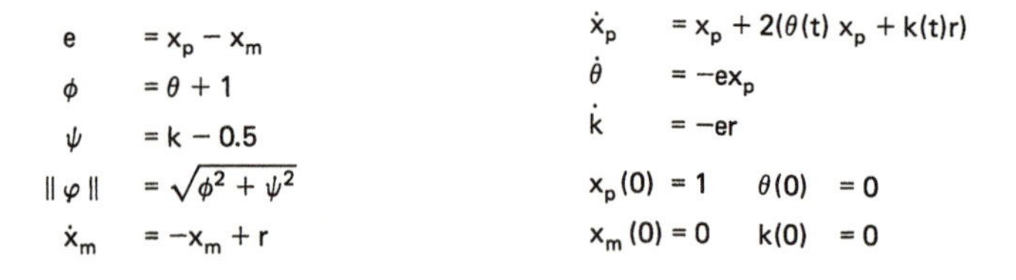

(a) $r(t) = 0.0$ (b) $r(t) = 5.0$ (c) $r(t) = 2\cos t + 3\cos 2t$

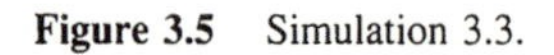

Figure 3.5 Simulation 3.3.

Control of Nonlinear Plants. As in Section 3.3.1, we show that the procedure adopted above to obtain stable adaptive control of the linear plant of Eq. (3.21) can also be extended to a class of nonlinear plants. This is indicated briefly below.

Let the nonlinear plant to be adaptively controlled be given by

$$\dot{x}_p = a_p x_p + \alpha f(x_p) + k_p u \tag{3.32}$$

where a_p, k_p and α are unknown, x_p as well as $f(x_p)$ can be measured, and $f(\cdot)$ is a smooth function of its argument with $f(0) = 0$. It is desired that x_p follow the output x_m of the linear reference model given in Eq. (3.17). We choose a control input of the form

$$u = \theta(t)x_p + \widehat{\alpha}(t)f\left(x_p\right) + k(t)r \tag{3.33}$$

and the adaptive laws for adjusting $\theta(t)$, $\widehat{\alpha}(t)$ and $k(t)$ are given by

$$\begin{aligned} \dot{\theta} &= -sgn(k_p)ex_p \\ \dot{\widehat{\alpha}} &= -sgn(k_p)ef(x_p) \\ \dot{k} &= -sgn(k_p)er. \end{aligned} \tag{3.34}$$

Using the same arguments as before, it can be shown that the overall system given by Eqs. (3.32)-(3.34) has bounded solutions and that $\lim_{t\to\infty} e(t) = 0$.

Simulation 3.4 Results similar to those in Simulation 3.3 are observed in the control of the nonlinear system $\dot{x}_p = x_p + 3x_p^3 + u$ with a reference model described by $\dot{x}_m = -x_m + r$. The convergence of the output error and parameter errors are shown in Fig. 3.6 for•(a) $r(t) \equiv 0$, (b) $r(t) \equiv 5$, and (c) $r(t) = 2 \cos t + 3 \cos 2t$.

Comment 3.1 Although the procedures adopted in the control problem for determining stable adaptive laws parallel those outlined in Section 3.3.1 for the identification problem, some of the differences in the assumptions made in the two cases are worth stating explicitly.

(i) In Eq. (3.5), the plant is assumed to be stable and, hence, a_p is negative. In the control problem, the plant, as described in Eq. (3.21), can be unstable.

(ii) In the identification problem, the principal objective is to estimate the parameters a_p and k_p of the plant. In contrast to this, the objective in the control problem is merely for the plant output to follow the output of the model asymptotically. Hence, the convergence of parameters to their desired values, which is central to the identification problem, is not critical in the control problem. In both cases, however, as shown in Chapter 6, the persistent excitation of the relevant input is needed to assure parameter convergence.

(iii) In the identification problem, the parameters of the identification model are adjusted. Any instability that can result, can manifest itself only in this model. In contrast to this, in the control problem, the controller parameters are adjusted in the plant feedback loop. Hence, the controlled process may become unstable, which from a practical standpoint, can be disastrous.

(iv) The structure of the identification model is at the discretion of the designer and is determined by the specific parametrization of the plant that is used. The identification of parameters is carried out in the context of this predetermined structure. Since the plant is assumed to be stable and the input bounded, the parametrization can be chosen in such a fashion that identification is carried out using known bounded signals. In contrast to this, only the reference input r can be assumed to be bounded in the control problem and the boundedness of all the signals in the adaptive loop has to be demonstrated using the specific adaptive law chosen.

From the analyses in Sections 3.3.1 and 3.3.2, the identification problem does not appear to be less difficult than the control problem, since essentially the same stability

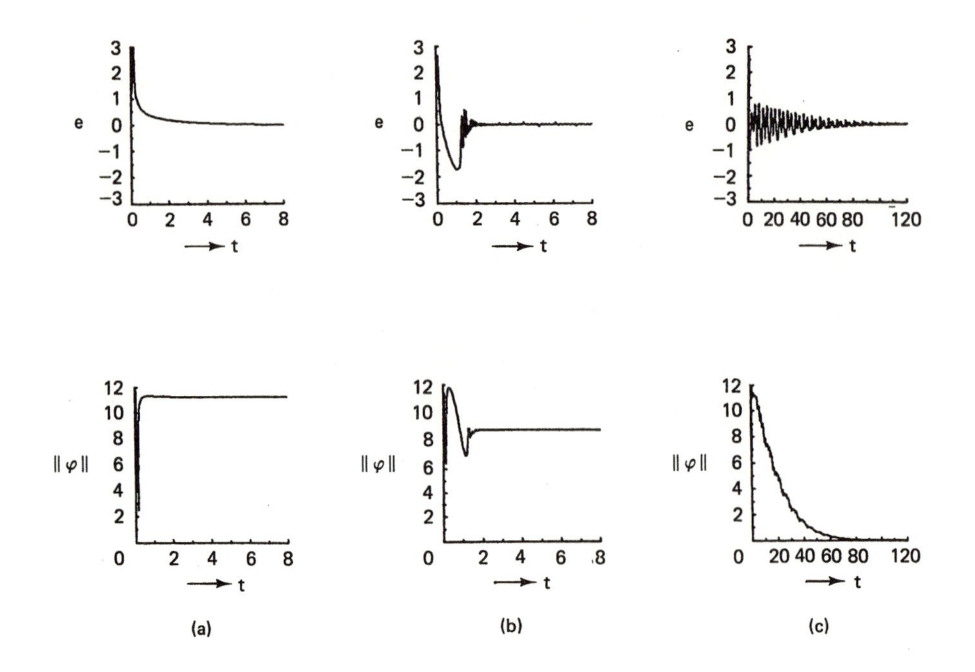

$$
\begin{aligned}
e &= x_p - x_m \\
\phi &= \theta + 2 \\
\tilde{\alpha} &= \hat{\alpha} - 3 \\
\|\varphi\| &= \sqrt{\phi^2 + \tilde{\alpha}^2} \\
\dot{x}_m &= -x_m + r \\
\dot{x}_p &= x_p + 3x_p^3 + u \\
u &= \theta(t)\, x_p + \hat{\alpha}(t) x_p^3 + r \\
\dot{\theta} &= -e x_p \\
\dot{\hat{\alpha}} &= -e x_p^3 \\
x_p(0) &= 1 \qquad \theta(0) = 0 \\
x_m(0) &= 0 \qquad \hat{\alpha}(0) = 0
\end{aligned}
$$

(a) $r(t) = 0.0$ (b) $r(t) = 5.0$ (c) $r(t) = 2\cos t + 3\cos 2t$

Figure 3.6 Simulation 3.4.

analysis was used in the two cases. However, this is no longer the case for higher order plants where the control problem is significantly more complex. This difference may be attributed to the different assumptions (i)-(iv) that have to be made in the two cases.

Comment 3.2 Error equations (3.25) and (3.29) can be rewritten as

$$\begin{aligned} \dot{e} &= a_m e + k_p\phi(e + x_m) + k_p\psi r \\ \dot{\phi} &= -sgn(k_p)e(e + x_m) \\ \dot{\psi} &= -sgn(k_p)er \end{aligned} \tag{3.35}$$

and represent the overall adaptive system. These equations are of third order and have a quadratic nonlinearity, and are time-varying when the reference input $r(t)$ varies with time. Before Theorem 2.5 is used to establish the boundedness of the solutions of these equations, it must be shown that the solutions exist. Since the right-hand side of Eq. (3.35) is continuous with respect to x and piecewise-continuous in t, the existence of a solution is assured for all $t \in [t_0, t_0 + \alpha]$ for some constant $\alpha > 0$. Since the Lyapunov function defined as in Eq. (3.30) ensures that all solutions of Eq. (3.35) belong to a compact set, their existence on $[t_0, \infty)$ is assured. The convergence of the output error to zero follows subsequently.

Comment 3.3 Contributions to adaptive control using the model-reference approach were made by numerous authors in the 1960s including Whitaker [8], Grayson [2], Parks [9], Shackloth and Butchart [12], Phillipson [11], and Monopoli [4]. The model reference approach was developed by Whitaker and his coworkers in the late 1950s where the parameters of a controller were adjusted to minimize an error function using a gradient approach. Around the same time, Grayson suggested a synthesis procedure based on the direct method of Lyapunov, so that the stability of the adaptive system would be assured from the outset. The method in [8] came to be known as the *M.I.T rule*. That such a method of adjusting the parameters might result in instability was pointed out by Parks and by Shackloth and Butchart who also suggested the synthesis procedure based on Lyapunov's direct method. Modifications and improvements were subsequently suggested by Phillipson, Monopoli, and others. A complete and rigorous analysis of adaptive identification and control of plants described by first-order differential equations was carried out in [5].

3.3.3 The Control Problem: Scalar Case (Indirect Control)

In Section 3.3.2, the control of an unknown plant was carried out by directly adjusting the control parameters in a feedback loop, based on the error between plant and model outputs. An alternative method is to estimate the plant parameters as described in Section 3.3.1 and to adjust the control parameters on the basis of such estimates. This procedure is called *indirect control* (refer to Section 1.4). We briefly show here that such a scheme would also result in a bounded control with the error $e(t)$ tending to zero asymptotically.

When the plant parameters a_p and k_p are known, it was shown in Section 3.3.2 that constant parameters θ^* and k^* exist, where

$$\theta^* \triangleq \frac{a_m - a_p}{k_p} \qquad \text{and} \qquad k^* \triangleq \frac{k_m}{k_p}.$$

These can be used to generate the input $u(t)$ as

$$u(t) = \theta^* x_p(t) + k^* r(t)$$

so that

$$\lim_{t \to \infty} |x_p(t) - x_m(t)| = 0.$$

In the direct control case, this motivated the choice of the controller structure given by Eq. (3.23). When the indirect control approach is used, the plant parameters a_p and k_p are estimated as $\widehat{a}_p(t)$ and $\widehat{k}_p(t)$ respectively, and the control parameters are adjusted, assuming that these correspond to the true values of the plant parameters. This motivates the use of a control input of the form (Fig. 3.7)

$$u(t) = \frac{a_m - \widehat{a}_p(t)}{\widehat{k}_p(t)} x_p(t) + \frac{k_m}{\widehat{k}_p(t)} r(t). \tag{3.36}$$

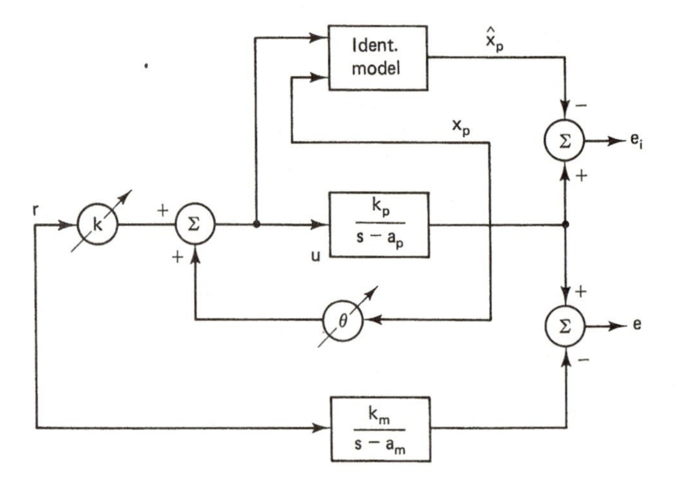

Figure 3.7 Indirect control: scalar case.

Hence, although an identification model is used to estimate $\widehat{a}_p(t)$ and $\widehat{k}_p(t)$, it is seen that an explicit reference model is not used for control purposes. Instead, the parameters a_m and k_m are used implicitly in Eq. (3.36) to determine the control parameters. This has the same form as Eq. (3.23) obtained by using direct control, if we define

$$\theta(t) \triangleq \frac{a_m - \widehat{a}_p(t)}{\widehat{k}_p(t)} \qquad \text{and} \qquad k(t) \triangleq \frac{k_m}{\widehat{k}_p(t)}.$$

As in Section 3.3.2, we consider two cases based on the prior information regarding k_p.

Case (i) k_p known. As in Section 3.3.1 we build an estimator of the form

$$\dot{\widehat{x}}_p = a_m \widehat{x}_p + \left(\widehat{a}_p(t) - a_m\right) x_p + k_p u. \tag{3.37}$$

Defining the output identification error and the parameter estimation error as

$$e_i(t) \triangleq x_p(t) - \widehat{x}_p(t) \qquad \text{and} \qquad \widetilde{a}_p(t) \triangleq \widehat{a}_p(t) - a_p,$$

we obtain the error equation

$$\dot{e}_i(t) = a_m e_i(t) - \widetilde{a}_p(t) x_p(t). \tag{3.38}$$

The adaptive identification law is chosen as described in Section 3.3.1 to be

$$\dot{\widehat{a}}_p(t) = \dot{\widetilde{a}}_p(t) = e_i(t) x_p(t). \tag{3.39}$$

The control input is then implemented as

$$u(t) = \frac{a_m - \widehat{a}_p(t)}{k_p} x_p(t) + \frac{k_m}{k_p} r(t). \tag{3.40}$$

The estimation scheme described by Eqs. (3.37) and (3.39) together with the control scheme of Eq. (3.40) can be shown to result in a globally stable adaptive system as well as asymptotic model-following.

Stability Analysis. A Lyapunov function candidate

$$V(e_i, \widetilde{a}_p) \triangleq \frac{1}{2}\left[e_i^2 + \widetilde{a}_p^2\right]$$

has a time derivative $\dot{V}(e_i, \widetilde{a}_p)$, which when evaluated along the trajectories of Eqs. (3.38) and (3.39) leads to the equation

$$\dot{V} = a_m e_i^2 \leq 0.$$

Hence, $e_i(t)$ and $\widetilde{a}_p(t)$ (and therefore $\widehat{a}_p(t)$) are uniformly bounded. Further, with the control input of the form in Eq. (3.40), the estimator can be expressed as

$$\dot{\widehat{x}}_p(t) = a_m \widehat{x}_p(t) + k_m r(t). \tag{3.41}$$

Choosing $\widehat{x}_p(t_0) = x_m(t_0)$, it follows from Eqs. (3.41) and (3.17) that

$$\widehat{x}_p(t) = x_m(t) \qquad \forall\, t \geq t_0$$

and, hence, $\widehat{x}_p(t)$ is uniformly bounded. Therefore, the output of the plant $x_p(t) = \widehat{x}_p(t) + e_i(t)$ is also uniformly bounded. As in Section 3.3.2, we obtain from Eq. (3.38) that $\dot{e}_i(t)$ is bounded and $e_i \in \mathcal{L}^2$. Therefore

$$\lim_{t\to\infty} |x_p(t) - x_m(t)| = \lim_{t\to\infty} |x_p(t) - \widehat{x}_p(t)| = 0.$$

The fact that the estimator equation (3.41) is identical to Eq. (3.17) in Section 3.3.2 and the adaptive laws are also identical implies that direct and indirect control are equivalent when k_p is known. Conditions under which such equivalence can be established for higher order plants are discussed in Chapter 5.

Case (ii) k_p unknown. The problem becomes considerably more difficult when k_p is unknown and an indirect approach is used to control the plant. This may be attributed to the fact that k_p has to be estimated on line as $\widehat{k}_p$ and the control input as given by Eq. (3.36) depends on $1/\widehat{k}_p(t)$. Hence, even though the parameter estimates are bounded, the control input may become unbounded as $\widehat{k}_p(t)$ assumes values close to zero. As shown in subsequent chapters, this problem is also encountered in more complex cases.

Let $sgn(k_p) = +1$ and let a lower bound on k_p be known so that

$$k_p \geq \underline{k}_p.$$

We choose an estimator of the form

$$\dot{\widehat{x}}_p = a_m \widehat{x}_p + \left[\widehat{a}_p(t) - a_m\right] x_p + \widehat{k}_p(t) u.$$

Adaptive laws for adjusting $\widehat{a}_p(t)$ and $\widehat{k}_p(t)$ are chosen as

$$\dot{\widehat{a}}_p = e_i x_p \tag{3.42}$$

and

$$\dot{\widehat{k}}_p = f(e_i, u, \underline{k}_p) \qquad \widehat{k}_p(t_0) > \underline{k}_p \tag{3.43}$$

where $e_i = x_p - \widehat{x}_p$, and

$$f(e_i, u, \underline{k}_p) = \begin{cases} e_i u & \text{if } \widehat{k}_p(t) > \underline{k}_p \\ e_i u & \text{if } \widehat{k}_p(t) = \underline{k}_p \text{ and } e_i u > 0 \\ 0 & \text{if } \widehat{k}_p(t) = \underline{k}_p \text{ and } e_i u \leq 0 \end{cases}$$

The adaptive law in Eq. (3.43) assures that $\widehat{k}_p(t) \geq \underline{k}_p$ for all $t \geq t_0$.

Stability Analysis. Defining the parameter errors $\widetilde{a}_p$ and $\widetilde{k}_p$ as

$$\widetilde{a}_p(t) \stackrel{\triangle}{=} \widehat{a}_p(t) - a_p, \qquad \text{and} \qquad \widetilde{k}_p(t) \stackrel{\triangle}{=} \widehat{k}_p(t) - k_p$$

the error equation is derived as

$$\dot{e}_i = a_m e_i - \widetilde{a}_p(t) x_p - \widetilde{k}_p(t) u. \tag{3.44}$$

Choosing a quadratic function

$$V = \frac{1}{2}\left(e_i^2 + \widetilde{a}_p^2 + \widetilde{k}_p^2\right)$$

its time derivative along the trajectories of Eqs. (3.42), (3.43), and (3.44) is given by

$$\dot{V} = a_m e_i^2 - e_i \widetilde{a}_p x_p - e_i \widetilde{k}_p u + \widetilde{a}_p e_i x_p + \widetilde{k}_p f(e_i, u, \underline{k}_p).$$

From the choice of $f(e_i, u, \underline{k}_p)$, it follows that

$$\dot{V} = \begin{cases} a_m e_i^2 & \text{if } \widehat{k}_p(t) > \underline{k}_p \\ a_m e_i^2 & \text{if } \widehat{k}_p(t) = \underline{k}_p \text{ and } e_i u > 0 \\ a_m e_i^2 - \widetilde{k}_p e_i u & \text{if } \widehat{k}_p(t) = \underline{k}_p \text{ and } e_i u \leq 0 \end{cases}$$

Hence, $\dot{V} \leq 0$ and e_i, $\widehat{a}_p$ and $\widehat{k}_p$ are bounded. Using similar arguments as in the previous sections, it can also be shown that $\lim_{t\to\infty} e_i(t) = 0$ and $\lim_{t\to\infty} |x_p(t) - x_m(t)| = 0$.

In the treatment above, the initial condition on $\widehat{k}_p(t_0)$ is restricted. The reader is referred to Problem 10 where an alternate method for adjusting $\widehat{k}_p$ which leads to global stability is suggested.

3.3.4 Indirect Control: Dynamic Adjustment of Control Parameters

The analysis of adaptive systems using the methods described in Sections 3.3.2 and 3.3.3 reveals that, even in simple systems, an asymmetry exists between the direct and indirect approaches. Although this was not present when k_p is assumed to be known, it becomes evident when k_p is unknown. Only the sign of k_p is needed to implement the adaptive law for direct control, whereas a knowledge of a lower bound on $|k_p|$ is also needed for indirect control. This seemingly minor difference raises fundamental questions regarding the relation between the two approaches. If it can be shown that knowledge of a lower bound on $|k_p|$ is necessary for indirect control of the simple system described in Section 3.3.3, it then follows that the two methods are essentially different. If stable adaptive control using an indirect approach can be demonstrated using only the sign of k_p, then the two methods require the same prior information at least in the simple case considered. Recently, it has been shown [1] that stable indirect control is possible without assuming a lower bound on $|k_p|$. The approach, which is based on adjusting the control parameters dynamically, is treated in the following section.

Let a plant be described by Eq. (3.21), where a_p and k_p are unknown constants. Let an estimator be described by the equation

$$\dot{\widehat{x}}_p = a_m \widehat{x}_p + \left[\widehat{a}_p(t) - a_m\right] x_p + \widehat{k}_p u. \tag{3.45}$$

$\widehat{a}_p(t)$ and $\widehat{k}_p(t)$ can be determined as described in the identification problem in Section 3.3.1 using the adaptive laws

$$\begin{aligned} \dot{\widehat{a}}_p &= e_i x_p \\ \dot{\widehat{k}}_p &= e_i u \end{aligned}$$

where $e_i = x_p - \widehat{x}_p$. At any instant t, our objective is to adjust $\theta(t)$ and $k(t)$ based on the estimates $\widehat{a}_p(t)$ and $\widehat{k}_p(t)$ of the plant parameters. Since it is desired that $\widehat{a}_p + \widehat{k}_p\theta$ and $\widehat{k}_p k$ tend asymptotically to a_m and k_m respectively, we define closed-loop estimation errors ϵ_θ and ϵ_k as

$$\begin{aligned}\epsilon_\theta &= \widehat{a}_p + \widehat{k}_p\theta - a_m \\ \epsilon_k &= \widehat{k}_p k - k_m.\end{aligned}$$

The control parameters $\theta(t)$ and $k(t)$ are adjusted using the adaptive laws

$$\begin{aligned}\dot{\theta} &= -sgn(k_p)\epsilon_\theta \\ \dot{k} &= -sgn(k_p)\epsilon_k\end{aligned} \tag{3.46}$$

to decrease the errors ϵ_θ and ϵ_k. To assure the stability of the overall system, the rules for adjusting the estimation parameters have to be modified as

$$\begin{aligned}\dot{\widehat{a}}_p &= e_i x_p - \epsilon_\theta \\ \dot{\widehat{k}}_p &= e_i u - \theta\epsilon_\theta - k\epsilon_k.\end{aligned} \tag{3.47}$$

The adaptive estimation laws of Eq. (3.47) together with the adaptive control laws of Eq. (3.46) determine the modified indirect adaptive scheme and assure the boundedness of all the signals in the system and that $\lim_{t\to\infty} |x_p(t) - x_m(t)| = 0$.

Stability Analysis. Let a function $V[e_i, \phi, \psi, \widetilde{a}_p, \widetilde{k}_p]$ be defined as

$$V = \frac{1}{2}\left[e_i^2 + |k_p|\left(\phi^2 + \psi^2\right) + \widetilde{a}_p^2 + \widetilde{k}_p^2\right].$$

The time derivative of V along the trajectories of the system described by Eqs. (3.45)-(3.47) can be expressed as

$$\dot{V} = a_m e_i^2 - \epsilon_\theta^2 - \epsilon_k^2 \leq 0.$$

Therefore, V is a Lyapunov function, and it follows that e_i, $\widehat{a}_p$, $\widehat{k}_p$, θ, and k are bounded. However, the boundedness of the signal x_p, $\widehat{x}_p$, or u does not follow directly.

The estimator equation can be rewritten as

$$\dot{\widehat{x}}_p(t) = [a_m + \epsilon_\theta(t)]\,\widehat{x}_p(t) + \epsilon_\theta(t)e_i(t) + k(t)\widehat{k}_p(t)r(t). \tag{3.48}$$

Since $\epsilon_\theta \in \mathcal{L}^2 \cap \mathcal{L}^\infty$, and $e_i, k, \widehat{k}_p$, and $r \in \mathcal{L}^\infty$, it follows that

$$\widehat{x}_p \in \mathcal{L}^\infty.$$

Hence, $x_p(= \widehat{x}_p + e_i) \in \mathcal{L}^\infty$. Since this implies that $\dot{e}_i$, $\dot{\epsilon}_\theta$, and $\dot{\epsilon}_k$ are bounded, we conclude that

$$\lim_{t\to\infty} e_i(t) = 0, \qquad \lim_{t\to\infty} \epsilon_\theta(t) = 0, \text{ and } \lim_{t\to\infty} \epsilon_k(t) = 0.$$

From Eq. (3.48), it also follows that $\lim_{t\to\infty} |\widehat{x}_p(t) - x_m(t)| = 0$ and hence x_p approaches x_m asymptotically.

The discussions in Sections 3.3.2-3.3.4 reveal the importance of the specific parametrization of the plant and the controller used as well as the adaptive laws chosen in the control problem. The results presented in this section indicate that, at least for a plant described by a scalar differential equation, the prior information needed for stable adaptive control is the same whether a direct or indirect approach is used. The equivalence between the two approaches in more general cases is considered in Chapter 5.

The importance of the prior information needed to solve simple adaptive problems is worth stressing here. Assumptions that appear mild in such cases have counterparts that are not reasonable in higher order and multivariable plants. For example, in all the cases considered in this section, it was assumed that the sign of k_p is known. The question can be raised whether this is necessary for determining stable adaptive laws. As shown by Nussbaum [7], stability can indeed be achieved without explicitly asking for such prior information, though the corresponding nonlinear control is considerably more complex. The implications of such developments are discussed further in Chapter 9.

3.3.5 Examples

Despite their simplicity, the mathematical models discussed in Section 3.3 are found to adequately describe many practical adaptive systems. In this section we include two examples selected from the published literature to illustrate the typical circumstances that necessitate the use of adaptive strategies.

The first example deals with the control of water levels in coupled tanks and was suggested in [3]. The second example, which is concerned with the control of motor speed in high speed cameras, was discussed in [10]. Although both articles consider a discrete-time representation of the systems under consideration, we study these examples using their continuous-time counterparts. In both cases, our interest is in the dynamical representation of the processes and the justification for the use of adaptive control. Since Chapter 11 contains detailed case studies of adaptive systems, we do not undertake a critical evaluation of the performance of the two examples considered here.

Example 3.2

Control of Water Level in a Pilot Unit. The pilot unit consists of two 10-liter tanks in series connected by a variable resistance. A third tank takes up some of the flow out of Tank 2 via a variable resistance (Fig. 3.8). Tank 1 is tapered at the bottom. The water level in Tank 1 is measured with a bubble system and controlled by manipulating the feed flow of water into Tank 1. The cylindrical section of the tank gives a linear response and the conical section a nonlinear response. As the level switches between the two sections, control becomes difficult since the process gain itself changes in these two

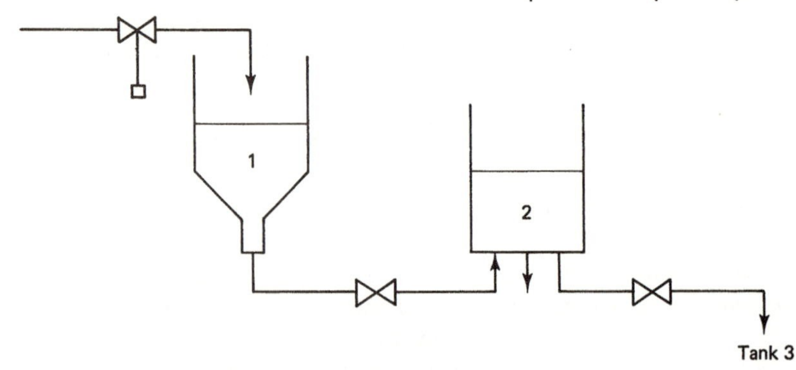

Figure 3.8 Control of water level in a pilot unit.

sections. Additional nonlinearities arise due to the control valve characteristic, changing line resistances and varying holdup in Tank 2. The aim of the control is to adjust the feed, so that the water level in Tank 1 is constant.

If a constant gain PI controller is used to control the tank water level, it is found that the set of gains required when the water level is in the linear cylindrical region is different from that in the nonlinear conical region. Hence a single set of gains is inadequate for both regions and an adaptive controller is called for. Taking into account the varying cross-section of Tank 1 but neglecting the effects due to the other nonlinearities, a first-order model is found to be adequate to describe the relation between the valve stem position ($u(t)$) and the water level in Tank 1 ($y(t)$). A mathematical description of the process can be given by

$$\dot{y} = f(y)u \tag{3.49}$$

where

$$f(y) = k_a \quad \text{in the cylindrical region}$$
$$= \frac{k_b}{y^2} \quad \text{in the conical region}$$

and k_a and k_b are positive constants. The nonlinear model in Eq. (3.49) can be further simplified as

$$\dot{x}_p = a_p x_p + k_p u$$

where a_p and k_p are constant gains that assume different values in each region and x_p denotes the deviation of the water level in Tank 1 about a nominal operating point, which is the level that separates the cylindrical and conical regions. Since the gains are fixed once the process enters each of the two regions, assuming that the controller adjusts itself sufficiently rapidly to the appropriate value in both cases, we can consider the process to have constant gains that are unknown. The desired response from the plant can be specified as

$$\dot{x}_m = a_m x_m + r$$

where r is a step input and $x_m(t)$ is the desired water level in Tank 1 at time t for all $t \geq 0$. The negative constant a_m is chosen so that the rise time of $x_m(t)$ is appropriate. The adaptive control problem is, therefore, to find a control input $u(t)$ for all $t \geq 0$ so that $\lim_{t\to\infty} |x_p(t) - x_m(t)| = 0$. The results of Section 3.3.2 indicate that a time-varying controller of the form

$$\begin{aligned} u &= \theta(t)x_p + k(t)r \\ \dot{\theta} &= -sgn(k_p)ex_p \\ \dot{k} &= -sgn(k_p)er \end{aligned} \tag{3.50}$$

where $e(t) = x_p(t) - x_m(t)$, can be implemented for stable adaptation; the results of Section 3.3.3 suggest that a controller described by

$$\begin{aligned} u &= \frac{a_m - \widehat{a}_p(t)}{\widehat{k}_p(t)} x_p + \frac{k_m}{\widehat{k}_p(t)} r \\ \dot{\widehat{a}}_p &= e_i x_p \\ \dot{\widehat{k}}_p &= e_i u \end{aligned} \tag{3.51}$$

where $e_i(t) = x_p(t) - \widehat{x}_p(t)$ can also be used provided adequate precautions are taken to prevent $\widehat{k}_p$ from taking values close to zero.

Example 3.3

Adaptive Control for High-Speed Cameras: High-speed photography is required to record events that occur in microsecond time frames. The controller of such a system is described in [10] and consists of a camera turbine rotor unit, a current-to-pressure transducer to regulate the gas turbine, and a controller that measures rotor speed. The output of the controller is the current to the transducer. The system however has many uncertainties. The turbine rotor can be any one of eighty different models that have different dynamic characteristics. The choice of the drive gas, helium, or nitrogen dramatically influences time responses. Each turbine exhibits inherent nonlinearities and uncertainties. The aim of the controller is to bring any rotor to a preselected speed and then regulate the speed to within a small error. The algorithm must also provide stable operation with minimal overshoot and settling time over the entire dynamic range of the rotor. These requirements, together with the uncertainties present, make a fixed controller inadequate for the purpose. An adaptive control scheme appears attractive for this application.

A rather crude model of the camera would be

$$\dot{x}_p = a_p x_p + k_p u$$

where $x_p(t)$ is the measured speed and $u(t)$ is the current at time t and constants a_p and k_p are unknown. A reference model

$$\dot{x}_m = a_m x_m + r$$

where $r(t) \equiv -a_m r^*$, r^* is the final desired speed, $a_m = -4/T_s$ and T_s is the desired settling time. The problem is again to determine a control u so that

$$\lim_{t\to\infty} x_p(t) = \lim_{t\to\infty} x_m(t).$$

The solution to this problem is provided by an adaptive controller of the form shown in Eq. (3.50) or (3.51).

3.4 STATE VARIABLES ACCESSIBLE

The procedure outlined in Section 3.3 for the stable identification and control of a linear plant of first order with unknown parameters can be extended directly to a higher order plant when its entire state vector is accessible [5]. This is a consequence of the well known result in control theory that a linear time-invariant controllable system can be stabilized using state feedback.

3.4.1 The Identification Problem: Vector Case

Consider a multiple-input multiple-output linear time invariant plant with unknown parameters. Let the plant be of nth order whose states are accessible and described by the vector differential equation

$$\dot{x}_p(t) \quad = \quad A_p x_p(t) + B_p u(t)$$

where A_p, B_p are unknown, $A_p \in \mathbb{R}^{n\times n}$, $B_p \in \mathbb{R}^{n\times p}$, A_p is asymptotically stable, and u is bounded. An estimator of the form

$$\dot{\widehat{x}}_p(t) \quad = \quad A_m \widehat{x}_p(t) + \left[\widehat{A}_p(t) - A_m\right] x_p(t) + \widehat{B}_p(t)u(t) \tag{3.52}$$

is chosen, where A_m is an asymptotically stable $(n \times n)$ matrix, $\widehat{A}_p(t)$ is an $n \times n$ matrix and $\widehat{B}_p(t)$ is an $n \times p$ matrix, both of whose elements can be adjusted at the discretion of the designer. If the state error and parameter errors are defined as

$$e(t) \triangleq \widehat{x}_p(t) - x_p(t), \qquad \Phi(t) \triangleq \widehat{A}_p(t) - A_p, \qquad \Psi(t) \triangleq \widehat{B}_p(t) - B_p,$$

then the error equations are given by

$$\dot{e}(t) \quad = \quad A_m e(t) + \Phi(t)x_p(t) + \Psi(t)u(t). \tag{3.53}$$

As in the scalar case discussed in Section 3.3, the problem is to adjust the elements of the matrices $\widehat{A}_p(t)$ and $\widehat{B}_p(t)$ or equivalently $\Phi(t)$ and $\Psi(t)$ so that the quantities $e(t), \Phi(t)$, $\Psi(t)$ tend to zero as $t \to \infty$. We choose the adaptive laws to be

$$\begin{aligned} \dot{\widehat{A}}_p(t) = \dot{\Phi}(t) &= -Pe(t)x_p^T(t) \\ \dot{\widehat{B}}_p(t) = \dot{\Psi}(t) &= -Pe(t)u^T(t) \end{aligned} \tag{3.54}$$

where P is an $n \times n$ symmetric positive-definite matrix ($P > 0$), which depends on A_m. It is shown in the following paragraphs that the equilibrium state ($e = 0, \Phi = 0, \Psi = 0$) of Eqs. (3.53) and (3.54) is uniformly stable in the large and $\lim_{t\to\infty} e(t) = 0$.

Stability Analysis. Let $V(e, \Phi, \Psi)$ be a Lyapunov function candidate of the form

$$V(e, \Phi, \Psi) = e^T P e + Tr\left(\Phi^T \Phi + \Psi^T \Psi\right) \tag{3.55}$$

where $Tr(A)$ denotes the trace of a matrix A. If P is a positive-definite matrix, $V(e, \Phi, \Psi)$ is positive-definite. The time derivative of Eq. (3.55) along the trajectories of Eq. (3.53) is given by

$$\dot{V} = e^T(PA_m + A_m^T P)e + 2e^T P\Phi x_p + 2e^T P\Psi u + Tr(2\dot{\Phi}^T\Phi + 2\dot{\Psi}^T\Psi).$$

Since A_m is an asymptotically stable matrix, the solution of the matrix equation

$$A_m^T P + PA_m = -Q_0 < 0,$$

where Q_0 is any $n \times n$ symmetric positive-definite matrix, yields a symmetric positive-definite matrix P (refer to Theorem 2.10). With the adaptive laws as defined in Eq. (3.54),

$$\begin{aligned}\dot{V} &= -e^T Q_0 e + 2e^T P\Phi x_p - 2\,Tr(x_p e^T P\Phi) + 2e^T P\Psi u - 2\,Tr(ue^T P\Psi) \\ &= -e^T Q_0 e \le 0\end{aligned}$$

since $Tr(ab^T) = b^T a$ where a and b are column vectors. Hence, the equilibrium state $(e = 0, \Phi = 0, \Psi = 0)$ of Eqs. (3.53) and (3.54) is globally stable. Also $e \in \mathcal{L}^2$ and $\dot{e}$ is bounded, which implies that

$$\lim_{t\to\infty} e(t) = 0.$$

Comment 3.4

(i) As in the scalar case, the analysis above only assures the asymptotic convergence of the state error to zero. The convergence of the parameters to their true values, which is the main aim of the identification procedure, depends on the persistent excitation of the input u and is discussed further in Chapter 6.

(ii) The convergence properties of the identification scheme are governed by the choice of the matrices A_m and Q_0, and the input u.

(iii) The procedure can be readily extended to the parameter identification of a stable nonlinear system of the form

$$\dot{x}_p = A_p x_p + C_p f(x_p) + B_p g(u) \tag{3.56}$$

where A_p, B_p, and C_p are unknown constant matrices, $f(\cdot)$ and $g(\cdot)$ are known smooth functions of their arguments such that Eq. (3.56) has bounded solutions for a bounded input u. The global stability of the equilibrium state of the error differential equations can be shown as in the scalar case (see Section 3.3).

The reader would have observed that the identification model used in Section 3.4.1 is a generalization of Model 2 described in Section 3.3.1. The generalization of Model 1 has the form

$$\dot{\widehat{x}}_p(t) = \widehat{A}_p(t)\widehat{x}_p(t) + \widehat{B}_p(t)u(t),$$

which leads to an error differential equation of the form

$$\dot{e}(t) = A_p e(t) + \Phi(t)\widehat{x}_p(t) + \Psi(t)u(t).$$

Although A_p is known to be asymptotically stable, its elements are unknown. Hence, a matrix P cannot be determined such that $A_p^T P + PA_p = -Q_0$ where Q_0 is a symmetric positive-definite matrix. Since P is explicitly used in the adaptive laws, this implies that Model 1 cannot be used for identification. Since the matrix A_m in Eq. (3.52) is known, it is clear why Model 2 can be implemented practically.

3.4.2 The Control Problem

The statement of the control problem of a general linear time-invariant plant of order n with unknown parameters and accessible state variables can be stated as follows: The plant to be adaptively controlled can be described by the differential equation

$$\dot{x}_p = A_p x_p + B_p u \tag{3.57}$$

where A_p is an unknown constant $n \times n$ matrix, B_p is an unknown constant $n \times p$ matrix, and (A_p, B_p) is controllable. The n-dimensional state x_p is assumed to be accessible. A reference model is specified by the linear time-invariant differential equation

$$\dot{x}_m = A_m x_m + B_m r \tag{3.58}$$

where A_m is an $n \times n$ asymptotically stable matrix, B_m is an $n \times p$ matrix, and r is a bounded reference input. It is assumed that $x_m(t)$, for all $t \geq t_0$, represents a desired trajectory for $x_p(t)$ to follow. The aim is to determine a method for controlling the plant so that

$$\lim_{t \to \infty} [x_p(t) - x_m(t)] = 0.$$

As in the scalar case, the solution to the control problem can be attempted under different assumptions regarding the prior information available concerning the plant. We briefly consider the following three cases:

Case (i). The control input $u(t) = r(t)$, $\forall\, t \geq t_0$, and the plant parameters a_{ij} and b_{ij}, which are the elements of the matrices A_p and B_p can be adjusted directly. We refer to this as the direct adjustment of plant parameters.

Case (ii). The matrix A_p is unknown while the matrix B_p is assumed to be known. The matrix B_m of the reference model, which is at the discretion of the designer, can be chosen to be identical to B_p. Alternately, B_m is chosen as $B_m = B_p Q^*$, where Q^* is a known constant matrix. The control input $u(t)$ to the plant is then generated using feedback as

$$u = \Theta(t)x_p + Q^* r$$

where Q^* depends on B_p. The feedback matrix Θ is adjusted adaptively.

Case (iii). In this case the matrices A_p and B_p are both assumed to be unknown and as in (ii), $u(t)$ is generated using feedback.

Since the stability analyses for cases (i) and (ii) are very similar to that given for the identification problem, we merely sketch the principal steps involved. Case (iii) however is considerably more difficult and only local stability can be assured. Hence, we discuss this problem in somewhat greater detail.

Case (i) Direct Adjustment of Plant Parameters. We assume that while the plant parameters, which are the elements of A_p and B_p, are unknown, their time derivatives can be adjusted using input-output data. If $u(t) = r(t)$, $\forall\ t \geq t_0$, the differential equations of the plant and model can be described by

$$\begin{aligned} \dot{x}_m &= A_m x_m + B_m r \qquad \text{(Model)} \\ \dot{x}_p &= A_p(t) x_p + B_p(t) r \qquad \text{(Plant).} \end{aligned}$$

As in the identification problem discussed earlier, the adaptive laws

$$\begin{aligned} \dot{A}_p &= P e x_p^T \\ \dot{B}_p &= P e r^T \end{aligned}$$

result in $\lim_{t\to\infty} e(t) = 0$, where $e(t) = x_p(t) - x_m(t)$, and the symmetric positive-definite matrix P is the solution of the matrix equation

$$A_m^T P + P A_m = -Q_0 \qquad Q_0 > 0.$$

Since this is the primary aim of control, the fact that the parameter errors may not tend to zero is not directly relevant here. However, as shown in Chapter 6, when the reference input is persistently exciting, $\|A_p(t) - A_m\|$ and $\|B_p(t) - B_m\|$ also tend to zero.

Case (ii) Feedback Control with B_p known. In the differential equation (3.57), the elements of the matrix A_p are assumed to be constant but unknown although those of B_p are constant and known. The control input u is chosen to be

$$u(t) = \Theta(t) x_p(t) + Q^* r(t)$$

where $\Theta(t) \in \mathbb{R}^{p\times n}$ consists of adjustable parameters and by definition, $Q^* \in \mathbb{R}^{p\times p}$ such that

$$B_p Q^* = B_m.$$

It is further assumed that a constant matrix Θ^* exists such that

$$A_p + B_p \Theta^* = A_m.$$

The error differential equations can then be expressed as

$$\dot{e} = A_m e + B_p \left[\Theta(t) - \Theta^*\right] x_p \tag{3.59}$$

and the aim is to determine an adaptive law for adjusting $\Theta(t)$ so that $\lim_{t\to\infty} e(t) = 0$.

Both the indirect and the direct approaches can be shown to result in stable adaptation. If the former approach is used, the estimate $\widehat{A}_p(t)$ of A_p at time t is determined using the procedure outlined in Section 3.3.3. Assuming $\widehat{A}_p(t)$ to be the true value of A_p, $\Theta(t)$ is computed so that

$$\widehat{A}_p(t) + B_p\Theta(t) \quad = \quad A_m. \tag{3.60}$$

If A_m in Eq. (3.52) is the same as the matrix A_m of the reference model, it follows that the output of the identification model is the same as $x_m(t)$, if the initial conditions are properly chosen. This results in the error differential equation (3.59).

Hence, both direct and indirect adaptive control methods yield the same adaptive law. While $\Theta(t)$ is adjusted directly in the former, its adjustment in the latter case depends on $\widehat{A}_p(t)$. In both cases, this can be expressed by the equation

$$\dot{\Theta}(t) = \dot{\Phi}(t) \quad = \quad -B_p^T P e x_p^T \tag{3.61}$$

where $\Phi(t) = \Theta(t) - \Theta^*$ and P is the symmetric positive-definite solution of the matrix equation $A_m^T P + PA_m = -Q_0$ where $Q_0 > 0$. This assures that the function $V(e, \Phi)$ defined by

$$V(e,\Phi) \ \stackrel{\triangle}{=}\ e^T P e + Tr\left[\Phi^T \Phi\right]$$

is a Lyapunov function for the system of Eqs. (3.59) and (3.61), and that $\lim_{t\to\infty}[x_p(t) - x_m(t)] = 0$.

Comment 3.5

(i) While, in theory, both direct and indirect methods can be used for asymptotic model following, the indirect approach requires the solution of the matrix algebraic equation (3.60) at every instant, and hence may be impractical. In contrast to this, the direct control method can be implemented relatively easily. Methods similar to those given in Section 3.3.4 using an indirect approach based on dynamic adjustment of control parameters and closed-loop estimation errors can be found in [1].

(ii) If B_p and B_m are vectors and (A_p, B_p) and (A_m, B_m) are in control canonical form, then the algebraic equation (3.60) can be solved at every instant. The parameter vector $\theta(t)$ then corresponds to the difference between the last rows of the matrices $\widehat{A}_p(t)$ and A_m. Hence, in this case, indirect control is also practically feasible.

(iii) The analysis is unchanged in both identification and control problems when an adaptive gain matrix Γ, which is symmetric and positive definite, is included in the adaptive law. If, for example, the adaptive law in Eq. (3.61) is modified as

$$\dot{\Phi} \quad = \quad -\Gamma B_p^T P e x_p^T$$

then global stability can be shown by replacing the term $\Phi^T\Phi$ in the Lyapunov function with $\Phi^T\Gamma^{-1}\Phi$.

(iv) The control of a class of nonlinear plants of the form in Eq. (3.56) can also be carried out exactly along the same lines as for the linear case described in this section. The assumption in such a case is that, using nonlinear feedback, the overall system can be made linear. For a given matrix B_p this places a constraint on the matrix C_p in Eq. (3.56). If a control input of the form

$$u(t) = \Theta(t)x_p(t) + \widehat{C}_p(t)f(x_p(t)) + Q^*r(t)$$

is used, it is assumed that in addition to matrices Θ^* and Q^*, a constant matrix G^* exists so that

$$B_pG^* = -C_p.$$

The matrices $\Theta(t)$ and $\widehat{C}_p(t)$ can be adjusted using the adaptive laws

$$\begin{aligned} \dot{\Theta} &= -B_p^T P e x_p^T \\ \dot{\widehat{C}}_p &= -B_p^T P e f^T(x_p) \end{aligned}$$

where

$$A_m^T P + P A_m = -Q_0, \qquad \text{and } Q_0 = Q_0^T > 0$$

resulting in $e(t)$ tending to zero as $t \to \infty$.

The cancellation of nonlinearities in the plant to be controlled using nonlinear feedback has been attempted in many applications, particularly in the field of robotics. The simulation study of a second-order nonlinear plant is described below.

Simulation 3.5 Consider a plant described by the second-order nonlinear differential equation

$$\ddot{x}_p - \dot{x}_p + \sin\,\dot{x}_p + 2x_p + 2\,\cos\,x_p = u, \tag{3.62}$$

and a reference model described by the equation

$$\ddot{x}_m + \dot{x}_m + x_m = r.$$

It is assumed that only the coefficients of the nonlinear terms $\sin\,\dot{x}_p$ and $\cos\,x_p$ in Eq. (3.62) are unknown. To achieve asymptotic model-following, the control input is implemented as

$$u(t) = -2\dot{x}_p + x_p + c_1(t)\cos\,x_p + c_2(t)\sin\,\dot{x}_p + r(t)$$

and $c_1(t)$ and $c_2(t)$ are adjusted as

$$\begin{aligned} \dot{c}_1 &= -(e_1 + 2e_2)\,\cos\,x_p \\ \dot{c}_2 &= -(e_1 + 2e_2)\sin\,\dot{x}_p \end{aligned}$$

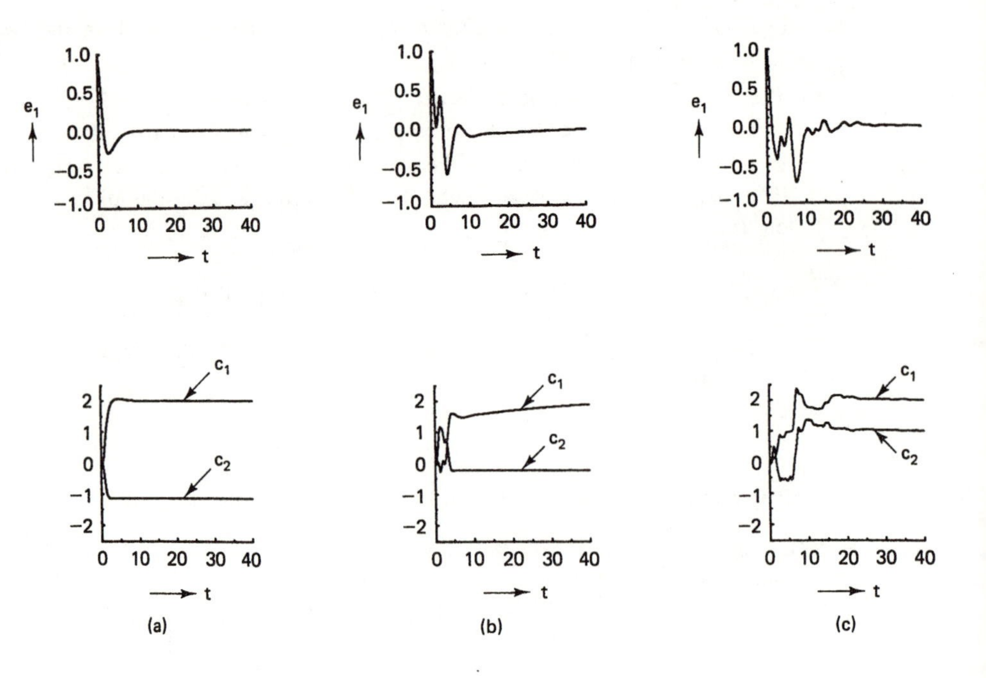

$$e_1 = x_p - x_m \qquad e_2 = \dot{x}_p - \dot{x}_m$$

$$\ddot{x}_m + \dot{x}_m + x_m = r$$

$$\ddot{x}_p - \dot{x}_p + \sin \dot{x}_p + 2x_p + 2\cos x_p = u$$

$$u = -2\dot{x}_p + x_p + c_1(t)\cos x_p + c_2(t)\sin \dot{x}_p + r$$

$$\dot{c}_1 = -(e_1 + 2e_2)\cos x_p$$

$$\dot{c}_2 = -(e_1 + 2e_2)\sin \dot{x}_p$$

$$c_1^* = 2$$

$$c_2^* = 1$$

$$x_p(0) = 1$$

$$\dot{x}_p(0) = 0$$

$$x_m(0) = 0$$

$$\dot{x}_m(0) = 0$$

$$c_1(0) = 0$$

$$c_2(0) = 0$$

(a) $r(t) = 0.0$ (b) $r(t) = 5.0$ (c) $r(t) = 2\cos t + 3\cos 2t$

Figure 3.9 Simulation 3.5.

where $e_1 = x_p - x_m$ and $e_2 = \dot{x}_p - \dot{x}_m$. Figure 3.9 shows the simulation results of this system. In Fig. 3.9(a)-(c), $e_1(t)$, $c_1(t)$ and $c_2(t)$ are displayed when the reference input is (a) $r(t) \equiv 0$, (b) $r(t) \equiv 5$, and (c) $r(t) = 2\cos t + 3\cos 2t$ respectively. While e_1 tends to zero in all the three cases, c_1 and c_2 tend to their true values only in case (c).

Case (iii) B_p unknown. The problem becomes considerably more complex when B_p is unknown. Since it is no longer possible to choose the matrix B_m of the reference model so that $B_m = B_p$, the design of the controller must be based on further assumptions concerning the structure of B_p. In particular, it is assumed that constant matrices Θ^* and Q^* exist such that

$$
\begin{aligned}
A_p + B_m\Theta^* &= A_m \\
B_pQ^* &= B_m.
\end{aligned}
$$

Two controller structures are shown in Fig. 3.10(a)-(b). An analysis of the adaptive system in Fig. 3.10(a) results in error equations that are nonlinear in the parameter errors so that the methods described thus far cannot be used to generate stable adaptive laws. Hence, we confine our attention to the structure shown in Fig. 3.10(b). The parameters of the feedforward gain matrix $Q(t)$ and the feedback gain matrix $\Theta(t)$ are adjusted using the available signals of the system. If $Q(t) \equiv Q^*$ and $\Theta(t) \equiv \Theta^*$, it follows that the transfer matrix of the plant together with the controller is identical to that of the model. As shown below, the adaptive laws derived for this problem assure

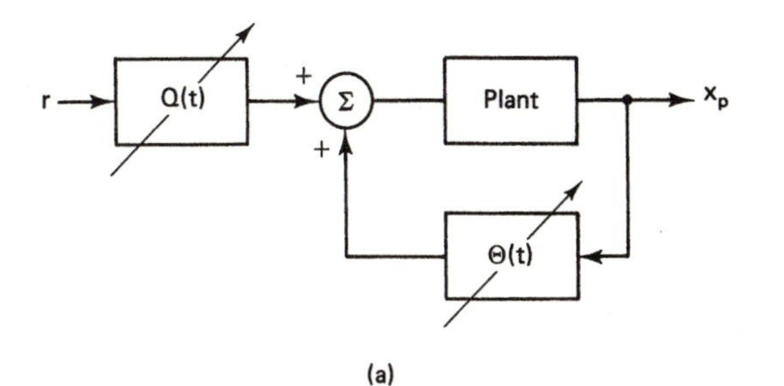

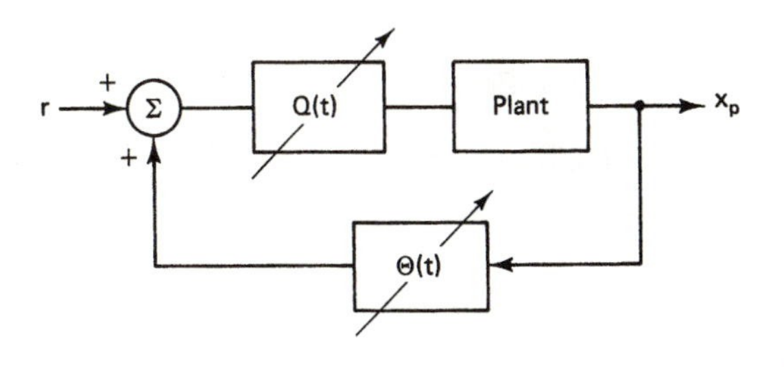

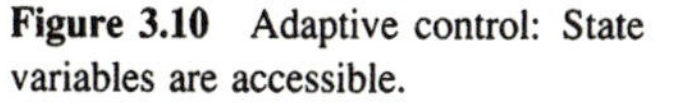
Figure 3.10 Adaptive control: State variables are accessible.

boundedness of parameters and convergence of output errors to zero only when the initial parameter values $Q(t_0)$ and $\Theta(t_0)$ lie in some bounded neighborhood of the desired values Q^* and Θ^*. This contrasts sharply with all the earlier sections where the results obtained hold for arbitrary initial conditions.

When the controller structure in Fig. 3.10(b) is used, the control input $u(t)$ can be expressed as

$$u(t) = Q(t)\Theta(t)x_p(t) + Q(t)r(t).$$

This yields an overall system described by

$$\dot{x}_p = \left[A_p + B_p Q(t)\Theta(t)\right] x_p + B_p Q(t) r. \tag{3.63}$$

From Eqs. (3.63) and (3.58), we obtain the error equations

$$\begin{aligned}
\dot{e} &= A_m e + \left[A_p + B_p\left(Q^* + Q(t) - Q^*\right)\Theta(t) - A_m\right] x_p \\
&\qquad + \left[B_p\left(Q(t) - Q^*\right)\right] r \\
&= A_m e + B_m\left[\Theta(t) - \Theta^*\right] x_p + B_m\left[{Q^*}^{-1} Q(t) - I\right]\Theta(t) x_p \\
&\qquad + B_m\left[{Q^*}^{-1} Q(t) - I\right] r \qquad (3.64) \\
&= A_m e + B_m \Phi(t) x_p + B_m \Psi(t) u
\end{aligned}$$

where

$$\Phi(t) \stackrel{\triangle}{=} \Theta(t) - \Theta^*, \quad \text{and} \quad \Psi(t) \stackrel{\triangle}{=} {Q^*}^{-1} - Q^{-1}(t).$$

Noting that Eq. (3.64) is of the same form as Eq. (3.59), where B_m is known, and x_p and u can be measured, we can now choose the adaptive laws for adjusting $\Phi(t)$ and $\Psi(t)$ as

$$\begin{aligned}
\dot{\Phi} &= -B_m^T P e x_p^T \\
\dot{\Psi} &= -B_m^T P e u^T
\end{aligned} \tag{3.65}$$

where $A_m^T P + P A_m = -Q_0 < 0$. This ensures the boundedness of solutions of Eqs. (3.64) and (3.65), which can be shown as follows. A scalar function

$$V = e^T P e + Tr(\Phi^T \Phi + \Psi^T \Psi) \tag{3.66}$$

has a time derivative along the trajectories of Eqs. (3.64) and (3.65) given by

$$\begin{aligned}
\dot{V} &= -e^T Q_0 e + 2e^T P B_m \Phi x_p + 2e^T P B_m \Psi u - Tr(2x_p e^T P B_m \Phi) - Tr(2u e^T P B_m \Psi) \\
&= -e^T Q_0 e \leq 0.
\end{aligned}$$

The adaptive controller may be implemented in terms of $\dot{\Theta}(t)$ and $\dot{Q}(t)$ as

$$\begin{aligned}
\dot{\Theta} &= -B_m^T P e x_p^T \\
\dot{Q} &= -Q B_m^T P e u^T Q.
\end{aligned}$$

The existence of the Lyapunov function V in Eq. (3.66) assures global stability in the $\{e, \Phi, \Psi\}$ space. However,

$$\Psi = \left[Q^{*^{-1}} - Q^{-1}(t)\right],$$

and our interest is in the parameter errors

$$\tilde{Q}(t) \triangleq Q(t) - Q^*.$$

Hence, only uniform stability, rather than uniform stability in the large, is implied in the $\{e, \Phi, \tilde{Q}\}$ space by the existence of the Lyapunov function V, since the latter is not radially unbounded in this space. This is shown in the following example.

Example 3.4

Consider a positive-definite function $V = e^2 + \psi^2$, which is radially unbounded in the (e, ψ) space. If

$$\psi \triangleq \frac{1}{q} - \frac{1}{q^*}, \qquad \text{and} \qquad \tilde{q} \triangleq q - q^*,$$

the level surfaces of $V(e, \tilde{q}) = c$ are given by (Fig. 3.11)

$$e^2 + \frac{\tilde{q}^2}{\left(\tilde{q} + q^*\right)^2 q^{*2}} = c$$

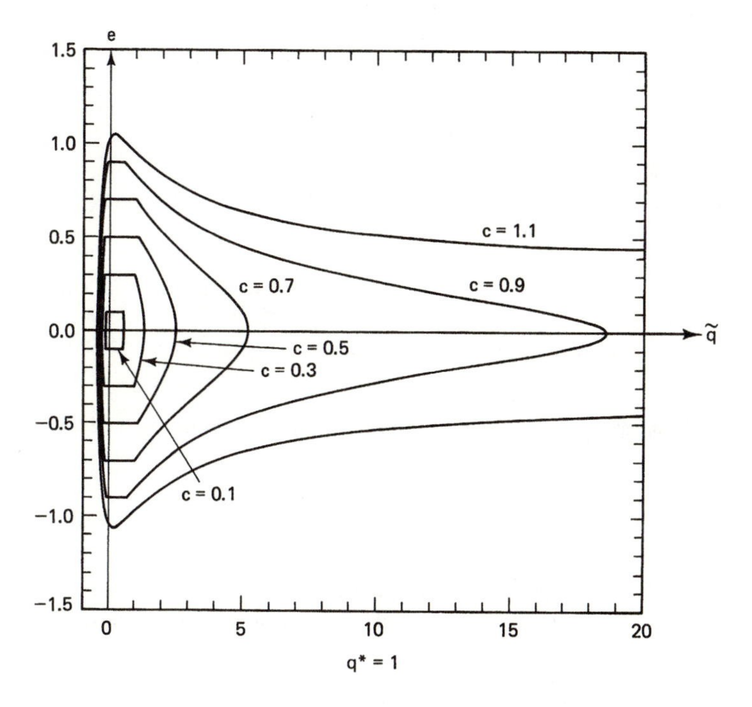

Figure 3.11 Level surfaces $V(e, \tilde{q}) = c$.

and are closed only for $c < 1/q^{*2}$. It is seen that $V(e,\widetilde{q})$ is not radially unbounded and, hence, does not satisfy assumption (iv) in Theorem 2.5.

3.5 SUMMARY

The identification and control of some simple adaptive systems are discussed in this chapter. Following the problem of identification of algebraic systems, the identification, indirect control, and direct control of a first-order plant with unknown parameters are discussed. In each case, given the error equation describing the output error between a plant and a model, adaptive laws are derived for adjusting the time derivatives of the relevant parameters. The existence of a Lyapunov function assures the boundedness of all the signals and the convergence of the output error to zero with time. Two examples of practical adaptive systems are included to illustrate situations where adaptive control strategies may be advantageous.

Higher order dynamical systems in which all the state variables of the plant can be measured are analyzed at the end of the chapter. In general, such problems are no more difficult than the corresponding problems described by first order differential equations. In all cases, the existence of constant feedforward and feedback gain matrices that assure perfect model following has to be assumed to ensure that the problems are mathematically tractable. When adaptation has to be carried out using the input and output of a plant, without explicit knowledge of its state, more interesting and difficult questions arise. These are discussed in Chapters 4 and 5.

PROBLEMS

Part I

1. If K is an $m \times n$ constant matrix with unknown elements, and $u(t)$ and $y(t)$ are time-varying vectors of appropriate dimensions with

$$Ku(t) = y(t),$$

determine (1) an algebraic method and (2) a dynamic method of estimating the elements of K.

2. If an asymptotically stable nonlinear plant is described by the scalar differential equation

$$\dot{x}_p = \sum_{i=1}^{N} a_i f_i(x_p, u, t)$$

where f_i are known bounded functions for $i = 1, \ldots, N$, so that x_p is bounded for a bounded input u, and $u(t)$ and $x_p(t)$ can be measured at each instant, describe a method of estimating the unknown parameters $a_1, \ldots, a_N$.

3. If ϕ is the parameter error vector, the methods described in this chapter result in

$$\lim_{t\to\infty} \dot{\phi}(t) = 0.$$

Does this imply that $\phi(t)$ tends to a constant? Justify your answer with a proof if the answer is yes and a counterexample if it is no.

4. When an identification or an adaptive control problem yields an error differential equation

$$\dot{e} = -e + \phi x_p,$$

the adaptive law $\dot{\phi} = -ex_p$ was chosen in Chapter 3. This makes $V(e, \phi) = (e_1^2 + \phi^2)/2$ a Lyapunov function since $V > 0$ and $\dot{V} = -e^2 \leq 0$. If the adaptive law is

$$\dot{\phi} = -\phi - ex_p,$$

$V(e, \phi)$ would still be a Lyapunov function with $\dot{V} = -\phi^2 - e^2 < 0$ assuring the uniform asymptotic stability in the large of the origin in the (e, ϕ) space. Why then is such an adaptive law not used?

Part II

5. If a nonlinear plant is described by the differential equation

$$\dot{x}_p = \sum_{i=1}^{n} a_i f_i(x_p) + bu$$

where $a_1, \ldots, a_n$ and b are unknown, but $sgn(b_p)$ is known, $f_1(\cdot), \ldots, f_n(\cdot)$ are known bounded functions, and x_p can be measured at each instant, design a controller to match x_p to the output of a time-invariant reference model which is asymptotically stable.

6. Let the error differential equation in an adaptive problem be $\dot{e} = -e + \phi x_p$. If the parameter error ϕ is adjusted using the adaptive law $\dot{\phi} = -\gamma(t) e x_p$, where $\gamma(t)$ is a time-varying gain, what are the conditions on $\gamma(t)$ so that the origin in the (e, ϕ) space is uniformly stable?

7. The reference model and a plant to be controlled are described by the scalar algebraic equations

$$y_m(t) = r(t)$$
$$y_p(t) = k_p u(t)$$

where $r : \mathbb{R}^+ \to \mathbb{R}$ is bounded and piecewise-continuous. k_p is an unknown constant whose sign is known. Determine a bounded control input u using a differentiator free controller so that

$$\lim_{t \to \infty} |y_p(t) - y_m(t)| = 0.$$

8. Assuming that a reference model described by the differential equation $\ddot{x}_m + 1.4\dot{x}_m + x_m = r$ can be constructed, show how $k(t)$ in Problem 6(d) in Chapter 1 can be adjusted so that $\lim_{t \to \infty}[x_p(t) - x_m(t)] = 0$.

9. In the system shown in Fig. 3.12, determine rules for adjusting the parameters k_1 and k_2 so that the output error e_1 tends to zero.

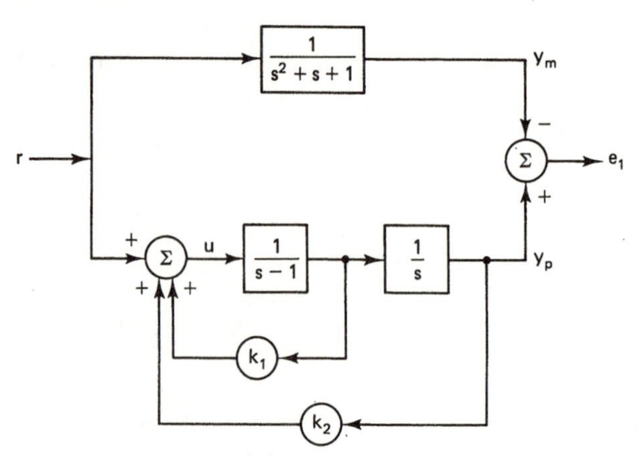

Figure 3.12 Adaptive control using state feedback.

10. Let the plant and model transfer functions be given by $k_p/(s-a_p)$ and $k_m/(s-a_m)$ respectively, where $k_m > 0$, $a_m < 0$ and k_p and a_p are unknown. In addition, $k_p \geq \underline{k}_p > 0$, where $\underline{k}_p$ is known. Let an indirect approach as described in Section 3.3.3 [case (ii)] be used to control the plant where, instead of Eq. (3.36) and Eq. (3.43), the following controller and adaptive law are used[1]:

$$
\begin{aligned}
u &= \frac{a_m - \widehat{a}_p}{\widehat{k}_p + f(\widehat{k}_p)} x_p + \frac{k_m}{\widehat{k}_p + f(\widehat{k}_p)} r \\
\dot{\widehat{k}}_p &= e_i u + f(\widehat{k}_p) \\
f(\widehat{k}_p) &= \begin{cases} \underline{k}_p - \widehat{k}_p & \text{if } \widehat{k}_p < \underline{k}_p \\ 0 & \text{if } \widehat{k}_p \geq \underline{k}_p. \end{cases}
\end{aligned}
$$

Show that the origin of the resultant error equations is globally stable.

Part III

11. In all the adaptive identification and control methods described in this chapter, the adaptive laws were expressed in terms of the time derivative of the parameters.

(a) Can stable adaptive laws be generated in which the second derivative of the parameters, with respect to time, are adjusted using the measured signals of the system?

(b) Can stable laws be generated in which the rth derivatives of the parameters with respect to time are adjusted? $(r > 2)$

12. In Problem 5, can you solve the problem if the reference model is described by a differential equation of the form

$$\dot{x}_m = f_1(x_m) + g_1(u)$$

where $f_1(\cdot)$ and $g_1(\cdot)$ are nonlinear functions?

13. In Section 3.3.4, it was shown that when indirect control is used, the control parameters can be adjusted dynamically to make the overall system stable without a knowledge of a lower bound on $|k_p|$. Can you determine an algebraic procedure for adjusting the control parameters by which the same results can be obtained?

[1] Suggested by Gerhard Kreisselmeier.

14. Using the direct or indirect approach, can you choose stable adaptive laws to adjust the controller and/or identifier parameters for the problem described in Section 3.3 when the sign of k_p is unknown?

REFERENCES

1. Duarte, M.A., and Narendra, K.S. *"Combined direct and indirect approach to adaptive control."* Technical Report No. 8711, Center for Systems Science, Yale University, New Haven, CT, Sept. 1987.
2. Grayson, L.P. "Design via Lyapunov's second method." *Proceedings of the 4th JACC*, Minneapolis, Minnesota, 1963.
3. Johnstone, R.M., Shah, S.L., Fisher, D.G., and Wan, R. "Experimental evaluation of hyperstable model reference adaptive control." In *Applications of adaptive control*, edited by K.S. Narendra and R.V. Monopoli, pp. 379–396, New York:Academic Press, 1980.
4. Monopoli, R.V. "Lyapunov's method for adaptive control-system design." *IEEE Transactions on Automatic Control* 12:334–335, 1967.
5. Narendra, K.S., and Kudva, P. "Stable adaptive schemes for system identification and control - Parts I & II." *IEEE Transactions on Systems, Man and Cybernetics* SMC-4:542–560, Nov. 1974.
6. Narendra, K.S., and Valavani, L.S. "Stable adaptive observers and controllers." *Proceedings of the IEEE* 64:1198–1208, August 1976.
7. Nussbaum, R.D. "Some remarks on a conjecture in parameter adaptive control." *Systems & Control Letters* 3:243–246, 1983.
8. Osburn, P.V., Whitaker, H.P., and Kezer, A. "New developments in the design of model reference adaptive control systems." In *Proceedings of the IAS 29th Annual Meeting*, New York, NY, 1961.
9. Parks, P.C. "Lyapunov redesign of model reference adaptive control systems." *IEEE Transactions on Automatic Control* 11:362–367, 1966.
10. Payne, A. "Adaptive control algorithms for high speed cameras." In *Proceedings of the CDC*, San Diego, CA, 1981.
11. Phillipson, P.H. "Design methods for model reference adaptive systems." *Proc. Inst. Mech. Engrs.* 183:695–700, 1969.
12. Shackcloth, B., and Butchart, R.L. "Synthesis of model reference adaptive systems by Lyapunov's second method." In *Proceedings of the IFAC Symposium on the Theory of Self-Adaptive Control Systems,"* Teddington, England, 1965.
13. Winsor, C.A., and Roy, R.J. "Design of model reference adaptive control systems by Lyapunov's second method." *IEEE Transactions on Automatic Control* 13:204, April 1968.

4

Adaptive Observers

4.1 INTRODUCTION

In Chapter 3, the adaptive identification and control of simple dynamical systems were discussed. The entire state vectors of such systems were assumed to be accessible for measurement. It was shown that stable adaptive laws could be generated when the parameters of the plant are constant but unknown. The theoretical basis for these results can be found in linear systems theory, where it is known that a controllable linear time-invariant plant can be stabilized by state feedback. In more realistic cases, where only the outputs of the plant rather than the state vector can be measured, these methods can no longer be applied directly. However, for linear time-invariant plants, it is well known that estimates of the state variables can be generated from the outputs using a Luenberger observer. It is also well established that by feeding back a linear combination of these estimates, the overall system can be stabilized. This in turn motivates efforts even in the adaptive case, when the parameter values are not known, to estimate the state variables, so that methods similar to those developed in Chapter 3 can be applied to stabilize the system. In this chapter, we focus our attention to the estimation of the state variables of an unknown plant; the use of the estimates to control the plant in a stable fashion is treated in Chapter 5. In both cases, it is found that the problems are significantly more complex than their counterparts when the plant parameters are known.

For a linear time-invariant plant with known parameters, the Luenberger observer that generates the state estimates, has for its inputs both the inputs and the available outputs of the plant. As shown in Section 4.2, a fixed observer structure can be chosen so that the error between the true state and the state of the observer tends to zero asymptotically. The problem is rendered more difficult in the adaptive case since the parameters are not known and hence the observer cannot be constructed. The best one can hope for is to use nominal values of the parameters in the observer to estimate the state variables, and use the latter in turn to update the estimates of the parameters. This immediately raises questions regarding the stability of the overall system. This is not surprising since the latter is nonlinear, with both parameter and state estimates for its state variables. Assuming that the "bootstrapping" procedure above can be shown to be stable, the estimates of the parameters and state variables can be obtained. Such an observer, which simultaneously estimates the parameters and state variables of a dynamical system, is referred to as an *adaptive observer*. This chapter is devoted to the analysis and synthesis of such observers.

4.2 THE LUENBERGER OBSERVER

The Luenberger observer, which is used for estimating the state variables of a linear time-invariant system with known parameters from the measured outputs, is a convenient starting point for the discussion of adaptive observers.

Consider a dynamic system described by the equations

$$\begin{aligned} \dot{x} &= Ax + Bu \qquad x(t_0) = x_0 \\ y &= Hx \end{aligned} \tag{4.1}$$

where A, B, and H are constant $(n \times n)$, $(n \times p)$, and $(m \times n)$ matrices whose elements are known. If only the input $u(t)$ and output $y(t)$ can be measured at every time instant t, the estimate $\widehat{x}(t)$ of the state $x(t)$ of the system can be obtained using an open-loop observer of the form

$$\begin{aligned} \dot{\widehat{x}} &= A\widehat{x} + Bu \qquad \widehat{x}(t_0) = \widehat{x}_0 \\ \widehat{y} &= H\widehat{x}. \end{aligned} \tag{4.2}$$

If the state error $\widetilde{x} = x - \widehat{x}$, then from Eqs. (4.1) and (4.2), $\widetilde{x}$ satisfies the differential equation

$$\dot{\widetilde{x}} = A\widetilde{x} \qquad \widetilde{x}(t_0) = x_0 - \widehat{x}_0 \stackrel{\Delta}{=} \widetilde{x}_0$$

whose solution is given by $\widetilde{x}(t) = \exp[A(t - t_0)]\,\widetilde{x}_0$. If A is asymptotically stable, the error $\widetilde{x}(t)$ tends to zero as $t \to \infty$. The rate of decrease of $\|\widetilde{x}(t)\|$ is dependent on the matrix A.

In the observer above, there is no feedback of information. If the matrix A is known only approximately, an error will exist between $x(t)$ and $\widehat{x}(t)$ even as $t \to \infty$. Further, even when A is known precisely, $\widetilde{x}(t) \to 0$ as $t \to \infty$ only if A is asymptotically stable. In view of these considerations, the following observer, known as the Luenberger observer, is preferred in practice. Such an observer is described by Eq. (4.3).

$$\begin{aligned} \dot{\widehat{x}} &= A\widehat{x} + L(y - \widehat{y}) + Bu \\ \widehat{y} &= H\widehat{x}. \end{aligned} \tag{4.3}$$

From Eq. (4.3), it is seen that the observer contains a feedback term that depends on the error between the true output and its estimate. The equations describing the error $\tilde{x}$ are given by

$$\dot{\tilde{x}} = (A - LH)\tilde{x}.$$

If (H, A) is observable, it is known [3] that the designer can choose L so that $(A - LH)$ is asymptotically stable. Then $\lim_{t\to\infty} \tilde{x}(t) = 0$ or the state estimate $\hat{x}(t)$ asymptotically approaches $x(t)$. By adjusting the matrix L and hence the eigenvalues of $A - LH$, the designer can also choose its rate of convergence.

Example 4.1

Let a plant be described by Eq. (4.1) where

$$A = \begin{bmatrix} -5 & 1 \\ -6 & 0 \end{bmatrix}; \qquad B = \begin{bmatrix} 1 \\ 1 \end{bmatrix}; \text{ and } H = [1, 0].$$

The plant is controllable and observable and has a transfer function

$$W_p(s) = \frac{s+1}{(s+2)(s+3)}.$$

To estimate the state of the plant, an observer of the form in Eq. (4.3) with $L = [10, 44]^T$ can be chosen so that the eigenvalues of $A - LH$ are -5 and -10.

Using the Luenberger observer, the true state of the system is attained even when the given system (that is, matrix A) is unstable. An attempt can therefore be made to stabilize the system by feeding back the estimate $\hat{x}(t)$ rather than $x(t)$. If (A, B) in Eq. (4.1) is controllable, it is known that a constant $(p \times n)$ matrix K exists such that $[A + BK]$ has desired eigenvalues. Assuming that u in Eq. (4.1) is chosen as

$$u(t) = K\hat{x}(t) + r(t),$$

where r is a p-dimensional reference input, the following equations describing the evolution of $x(t)$, $\hat{x}(t)$, and $\tilde{x}(t)$ respectively, can be obtained:

$$\begin{aligned} \dot{x} &= [A + BK]x - BK\tilde{x} + Br && \text{(System)} \\ \dot{\hat{x}} &= [A + BK]\hat{x} + LH\tilde{x} + Br && \text{(Observer)} \\ \dot{\tilde{x}} &= [A - LH]\tilde{x} && \text{(Error).} \end{aligned}$$

Hence the system is stabilized and $\tilde{x}(t)$ tends to zero asymptotically.

From the discussion above it is clear that if the given system is unstable, observability of Eq. (4.1) is needed to design a stable observer, while its controllability is required to stabilize the system using feedback.

Reduced Order Observer. When some of the state variables of the system can be measured and only the rest need to be estimated, a reduced order Lunberger observer can be used to generate these estimates. For example, if $x^T = [x_1^T \vdots x_2^T]$, $x_1 : \mathbb{R}^+ \to \mathbb{R}^{n_1}$, and $x_2 : \mathbb{R}^+ \to \mathbb{R}^{n_2}$, assuming that the n_1-dimensional vector x_1 can be measured, an observer can be designed to estimate only the n_2 dimensional vector x_2. Let such a dynamical system be described by the equations

$$\begin{aligned}\dot{x}_1 &= A_{11}x_1 + A_{12}x_2 + B_1 u \\ \dot{x}_2 &= A_{21}x_1 + A_{22}x_2 + B_2 u \\ y &= \left[I \mid 0\right] x = x_1\end{aligned}$$

where $A_{ij}(i, j = 1, 2)$ are constant matrices of appropriate dimensions and $I \in \mathbb{R}^{n_1 \times n_1}$ is an identity matrix. If the entire state vector $x(t)$ is to be determined using only the output $x_1(t)$, it is necessary that the given system be observable. Assuming that such is indeed the case, it can be shown [5] that there exists an $(n_2 \times n_1)$ matrix L such that the eigenvalues of the matrix $(A_{22} - LA_{12})$ can be arbitrarily chosen. This suggests the transformation

$$z \triangleq x_2 - Lx_1 = x_2 - Ly$$

where the new variable z satisfies the differential equation

$$\dot{z} = (A_{22} - LA_{12})\, z + (A_{22} - LA_{12})\, Ly + (A_{21} - LA_{11})\, y + (B_2 - LB_1)\, u. \quad (4.4)$$

By choosing L appropriately, $(A_{22} - LA_{12})$ can be made asymptotically stable in Eq. (4.4) while all the other terms depend on the known signals u and y. Hence an n_2^{th} order observer for estimating z can be designed, where the state $\widehat{z}$ of the observer satisfies the differential equation

$$\dot{\widehat{z}} = (A_{22} - LA_{12})\, \widehat{z} + (A_{22} - LA_{12})\, Ly + (A_{21} - LA_{11})\, y + (B_2 - LB_1)\, u. \quad (4.5)$$

From Eqs. (4.4) and (4.5), it follows that the state error $\tilde{z} = z - \widehat{z}$ satisfies the equation

$$\dot{\tilde{z}} = (A_{22} - LA_{12})\, \tilde{z}.$$

Hence, $\tilde{z}$ tends to zero as $t \to \infty$ for arbitrary initial conditions, or $\lim_{t\to\infty} [z(t) - \widehat{z}(t)] = 0$. Defining

$$\widehat{x}_2 = \widehat{z} + Ly$$

it follows that

$$\lim_{t\to\infty} [\widehat{x}_2(t) - x_2(t)] = \lim_{t\to\infty} [\widehat{z}(t) - z(t)] = 0.$$

Example 4.2

Let the plant be described as in Example 4.1 where

$$\begin{aligned}\dot{x}_1 &= -5x_1 + x_2 + u; \qquad y = x_1 \\ \dot{x}_2 &= -6x_1 + u.\end{aligned}$$

A reduced-order observer can be chosen with $L = 5$ as

$$\begin{aligned}\dot{\widehat{z}} &= -5\widehat{z} - 6y - 4u \\ \widehat{x}_2 &= \widehat{z} + 5y.\end{aligned}$$

The state error $\widetilde{x}_2 = \widehat{x}_2 - x_2$ satisfies the equation $\dot{\widetilde{x}}_2 = -5\widetilde{x}_2$, and, hence, decays exponentially.

4.3 ADAPTIVE OBSERVERS

As mentioned in the introduction, after the adaptive identification and control problems of linear time-invariant systems whose entire state-vectors are accessible were resolved, attention shifted to the design of adaptive observers. Since the Luenberger observer was well established as an efficient method of estimating the state variables of a known observable system using input-output data, it was generally felt that an observer with adjustable parameters could be designed to estimate the state of a linear time-invariant system with unknown parameters as well. However, it soon became evident that assuring the stability of such an observer was far from straightforward. The structure of the observer as well as the adaptive laws for updating its parameters had to be chosen judiciously for this purpose. This was accomplished in [2,8,12,14]. In 1977, an alternate method of generating the estimates of the states and the parameters of the plant was suggested [7] where the adaptive algorithms ensured faster convergence of the parameter estimates under certain conditions. In view of the importance of adaptive observer theory in the history of adaptive systems, we discuss the various schemes suggested and their modifications in the sections following.

4.3.1 Statement of the Problem

A single-input single-output linear time-invariant plant of order n with unknown parameters is described by the differential equation

$$\dot{x}_p = Ax_p + bu \qquad y_p = h^T x_p$$

where u and y_p are scalar input and output functions respectively, u is a piecewise-continuous uniformly bounded function of time, and x_p is the state vector of dimension n. It is assumed that the plant is observable. The problem is to construct an adaptive observer, which estimates both the parameters of the plant as well as the state vector x_p, using only input-output data.

Although the triple (h^T, A, b) contains $n^2 + 2n$ elements, only $2n$ parameters are needed to uniquely determine the input-output relationship. n of these correspond to the matrix A while the remaining n are determined jointly by h and b. In terms of the rational transfer function of the given plant, the $2n$ parameters correspond to the coefficients of the denominator and numerator polynomials. Hence, the given plant is parametrized in such a fashion that the unknown parameters are realized explicitly as the

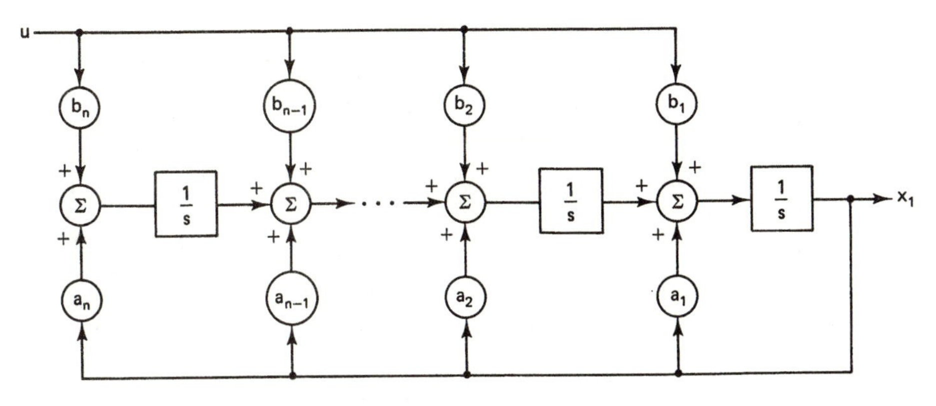

Figure 4.1 Canonical form in [2].

elements of two n-dimensional vectors. Although a large number of such representations are possible, only some of them can be used in the design of adaptive observers, since the parametrization together with the adaptive law determines whether or not the adaptive observer will be stable.

A minimal observer is first considered where the order of the observer is the same as that of the plant. Following a description of the structure of the fixed observer, we shall discuss the adaptive observer and the corresponding adaptive laws.

4.3.2 Minimal Realization of Adaptive Observers

Any single-input single-output linear time-invariant plant with unknown parameters, which is observable, can be described by a differential equation of the form [2]

$$\begin{aligned} \dot{x}_p &= \left[a|\overline{A}\right] x_p + bu \\ y_p &= h^T x_p \;=\; x_1 \end{aligned} \tag{4.6}$$

where $a^T = [a_1, a_2, \ldots, a_n]$, $b^T = [b_1, b_2, \ldots, b_n]$, the $(n \times (n-1))$ matrix $\overline{A}$ is known, and $h^T = [1, 0, \ldots, 0]$. The vectors a and b represent the unknown parameters of the plant and specify it completely. The objective is to estimate these parameters as well as the state x_p of the plant from input-output data.

In many of the early papers, different minimal representations were used by choosing different forms for the matrix $\overline{A}$. For example, Carroll and Lindorff [2] used the canonical form

$$\overline{A} = \left[\frac{I}{0}\right]$$

where $I \in \mathbb{R}^{(n-1)\times(n-1)}$ is an identity matrix, so that the plant can be represented as shown in Fig. 4.1.

Luders and Narendra chose the canonical form [13]

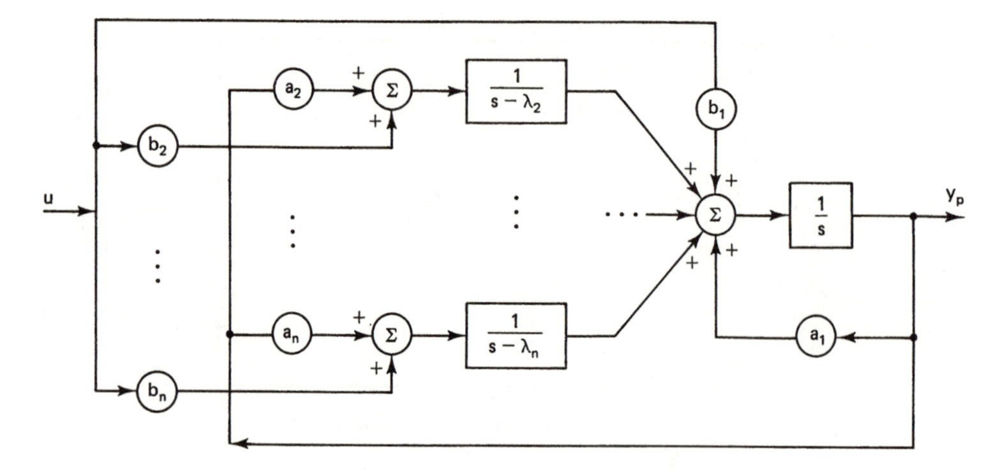

Figure 4.2 Canonical form in [13].

$$\overline{A} = \begin{bmatrix} 1 & \cdots & 1 \\ \lambda_2 & & 0 \\ & \ddots & \\ 0 & & \lambda_n \end{bmatrix}$$

which leads to the representation of the plant as shown in Fig. 4.2.

The corresponding transfer functions of the plant, using the representations in Figs. 4.1 and 4.2, are respectively,

$$W_p(s) = \frac{\sum_{i=1}^{n} s^{n-i} b_i}{s^n - \sum_{i=1}^{n} s^{n-i} a_i}; \tag{4.7}$$

and

$$W_p(s) = \frac{b_1 + \sum_{i=2}^{n} \frac{b_i}{s-\lambda_i}}{(s-a_1) - \sum_{i=2}^{n} \frac{a_i}{s-\lambda_i}}. \tag{4.8}$$

Equations (4.7) and (4.8) indicate that the transfer function of the plant can be completely specified by the choice of the $2n$ elements $a_i, b_i (i = 1, 2, \ldots, n)$ of the vectors a and b. Hence, parameter identification of a given plant using these representations involve the identification of the vectors a and b.

Example 4.3

For the plant defined in Example 4.1 the representation in Fig. 4.1 has the values $b_1 = 1$, $b_2 = 1$, $a_1 = -5$, and $a_2 = -6$, while that in Fig. 4.2 is specified by $b_1 = 1$, $b_2 = 0$, $a_1 = -4$, and $a_2 = -2$, where $\lambda_2 = -1$.

A representation of the plant, equivalent to that given in Eq. (4.6), is described by the equations

$$\begin{aligned} \dot{x}_p &= \left[-k\middle|\overline{A}\right] x_p + g y_p + bu \\ y_p &= h^T x_p = x_1 \end{aligned} \tag{4.9}$$

where k is a known constant vector such that $K = [-k|\overline{A}]$ is an asymptotically stable matrix and $g = k + a$. The unknown vectors are g and b respectively. This representation is desirable in adaptive observers since the state vector x_p of the system can be considered as the output of a known stable system with an input $gy_p + bu$, where g and b are unknown, but the signals $u(t)$ and $y_p(t)$ can be measured at time t. To estimate g and b in Eq. (4.9), the following structure is a natural one for the adaptive observer:

$$\begin{aligned} \dot{\widehat{x}}_p(t) &= K\widehat{x}_p(t) + \widehat{g}(t)y_p(t) + \widehat{b}(t)u(t) \\ \widehat{y}_p(t) &= h^T\widehat{x}_p(t) \end{aligned} \tag{4.10}$$

where $\widehat{g}(t)$ and $\widehat{b}(t)$ are the estimates of the parameter vectors g and b respectively, and $\widehat{x}_p(t)$ is the estimate of $x_p(t)$. Hence, the error differential equation can be derived as

$$\dot{e} = Ke + \phi y_p + \psi u \qquad e_1 = h^T e \tag{4.11}$$

where

$$e(t) \stackrel{\Delta}{=} \widehat{x}_p(t) - x_p(t), \qquad e_1(t) \stackrel{\Delta}{=} \widehat{y}_p(t) - y_p(t),$$

$$\phi(t) \stackrel{\Delta}{=} \widehat{g}(t) - g, \qquad \text{and } \psi(t) \stackrel{\Delta}{=} \widehat{b}(t) - b.$$

The aim is to determine the adaptive laws for adjusting $\dot{\phi}$ and $\dot{\psi}$ (or equivalently $d/dt\{\widehat{g}(t)\}$ and $d/dt\{\widehat{b}(t)\}$) so that both parameter and state errors tend to zero asymptotically.

The motivation for the representation of the plant as well as the observer, as given in Eqs. (4.9) and (4.10) respectively, can be traced to the identification model described in Chapter 3, which was used to identify an unknown plant whose state variables are accessible. The error equation (4.11) is seen to be similar to Eq. (3.7b). However, the same procedure can no longer be used for the generation of simple adaptive laws, since the entire state $e(t)$ is not accessible in the present case. The problem is rendered difficult by the fact that the estimates of the $2n$ unknown parameters have to be updated using only the two scalar signals $u(t)$ and $y_p(t)$. Consequently, to provide the required flexibility, two additive auxiliary vector inputs $v_1(t)$ and $v_2(t)$ are introduced in the adaptive observer. This modifies the adaptive observer equations as follows:

$$\begin{aligned} \dot{\widehat{x}}_p(t) &= K\widehat{x}_p(t) + \widehat{g}(t)y_p(t) + \widehat{b}(t)u(t) + v_1(t) + v_2(t) \\ \widehat{y}_p(t) &= h^T\widehat{x}_p(t). \end{aligned}$$

The error equations are given by

$$\dot{e} = Ke + \phi y_p + \psi u + v_1(t) + v_2(t) \qquad e_1 = h^T e. \tag{4.12}$$

With the additional freedom provided by the vectors $v_1(t)$ and $v_2(t)$, adaptive laws similar to those derived in Chapter 3 can also be obtained for the adaptive observer. It is based on the fact that by the proper choice of v_1 and v_2, the output $e_1(t)$ of the error equation (4.12) may be made identical to that of a system for which stable adaptive laws can be established. Theorem 4.1 is found to be useful in determining the latter.

Theorem 4.1 Let a dynamical system be represented by the controllable and observable triple (h_1^T, A_1, b_1) where A_1 is an asymptotically stable matrix, and

$$W(s) \triangleq h_1^T (sI - A_1)^{-1} b_1 \text{ is strictly positive real.}$$

Let the elements of a vector $\omega(t)$ be bounded piecewise-continuous functions. Then the origin of the system of differential equations

$$\dot{\epsilon} = A_1\epsilon + b_1\phi^T\omega \qquad \epsilon_1 = h_1^T\epsilon \tag{4.13}$$

$$\dot{\phi} = -\epsilon_1\omega \tag{4.14}$$

is uniformly stable in the large.

Proof. Consider the following quadratic function $V(\epsilon, \phi)$ as a Lyapunov function candidate:

$$V(\epsilon, \phi) = \epsilon^T P\epsilon + \phi^T\phi$$

where P is a suitably chosen symmetric positive-definite matrix. Evaluating the time derivative of V along the trajectories of the system described by Eqs. (4.13) and (4.14), we obtain

$$\dot{V} = \epsilon^T \left(A_1^T P + PA_1\right)\epsilon + 2\left[\phi^T\omega \cdot \epsilon^T Pb_1 - \phi^T\omega \cdot \epsilon_1\right].$$

We recall from the Lefschetz version of the Kalman-Yakubovich lemma presented in Chapter 2 (Lemma 2.3) that if $h_1^T[sI - A_1]^{-1}b_1$ is strictly positive real, then there exist symmetric positive-definite matrices P and Q such that the matrix and vector equations

$$\begin{aligned} A_1^T P + PA_1 &= -Q \\ Pb_1 &= h_1 \end{aligned}$$

are simultaneously satisfied. If such a matrix P is used in the definition of the function $V(\epsilon, \phi)$, we obtain

$$\begin{aligned} \dot{V} &= -\epsilon^T Q\epsilon + 2\epsilon^T h_1\phi^T\omega - 2\epsilon_1\phi^T\omega \\ &= -\epsilon^T Q\epsilon \le 0, \end{aligned}$$

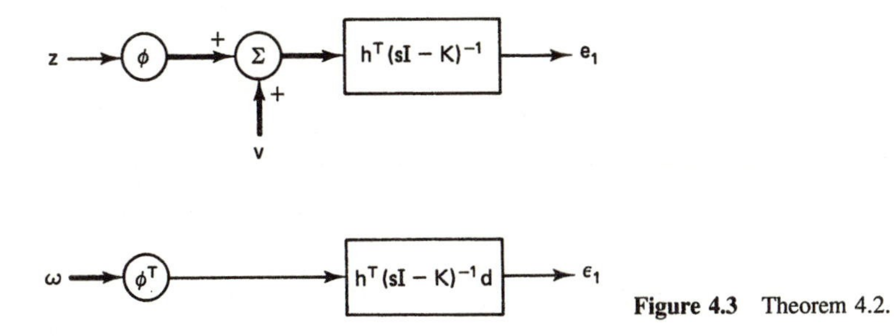

Figure 4.3 Theorem 4.2.

that is, $\dot{V}$ is negative-semidefinite and the origin of the system described by Eqs. (4.13) and (4.14) is uniformly stable in the large.

If, in addition to the conditions of Theorem 4.2, $\omega(t)$ is also bounded, then $\dot{\epsilon}$ is bounded. Since $\dot{V} = -\epsilon^T Q\epsilon$, $\epsilon \in \mathcal{L}^2$. Hence, from Lemma 2.12, we have $\lim_{t\to\infty} \epsilon(t) = 0$.

Theorem 4.1 implies that if the error equation is of the form described in Eq. (4.13), where ϕ corresponds to the unknown parameter error, then the adaptive law for adjusting the parameter can be chosen according to Eq. (4.14). Theorem 4.2 provides the conditions on the auxiliary inputs $v_1(t)$ and $v_2(t)$ in Eq. (4.12) under which Eqs. (4.12) and (4.13) have the same outputs asymptotically. With such a choice of v_1 and v_2, the same adaptive laws in Eq. (4.14) can also be used in the adaptive observer.

Theorem 4.2 Let (h^T, K) be an observable pair where $K \in \mathbb{R}^{n\times n}$ is asymptotically stable. Given bounded piecewise-continuous functions of time $z : \mathbb{R}^+ \to \mathbb{R}$ and $\dot{\phi} : \mathbb{R}^+ \to \mathbb{R}^n$, a vector $d \in \mathbb{R}^n$ and functions $v, w : \mathbb{R}^+ \to \mathbb{R}^n$ can be determined such that the systems

$$\dot{e} = Ke + \phi z + v \qquad e_1 = h^T e \tag{4.15}$$

$$\dot{\epsilon} = K\epsilon + d\phi^T\omega \qquad \epsilon_1 = h^T\epsilon \tag{4.16}$$

have the same outputs as $t \to \infty$.

Proof. Without loss of generality, the proof is established for the pair (h^T, K) in observer canonical form:

$$K = \left[\begin{array}{c|c} -k & \begin{array}{c} I \\ \hline 0 \end{array} \end{array}\right] \qquad \text{and} \qquad h^T = (1\,0\,0\cdots 0). \tag{4.17}$$

(i) The vector $d^T = [1, d_2, \ldots, d_n]$ is chosen so that $h^T(sI - K)^{-1}d$ is strictly positive real.

(ii) Let $G_i(s)$ be the transfer function

$$G_i = \left[\frac{s^{n-i}}{s^{n-1} + d_2 s^{n-2} + \cdots + d_n}\right] \qquad i = 1, 2, \ldots, n,$$

and let

$$G^T(s) = [G_1(s), G_2(s), \ldots, G_n(s)].$$

If $\omega^T = [\omega_1, \omega_2, \ldots, \omega_n]$, let $\omega_i(t)$ be defined by $\omega_i(t) = G_i(s)z(t)$[1]. Since $d^T G(s) = 1$, it follows that when ϕ is a constant vector, by choosing $v(t) \equiv 0$,

$$h^T[sI - K]^{-1}\phi = \left\{h^T[sI - K]^{-1}d\right\}\phi^T G(s)$$

or, the transfer function from z to e_1 is identical to that from z to ϵ_1.

(iii) Define

$$v^T(t) = -\dot{\phi}^T(t)\,[0, A_2\omega(t), \ldots, A_n\omega(t)]$$

where $A_m \in \mathbb{R}^{n \times n}$, $m = 1, \ldots, n$, given by

$$A_m = \begin{bmatrix} 0 & -d_m & -d_{m+1} & & \cdots & & -d_n & 0 & \cdots & \cdots & 0 \\ 0 & 0 & -d_m & & \cdots & & -d_{n-1} & -d_n & 0 & \cdots & 0 \\ 0 & \cdots & 0 & \ddots & & & & & \ddots & & \\ 0 & & \cdots & 0 & \ddots & & & & & \ddots & \\ 0 & & \cdots & & 0 & -d_m & -d_{m+1} & \cdots & & & -d_n \\ 0 & 1 & d_2 & d_3 & \cdots & d_{m-1} & 0 & 0 & & \cdots & 0 \\ \vdots & \ddots & \ddots & \ddots & & & & \ddots & 0 & \cdots & 0 \\ 0 & \cdots & 0 & 1 & d_2 & d_3 & & \cdots & & & d_{m-1} \end{bmatrix}$$

With the choice above of d, ω, and v in (i)-(iii), it can be shown that

$$\sum_{i=1}^{n} s^{n-i}\left[\phi_i z + v_i - d_i\phi^T\omega\right] = 0.$$

This in turn implies from Eq. (4.17) that, for any differentiable function $\phi(t)$,

$$h^T\,[sI - K]^{-1}\left[\phi z + v - d\phi^T\omega\right] = \sum_{i=1}^{n}\frac{s^{n-i}}{k(s)}\left[\phi_i z + v_i - d_i\phi^T\omega\right] = 0$$

where $k(s)$ is the characteristic polynomial of K. Since K is an asymptotically stable matrix, for arbitrary initial conditions of the two differential equations (4.15) and (4.16), $\lim_{t\to\infty}|e_1(t) - \epsilon_1(t)| = 0$.

In the proof of Theorem 4.2, it is seen that the vector d is chosen to make the system described by the triple $\{h^T, K, d\}$ strictly positive real. The transfer functions $G_i(s)$, obtained using d, along with z define the vector $\omega(t)$, and $v(t)$ is in turn determined by $\dot{\phi}(t)$ and $\omega(t)$. In the adaptive problems where this theorem finds application, $\dot{\phi}(t)$ is specified by the adaptive law and is consequently known.

[1] The argument s denotes the differential operator d/dt.

In the adaptive observer described in Eq. (4.12), the objective is to choose v_1 and v_2 in such a manner that stable adaptive laws can be given. This is accomplished by first defining the vectors $\omega^1(t)$ and $\omega^2(t)$ as

$$G(s)u(t) = \omega^1(t); \qquad G(s)y_p(t) = \omega^2(t)$$

and the vectors $v_1(t)$ and $v_2(t)$ in turn as

$$\begin{aligned} v_1^T &= \dot{\psi}^T \left[0, A_2\omega^1, \ldots, A_n\omega^1\right] \\ v_2^T &= \dot{\phi}^T \left[0, A_2\omega^2, \ldots, A_n\omega^2\right]. \end{aligned}$$

By Theorem 4.2, it follows that the output $e_1(t)$ of Eq. (4.12) approaches $\epsilon_1(t)$ asymptotically, where the latter is the solution of

$$\dot{\epsilon} = K\epsilon + d\left(\psi^T\omega^1 + \phi^T\omega^2\right) \qquad \epsilon_1 = h^T\epsilon.$$

From Theorem 4.1, if ω^1 and ω^2 are bounded and the adaptive laws are chosen as

$$\begin{aligned} \dot{\psi} &= -\epsilon_1\omega^1 \\ \dot{\phi} &= -\epsilon_1\omega^2, \end{aligned}$$

then $\lim_{t\to\infty} \epsilon_1(t) = 0$. Hence, the adaptive laws

$$\begin{aligned} \dot{\psi} &= -e_1\omega^1 \\ \dot{\phi} &= -e_1\omega^2 \end{aligned}$$

result in the uniform stability in the large of the origin of Eq. (4.12) and in $\lim_{t\to\infty} e_1(t) = 0$. If, in addition, u is persistently exciting with sufficient number of frequencies (refer to Chapter 6), it can be shown that $\phi(t)$, $\psi(t)$ and therefore $e(t)$ tend to zero asymptotically. The structure of the adaptive observer is shown in Fig. 4.4.

4.3.3 Nonminimal Realization of Adaptive Observers

The observer presented in the previous section is unwieldy from the point of view of implementation due to the complexity of the signals v_1 and v_2. The question arises as to whether it is possible to eliminate these signals by choosing a suitable representation for the plant triple (h^T, A, b). It is found that a simpler adaptive observer can be constructed using a nonminimal representation of the plant. While several nonminimal structures have evolved over the years, we present only the original structure suggested in [12] and one of its more recent modifications [7]. Theorem 4.3 provides the basis for such a representation.

Theorem 4.3 [**Representation 1.**] Any controllable and observable n^{th} order single-input single-output LTI system

$$\begin{aligned} \dot{x}_p &= Ax_p + bu \\ y_p &= h^T x_p \end{aligned} \tag{4.18}$$

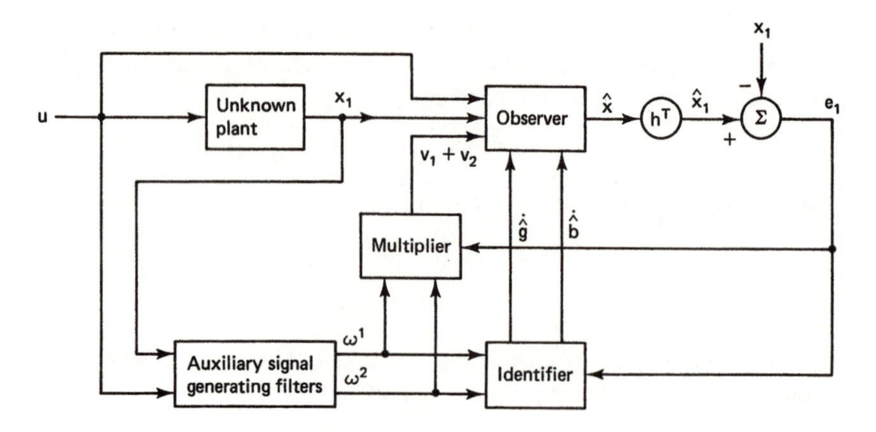

Figure 4.4 The minimal adaptive observer.

is input-output equivalent to the LTI system described by the differential equations

$$\begin{aligned} \dot{x}_1 &= -\lambda x_1 + \theta^T \omega \\ \dot{\omega}_1 &= \Lambda \omega_1 + \ell u \\ \dot{\omega}_2 &= \Lambda \omega_2 + \ell y_p \\ y_p &= x_1 \end{aligned} \tag{4.19}$$

when the parameter vector θ is suitably chosen. θ and ω are defined as

$$\theta \triangleq \begin{bmatrix} c_0 \\ \bar{c} \\ d_0 \\ \bar{d} \end{bmatrix} \quad \text{and} \quad \omega \triangleq \begin{bmatrix} u \\ \omega_1 \\ y_p \\ \omega_2 \end{bmatrix},$$

$\lambda > 0$ is a scalar, (Λ, ℓ) is controllable, and Λ is an $(n-1) \times (n-1)$ asymptotically stable matrix. Such a representation is shown in Fig. 4.5.

Proof. Defining the n-vectors

$$\theta_1 \triangleq \begin{bmatrix} c_0 \\ \bar{c} \end{bmatrix}, \ \theta_2 \triangleq \begin{bmatrix} d_0 \\ \bar{d} \end{bmatrix}, \ \overline{\omega}_1 \triangleq \begin{bmatrix} u \\ \omega_1 \end{bmatrix}, \ \text{and } \overline{\omega}_2 \triangleq \begin{bmatrix} y_p \\ \omega_2 \end{bmatrix};$$

the transfer function from u to $\theta_1^T \overline{\omega}_1$ is given by

$$\bar{c}^T [sI - \Lambda]^{-1} \ell + c_0 \triangleq \frac{P(s)}{R(s)}$$

where $P(s)$ is an $(n-1)^{\text{th}}$ degree polynomial in s and $R(s)$ is the characteristic polynomial of the asymptotically stable matrix Λ. Similarly, the transfer function from y_p to $\theta_2^T \overline{\omega}_2$ is

$$\bar{d}^T [sI - \Lambda]^{-1} \ell + d_0 \triangleq \frac{Q(s)}{R(s)}$$

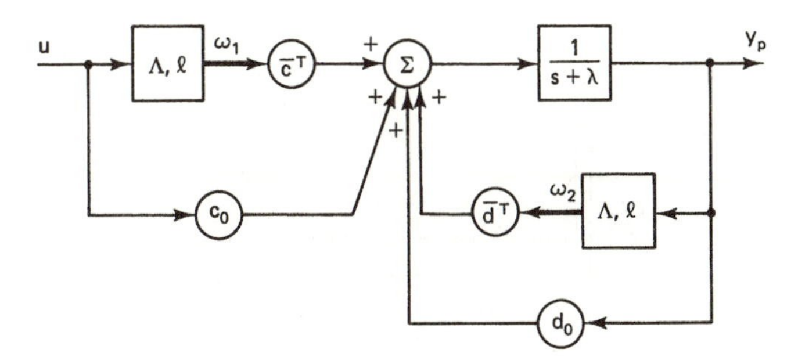

Figure 4.5 Representation 1.

where $Q(s)$ is an $(n-1)^{\text{th}}$ degree polynomial in s. The elements of θ_1 and θ_2, respectively, determine the coefficients of the polynomials $P(s)$ and $Q(s)$. The transfer function from the input u to the output y_p can be expressed as

$$\begin{aligned} W_p(s) &= \frac{P(s)}{R(s)} \cdot \frac{\dfrac{1}{s+\lambda}}{1 - \dfrac{Q(s)}{R(s)(s+\lambda)}} \\ &= \frac{P(s)}{(s+\lambda)R(s) - Q(s)} \triangleq \frac{Z_p(s)}{R_p(s)}. \end{aligned}$$

Since $R(s)$ is of degree $(n-1)$, it follows that the denominator polynomial is monic and of degree n. Further, the n coefficients of the numerator polynomial $Z_p(s)$ and the denominator polynomial $R_p(s)$ can be chosen arbitrarily by appropriately choosing the vectors θ_1 and θ_2, respectively. Hence it follows that any linear time-invariant plant can be parametrized as shown in Eq. (4.19). The parameter vector θ_1 is determined by the zeros of the transfer function $W_p(s)$ while the vector θ_2 is determined by its poles.

If the state of the nonminimal representation in Eq. (4.19) is defined as $x_{np} = [x_1, \omega_1^T, \omega_2^T]^T$, then there exists a constant matrix $C \in \mathbb{R}^{n\times(2n-1)}$ such that

$$x_p = Cx_{np}. \tag{4.20}$$

(See Problem 5). The matrix C is however unknown, since it depends on the parameters of the plant.

An obvious choice for the adaptive observer, which estimates the parameter θ as well as the state x_{np}, is a dynamical system with the same structure as representation 1 but with adjustable parameters. This is described by the following equations (Fig. 4.6):

$$\begin{aligned} \dot{\hat{x}}_1 &= -\lambda\hat{x}_1 + \hat{\theta}^T\hat{\omega} \\ \dot{\hat{\omega}}_1 &= \Lambda\hat{\omega}_1 + \ell u \\ \dot{\hat{\omega}}_2 &= \Lambda\hat{\omega}_2 + \ell\hat{y}_p \\ \hat{y}_p &= \hat{x}_1 \end{aligned}$$

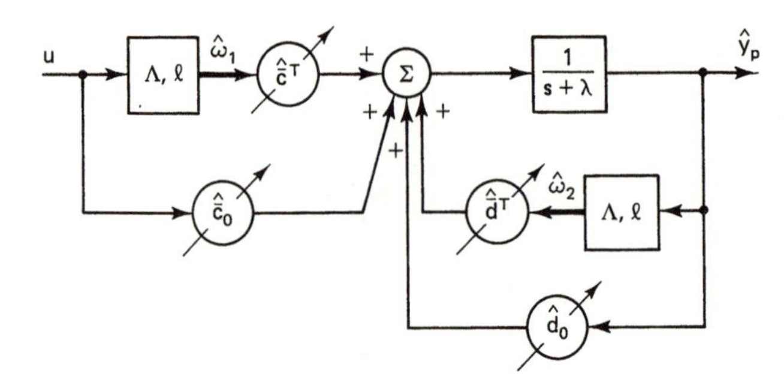

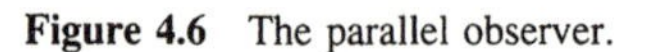
Figure 4.6 The parallel observer.

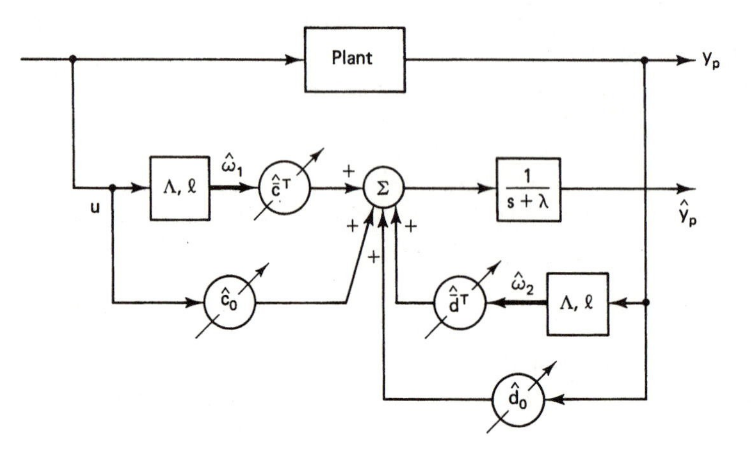

Figure 4.7 The series-parallel observer - Representation 1.

where $\widehat{\theta}(t) = [\widehat{c}_0(t), \widehat{\overline{c}}^T(t), \widehat{d}_0(t), \widehat{\overline{d}}^T(t)]^T$ and $\widehat{\omega}(t) = [u(t), \widehat{\omega}_1^T(t), \widehat{y}_p(t), \widehat{\omega}_2^T(t)]^T$ are estimates of θ and $\omega(t)$ at time t, respectively. Such an observer is referred to as the parallel observer [9] and the approach described is generally referred to as the output error approach. At the time of the writing of this book, a stable method of estimating the parameters and state variables of a given stable plant using a parallel observer had not been demonstrated. However, some authors [4,10] have shown that if some prior information concerning the location of the plant parameters is available, stable adaptive laws for adjusting the observer parameters can be realized.

The Series-Parallel Observer. In 1973, Luders and Narendra suggested a modification of the parallel observer [12], which would admit the generation of stable adaptive laws for estimating the unknown parameters of the plant. Their main contribution was the use of the actual plant output y_p rather than the estimate $\widehat{y}_p$ in the generation of the vector $\widehat{\omega}_2$. This yields a structure for the adaptive observer, shown in Fig. 4.7, and described by the equations

$$\begin{aligned}\dot{\widehat{\omega}}_1 &= \Lambda\widehat{\omega}_1 + \ell u\\ \dot{\widehat{\omega}}_2 &= \Lambda\widehat{\omega}_2 + \ell y_p\\ \dot{\widehat{x}}_1 &= -\lambda\widehat{x}_1 + \widehat{\theta}^T\widehat{\omega}\\ \widehat{y}_p &= \widehat{x}_1\end{aligned} \tag{4.21}$$

where $\Lambda \in \mathbb{R}^{(n-1)\times(n-1)}$ is asymptotically stable, and

$$\widehat{\theta}(t) = \begin{bmatrix}\widehat{c}_0(t)\\ \widehat{\bar{c}}(t)\\ \widehat{d}_0(t)\\ \widehat{\bar{d}}(t)\end{bmatrix}, \qquad \widehat{\omega}(t) = \begin{bmatrix}u(t)\\ \widehat{\omega}_1(t)\\ y_p(t)\\ \widehat{\omega}_2(t)\end{bmatrix}. \tag{4.22}$$

Since a part of the observer related to the output y_p is in series with the plant and the part related to the input u is in parallel with it, the observer is generally referred to as a series-parallel observer [9]. The approach based on such an observer has also been referred to as the equation error approach.

In the following paragraphs, adaptive laws for adjusting $\widehat{\theta}(t)$ are devised such that all the signals in the system remain bounded while the error between plant and observer outputs tends to zero asymptotically with time.

Stability Analysis. Let $\widehat{x}_{np} = [\widehat{y}_p, \widehat{\omega}_1^T, \widehat{\omega}_2^T]^T$ be the state of the series-parallel observer. From Eqs. (4.19) and (4.21), it follows that ω_i and $\widehat{\omega}_i$, $i = 1, 2$, satisfy the same differential equations and hence differ only by exponentially decaying terms due to initial conditions. The outputs y_p and $\widehat{y}_p$ of the plant and the observer, respectively, satisfy the first-order equations

$$\begin{aligned}\dot{y}_p(t) = \dot{x}_1(t) &= -\lambda x_1(t) + \theta^T\omega(t)\\ \dot{\widehat{y}}_p(t) = \dot{\widehat{x}}_1(t) &= -\lambda\widehat{x}_1(t) + \widehat{\theta}^T(t)\widehat{\omega}(t)\end{aligned} \tag{4.23}$$

where $\widehat{\theta}(t)$ and $\widehat{\omega}(t)$ are as defined in Eq. (4.22). If $\widetilde{\omega} = \widehat{\omega} - \omega$, it follows from Eqs. (4.19) and (4.21) that

$$\dot{\widetilde{\omega}} = \begin{bmatrix}\Lambda & 0\\ 0 & \Lambda\end{bmatrix}\widetilde{\omega} \triangleq \overline{\Lambda}\widetilde{\omega}.$$

Hence, $\lim_{t\to\infty}\widetilde{\omega}(t) = 0$. If the output error and the parameter error are defined as

$$e_1(t) \triangleq \widehat{y}_p(t) - y_p(t) \qquad \text{and} \qquad \phi(t) \triangleq \widehat{\theta}(t) - \theta,$$

it follows from Eq. (4.23) that $e_1(t)$ satisfies the first-order equation

$$\dot{e}_1(t) = -\lambda e_1(t) + \phi^T(t)\widehat{\omega}(t) + \theta^T\widetilde{\omega}(t). \tag{4.24}$$

Since θ is a constant vector and $\widetilde{\omega}$ decays exponentially, the last term on the right-hand side of Eq. (4.24) is not found to affect the stability analysis as shown below.

Define a Lyapunov function candidate as

$$V(e_1, \phi, \widetilde{\omega}) = \frac{1}{2}\left(e_1^2(t) + \phi^T(t)\phi(t) + \beta\widetilde{\omega}^T(t)P\widetilde{\omega}(t)\right)$$

where P is a symmetric positive-definite matrix such that $\overline{\Lambda}^T P + P\overline{\Lambda} = -Q < 0$ and β is a positive scalar constant. The time derivative $\dot{V}(e_1, \phi, \widetilde{\omega})$ along the trajectories of Eq. (4.24) can be evaluated as

$$\begin{aligned} \dot{V}(e_1, \phi, \widetilde{\omega}) &= -\lambda e_1^2(t) + e_1(t)\phi^T(t)\widehat{\omega}(t) + e_1(t)\theta^T\widetilde{\omega}(t) + \phi^T(t)\dot{\phi}(t) \\ &\quad - \frac{\beta}{2}\widetilde{\omega}^T(t)Q\widetilde{\omega}(t). \end{aligned} \tag{4.25}$$

As described in Chapter 3, if the adaptive law for updating the parameter vector $\widehat{\theta}(t)$ is chosen as

$$\dot{\widehat{\theta}} = \dot{\phi} = -e_1\widehat{\omega}, \tag{4.26}$$

Eq. (4.25) becomes

$$\dot{V}(e_1, \phi, \widetilde{\omega}) = -\lambda e_1^2(t) + e_1(t)\theta^T\widetilde{\omega}(t) - \frac{\beta}{2}\widetilde{\omega}^T(t)Q\widetilde{\omega}(t).$$

If β is chosen to be greater than $2\|\theta\|^2/(\lambda\lambda_Q)$, where λ_Q is the minimum eigenvalue of Q, then $\dot{V}(e_1, \phi) \le 0$ so that the origin of Eqs. (4.24) and (4.26) is uniformly stable in the large.

From the discussion above, it is clear that the effect of exponentially decaying terms of the form $\widetilde{\omega}(t)$ can be taken into account in the stability analysis by introducing an additional term in the Lyapunov function. Hence, in the chapters following, we shall neglect exponentially decaying terms due to initial conditions wherever they do not affect the stability arguments.

Since $\int_{t_0}^{\infty} \dot{V} dt < \infty$, it follows that $e_1 \in \mathcal{L}^2$. If we assume that the plant is asymptotically stable, then it follows that ω and therefore $\widehat{\omega}$ are bounded. Hence we can conclude from Eq. (4.24) that $\dot{e}_1$ is bounded. Using Lemma 2.12, it follows that

$$\lim_{t\to\infty} e_1(t) = 0, \qquad \lim_{t\to\infty} [\widehat{x}_{np}(t) - x_{np}(t)] = 0.$$

In addition, $\lim_{t\to\infty} \dot{\widehat{\theta}}(t) = 0$. Since $x_p(t) = Cx_{np}(t)$ by Eq. (4.20), the estimate $\widehat{x}_p(t)$ can be computed as $\widehat{x}_p(t) = \widehat{C}x_{np}(t)$, where $\widehat{C}(t)$ is the estimate of C and depends on the plant parameter estimate $\widehat{\theta}(t)$.

The nature of the convergence of the parameter estimates is more complex and depends on the nature of the reference input u, the plant output y_p and consequently the signal $\widehat{\omega}$ in Eq. (4.21). If $\widehat{\omega}$ is persistently exciting in $\mathbb{R}^{2n}$, it can be shown that

$\lim_{t\to\infty}\widehat{\theta}(t) = \theta$, so that the parameter estimates converge to their true values and $\widehat{x}_p(t)$ converges to $x_p(t)$. The definition of persistent excitation and the conditions on the input that assure parameter convergence are discussed in Chapter 6.

The main features of the adaptive observer described thus far may be summarized as follows:

(i) Using a nonminimal representation of the plant, the output y_p can be described as the output of a first-order differential equation whose input is a linear combination of $2n$ signals. The coefficients of these signals represent the unknown parameters of the plant.

(ii) Using the input and the output of the plant, the adaptive observer is parametrized in such a fashion that the output of the observer is also described by a first-order differential equation similar to that in (i). The parameters of the observer can be considered to be the adjustable coefficients of the vector $\widehat{\omega}$.

(iii) The uniform stability of the origin of the error equation (4.24) can be derived without any assumptions on the stability of the plant. However, if the plant is stable, the convergence of the error e_1 to zero can also be concluded.

(iv) The parameters of the observer are adjusted using stable adaptive laws so that the error between plant and observer outputs tends to zero. The convergence of the parameters to the desired values however depends on the persistent excitation of the input signals (see Chapter 6).

(v) The prior information required of the plant to build the observer is its order n. This is in contrast to the control problem treated in Chapter 5, where more information is needed to build a stable controller. Also, the underlying error equations (4.24) and (4.26) are linear, in contrast to the error equations in the control problem, which are nonlinear. Hence, all the stability properties of the observer hold in the large.

(vi) If the number of zeros m of $W_p(s)$ is known and is less than $n-1$, then $\widehat{\theta}_1$ can be chosen to be of dimension $m+1$ and $\widehat{\overline{\omega}}_1(t) = \widehat{\omega}_1(t) = [1/R(s), \ldots, s^m/R(s)]^T u(t)$, where $R(s)$ is a Hurwitz polynomial of degree $n-1$. This reduces the total number of adjustable parameters to $n+m+1$.

(vii) When the plant is partially known, a reduced order adaptive observer can be constructed in some cases. For example, if $W_p(s) = W(s)\overline{W}_p(s)$, where $W(s)$ is known and the order of $\overline{W}_p(s) = \overline{n} < n$, then an observer of order $2\overline{n}-1$ suffices. The state of the observer in this case is given by $[\widehat{y}_p, \widehat{\omega}_1^T, \widehat{\omega}_2^T]^T$ where

$$\widehat{\omega}_1 = \begin{bmatrix} \frac{1}{R(s)} \\ \frac{s}{R(s)} \\ \vdots \\ \frac{s^{\overline{n}-2}}{R(s)} \end{bmatrix} W(s)u, \qquad \widehat{\overline{\omega}}_1 = \begin{bmatrix} W(s)u \\ \widehat{\omega}_1 \end{bmatrix}, \qquad \widehat{\omega}_2 = \begin{bmatrix} \frac{1}{R(s)} \\ \frac{s}{R(s)} \\ \vdots \\ \frac{s^{\overline{n}-2}}{R(s)} \end{bmatrix} y_p$$

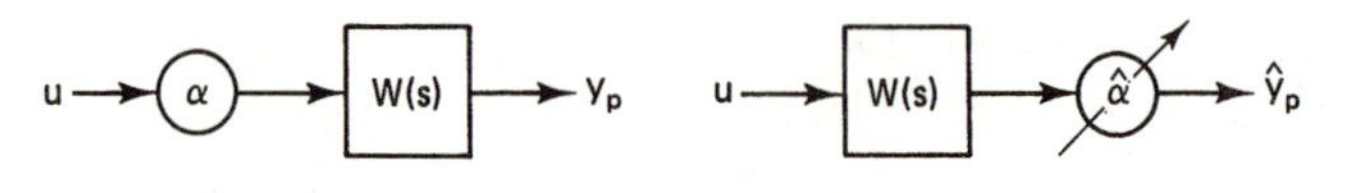

Figure 4.8 A modified adaptive observer.

and $R(s)$ is a Hurwitz polynomial of degree $\overline{n}-1$. The estimate of the plant output can be determined using Eq. (4.21), where $\widehat{\theta}$ is of dimension $2\overline{n}$.

4.3.4 Modified Adaptive Observer

An alternate method of parametrizing the plant for the simultaneous identification of the state vector as well as the parameters was suggested by Kreisselmeier [7]. The approach used is different from that described in the previous section, where the parameters of a Luenberger observer were adjusted. The motivation for Kreisselmeier's observer can be explained in terms of the following simple example.

Consider a plant that has the structure shown in Fig. 4.8(a), where $W(s)$ is a known asymptotically stable transfer function and α is an unknown scalar parameter. If $u(t)$ and $y_p(t)$ can be measured at every instant of time t, the objective is to estimate the value of the parameter α. If an observer structure identical to that given in Fig. 4.8(a) with α replaced by its estimate $\widehat{\alpha}$ is used, then the estimate $\widehat{y}_p(t)$ of the output is given by

$$\widehat{y}_p(t) = W(s)\widehat{\alpha}(t)u(t).$$

No method currently exists for adjusting $\widehat{\alpha}$ with such an observer structure, if $W(s)$ is not strictly positive real. This problem is avoided by choosing an observer structure as shown in Fig. 4.8(b), where the adjustable parameter $\widehat{\alpha}(t)$ follows the transfer function $W(s)$. This simplifies the analysis substantially. Denoting $W(s)u(t) = \omega(t)$, the equation satisfied by the output error $e_1(t) = \widehat{y}_p(t) - y_p(t)$ is given by

$$e_1(t) = \left[\widehat{\alpha}(t) - \alpha\right] \omega(t).$$

Hence, $\widehat{\alpha}(t)$ can be updated using the adaptive law

$$\dot{\widehat{\alpha}}(t) = -e_1(t)\omega(t).$$

This idea can be directly generalized to the multiparameter case as follows. If the unknown plant can be represented in the form given in Eq. (4.18), where A is an asymptotically stable matrix, A, and h are known, and the elements of the vector b are unknown, then the latter can be estimated as shown below:

Let e_i represent a unit vector $[0 \cdots 0\ 1\ 0 \cdots 0]^T$, which contains unity as its ith element. Since the vector b can be expressed as

$$b = \sum_{i=1}^{n} b_i e_i,$$

by linearity, the state x_p can be generated as a linear combination of the vectors ξ_i where

$$\dot{\xi}_i(t) = A\xi_i(t) + e_i u(t), \qquad \xi_i(t_0) = 0.$$

Then

$$x_p(t) = [\xi_1(t),\ \xi_2(t), \ldots, \xi_n(t)]\, b + \exp\{A(t-t_0)\}x_p(t_0)$$

and

$$y_p(t) = z_1^T(t)b + h^T \exp\{A(t-t_0)\}x_p(t_0)$$

where

$$\overline{\xi}(t) \triangleq [\xi_1(t),\ \xi_2(t), \ldots, \xi_n(t)], \qquad z_1(t) \triangleq \overline{\xi}(t)h.$$

Since $\overline{\xi}(t)$ can be measured, an estimate $\widehat{y}_p(t)$ of the output $y_p(t)$ is defined as

$$\widehat{y}_p(t) = z_1^T(t)\widehat{b}(t)$$

where $\widehat{b}(t)$ is an estimate of b at time t. If $\widehat{b}(t)$ is adjusted according to the adaptive law

$$\dot{\widehat{b}}(t) = -\left[\widehat{y}_p(t) - y_p(t)\right] z_1(t),$$

it follows that $\lim_{t\to\infty} |y_p(t) - \widehat{y}_p(t)| = 0$.

We now consider the identification of the unknown parameters of a plant, using only input and output measurements, based on the procedure above. The plant is parametrized in the form

$$\begin{aligned} \dot{x}_p &= Kx_p + gy_p + bu \\ y_p &= h^T x_p \end{aligned}$$

where

$$K \triangleq \left[\begin{array}{c|c} -k & \begin{array}{c} I \\ \hline 0 \end{array} \end{array} \right], \qquad h^T = [1\,0\cdots 0],$$

and the vector k is chosen so that the matrix K is asymptotically stable. The identification of the plant now involves the determination of the unknown vectors g and b from the input-output data. Following the procedure described earlier, $2n$ vectors ξ_i are generated as

$$\begin{aligned} \dot{\xi}_i(t) &= K\xi_i(t) + e_i u(t), & \xi_i(t_0) &= 0 \\ \dot{\xi}_{i+n}(t) &= K\xi_{i+n}(t) + e_i y_p(t), & \xi_{i+n}(t_0) &= 0 \end{aligned} \tag{4.27}$$

for $i = 1, 2, \ldots, n$. The state and the output of the plant can be written as

$$x_p(t) = \overline{\xi}(t)p + \exp\{K(t-t_0)\}x_p(t_0)$$

$$y_p(t) = z_2^T(t)p + h^T \exp\{K(t-t_0)\}x_p(t_0)$$

where

$$\bar{\xi}(t) \triangleq [\xi_1(t), \xi_2(t), \ldots, \xi_{2n}(t)]$$
$$z_2(t) \triangleq \bar{\xi}(t)h$$
$$p \triangleq [b^T, g^T]^T.$$

We define an estimate $\widehat{y}_p(t)$ of $y_p(t)$ as

$$\widehat{y}_p(t) = h^T\widehat{x}_p(t) = z_2^T(t)\widehat{p}(t)$$

where $\widehat{p}(t)$ is the estimate of the parameter vector p at time t. If $\widehat{p}(t)$ is updated using the adaptive law

$$\dot{\widehat{p}}(t) = -\left[\widehat{y}_p(t) - y_p(t)\right] z_2(t)$$

from the arguments used thus far in this section, it follows that $\lim_{t\to\infty} |\widehat{y}_p(t) - y_p(t)| = 0$. Also, if the input to the plant is such that all the modes of the plant are excited, that is, the vector $\widehat{\omega}$ is persistently exciting in $\mathbb{R}^{2n}$, then the parameter estimates can also be shown to converge to their true values (see Chapter 6).

Equation (4.27) indicates that the order of the observer is $2n^2$. However, from the particular form of K, it follows that

$$(sI - K^T)^{-1}e_1 = \frac{1}{\lambda(s)}[s^{n-1}, \ldots, s, 1]^T$$

where $\lambda(s)$ is the characteristic polynomial of K. Hence,

$$(sI - K)^{-1}e_i = T_i(sI - K^T)^{-1}e_1, \qquad i = 1, \ldots, n$$

where T_i are constant matrices whose elements are the coefficients of the numerator polynomials of $(sI-K)^{-1}e_i$ (See Problem 6). Therefore Eq. (4.27) can also be expressed as

$$\begin{aligned}
\dot{v}_1 &= K^T v_1 + e_1 u \qquad v_1(t_0) = 0 \\
\dot{v}_2 &= K^T v_2 + e_1 y_p \qquad v_2(t_0) = 0 \\
\xi_i &= T_i v_1, \qquad \xi_{i+n} = T_i v_2 \qquad i = 1, \ldots, n.
\end{aligned} \tag{4.28}$$

The $2n$th order observer in Eq. (4.28) therefore suffices to generate the estimate of the output.

Adaptive Observer (Representation 2). After the adaptive observer described in the preceding section (representation 1) was proposed, several variations of the basic structure were also suggested in [12]. Following the work of Kreisselmeier reported in this section, a version of the adaptive observer based on a different parametrization of an LTI plant has found wide application (Fig. 4.9). In this case, the plant can be represented as a $2n^{\text{th}}$ order system with $2n$ parameters. Denoting the parameter vector θ and the vector ω as

$$\theta \triangleq \begin{bmatrix} c \\ d \end{bmatrix} \qquad \omega \triangleq \begin{bmatrix} \omega_1 \\ \omega_2 \end{bmatrix},$$

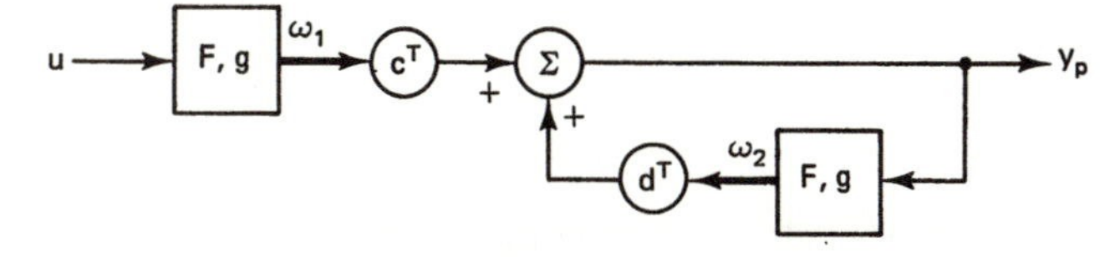

Figure 4.9 Representation 2.

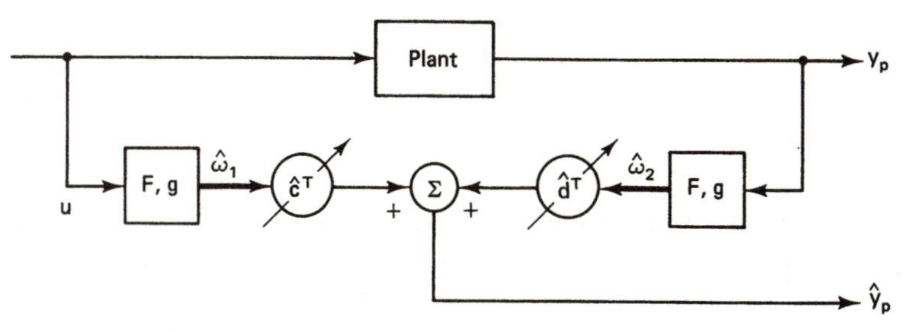

Figure 4.10 The adaptive observer - Representation 2.

the plant can be described by the equations

$$
\begin{aligned}
\dot{\omega}_1 &= F\omega_1 + gu \\
\dot{\omega}_2 &= F\omega_2 + gy_p \\
y_p &= \theta^T\omega \;=\; c^T\omega_1 + d^T\omega_2
\end{aligned} \tag{4.29}
$$

where F is an arbitrary $n \times n$ asymptotically stable matrix, and (F, g) is controllable.

The essential difference between the parametrization used here and that used earlier lies in the fact that the output is directly expressed as a linear combination of $2n$ accessible signals. This leads to a different error equation and a consequent modification in the stability analysis that follows.

The observer structure corresponding to Eq. (4.29) is shown in (Fig. 4.10) and is described by the equations

$$
\begin{aligned}
\dot{\widehat{\omega}}_1 &= F\widehat{\omega}_1 + gu \\
\dot{\widehat{\omega}}_2 &= F\widehat{\omega}_2 + gy_p \\
\widehat{y}_p &= \widehat{\theta}^T\widehat{\omega}
\end{aligned}
$$

where $\widehat{\theta}^T(t) = [\widehat{c}^T(t), \widehat{d}^T(t)]$ and $\widehat{\omega}^T(t) = [\widehat{\omega}_1^T(t), \widehat{\omega}_2^T(t)]$. The problem is to estimate the parameter vector θ based on input-output data, or alternately to determine adaptive laws for updating the parameter estimate $\widehat{\theta}(t)$ so that $\lim_{t\to\infty}\widehat{\theta}(t) = \theta$.

Stability Analysis. If

$$
\phi \stackrel{\triangle}{=} \widehat{\theta} - \theta, \qquad \widetilde{\omega} \stackrel{\triangle}{=} \widehat{\omega} - \omega, \text{ and } e_1 \stackrel{\triangle}{=} \widehat{y}_p - y_p, \tag{4.30}
$$

then the error e_1 satisfies the equation

$$e_1 = \phi^T \widehat{\omega} + \varepsilon \tag{4.31}$$

where $\varepsilon = \theta^T \widetilde{\omega}$. The adaptive law in this case is described by

$$\dot{\phi} = -e_1 \widehat{\omega} = -\widehat{\omega}\widehat{\omega}^T \phi - \widehat{\omega}\varepsilon. \tag{4.32}$$

Since the sensitivity error $\widetilde{\omega}$ satisfies the differential equation

$$\dot{\widetilde{\omega}} = \begin{bmatrix} F & 0 \\ 0 & F \end{bmatrix} \widetilde{\omega} \stackrel{\Delta}{=} \overline{F}\widetilde{\omega} \tag{4.33}$$

where F is an asymptotically stable matrix, $\widetilde{\omega}$ tends to zero as $t \to \infty$. Therefore $\varepsilon \to 0$ as $t \to \infty$. Hence, as mentioned earlier, we ignore the terms due to $\widetilde{\omega}$ in the stability analysis that follows. If a Lyapunov function candidate is chosen as

$$V = \frac{1}{2}\phi^T \phi$$

the time derivative along the trajectories of Eqs. (4.32) and (4.33) leads to

$$\dot{V} = -\left(\phi^T \widehat{\omega}\right)^2 \leq 0.$$

Hence, $\|\phi(t)\|$ is bounded for all $t \geq t_0$.

As in the previous representation, if we make a further assumption that the plant is stable, it follows that $\widehat{\omega}$ is bounded, and hence e_1 is bounded. We also have $\dot{\phi}$ and $\dot{\omega}$ bounded, and hence $\dot{e}_1$ is bounded. Therefore, as in Section 4.3.3, we can conclude that

$$\lim_{t \to \infty} e_1(t) = 0.$$

Further, if the input is persistently exciting,

$$\lim_{t \to \infty} \widehat{\theta}(t) = \theta \qquad \text{and} \qquad \lim_{t \to \infty} [\widehat{x}_p(t) - x_p(t)] = 0.$$

4.4 DESIGN OF ADAPTIVE OBSERVERS

The primary objective of the stability approach in the design of adaptive systems is to determine stable adaptive laws for the adjustment of parameters, so that all the signals of the system remain bounded. Once this is accomplished, the choice of suitable values for the fixed parameters in the adaptive law is undertaken so that the behavior of the overall adaptive system meets performance specifications. In all the adaptive observers described in the previous section, we were mainly concerned with different parametrizations of the observer which led to stable adaptive laws. However, the conditions under which parameters tend to their true values and the rate at which they would converge have

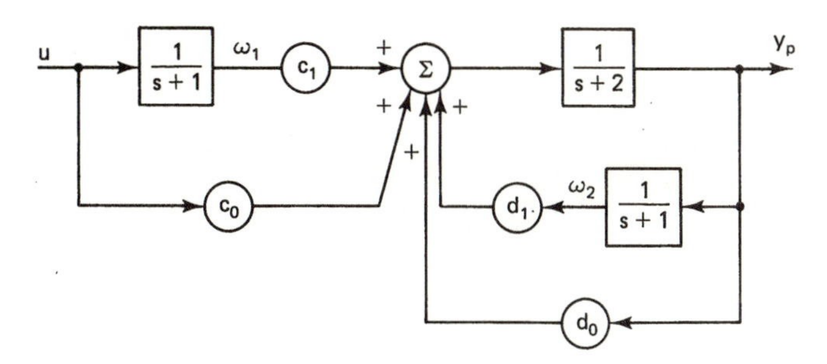

Figure 4.11 Representation 1.

not been described thus far. The choice of the input u, the observer structure, and the adaptive gains are of importance when adaptive observers are designed for practical applications. In this section, we present some simulation studies on simple plants using the methods discussed earlier. These studies serve to introduce the reader to some of the design considerations above. The theoretical basis for the choice of inputs and adaptive gains is discussed in detail in Chapters 6 and 8.

Unless mentioned otherwise, initial conditions were chosen to be identically zero in all simulations.

4.4.1 Representation 1

Simulation 4.1 The transfer function of a plant to be identified was chosen as

$$W_p(s) = \frac{s+1}{(s+2)(s+3)}. \tag{4.34}$$

Using representation 1 (Fig. 4.5), it can be parametrized as shown in Fig. 4.11. The parameters c_0, c_1, d_0, and d_1, can be calculated as

$$c_0 = 1, \qquad c_1 = 0, \qquad d_0 = -2, \qquad d_1 = -2.$$

To identify these parameters, an observer described in Section 4.3.3 (Fig. 4.7) was simulated with four parameters $\widehat{c}_0$, $\widehat{c}_1$, $\widehat{d}_0$, and $\widehat{d}_1$, which were adjusted according to the adaptive laws

$$\dot{\widehat{c}}_0 = -\gamma_1 e_1 u; \qquad \dot{\widehat{c}}_1 = -\gamma_2 e_1 \omega_1; \qquad \dot{\widehat{d}}_0 = -\gamma_3 e_1 y_p; \qquad \dot{\widehat{d}}_1 = -\gamma_4 e_1 \omega_2.$$

The experiments performed on the overall system, the values of the adaptive gains γ_i, and the corresponding behavior of the adaptive observer are given below:

(a) With $u(t) = 5 \cos t$, the evolution of the parameter values as well as the output error are shown in Fig. 4.12-(a). All the adaptive gains were chosen to be unity. The output error is seen to tend to zero, even while the parameter values tend to constant values that are not the same as those of the plant. The estimated plant transfer function, however, is seen to have the same amplitude and phase shift as that of the plant transfer function at the frequency 1 rad/sec.

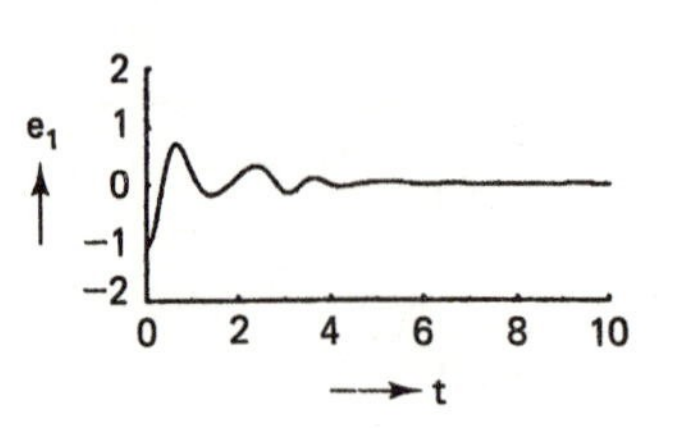

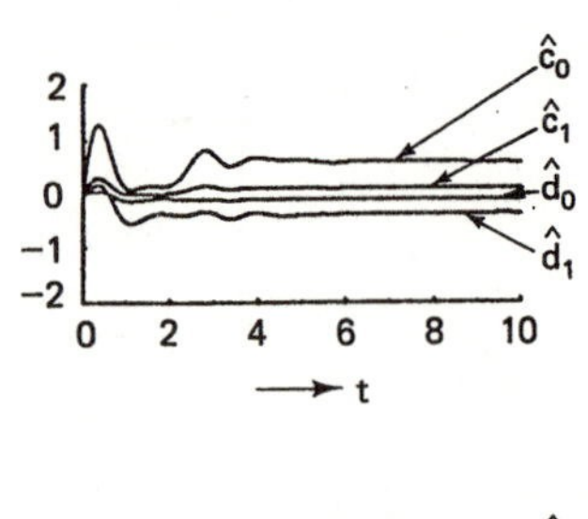

(a)

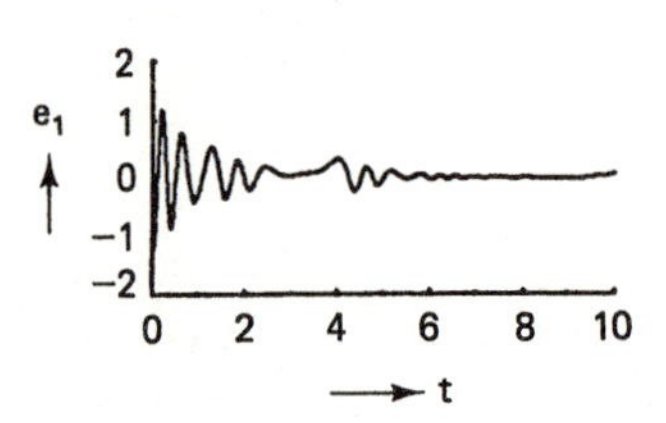

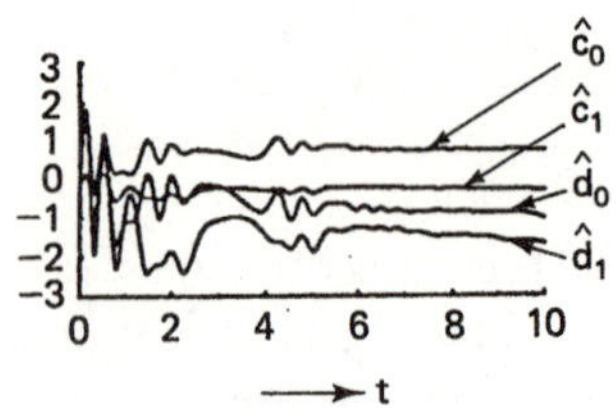

(b)

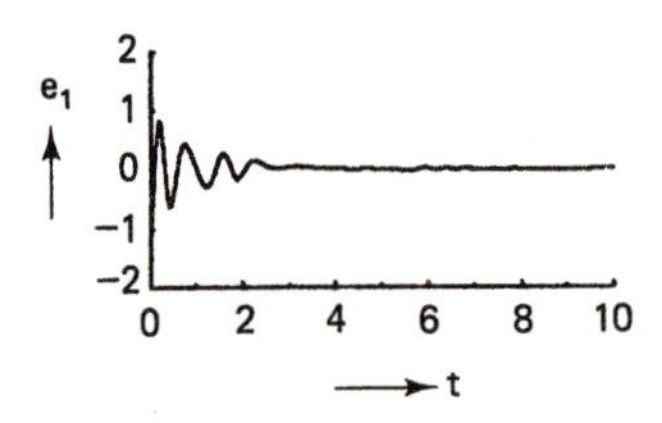

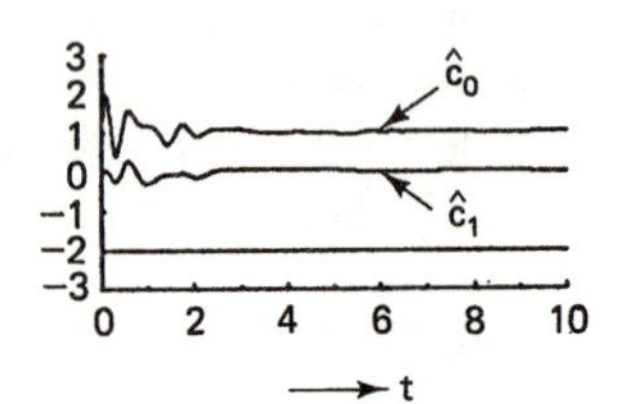

(c)

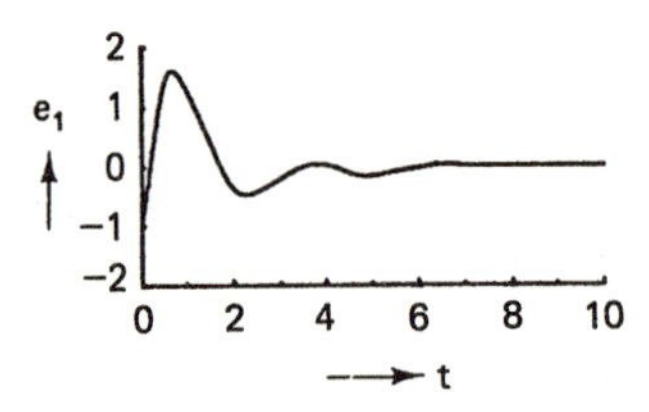

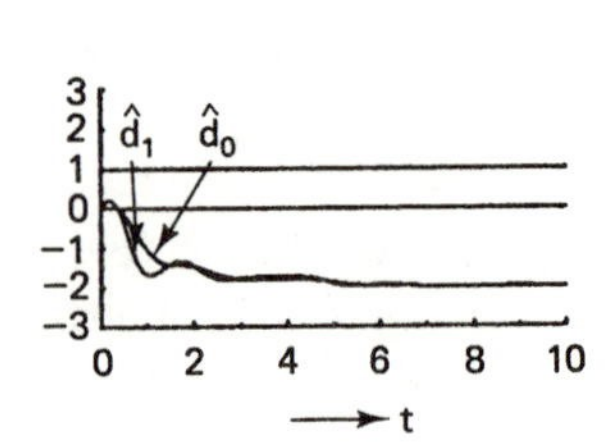

(d)

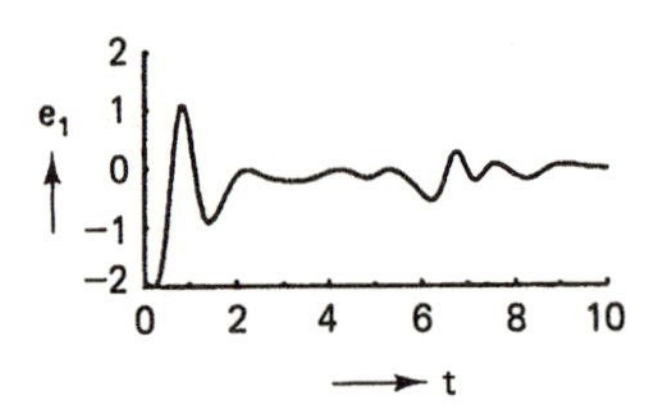

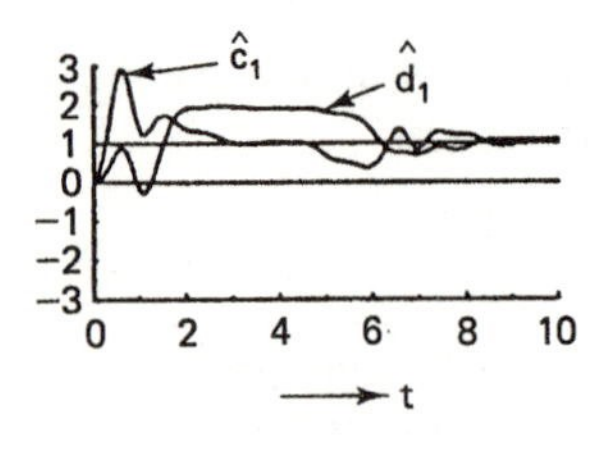

(e)

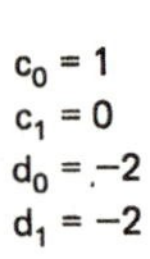

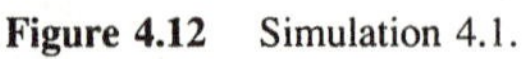

Figure 4.12 Simulation 4.1.

(b) When the input was chosen as $u(t) = 5 \cos\, t{+}10 \cos\, 2t$, all the parameter estimates approach their true values (Fig. 4.12(b)). The adaptive gains $\gamma_1 = \gamma_2 = 1$, $\gamma_3 = 5$, and $\gamma_4 = 20$ resulted in the fastest response.

(c) Figure 4.12(c) shows the nature of the convergence of parameters $\widehat{c}_0(t)$ and $\widehat{c}_1(t)$ when d_0 and d_1 (that is, the poles of the plant) were assumed to be known and the same input $u(t)$ as in the preceding entry was used, with $\gamma_1 = \gamma_2 = 1$.

(d) Figure 4.12(d) gives the response of the observer when c_0 and c_1 were assumed to be known and only $\widehat{d}_0(t)$ and $\widehat{d}_1(t)$ were adjusted. The input $u(t)$ was again chosen as $u(t) = 5 \cos\, t + 10 \cos\, 2t$ with $\gamma_3 = \gamma_4 = 1$.

(e) In Fig. 4.12(e) the performance of the observer is shown when the numerator and denominator terms of the plant transfer function are both known partially, that is, with the transfer function $(s + a)/(s^2 + 3s + b)$ where a and b are unknown. This implies that $c_0 = 1$, $c_1 = a - 1$, $d_0 = 0$, and $d_1 = 2 - b$, and therefore only two parameters $\widehat{c}_1$ and $\widehat{d}_1$ need to be adjusted. The evolution of the output error and parameter estimates is shown in Fig. 4.12(e) assuming that $a = 2$, $b = 1$, and $u(t) = 5 \cos\, t + 10 \cos\, 2t$. The adaptive gains were chosen as $\gamma_2 = \gamma_4 = 1$.

The results of the simulation study above can be summarized as follows:

(i) Although the output error converges to zero for all inputs, the parameter errors may tend to constant values. When the input contains at least two distinct frequencies, all four parameters tend to their true values.

(ii) When either the poles or the zeros of the plant transfer function are known, some of the parameters of the observer can be fixed. Convergence of the remaining parameters is rapid in this case. Similar behavior is also observed when partial information is available regarding both the numerator and denominator polynomials of the plant transfer function. Convergence of the parameters of the observer is found to be relatively slow when all the parameters are unknown.

(iii) Although considerable freedom exists in the choice of the adaptive gains, in practice it is found that determining the optimal values of $\gamma_i (i = 1, 2, 3, 4)$ is a tedious task. This is mainly due to the stiff nature of the error differential equations, which results in rapid convergence for some initial conditions, but in very slow convergence for others. These studies reveal the need for time-varying adaptive gains, where such gains are based on the specific inputs used to excite the plant [6].

4.4.2 Representation 2

Simulation 4.2 The identification of the same plant [refer to Eq. (4.34)] considered in Simulation 4.1 was carried out using representation 2. Using such a parametrization, the plant has a representation as shown in Fig. 4.13. The four unknown parameter values are

$$c_0 = 0, \qquad c_1 = 1, \qquad d_0 = -2, \qquad d_1 = 0.$$

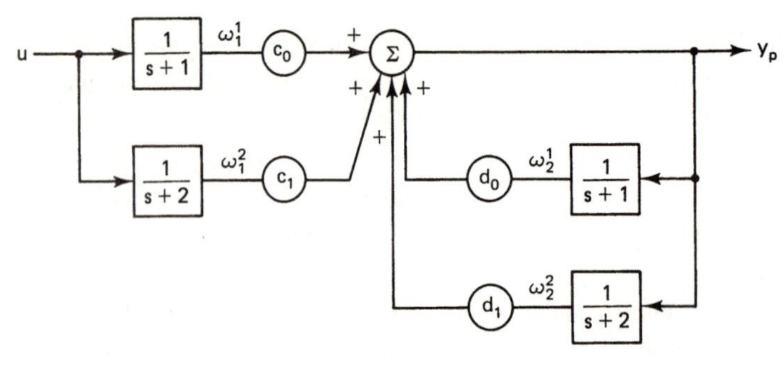

Figure 4.13 Representation 2.

The observer therefore has four adjustable parameters: $\widehat{c}_0$, $\widehat{c}_1$, $\widehat{d}_0$, and $\widehat{d}_1$. These parameters were adjusted according to the adaptive laws

$$\dot{\widehat{c}}_0 = -\gamma_1 e_1 \omega_1^1; \qquad \dot{\widehat{c}}_1 = -\gamma_2 e_1 \omega_1^2; \qquad \dot{\widehat{d}}_0 = -\gamma_3 e_1 \omega_2^1; \qquad \dot{\widehat{d}}_1 = -\gamma_4 e_1 \omega_2^2$$

with $\gamma_i > 0, i = 1, 2, 3, 4$. Figure 4.14(a)-(d) shows the results of the simulation experiments that were performed on the overall system. As in Simulation 4.1, the adaptive gains $\gamma_i, i = 1 - 4$, were set to unity except in case (b).

(a) With $u(t) = 5 \cos t$, the evolution of the parameter values as well as the output error are shown in Fig. 4.14(a). As in Simulation 4.1, the output error is seen to tend to zero even while the parameter values tend to constant values that are not the same as those of the plant. Once again, the estimated plant transfer function is such that it has the same amplitude and phase shift as the plant at the input frequency $\omega = 1$. The latter would account for the output error tending to zero asymptotically.

(b) When the input was chosen to be $u(t) = 5 \cos t + 10 \cos 2t$, all the parameter estimates approach their true values [Fig. 4.14(b)]. The values of the adaptive gains in this case were $\gamma_1 = \gamma_2 = 5$, $\gamma_3 = \gamma_4 = 10$.

(c) Figure 4.14(c) shows the nature of the convergence of parameters $\widehat{c}_0(t)$ and $\widehat{c}_1(t)$ when d_0 and d_1 (that is, the poles of the plant) are assumed to be known and the same input in the entry above is used.

(d) Figure 4.14(d) gives the response of the observer when c_0 and c_1 were assumed to be known and only $\widehat{d}_0(t)$ and $\widehat{d}_1(t)$ were adjusted. When both the numerator and denominator polynomials of the plant transfer function were only partially known, the parameters of the observer rapidly approach their true values.

4.4.3 Hybrid Algorithms

In all the adaptive observers considered thus far, the estimates of the parameters are adjusted continuously using input-output data. However, for robustness of estimation, it may be desirable to update the estimates only at specific instants of time t_i, where $\{t_i\}$ is an unbounded monotonic sequence, and $\sup_i[t_{i+1} - t_i] < \infty$. This implies that while

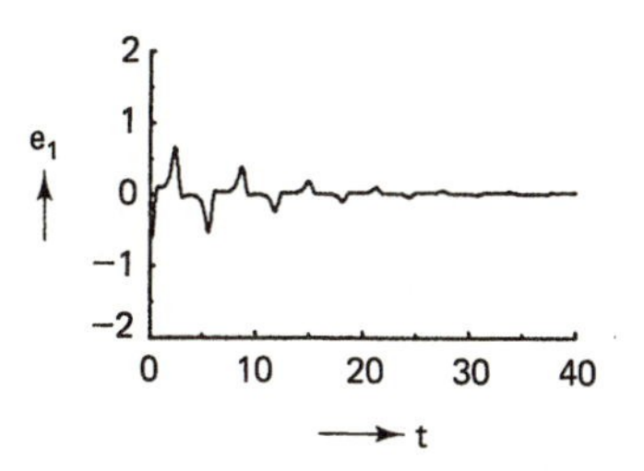

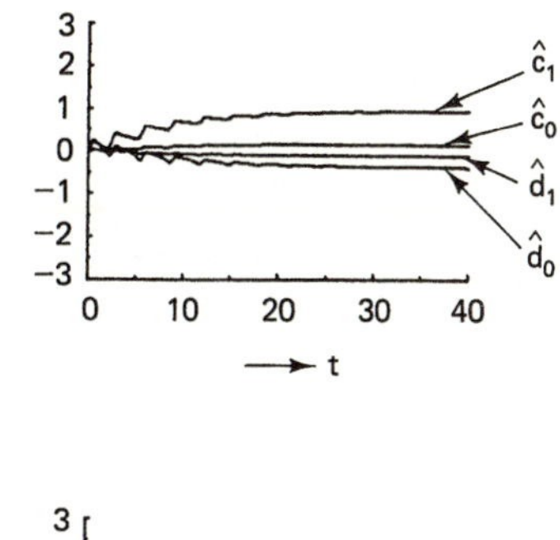

(a)

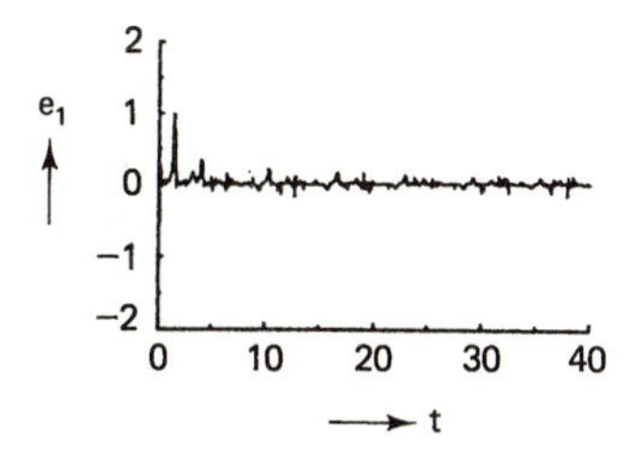

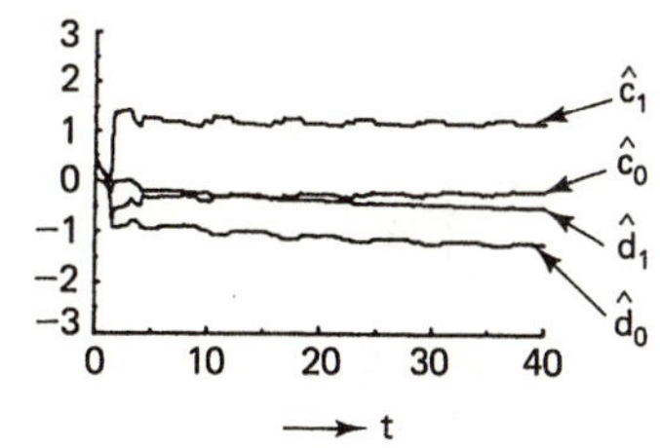

(b)

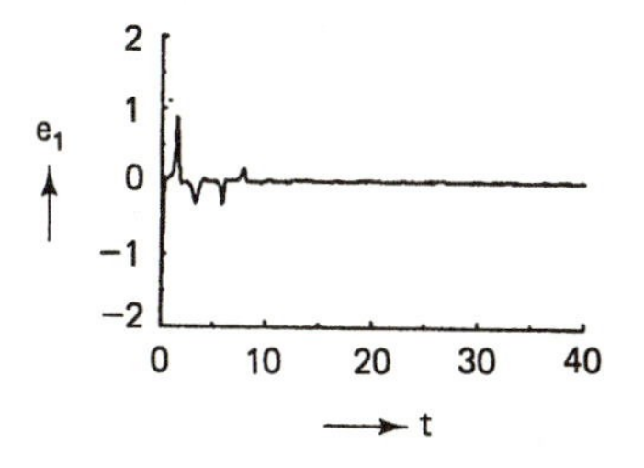

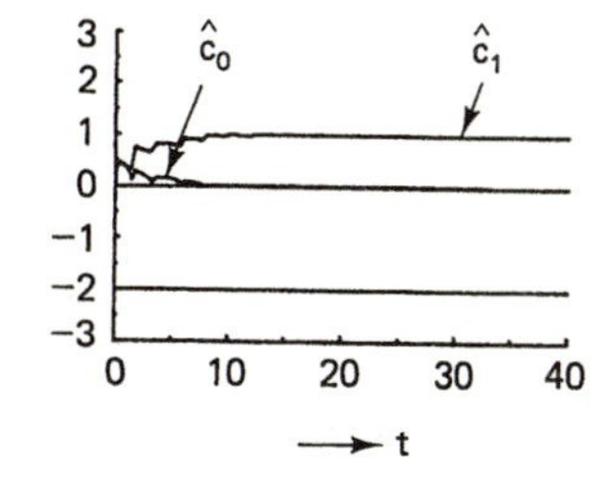

(c)

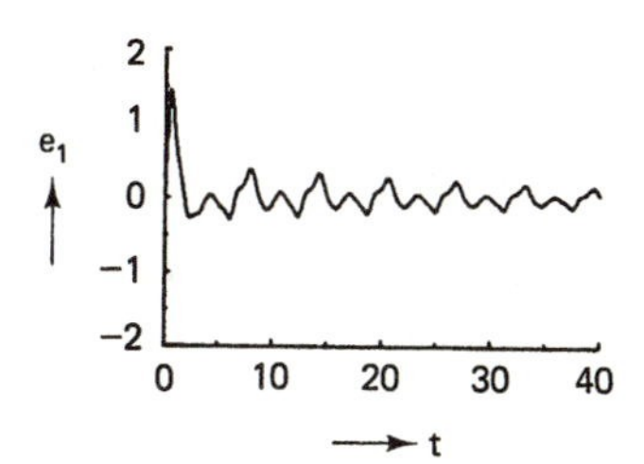

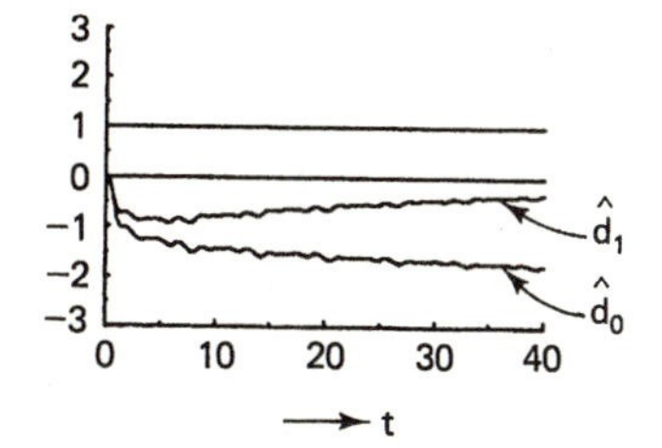

(d)

$c_0 = 0$
$c_1 = 1$
$d_0 = -2$
$d_1 = 0$

Figure 4.14 Simulation 4.2.

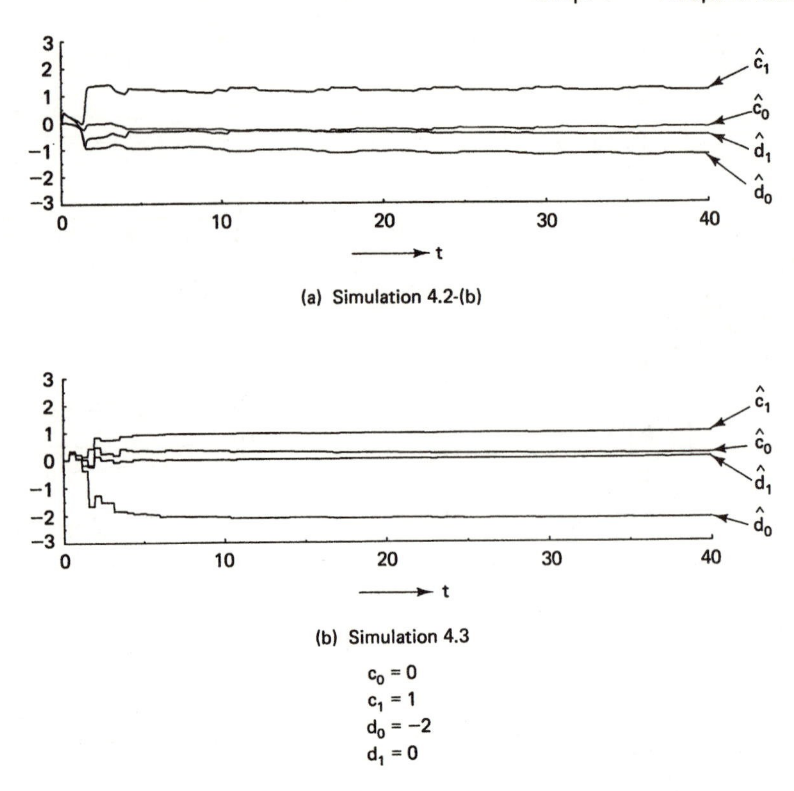

Figure 4.15 Hybrid algorithm.

all the signals of the system are piecewise-continuous, the parameters are adjusted in a discrete fashion. For example, given the plant $W_p(s)$ whose parametrization leads to the equation $\omega^T(t)\theta = y_p(t)$ (using representation 2), a discrete stable adaptive law is used to update the parameters $\widehat{\theta}(t)$ of the adaptive observer. This is given by

$$\phi_{k+1} \quad = \quad \phi_k - \Gamma R_k \phi_k \tag{4.35}$$

where

$$\phi_k \stackrel{\triangle}{=} \phi(t_k) = \theta(t_k) - \theta, \;\; T_k = t_{k+1} - t_k, \;\; R_k \stackrel{\triangle}{=} \frac{1}{T_k}\int_{t_k}^{t_{k+1}} \frac{\omega(\tau)\omega^T(\tau)}{1+\omega^T(\tau)\omega(\tau)}d\tau,$$

and Γ is a matrix in $\mathbb{R}^n$ chosen so that $[I - \Gamma R_k]$ has eigenvalues less than or equal to unity. A detailed stability analysis of such hybrid systems is given in Chapter 8. However, in view of its practical importance, we include some simulation results here.

Simulation 4.3 With the same plant as in Simulations 4.1 and 4.2, an observer similar to that described in Simulation 4.2 was simulated. The adaptive law in Eq. (4.35) was used to adjust the parameters with $T = 0.5$ and $u(t) = 5\cos t + 10\cos 2t$. The evolution of $\widehat{\theta}(t)$ is shown in Fig. 4.15(b) with $\Gamma = diag[1, 1, 10, 10]$. In Fig. 4.15(a), the results of Simulation 4.2(b) is repeated for comparison purposes. It is seen that $\widehat{\theta}(t)$ converges to its true value faster with the hybrid algorithm.

4.4.4 Time-Varying Adaptive Gains

In Chapter 3, the adaptive identification and control of simple first-order systems, as well as systems whose state vectors are completely accessible, were discussed. In all cases, for ease of exposition, the adaptive law was initially assumed to have a gain of unity. Later, it was shown that a positive scalar gain γ or a constant symmetric positive-definite matrix Γ can be used in the adaptive law. In each case the Lyapunov function candidate is suitably modified to include the adaptive gain so that its time derivative $\dot{V}$ is negative-semidefinite. In this section, we outline the procedure for including a time-varying gain matrix in the adaptive law.

Our starting point is the error equation (4.31) in which the output error is linearly related to the parameter errors. Neglecting the term due to initial conditions, the error equation becomes

$$\phi^T \omega \;=\; e_1. \tag{4.36}$$

A Lyapunov function candidate $V(\phi) = \phi^T \phi/\gamma$ results in $\dot{V} = 2\phi^T \dot{\phi}/\gamma$. This suggests an adaptive law $\dot{\phi} = -\gamma e_1 \omega$, so that $\dot{V} = -2e_1^2 \leq 0$. Similarly, a scalar function $V(\phi) = \phi^T \Gamma^{-1} \phi$ together with the adaptive law $\dot{\phi} = -\Gamma e_1 \omega$, yields $\dot{V} = -2e_1^2 \leq 0$. Let the gain matrix Γ be time-varying, whose time derivative is defined by the differential equation $d/dt(\Gamma^{-1}) = \omega\omega^T$, or

$$\dot{\Gamma} \;=\; -\Gamma \omega \omega^T \Gamma, \qquad \Gamma(t_0) = \Gamma^T(t_0) = \Gamma_0 > 0. \tag{4.37}$$

Let the adaptive law have the form

$$\dot{\phi} \;=\; -\gamma \Gamma(t) e_1 \omega \qquad \gamma > \frac{1}{2}. \tag{4.38}$$

A Lyapunov function candidate of the form $\phi^T \Gamma^{-1} \phi$ has a time derivative $\dot{V}$ along the trajectories of Eqs. (4.37) and (4.38) given by

$$\dot{V} \;=\; -(2\gamma - 1)e_1^2 \;\leq\; 0.$$

Hence, the origin of Eq. (4.38) is stable in the large. Since $\Gamma(t)$ can tend to zero, V is not decrescent, and hence uniform stability of the origin cannot be concluded. Equations (4.37) and (4.38) can be considered as the continuous-time analogs of the least squares algorithm, which is well known in discrete-time plants.

Simulation 4.4 indicates the improvement in performance by including a time-varying gain matrix in the adaptive law.

Simulation 4.4 With the plant and observer as in Simulation 4.2, the adaptive law in Eq. (4.38), with a time-varying gain matrix as in Eq. (4.37), was used to adjust the parameters of the observer. The response of $\widehat{\theta}(t)$ for an input $u(t) = 5 \cos t + 10 \cos 2t$ is shown in Fig. 4.16. The initial value of Γ at time $t = 0$ was chosen to be 100.

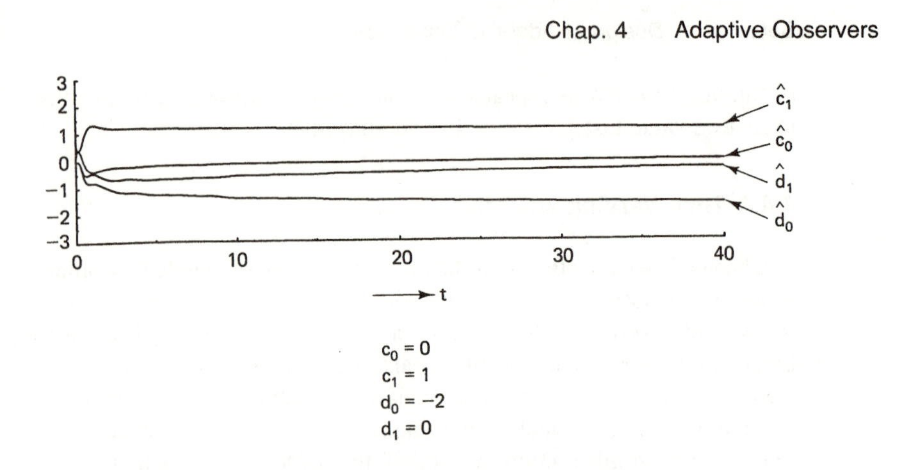

Figure 4.16 Adaptive observer with time-varying adaptive gains.

4.4.5 Integral Algorithm

An alternate scheme that is often used to ensure fast convergence of parameter estimates is one that is referred to as the integral algorithm [7]. Detailed descriptions of this algorithm as well as the one discussed in the previous section are given in Chapter 6 where their convergence properties are examined more carefully. In this section, we state the adaptive law for the sake of completeness. The brief description that follows is our interpretation of what was suggested by Kreisselmeier in [7].

We define an error $e_1 : \mathbb{R}^+ \times \mathbb{R}^+ \to \mathbb{R}$ as

$$e_1(t,\tau) \quad \stackrel{\triangle}{=} \quad \phi^T(t)\omega(\tau) \qquad \tau \leq t$$

where $\phi(t)$ is the parameter error at time t, and $\omega(\tau)$ is the vector of sensitivity functions evaluated at all the past instants of time $\tau \leq t$. Hence, $e_1(t,\tau)$ represents the error that would have resulted at time τ if the parameter estimate at time t had been used. If the rule for the adjustment of ϕ is determined so that $\int_{t_0}^{t} e_1^2(t,\tau)\, d\tau$ is minimized, we obtain

$$\dot{\phi} \quad = \quad -\gamma \int_{t_0}^{t} e_1(t,\tau)\omega(\tau)d\tau \qquad \gamma > 0. \tag{4.39}$$

The minimization of $\int_{t_0}^{t} e_1^2(t,\tau)\, d\tau$ implies that the parameter $\widehat{\theta}(t)$ chosen at time t minimizes the integral square of the error on all past data, that is, the interval $[t_0, t]$. As $t \to \infty$, the integral in Eq. (4.39) may become unbounded. Hence, to keep the time derivative of ϕ bounded, the performance criterion is modified as the weighted integral norm

$$\int_{t_0}^{t} exp\{-q(t-\tau)\}e_1^2(t,\tau)d\tau \qquad q > 0.$$

This yields the adaptive law

$$\dot{\phi}(t) \quad = \quad -\gamma\Gamma(t)\phi(t), \tag{4.40}$$

$$\dot{\Gamma}(t) \quad = \quad -q\Gamma(t) + \omega(t)\omega^T(t) \qquad \Gamma(t_0) \;=\; 0, \tag{4.41}$$

which assures the uniform asymptotic stability of the origin of Eq. (4.40) (see Chapter 6).

Although Eqs. (4.40) and (4.41) are convenient from the point of view of stability analysis, Eq. (4.40) cannot be implemented since $\phi(t)$ is unknown. But the error $e_1(t,\tau)$ can be rewritten as

$$\begin{aligned} e_1(t,\tau) &= \widehat{\theta}^T(t)\omega(\tau) - \theta^T\omega(\tau) && \text{from (4.30)},\\ &= \widehat{\theta}^T(t)\omega(\tau) - y_p(\tau) && \text{from (4.29)}. \end{aligned}$$

Hence the adaptive laws in Eqs. (4.40) and (4.41) can be expressed as

$$\begin{aligned} \dot{\phi}(t) &= -\gamma\Gamma(t)\widehat{\theta}(t) + \gamma\delta(t) \\ \dot{\Gamma}(t) &= -q\Gamma(t) + \omega(t)\omega^T(t) & \Gamma(t_0) &= 0 \\ \dot{\delta}(t) &= -q\delta(t) + \omega(t)y_p(t) & \delta(t_0) &= 0. \end{aligned} \tag{4.42}$$

The adaptive law can therefore be implemented using Eq. (4.42). The discrete analog of the weighting term $\exp(-qt)$ is referred to as the *forgetting factor* in the literature on discrete-time plants [1].

4.4.6 Multiple Models

An alternative method of achieving rapid convergence is by the use of multiple models [11]. If the input u and output y of a linear time-invariant plant with a transfer function $W_p(s)$ are related by the equation

$$W_p(s)u(t) = y(t), \tag{4.43}$$

an input $u_1(t) = W_1(s)u(t)$ results in the corresponding output $y_1(t)$ given by $W_1(s)y(t)$, if $W_1(s)$ is an asymptotically stable transfer function with zero initial conditions. Similarly, by using asymptotically stable transfer functions $W_2(s), \ldots, W_{2n-1}(s)$, $2n-1$ input-output pairs $\{u_i(t), y_i(t)\}$, $i = 1, \ldots, 2n-1$ can be generated so that $W_i(s)u(t) = u_i(t)$ and $W_i(s)y(t) = y_i(t)$. By the results described in Section 4.3, the plant can be parametrized in such a fashion that Eq. (4.43) can also be expressed in the form

$$\omega^T(t)\theta = y(t).$$

Using the same parametrization of the plant as before, along with the $2n$ input-output pairs, leads to the equation

$$\Omega^T(t)\theta = Y(t)$$

where

$$\Omega(t) \triangleq [\omega,\ W_1(s)\omega,\ \ldots, W_{2n-1}\omega] \qquad Y \triangleq [y,\ y_1,\ \ldots, y_{2n-1}]^T.$$

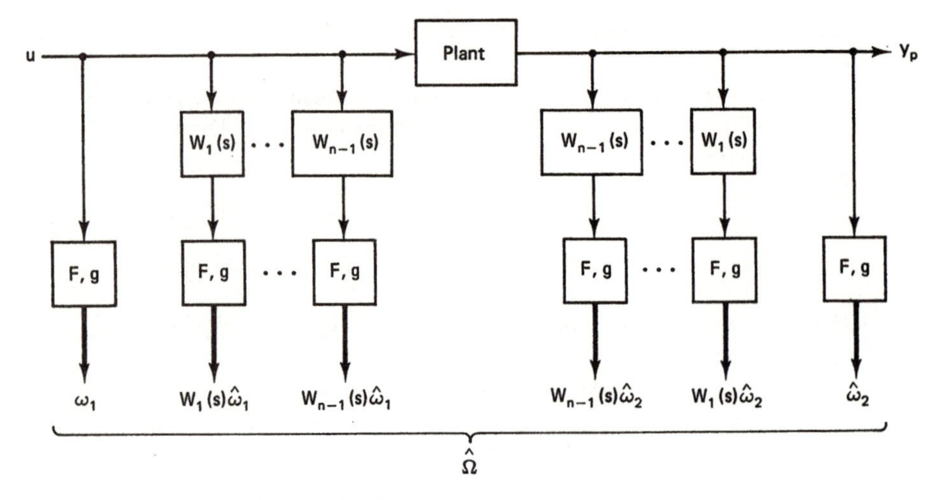

Figure 4.17 The observer with multiple models.

This leads to an observer described by the equation (Fig. 4.17)

$$\widehat{\Omega}^T(t)\widehat{\theta}(t) = \widehat{Y}(t) \tag{4.44}$$

and the adaptive law for adjusting the parameter $\widehat{\theta}$ is given by

$$\dot{\widehat{\theta}}(t) = -\gamma\left[\widehat{\Omega}(t)\widehat{\Omega}^T(t)\right]\widehat{\theta}(t) + \gamma\left[\widehat{\Omega}(t)Y(t)\right].$$

The parameter error differential equation is then derived as (ignoring exponentially decaying terms due to initial conditions)

$$\dot{\phi}(t) = -\gamma\widehat{\Omega}(t)\widehat{\Omega}^T(t)\phi(t). \tag{4.45}$$

If the columns of $\widehat{\Omega}(t)$ are independent, $\widehat{\Omega}(t)\widehat{\Omega}^T(t)$ is a positive-definite matrix. Hence, arbitrarily fast convergence rates can be realized in theory by making the adaptive gain γ sufficiently large.

The following simulation results reveal that, in practice, due to the skewed nature of the matrix $\widehat{\Omega}\widehat{\Omega}^T$, large values of the input and γ may be needed for fast convergence.

Simulation 4.5 The plant to be identified was chosen as

$$W_p(s) = \frac{s+1}{(s+2)(s+3)}.$$

If $\omega_1^T = [1/(s+1)u,\ 1/(s+2)u,\ 1/(s+1)y_p,\ 1/(s+2)y_p]$, then $\theta^T = [0,\ 1,\ -2,\ 0]$. Vectors ω_2, ω_3 and $\omega_4 \in \mathbb{R}^4$ were chosen as $\omega_2 = [1/(s+2)]\cdot\omega_1$, $\omega_3 = [1/(s+3)]\cdot\omega_1$ and $\omega_4 = [s/(s+3)]\cdot\omega_1$, and the matrix Ω as $\Omega = [\omega_1,\ \omega_2,\ \omega_3,\ \omega_4]$. An observer was constructed as described in Eq. (4.44) with the adaptive law as in Eq. (4.45). The

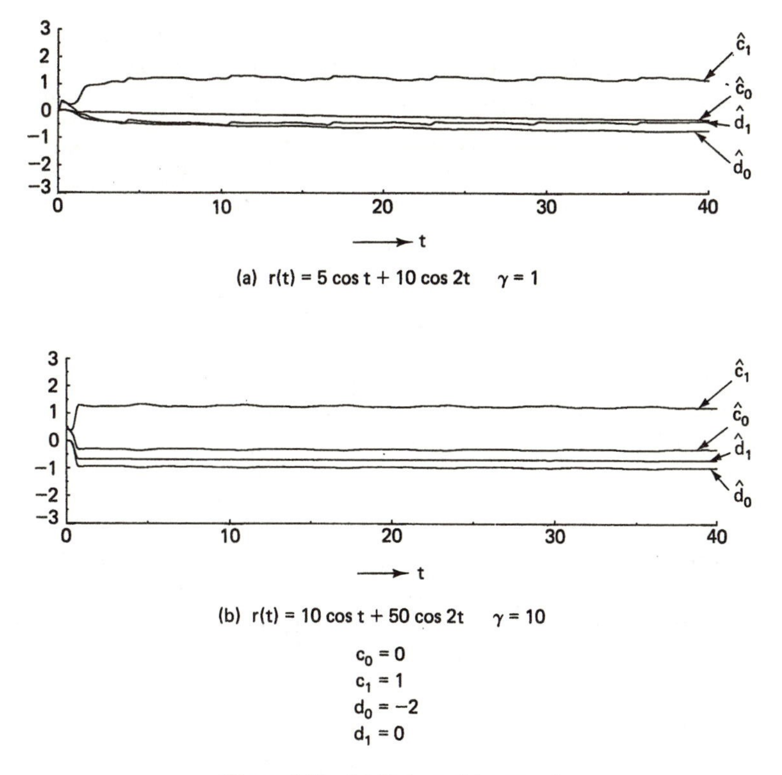

Figure 4.18 Multiple model approach.

evolution of $\|e(t)\| = \|\widehat{Y}(t) - Y(t)\|$ and $\widehat{\theta}(t)$ are shown in Fig. 4.18. In Fig. 4.18(a), $u(t) = 5\cos t + 10\cos 2t$ and $\gamma = 1$ and in Fig. 4.18(b), $u(t) = 10\cos t + 20\cos 2t$ and $\gamma = 10$. It is seen that the parameter estimates converge to their true values more rapidly as γ is increased.

4.4.7 Adaptive Algorithms: An Overview

In Chapter 3 the identification problem was first posed as an algebraic problem in which a parameter vector θ had to be estimated using input-output data. If u is an input vector function, and y_p the corresponding scalar output function, we have the linear equation

$$u^T(t)\theta = y_p(t) \qquad \forall\, t \geq t_0 \tag{4.46}$$

where θ is an unknown n-dimensional vector, $u : \mathbb{R}^+ \to \mathbb{R}^n$ and $y_p : \mathbb{R}^+ \to \mathbb{R}$. From Sections 4.3.3 and 4.3.4, it follows that the plant whose parameters need to be identified using input-output data can also be represented in the form of Eq. (4.46). Therefore, the problem of parameter and state estimation can be reformulated as the problem of estimating θ in Eq. (4.46). Equation (4.46) can be premultiplied on either side by $u(t)$ to obtain

$$u(t)u^T(t)\theta = u(t)y_p(t) \tag{4.47}$$

and represents the basic form used to estimate θ. As mentioned in Chapter 3, this generally involves the minimization of a performance criterion and requires the inversion of a suitable matrix. For example, the minimization of the integral square error

$$\int_{t_0}^{t} \left[y_p(\tau) - \widehat{\theta}^T(t)u(\tau)\right]^2 d\tau$$

leads to the estimate

$$\widehat{\theta}(t) = \left[\int_{t_0}^{t} u(\tau)u^T(\tau)d\tau\right]^{-1} \int_{t_0}^{t} u(\tau)y_p(\tau)d\tau.$$

The algebraic equation (4.47) is also the starting point of all the iterative identification methods described in this chapter and Chapter 3. The adaptive law for adjusting $\widehat{\theta}(t)$ consists of a differential equation of the form

$$\dot{\widehat{\theta}} = f_1(u, y_p, \widehat{y}_p, \widehat{\theta}). \tag{4.48}$$

The right-hand side of Eq. (4.48) is chosen, using only the accessible signals in the system, such that the equilibrium state is stable and corresponds to the desired constant value θ. Alternately, if $\phi(t) = \widehat{\theta}(t) - \theta$, the differential equation (4.48) can also expressed as

$$\dot{\phi} = f_2(u, y_p, \widehat{y}_p, \phi) \tag{4.49}$$

where $\phi = 0$ is an equilibrium state that is stable. When the input u satisfies certain (persistent excitation) conditions, the equilibrium state $\phi = 0$ is also asymptotically stable. The various algorithms described in this chapter can be viewed as different realizations of the differential equation (4.48) [or Eq. (4.49)] so that this can be assured. Hence, if the initial estimate of $\theta(t)$ at time $t = t_0$ is chosen arbitrarily, $\widehat{\theta}(t)$ approaches θ asymptotically in all the cases described.

Since the desired parameter vector θ is a constant, different equations can be generated by premultiplying both sides of Eq. (4.46) by a suitable time-varying function $r(t)$. Then the input-output relation has the form

$$r(t)u^T(t)\theta = r(t)y_p(t) \tag{4.50}$$

and the estimator is chosen as

$$r(t)u^T(t)\widehat{\theta}(t) = r(t)\widehat{y}_p(t).$$

More generally, using a linear operator $\mathcal{L}$ on both sides of Eq. (4.50), the input-output relation is given by

$$\mathcal{L}\left[r(t)u^T(t)\theta\right] = \mathcal{L}[r(t)y_p(t)].$$

The corresponding estimator has the form

$$\mathcal{L}\left[r(t)u^T(t)\right]\widehat{\theta}(t) = \mathcal{L}[r(t)\widehat{y}_p(t)].$$

The adaptive law for the adjustment of $\widehat{\theta}(t)$ is then chosen as

$$\dot{\widehat{\theta}}(t) = -\mathcal{L}\left[r(t)u^T(t)\right]\widehat{\theta}(t) + \mathcal{L}\left[r(t)y_p(t)\right]$$

and, hence, the parameter error $\phi = \widehat{\theta} - \theta$ satisfies the differential equation

$$\dot{\phi}(t) = -\mathcal{L}\left[r(t)u^T(t)\right]\phi(t). \tag{4.51}$$

It is immediately evident that $\phi = 0$ is an equilibrium state of the differential equation (4.51). The operator $\mathcal{L}$ and $r(t)$ must be chosen in such a fashion that this equilibrium state is stable (and uniformly asymptotically stable if the input is persistently exciting). In the following paragraphs, we consider, in succession, the different adaptive algorithms described in this chapter and the corresponding operator $\mathcal{L}$ and function r used in each case.

(a) When $\mathcal{L}$ is the unity operator and $r(t) = u(t)$, $\gamma u(t)$, or $\Gamma u(t)$ (where γ is a positive scalar constant, and Γ is a constant symmetric positive-definite matrix), the adaptive law has the form $\dot{\phi} = -e_1 u$, $\dot{\phi} = -\gamma e_1 u$, or $\dot{\phi} = -\Gamma e_1 u$, respectively. The origin of $\dot{\phi} = -\Gamma e_1 u$ was shown to be stable in Section 4.4.4 by demonstrating that $V(\phi) = \phi^T\Gamma^{-1}\phi$ is a Lyapunov function.

(b) Consider the case where $r(t) = u(t)$ and $\mathcal{L}$ is the integral operator. The input-output relation in this case is given by

$$\int_{t_0}^{t} u(\tau)u^T(\tau)d\tau\,\theta = \int_{t_0}^{t} u(\tau)y_p(\tau)d\tau$$

and yields the (integral) adaptive law

$$\dot{\widehat{\theta}}(t) = -\int_{t_0}^{t} u(\tau)u^T(\tau)d\tau\,\widehat{\theta}(t) + \int_{t_0}^{t} u(\tau)y_p(\tau)d\tau.$$

The parameter error then satisfies the differential equation

$$\dot{\phi}(t) = -\int_{t_0}^{t} u(\tau)u^T(\tau)d\tau\,\phi(t). \tag{4.52}$$

The origin of Eq. (4.52) is stable since the integral is always positive-semidefinite. It can also be shown to be u.a.s. if u is persistently exciting since the integral then becomes positive-definite.

The main drawback of the approach in (b) is that the integral can become unbounded as $t \to \infty$. The hybrid algorithm, the use of a time-varying adaptive gain, the integral algorithm, and the use of multiple inputs that were discussed in this section, may all be considered as efforts to circumvent this difficulty.

(c) In the hybrid algorithm mentioned in Section 4.4.3, $\widehat{\theta}(t)$ is adjusted only at finite intervals. If $\widehat{\theta}(t_i)$ is the estimate at the beginning of a finite interval $[t_i, t_i+T]$, the adaptive law has the form

$$\widehat{\theta}(t_i+T)-\widehat{\theta}(t_i) = -\frac{\Gamma}{T}\int_{t_i}^{t_i+T}\frac{u(\tau)u^T(\tau)}{1+u^T(\tau)u(\tau)}d\tau\,\widehat{\theta}(t_i)+\frac{\Gamma}{T}\int_{t_i}^{t_i+T}\frac{u(\tau)y_p(\tau)}{1+u^T(\tau)u(\tau)}d\tau.$$

Hence, $r(t)=u(t)/[1+u^T(t)u(t)]$ and $\mathcal{L}$ corresponds to the integral between two finite limits. The stability properties of such equations are discussed in Chapter 8.

(d) If $\mathcal{L}$ is unity but $r(t)=S(t)u(t)$ where $S(t)$ is a time-varying matrix, the adaptive law with a time-varying gain results. The adaptive law in this case has the form

$$\dot{\widehat{\theta}}(t) = -S(t)u(t)u^T(t)\widehat{\theta}(t)+S(t)u(t)y_p(t)$$

resulting in the error differential equation

$$\dot{\phi}(t) = -S(t)u(t)u^T(t)\phi(t). \tag{4.53}$$

To assure the stability of Eq. (4.53), $S(t)$ is chosen as

$$S(t) = \left[\int_{t_0}^{t}u(\tau)u^T(\tau)d\tau\right]^{-1}$$

so that $\phi^T(t)S^{-1}(t)\phi(t)$ is positive-definite with a negative-semidefinite time derivative along the trajectories of Eq. (4.53), as shown in Section 4.4.4. Unlike the adaptive law in (b), this adaptive law can be used in practice if

$$\int_{t_0}^{t}uu^T\,d\tau > 0 \qquad \forall\ t\geq t_1,$$

since $S(t)$ acts as a normalizing factor and all the relevant signals are bounded.

(e) If $r(t)=u(t)$ and the linear operator $\mathcal{L}=1/(s+q)$ where q is a positive scalar constant, the integral algorithm in Section 4.4.5 results. In this case, the adaptive law can be shown to have the form

$$\begin{aligned}\dot{\widehat{\theta}} &= -\gamma\Gamma\widehat{\theta}+\gamma\delta & \gamma&>0\\ \dot{\Gamma} &= -q\Gamma+uu^T & \Gamma(t_0)&=0\\ \dot{\delta} &= -q\delta+uy_p & \delta(t_0)&=0\end{aligned}$$

where $\Gamma(t)$ and $\delta(t)$ are bounded.

(f) The multiple model approach described in Section 4.4.6 is somewhat different from the other adaptive laws in that it cannot be derived as a special case of the general law described in the introduction to this section. Instead the roles of r and $\mathcal{L}$ are reversed as shown below. $\mathcal{L}$ is a matrix operator defined as

$$\mathcal{L}u \triangleq I[u\,0\cdots0]+W_1(s)I[0\,u\cdots0]+\cdots+W_{n-1}(s)I[0\cdots0\,u] \triangleq U$$

where I is an $n \times n$ identity matrix and $W_i(s)$, $i = 1, \ldots, n-1$ are $n-1$ distinct asymptotically stable transfer functions. Using Eq. (4.46), we can express the input-output relation as

$$U^T\theta = Y. \tag{4.54}$$

The ith column of the matrix U is $W_{i-1}u$ and the ith element of the vector Y is $W_p(s)W_{i-1}(s)u$, where $W_p(s)$ is the transfer function of the plant. Choosing $r(t) = U(t)$ and premultiplying both sides of Eq. (4.54) by $r(t)$, we obtain

$$UU^T\theta = UY.$$

The adaptive law for updating $\widehat{\theta}(t)$ can be chosen as

$$\dot{\widehat{\theta}} = -\gamma UU^T\widehat{\theta} + \gamma UY \qquad \gamma > 0.$$

The origin $\phi = 0$ of the error equation

$$\dot{\phi} = -\gamma UU^T\phi$$

can be shown to be uniformly stable in the large, and u.a.s.l. if u is persistently exciting.

4.5 SUMMARY

The design of adaptive observers that simultaneously estimate the state variables and parameters of linear time-invariant plants is discussed in this chapter. The Luenberger observer, which can be used to estimate the state variables of the plant when the parameters are known, forms a natural starting point. The design of an adaptive observer includes the choice of a suitable parametrization of the plant as well as stable adaptive laws for the adjustment of its parameters. Minimal order observers are first discussed and are shown to be stable but unwieldy. Nonminimal observers are then discussed, which are shown to have a simpler structure. Using two different nonminimal representations of the plant, adaptive observers are constructed whose parameters can be adjusted using adaptive laws that assure stability in the large. Design issues related to the convergence of parameters are studied mainly through simulations, and various modifications of the adaptive laws are suggested to improve the rate of convergence.

PROBLEMS

Part I

1. A linear time-invariant system has the transfer function

$$W(s) = \frac{b_1 s + b_2}{s^2 + a_1 s + a_2}.$$

Represent the plant in the following canonical forms:

(a) $\dot{x} = Ax + bu; \qquad y = h^T x$ where $a = [a_1, \ldots, a_n]^T$,

$$A = \left[\begin{array}{c|c} -a & \begin{array}{c} I \\ -- \\ 0 \end{array} \end{array} \right]$$

and $h^T = [1, 0, \ldots, 0]$.

(b) h is the same as in (a), $\alpha = [\alpha_1, \ldots, \alpha_n]^T$,

$$A = \left[\begin{array}{c|ccccc} & 1 & 1 & 1 & \cdots & 1 \\ -\alpha & & & & & \\ & & & \Lambda & & \end{array} \right],$$

where $\Lambda = \lambda I$.

(c) nonminimal representation 1.

(d) nonminimal representation 2.

2. Determine an adaptive observer for identifying

(a) the unknown parameter vectors a and b in Problem 1(a).

(b) the unknown parameters in Problem 1(d).

3. Determine the structure of an adaptive observer to estimate the unknown parameters a_1 and a_2 of the transfer function $W(s)$ in Problem 1 when b_1 and b_2 are known.

Part II

4. For the plant described in Problem 1, design a minimal order observer. In particular, determine ω, the auxiliary signals v_1 and v_2, and the adaptive laws for adjusting the parameters of the observer.

5. A nonminimal observer of the form given in Problem 1(d) is used to estimate the state x of a dynamical system. While $x(t) \in \mathbb{R}^n$, the state of the observer $x_{np}(t) \in \mathbb{R}^{2n}$. Show that

$$x(t) = Cx_{np}(t)$$

where C is a constant $(n \times 2n)$ matrix.

6. Show that

$$(sI - K)^{-1} e_i = T_i (sI - K^T)^{-1} e_1, \qquad i = 1, \ldots, n$$

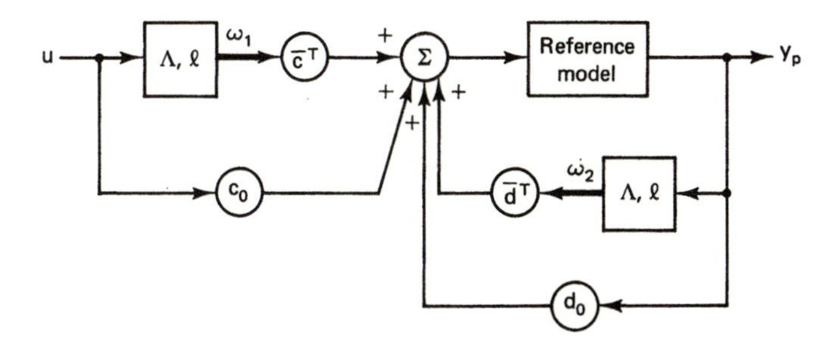

Figure 4.19 A nonminimal representation of an LTI plant.

where e_i is a unit vector $[0 \cdots 0\ 1\ 0 \cdots 0]^T$, which contains unity as its ith element and T_i are constant matrices whose elements are the coefficients of the numerator polynomials of $(sI - K)^{-1}e_i$.

7. A linear time-invariant plant has a transfer function $W_p(s)$. The reference model chosen has a transfer function $W_m(s)$ with the same relative degree as the plant. Show that the plant can be parametrized as shown in Fig. 4.19.

Part III

8. If an LTI plant with unknown parameters is unstable, can its states and parameters be estimated in a stable fashion? Justify your answer.

9. In Chapter 3, adaptive laws for controlling an LTI plant with unknown parameters were given when its entire state vector can be measured. In this chapter, adaptive observers for estimating the state of an unknown plant from input-output data were described. Using the results of the two chapters, can an unknown plant be controlled in a stable fashion?

REFERENCES

1. Åström, K.J., Borisson, U., Ljung, L., and Wittenmark, B. "Theory and applications of self-tuning regulators." *Automatica* 13:457–476, 1977.
2. Carroll, R., and Lindorff, D.P. "An adaptive observer for single-input single-output linear systems." *IEEE Transactions on Automatic Control* 18:428–435, Oct. 1973.
3. Chen, C.T. *Introduction to linear system theory.* New York, NY:Holt, Rinehart and Winston, Inc., 1970.
4. Dugard, L., and Goodwin, G.C. "Global convergence of Landau's 'output error with adjustable compensator' adaptive algorithm." *IEEE Transactions on Automatic Control* 30:593–595, 1985.
5. Gopinath, B. "On the control of linear multiple input-output systems." *Bell Syst. Tech. Journal* 50:1063–1081, March 1971.
6. Kim, C. *Convergence studies for an improved adaptive observer.* Ph.D. thesis, University of Connecticut, 1975.
7. Kreisselmeier, G. "Adaptive observers with exponential rate of convergence." *IEEE Transactions on Automatic Control* 22:2–8, 1977.

8. Kudva, P., and Narendra, K.S. "Synthesis of an adaptive observer using Lyapunov's direct method." *Int. Journal of Control* 18:1201–1210, 1973.
9. Landau, I.D. *Adaptive control.* New York:Marcel Dekker, Inc., 1979.
10. Landau, I.D. "Elimination of the real positivity condition in the design of parallel MRAS." *IEEE Transactions on Automatic Control* 23:1015–1020, 1978.
11. Lion, P.M. "Rapid identification of linear and nonlinear systems." *AIAA Journal* 5:1835–1842, 1967.
12. Luders, G., and Narendra, K.S. "A new canonical form for an adaptive observer." *IEEE Transactions on Automatic Control* 19:117–119, April 1974.
13. Luders, G., and Narendra, K.S. "An adaptive observer and identifier for a linear system." *IEEE Transactions on Automatic Control* 18:496–499, Oct. 1973.
14. Narendra, K.S., and Kudva, P. "Stable adaptive schemes for system identification and control - Parts I & II." *IEEE Transactions on Systems, Man and Cybernetics* SMC-4:542–560, Nov. 1974.

5

The Control Problem

5.1 INTRODUCTION

The control of dynamical systems, whose characteristics are not completely known, has been the goal of control theorists and practitioners ever since the term adaptive control was first introduced twenty-five years ago. This chapter, which is devoted entirely to the discussion of the control problem, naturally occupies a central place in this book. As stated in Chapter 1, to keep the problem analytically tractable, its scope has to be restricted to include mainly linear time-invariant plants with unknown parameters. It is well known from linear theory that when the parameters of such systems are known, the theoretical basis of control using only input-output data is found in the separation principle [1]. According to this principle, if the state estimates of the system are determined using a Luenberger observer, a linear combination of these estimates can be fed back to control it in a stable fashion (refer to Chapter 4). Hence, the control problem consists of realizing an observer and a feedback controller independently.

In the early days of adaptive control it was felt that a similar separation principle could be derived in the adaptive context as well. The results reported in Chapter 3, for example, indicate that linear time-invariant plants with unknown parameters can be adaptively controlled in a stable fashion when their state vectors can be measured. The contents of Chapter 4 revealed that state variables of such plants can be reconstructed from input-output data using stable adaptive observers. Hence, attempts were made to solve the adaptive control problem by combining the results of Chapters 3 and 4. Contrary to expectations, it was soon realized that the problem was more involved. The principal difficulty lay in the assumption, made in the design of the adaptive observer, that the input and the output of the observed plant are uniformly bounded. This assumption is not valid in the control problem, since it is precisely the question of stability of the overall feedback system that has to be resolved in this case. The period 1975-1980 consequently witnessed intense activity in the area and when the problem was finally resolved in 1980, it represented an important step in the evolution of the field of adaptive control.

As in the previous chapters, the solution to the problem above can also be conveniently divided into algebraic and analytic parts. The choice of the reference model as well as the controller structure is important in assuring the existence of a solution to the algebraic part of the problem. A linear time-invariant controller exists so that the transfer function of the plant, together with the controller, is identical to that of the reference model, resulting in the error between the plant and model outputs approaching zero asymptotically. Since the plant parameters are assumed to be unknown, the control parameters are only known to exist but cannot be determined a priori. The analytic part then deals with the manner in which the control parameters are to be adjusted so they evolve toward the desired values, and the overall system has globally bounded solutions.

The adaptive control problem for the case where the relative degree n^* of the plant [(number of poles − number of zeros) of the plant transfer function] is unity was solved by many authors in the 1970s. In Section 5.3, we treat this in some detail. The motivation for the choice of the controller structure as well as the adaptive laws is relatively transparent in this case and is based on the positive realness of the transfer function of the reference model. In contrast to this, the general adaptive control problem of a plant with $n^* \geq 2$ is considerably more complex. It is therefore not surprising that, for several years after the solution to the problem with $n^* = 1$ was obtained, this general case remained unresolved. Modifications in the structure of the controller, as well as changes in the adaptive law, were needed for a complete solution.

5.2 STATEMENT OF THE PROBLEM

The problem of controlling a plant P with unknown parameters can be conveniently divided into regulation and tracking problems. In the former, the main objective is to stabilize P, using input-output data, around a fixed operating point, while in the latter, the aim is to make the model output follow a desired signal asymptotically. These may be stated as follows:

Regulation. Let P be represented by a linear time-invariant differential equation

$$\begin{aligned} \dot{x}_p &= A_p x_p + b_p u \\ y_p &= h_p^T x_p \end{aligned} \tag{5.1}$$

where $u : \mathbb{R}^+ \to \mathbb{R}$ is the input, $y_p : \mathbb{R}^+ \to \mathbb{R}$ is the output, and $x_p : \mathbb{R}^+ \to \mathbb{R}^n$ is the n-dimensional state vector of the plant. Let the triple

$$\{h_p^T, A_p, b_p\} \quad \text{be observable and controllable} \tag{5.2}$$

and have unknown elements. Using only input-output data, determine a control function u, using a differentiator free controller, which stabilizes the overall system.

Tracking. Assuming that $y_m(t)$ is a uniformly bounded desired output signal, determine a bounded input u in Eq. (5.1) using a differentiator free controller such that $\lim_{t\to\infty}|y_p(t) - y_m(t)| = 0$.

The regulation problem, as stated above, calls for a nontrivial solution only when the plant is unstable. Hence, the instability of the matrix A_p can also be explicitly assumed in the statement of the problem. It should be added that, even if the matrix A_p is stable, the overall system may become unstable as the adaptation proceeds. Hence, the two problems are equivalent, if it is assumed that the designer is not aware of the stability or instability of A_p when the adaptive process is started. If, however, the designer is aware that A_p is stable, the nature of the problem changes significantly (See Comment 5.9). We shall focus our attention on the former problem in the sections following.

For the regulation problem, a stabilizing controller is known to exist when the plant parameters are known. In contrast to this, the existence of a realizable controller to achieve perfect tracking is not evident immediately. More precisely, the class of functions y_m for which the requirement that $\lim_{t\to\infty}|y_p(t) - y_m(t)| = 0$ can be satisfied must be specified, if the tracking problem is to be well posed. This is an important question in its own right and was discussed in Chapter 1. To make it analytically tractable we assume that the desired signal y_m is the output of a reference model with a transfer function $W_m(s)$ whose input r is a uniformly bounded piecewise continuous function of time, and determine conditions on $W_m(s)$ so that perfect tracking is possible. It is well known, from the theory of linear systems, that to achieve perfect tracking, the relative degree of the reference model must be greater than or equal to that of the plant. Hence, the tracking problem can be specialized to the model reference adaptive control (MRAC) problem stated below:

MRAC. Let a linear time-invariant reference model have an asymptotically stable transfer function $W_m(s)$ with an input-output pair $\{r(\cdot), y_m(\cdot)\}$ where r is a uniformly bounded piecewise-continuous function of time. Let a linear time-invariant plant be described by Eq. (5.1) and have a transfer function

$$W_p(s) = k_p \frac{Z_p(s)}{R_p(s)} \tag{5.3}$$

where

(i) the sign of the high frequency gain k_p,

(ii) an upper bound on the order n, and

(iii) the relative degree n^* of $W_p(s)$ are known, and (I)

(iv) the zeros of the monic polynomial $Z_p(s)$ lie in $\mathbb{C}^-$.

Let the relative degree n_m^* of $W_m(s)$ be greater than or equal to that of the plant, that is, $n_m^* \geq n^*$. Determine a differentiator free controller that generates a bounded control input u, so that all the signals in the closed loop system remain bounded and

$$\lim_{t\to\infty} |y_p(t) - y_m(t)| = 0.$$

When $r(t) \equiv 0$ for all $t \geq t_0$ and the initial conditions of the reference model are also zero, $y_m(t) \equiv 0$. The tracking problem described in Section 5.3 is then specialized to the regulation problem. Hence, the methods developed in the following sections for the former are applicable to the latter problem as well.

As stated earlier, the adaptive systems discussed here, as in the previous chapters, consist of algebraic and analytic parts. Assumptions (ii) and (iii) in (I) are found to be relevant for the former and assure the existence of a solution. Assumptions (i) and (iv) are related to the analytic part and are required to generate stable adaptive laws.

Even the statement of the MRAC problem shows a certain amount of hindsight. In fact, prior to its solution in 1980, the assumptions (i)-(iv) were not stated explicitly. It was only after the problem was resolved that their importance was realized. They are included here in the statement of the problem to emphasize the importance of prior information in the search for a solution. The implications of (I) will become transparent during the course of the discussions in Sections 5.3 and 5.4, where the solutions to the MRAC problem for $n^* = 1$ and $n^* \geq 2$ are presented. Although they are found to be sufficient to assure global stability of the system, whether they are also necessary is discussed in Chapter 9.

5.3 ADAPTIVE CONTROL OF PLANTS WITH RELATIVE DEGREE UNITY

Let the plant P to be controlled be described by the differential equation (5.1) and have a transfer function $W_p(s)$ given by Eq. (5.3), where $Z_p(s)$ and $R_p(s)$ are monic polynomials of order $n-1$ and n respectively[1]. From the assumption in Eq. (5.2) it also follows that $Z_p(s)$ and $R_p(s)$ are coprime polynomials. The reference model is assumed to be of order n with a strictly positive real transfer function

$$W_m(s) = k_m \frac{Z_m(s)}{R_m(s)} \tag{5.4}$$

[1]The same analysis applies when n denotes an upper bound on the order of the plant [see Section 5.7(e)].

and $Z_m(s)$ and $R_m(s)$ are monic Hurwitz polynomials of degrees $n-1$ and n respectively, and k_m is positive. The objective, as stated in Section 5.2, is to determine the input u to the plant so that its output $y_p(t)$ approaches $y_m(t)$ asymptotically.

The simplest solution to the problem above would consist of a cascade controller so that

$$u(t) = W_c(s)r(t)$$

where

$$W_c(s) = \frac{k_m}{k_p} \cdot \frac{Z_m(s)R_p(s)}{Z_p(s)R_m(s)},$$

and the resultant proper transfer function would be identical to that of the reference model. This solution is feasible only when $R_p(s)$ is a Hurwitz polynomial, so that any pole-zero cancellation occurs in $\mathbb{C}^-$. However, it suffers from the usual drawbacks of open-loop control. Hence a controller structure with feedback, which becomes mandatory when $R_p(s)$ is not Hurwitz, has to be used in practical design. In the following paragraphs, we consider three simple specific cases [15] where the controller structure and adaptive laws can be determined in a straightforward fashion. These cases correspond to different prior information available to the designer regarding the plant and are given below:

(i) The transfer function of the plant is known except for the high frequency gain k_p, and $Z_p(s) = Z_m(s)$, $R_p(s) = R_m(s)$.

(ii) The denominator polynomial $R_p(s)$ is known, $R_p(s) = R_m(s)$ but the numerator polynomial $k_pZ_p(s)$ is unknown. $Z_p(s)$ is however known to be Hurwitz.

(iii) The monic polynomial $Z_p(s)$ is Hurwitz, known, and $Z_p(s) = Z_m(s)$, while k_p and $R_p(s)$ are unknown.

The adaptive laws for the special cases above, as well as the general case discussed in Section 5.3.4, when $n^* = 1$, can be derived using Lemma 5.1:

Lemma 5.1 Consider a dynamical system described by the equation

$$\begin{aligned} \dot{x}(t) &= Ax(t) + b\phi^T(t)\mathrm{w}(t); \qquad y = h^T x(t) \\ z_1(t) &= ky(t) \end{aligned} \tag{5.5}$$

where (A, b) is stabilizable, (h^T, A) is detectable, and

$$h^T(sI - A)^{-1}b \triangleq H(s) \text{ is a strictly positive real transfer function.}$$

Further let $\phi : \mathbb{R}^+ \to \mathbb{R}^m$ be a vector of adjustable parameters, k an unknown constant with a known sign, and w: $\mathbb{R}^+ \to \mathbb{R}^m$ and $z_1 : \mathbb{R}^+ \to \mathbb{R}$ time-varying functions that can be measured. If $\phi(t)$ is adjusted as

$$\dot{\phi}(t) = -sgn(k)z_1(t)\mathrm{w}(t) \tag{5.6}$$

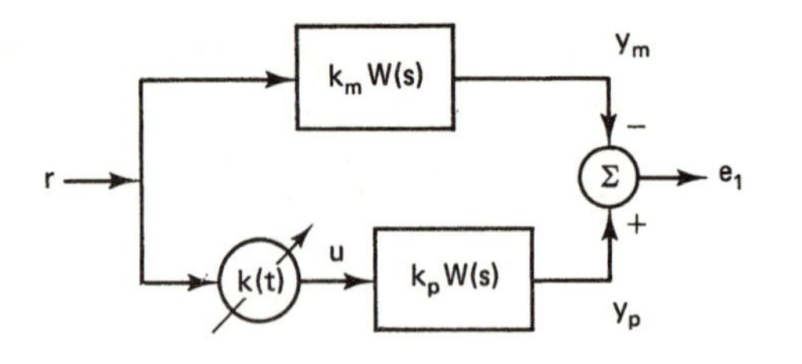

Figure 5.1 k_p unknown.

then the equilibrium state ($x = 0, \phi = 0$) of Eqs. (5.5) and (5.6) is uniformly stable in the large.

Proof. Since $H(s)$ is SPR, it follows from Lemma 2.4 (MKY) that, given a matrix $Q = Q^T > 0$, there exists a $P = P^T > 0$ such that

$$\begin{aligned} A^T P + PA &= -Q \\ Pb &= h. \end{aligned}$$

Let V be a positive-definite function of the form

$$V = x^T P x + (1/|k|)\phi^T \phi.$$

Its time derivative $\dot{V}$, evaluated along the solutions of Eqs. (5.5) and (5.6), leads to the equation

$$\begin{aligned} \dot{V} &= x^T(PA + A^T P)x + 2x^T P b \phi^T \mathrm{w} - 2\phi^T y \mathrm{w} \\ &= -x^T Q x \leq 0. \end{aligned}$$

Therefore the origin of Eqs. (5.5) and (5.6) is globally stable.

Comment 5.1 In Lemma 5.1 we only require that (A, b) be stabilizable and (h^T, A) be detectable. This implies that uncontrollable and unobservable modes are stable. Hence, if the system in Eq. (5.5) is of order n, the transfer function $H(s)$ is of order $\leq n$. Lemma 2.4, which assures the existence of an $(n \times n)$ matrix P, is needed in this case. If, however (h^T, A, b) is controllable and observable, Lemma 2.3 (LKY) can be applied. In the three cases discussed below, Lemma 2.3 is adequate for case (i), while Lemma 2.4 is needed in the last two cases.

5.3.1 Case (i) k_p Unknown

Let the transfer functions of the plant and the reference model differ only in their high frequency gains. An adjustable parameter $k(t)$ is included in series with the plant as shown in Fig. 5.1 and the control input into the plant is chosen as

$$u(t) = k(t)r(t).$$

If $W(s) = Z_m(s)/R_m(s)$, the outputs y_p and y_m of the plant and the model respectively can be expressed as

$$\begin{aligned} y_p(t) &= W(s)k_p k(t)r(t) \\ y_m(t) &= W(s)k_m r(t). \end{aligned}$$

Defining the error between the plant and model outputs as e_1 and the parameter error as ψ, we have

$$e_1(t) \stackrel{\Delta}{=} y_p(t) - y_m(t) \qquad \text{and} \qquad \psi(t) \stackrel{\Delta}{=} k(t) - \frac{k_m}{k_p}.$$

The error equation can then be derived as

$$e_1(t) = W(s)k_p\psi(t)r(t). \tag{5.7}$$

Since $W_m(s)$ is strictly positive real and k_m is positive, $W(s)$ is also strictly positive real. Hence Lemma 5.1 can be used to obtain the adaptive law

$$\dot{k}(t) = -sgn(k_p)e_1(t)r(t),$$

and it follows that the state $e(t)$ of the dynamical system described in Eq. (5.7) and $\phi(t)$ are bounded for all $t \geq t_0$. Since r is uniformly bounded, it follows from Eq. (5.7) that $\dot{e}$ is bounded. Hence, from Lemma 2.12, we have $\lim_{t\to\infty} |y_p(t) - y_m(t)| = 0$.

Comment 5.2 The problem considered in this section is similar to the identification problem of a single parameter considered in Section 4.3. The role of the plant and the model are, however, reversed. In the present context the adjustable parameter $k(t)$ is in series with the plant and, hence, unlike in Section 4.3, $sgn(k_p)$ is needed for the solution.

5.3.2 Case (ii) Unknown Zeros

The plant to be controlled is described by the differential equation (5.1) and has a transfer function $W_p(s)$ given by Eq. (5.3). The denominator polynomial $R_p(s)$ is assumed to be Hurwitz and known, while the numerator polynomial $k_p Z_p(s)$ is Hurwitz but unknown. A reference model is chosen so that $R_m(s) = R_p(s)$, that is, the poles of the model are the same as those of the plant. Since the zeros of the plant transfer function cannot be altered by state feedback, the controller is chosen as a prefilter in series with the plant. The aim is to parametrize the prefilter in such a fashion that its transfer function can be made equal to

$$W_c^*(s) \stackrel{\Delta}{=} \frac{k_m}{k_p}\frac{Z_m(s)}{Z_p(s)} \tag{5.8}$$

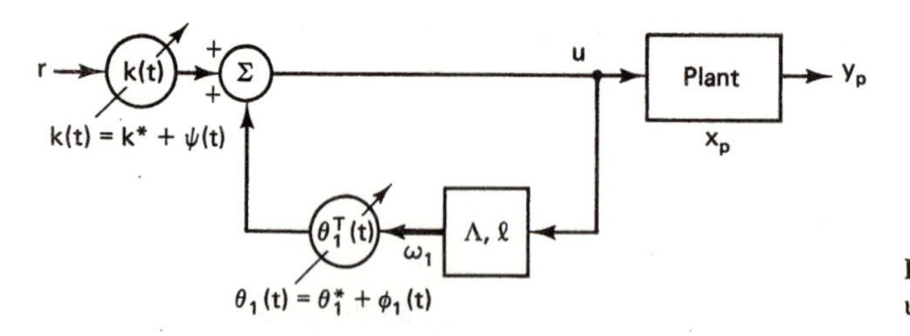

Figure 5.2 The adaptive controller with unknown zeros.

for a suitable choice of the control parameters. This implies that the plant together with the controller has an asymptotically stable transfer function with n poles, while $n-1$ zeros of the plant transfer function in $\mathbb{C}^-$ are cancelled by the $n-1$ poles of $W_c^*(s)$. Hence, the overall system is of dimension $2n-1$, of which $n-1$ modes are unobservable, but stable.

Although many realizations of the controller structure are possible, the one chosen must permit stable adaptive laws to be designed. Since the controller must be differentiator free, have a transfer function with predetermined zeros, and enable the poles to be located arbitrarily, it is chosen to have the structure shown in Fig. 5.2. As seen from the figure, the controller consists of a single parameter $k(t)$ in series with a feedback loop that contains a unity gain in the feedforward path and a dynamical system in the feedback path. The differential equation representing the controller is therefore given by

$$\begin{aligned} \dot{\omega}_1(t) &= \Lambda\omega_1(t) + \ell u(t) \\ u(t) &= \theta_1^T(t)\omega_1(t) + k(t)r(t) \end{aligned} \tag{5.9}$$

where Λ is an $(n-1)\times(n-1)$ asymptotically stable matrix, $\det(sI-\Lambda) = \lambda(s)$, and (Λ, ℓ) is controllable. When $k(t) \equiv k_c$ and $\theta(t) \equiv \theta_{1c}$, where $k_c \in \mathbb{R}$ and $\theta_{1c} \in \mathbb{R}^{n-1}$, the transfer function of the controller can be expressed as

$$W_c(s) = k_c \frac{\lambda(s)}{\lambda(s) - C(s)}$$

where

$$\frac{C(s)}{\lambda(s)} = \theta_{1c}^T (sI - \Lambda)^{-1} \ell.$$

Let Λ be chosen so that $\lambda(s) = Z_m(s)$. Since the polynomial $C(s)$ depends on the vector θ_{1c}, a specific value θ_1^* of θ_{1c} can be chosen such that the corresponding value of $C(s)$ is $C^*(s)$ where $C^*(s) = Z_m(s) - Z_p(s)$. If $k^* = k_m/k_p$, then by choosing $k_c = k^*$ and $\theta_{1c} = \theta_1^*$, the desired transfer function in Eq. (5.8) of the controller can be realized.

With these parameter values, let the reference model be expressed as shown in Fig. 5.3 where the signals corresponding to $\omega_1(t)$ and $x_p(t)$ are given by $\omega_1^*(t)$ and $x_p^*(t)$, respectively. The analysis above confirms the existence of a constant vector $\overline{\theta}_1^*$ defined as

$$\overline{\theta}_1^* \triangleq [k^*, \theta_1^{*T}]^T$$

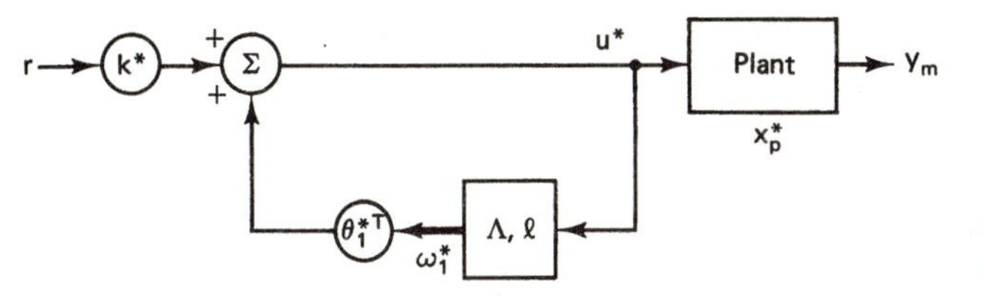

Figure 5.3 Nonminimal representation of the model with unknown zeros.

such that the objective of the control problem can be realized when the plant parameters are known. In the adaptive case, the parameters k_c and θ_{1c} are replaced by time-varying parameters $k(t)$ and $\theta_1(t)$, and the aim is to determine the rules by which they are to be adjusted so as to result in a stable system. Defining

$$\psi(t) \triangleq k(t) - k^*, \qquad \phi_1(t) \triangleq \theta_1(t) - \theta_1^*, \qquad \overline{\theta}_1(t) \triangleq [k(t), \theta_1(t)]^T,$$
$$\overline{\phi}_1(t) \triangleq [\psi(t), \phi_1^T(t)]^T, \quad \text{and} \quad \overline{\omega}_1(t) \triangleq [r(t), \omega_1^T(t)]^T,$$

the equations describing the overall system and the model can be expressed respectively as

$$\begin{bmatrix} \dot{x}_p \\ \dot{\omega}_1 \end{bmatrix} = \begin{bmatrix} A_p & b_p\theta_1^{*T} \\ 0 & \Lambda + \ell\theta_1^{*T} \end{bmatrix} \begin{bmatrix} x_p \\ \omega_1 \end{bmatrix} + \begin{bmatrix} k^* b_p \\ k^*\ell \end{bmatrix} r + \begin{bmatrix} b_p \\ \ell \end{bmatrix} \overline{\phi}_1^T \overline{\omega}_1$$
$$y_p = [h_p^T, 0]x, \qquad x \triangleq [x_p^T, \omega_1^T]^T \tag{5.10}$$

and

$$\begin{bmatrix} \dot{x}_p^* \\ \dot{\omega}_1^* \end{bmatrix} = \begin{bmatrix} A_p & b_p\theta_1^{*T} \\ 0 & \Lambda + \ell\theta_1^{*T} \end{bmatrix} \begin{bmatrix} x_p^* \\ \omega_1^* \end{bmatrix} + \begin{bmatrix} k^* b_p \\ k^*\ell \end{bmatrix} r \tag{5.11}$$
$$y_m = [h_p^T, 0]x_{mn}, \qquad x_{mn} \triangleq [x_p^{*T}, \omega_1^{*T}]^T.$$

From Eqs. (5.10) and (5.11), the equations describing the error between the system and the model are given by

$$\dot{e} = A_{mn}e + b_{mn}\overline{\phi}_1^T\overline{\omega}_1; \qquad e_1 = h_{mn}^T e$$

where

$$A_{mn} = \begin{bmatrix} A_p & b_p\theta_1^{*T} \\ 0 & \Lambda + \ell\theta_1^{*T} \end{bmatrix}, \quad b_{mn} = \begin{bmatrix} b_p \\ \ell \end{bmatrix}, \quad h_{mn} = \begin{bmatrix} h_p \\ 0 \end{bmatrix},$$

$e = x - x_{mn}$ and $e_1 = y_p - y_m$. From Fig. 5.3, it follows that

$$h_{mn}^T(sI - A_{mn})^{-1}b_{mn}k^* = W_m(s).$$

Therefore we have

$$e_1 = \frac{k_p}{k_m} W_m(s)\overline{\phi}_1^T\overline{\omega}_1. \tag{5.12}$$

Equation (5.12) is in the standard form treated in Lemma 5.1 and, hence, an adaptive law

$$\dot{\overline{\phi}}_1 = -sgn(k_p)e_1\overline{\omega}_1$$

assures the boundedness of $e(t)$ and $\overline{\phi}_1(t)$ for all $t \geq t_0$.

When $\overline{\theta}_1(t) \equiv \overline{\theta}_1^*$, from the controller structure, it follows that $\overline{\omega}_1^*(t)$ is bounded. Further, since $e(t)$ is bounded, $\overline{\omega}_1(t)$ is also bounded. Therefore it can once again be shown that $\lim_{t\to\infty} e_1(t) = 0$.

Comment 5.3 The plant together with the adaptive controller in Fig. 5.2 can be expressed in terms of the LTI system shown in Fig. 5.3 together with an input $\overline{\phi}_1^T \overline{\omega}_1$ at the summer. This yields the error equation (5.12).

Comment 5.4 The choice of the controller ensures that the poles of the controller cancel the zeros of the plant to obtain the transfer function of the reference model. Hence, the zeros of the plant should necessarily lie in $\mathbb{C}^-$ so that no unstable pole-zero cancellations can take place. The assumption that $Z_p(s)$ be Hurwitz is found to be common to all the adaptive systems discussed in this chapter.

Comment 5.5 The controller described in Eq. (5.9) has a feedback structure. To avoid algebraic loops, the dynamical system in the feedback path should have a strictly proper transfer function (for constant values of θ_1). This accounts for the fact that although the coefficients of the denominator polynomial of $W_c(s)$ are determined by the parameters in the feedback path, its high frequency gain is realized in the forward path.

5.3.3 Case (iii) Unknown Poles

In this section, we consider a plant whose transfer function $W_p(s)$ is given in Eq. (5.3) with $Z_p(s) = Z_m(s)$, a specified Hurwitz polynomial of degree $n-1$, and $R_p(s)$ is an nth degree monic polynomial with unknown coefficients. Since the roots of $R_p(s)$ can lie in $\mathbb{C}^+$, a cascade controller of the form of Eq. (5.9), considered in the previous section, cannot be used. Instead, the plant is stabilized using a feedback controller defined by

$$\begin{aligned} \dot{\omega}_2(t) &= \Lambda\omega_2(t) + \ell y_p(t) \\ u(t) &= k(t)r(t) + \theta_0(t)y_p(t) + \theta_2^T(t)\omega_2(t) \end{aligned} \tag{5.13}$$

where $\theta_0 : \mathbb{R}^+ \to \mathbb{R}$, $\omega_2, \theta_2 : \mathbb{R}^+ \to \mathbb{R}^{n-1}$, Λ is an $(n-1) \times (n-1)$ asymptotically stable matrix with $\lambda(s)$ as its characteristic polynomial, and (Λ, ℓ) is controllable (Fig. 5.4). When $\theta_0(t) \equiv \theta_{0c}$ and $\theta_2(t) \equiv \theta_{2c}$, where θ_{0c} and θ_{2c} are constants, the transfer function of the feedback controller can be written as

$$W_c(s) = \theta_{0c} + \theta_{2c}^T(sI - \Lambda)^{-1}\ell \triangleq \frac{D(s)}{\lambda(s)}.$$

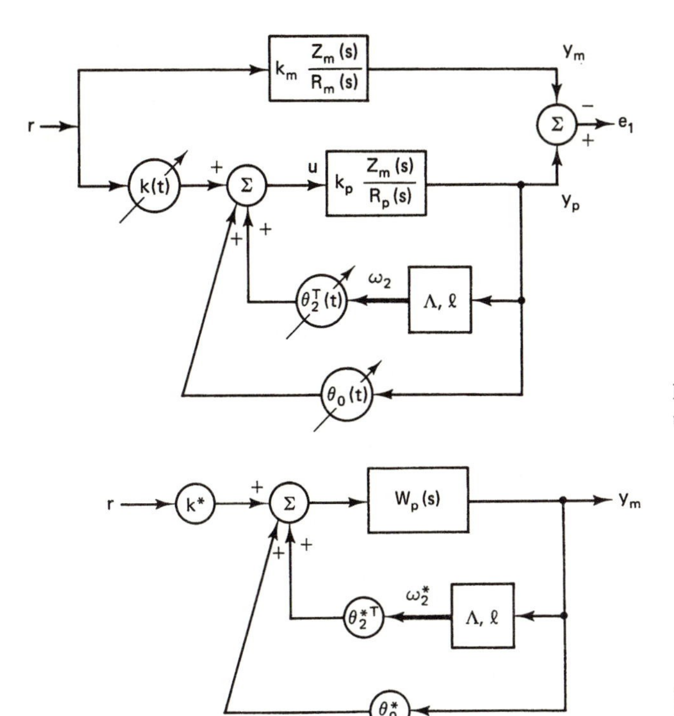

Figure 5.4 The adaptive controller with unknown poles.

Figure 5.5 Nonminimal representation of the model with unknown poles.

If $k(t) \equiv k_c$, where k_c is a constant, the overall transfer function $W_o(s)$ of the plant together with the controller is then given by

$$W_o(s) \triangleq k_c \frac{W_p(s)}{1 - W_p(s)W_c(s)} = \frac{k_c k_p Z_m(s)}{R_p(s)\left[1 - k_p \frac{Z_m(s)}{R_p(s)} \cdot \frac{D(s)}{\lambda(s)}\right]}.$$

Since the matrix Λ [and hence the polynomial $\lambda(s)$] is at the discretion of the designer, it is chosen so that $\lambda(s) = Z_m(s)$. Since $R_p(s)$ and $R_m(s)$ are assumed to be monic, there exists a parameter vector $[\theta_0^*, \theta_2^{*T}]$ such that if $\theta_{0c} = \theta_0^*$ and $\theta_{2c} = \theta_2^*$, then $D(s) = D^*(s)$ and

$$R_p(s) - k_p D^*(s) = R_m(s).$$

If in addition,

$$k_c = k^* \triangleq \frac{k_m}{k_p},$$

the closed-loop transfer function $W_o(s)$ can be made identical to the transfer function $W_m(s)$ (Fig. 5.5). Once again, it is to be noted that $n-1$ hidden modes resulting from pole-zero cancellation are stable, since $Z_m(s)$ is a Hurwitz polynomial.

In the adaptive case, when k, θ_0, and θ_2 are time-varying as in Eq. (5.13), the parameter errors are defined as $\psi(t) = k(t) - k^*$, $\phi_0(t) = \theta_0(t) - \theta_0^*$, and $\phi_2(t) = \theta_2(t) - \theta_2^*$.

For convenience, we shall represent the entire $(n+1)$-dimensional control parameter error vector as $\overline{\phi}_2(t)$, where

$$\overline{\phi}_2(t) = [\psi(t),\ \phi_0(t),\ \phi_2^T(t)]^T.$$

Using the same procedure as in Section 5.3.2, the output of the plant can then be expressed as

$$y_p(t) = \frac{k_p}{k_m} W_m(s)\left[k^* r(t) + \overline{\phi}_2^T(t)\overline{\omega}_2(t)\right]$$

where $\overline{\omega}_2(t) = [r(t),\ y_p(t),\ \omega_2^T(t)]^T$. Since $y_m(t) = W_m(s)r(t)$, the output error $e_1(t) = y_p(t) - y_m(t)$ satisfies the differential equation

$$e_1(t) = \frac{k_p}{k_m} W_m(s)\overline{\phi}_2^T(t)\overline{\omega}_2(t). \tag{5.14}$$

By Lemma 5.1, we infer that adaptive laws of the form

$$\begin{aligned} \dot{k}(t) = \dot{\psi}(t) &= -sgn(k_p)e_1(t)r(t) \\ \dot{\theta}_0(t) = \dot{\phi}_0(t) &= -sgn(k_p)e_1(t)y_p(t) \\ \dot{\theta}_2(t) = \dot{\phi}_2(t) &= -sgn(k_p)e_1(t)\omega_2(t) \end{aligned} \tag{5.15}$$

assure that the solutions of Eqs. (5.14) and (5.15) are bounded. Since e_1 as well as the output of the reference model are bounded, it follows that y_p is bounded. Hence, the $(n-1)$-dimensional vector ω_2 is bounded. As in all the previous cases, we can once again show that $\lim_{t\to\infty} |y_p(t) - y_m(t)| = 0$.

The adaptive laws suggested in Sections 5.3.1-5.3.3 are summarized in Table 5.1.

5.3.4 General Case ($n^* = 1$)

The discussions of the three special cases in the previous sections now permit us to present the controller structure and the solution to the adaptive control of a single-input single-output (SISO) plant whose relative degree is unity [14]. We assume as in the three special cases in Sections 5.3.1-5.3.3, that the plant is described by the equations

$$\dot{x}_p = A_p x_p + b_p u; \qquad y_p = h_p^T x_p$$

where the constant matrix A_p and vectors b_p and h_p have unknown elements. If the transfer function of the plant is given by

$$W_p(s) = k_p \frac{Z_p(s)}{R_p(s)} = h_p^T \left(sI - A_p\right)^{-1} b_p,$$

TABLE 5.1

Unknown quantities	Adjustable parameter	Fixed parameter	Adaptive law	Desired parameter
k_p	k	$\theta_1 = \theta_2 = 0$ $\theta_0 = 0$	$\dot{k} = -sgn(k_p)e_1 r$	k^*
k_p, zeros	k, θ_1	$\theta_2 = 0$ $\theta_0 = 0$	$\dot{k} = -sgn(k_p)e_1 r$ $\dot{\theta}_1 = -sgn(k_p)e_1\omega_1$	k^*, θ_1^*
k_p, poles	k, θ_0, θ_2	$\theta_1 = 0$	$\dot{k} = -sgn(k_p)e_1 r$ $\dot{\theta}_0 = -sgn(k_p)e_1 y_p$ $\dot{\theta}_2 = -sgn(k_p)e_1\omega_2$	$k^*, \theta_0^*, \theta_2^*$

$$k^* = \frac{k_m}{k_p}, \qquad \theta_1^{*T}(sI - \Lambda)^{-1}\ell = \frac{Z_m(s) - Z_p(s)}{Z_m(s)}, \qquad \theta_0^* + \theta_2^{*T}(sI - \Lambda)^{-1}\ell = \frac{R_p(s) - R_m(s)}{k_p Z_m(s)}.$$

then by the assumption above, k_p, as well as the $(2n - 1)$ coefficients of the monic polynomials $Z_p(s)$ and $R_p(s)$ are unknown. We further assume that $Z_p(s)$ is a Hurwitz polynomial and that the sign of the high frequency gain k_p is known.

The reference model is chosen to have a strictly positive real transfer function $W_m(s)$, defined in Eq. (5.4), and has an input-output pair $\{r(\cdot), y_m(\cdot)\}$. The high frequency gain k_m is assumed to be positive. The aim of the adaptive control problem, as in the previous cases, is to determine a bounded control input u to the plant using a differentiator free controller so that $\lim_{t\to\infty} |y_p(t) - y_m(t)| = 0$.

The Concept of a Tuner. In the three simple cases that we considered in Sections 5.3.1-5.3.3, the controller structure was chosen so that constant values of the controller parameter θ exist for which perfect regulation or tracking was achieved asymptotically. The adjustment of the parameter vector is carried out using the adaptive laws. The generation of the control input u can therefore be conveniently described in terms of a controller and a tuner [7], [10]. The former is linear and time-invariant for constant values of the control parameter vector θ and the latter determines how $\theta(t)$ is to be adjusted based on input-output data. Such a representation is shown in Fig. 5.6.

The representation above is found to be satisfactory for most of the adaptive methods described in the chapters following where the subsystem generating the control input can be conveniently divided into a controller and a tuner. The structure of the controller is dictated by linear systems theory, while the design of the tuner is based on stability considerations. Obviously, if more general nonlinear controllers are used in the future, such a division may not be fruitful.

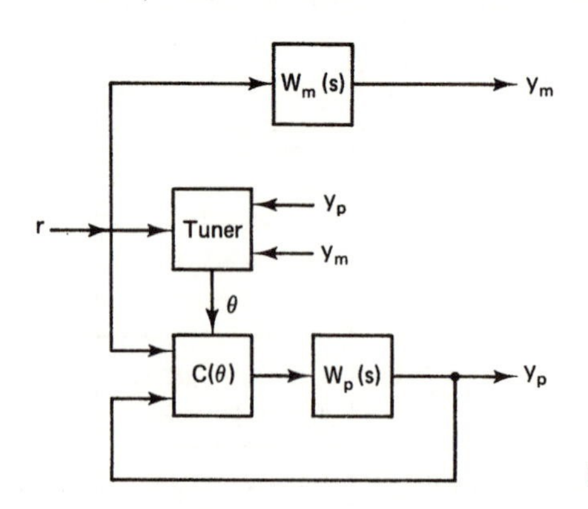

Figure 5.6 The adaptive tuner.

The controller structure chosen in this case (Fig. 5.7) includes all the elements used in Sections 5.3.1-5.3.3. In particular, it contains a gain $k(t)$ used in Section 5.3.1, the feedforward control loop with the parameter vector $\theta_1(t)$ as in Section 5.3.2, and the feedback controller with the parameters $\theta_0(t)$ and $\theta_2(t)$ as in Section 5.3.3. The controller is described completely by the differential equation

$$\begin{aligned} \dot{\omega}_1(t) &= \Lambda\omega_1(t) + \ell u(t) \\ \dot{\omega}_2(t) &= \Lambda\omega_2(t) + \ell y_p(t) \\ \omega(t) &\stackrel{\Delta}{=} \left[r(t),\ \omega_1^T(t),\ y_p(t),\ \omega_2^T(t)\right]^T \\ \theta(t) &\stackrel{\Delta}{=} \left[k(t), \theta_1^T(t), \theta_0(t), \theta_2^T(t)\right]^T \\ u(t) &= \theta^T(t)\omega(t) \end{aligned} \tag{5.16}$$

where $k : \mathbb{R}^+ \to \mathbb{R}$, $\theta_1, \omega_1 : \mathbb{R}^+ \to \mathbb{R}^{n-1}$, $\theta_0 : \mathbb{R}^+ \to \mathbb{R}$, $\theta_2, \omega_2 : \mathbb{R}^+ \to \mathbb{R}^{n-1}$, Λ is an asymptotically stable matrix and $det[sI - \Lambda] = \lambda(s)$.

It follows that when the control parameters $k(t)$, $\theta_1(t)$, $\theta_0(t)$, and $\theta_2(t)$ assume constant values k_c, θ_{1c}, θ_{0c}, and θ_{2c}, respectively, the transfer functions of the feedforward and the feedback controllers are respectively

$$\frac{\lambda(s)}{\lambda(s) - C(s)} \quad \text{and} \quad \frac{D(s)}{\lambda(s)},$$

and the overall transfer function of the plant together with the controller can be expressed as

$$W_o(s) = \frac{k_c k_p Z_p(s)\lambda(s)}{(\lambda(s) - C(s))\, R_p(s) - k_p Z_p(s) D(s)}. \tag{5.17}$$

As in Sections 5.3.1-5.3.3, $\lambda(s)$ is a monic polynomial of degree $n-1$ and $C(s)$ and $D(s)$ are polynomials of degree $n-2$ and $n-1$ respectively. The parameter vector θ_1 determines the coefficients of $C(s)$ while θ_0 and θ_2 together determine those of $D(s)$.

Let $C^*(s)$ and $D^*(s)$ be polynomials in s such that

$$\lambda(s) - C^*(s) = Z_p(s), \qquad R_p(s) - k_p D^*(s) = R_m(s).$$

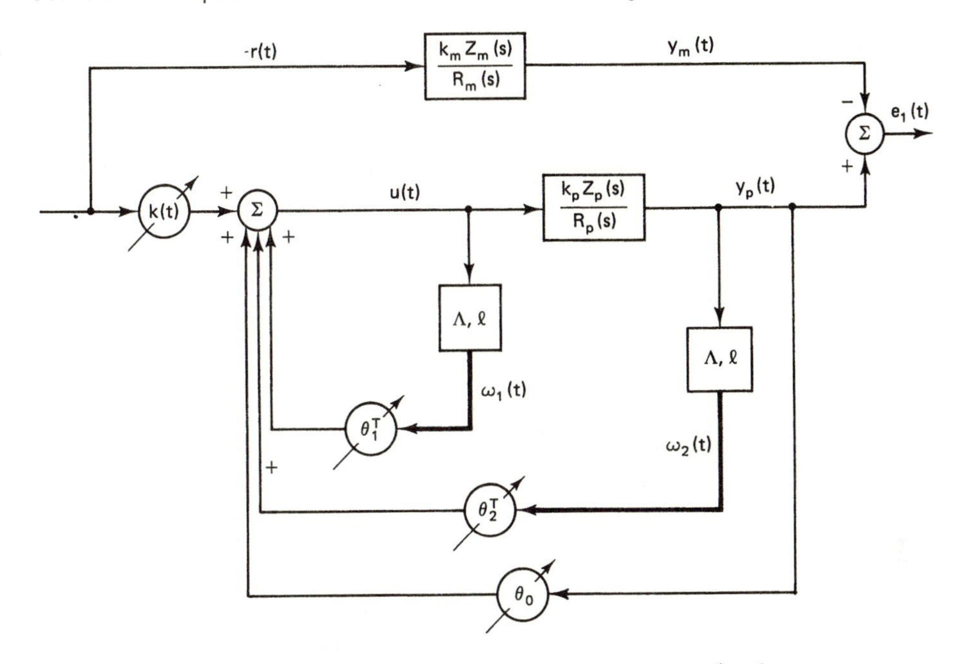

Figure 5.7 The adaptive controller: The general case ($n^* = 1$).

Further let $\lambda(s) = Z_m(s)$. Then scalars k^*, θ_0^* and vectors θ_1^* and θ_2^* exist such that

$$k^* = \frac{k_m}{k_p}, \quad \theta_1^{*T}(sI - \Lambda)^{-1}\ell = \frac{C^*(s)}{\lambda(s)}, \text{ and } \theta_0^* + \theta_2^{*T}(sI - \Lambda)^{-1}\ell = \frac{D^*(s)}{\lambda(s)}.$$

Choosing $\theta(t) \equiv \theta^*$, where θ^* is defined as

$$\theta^* \triangleq \left[k^*, \theta_1^{*T}, \theta_0^*, \theta_2^{*T}\right]^T,$$

the transfer function $W_o(s)$ in Eq. (5.17) becomes

$$W_o(s) = k_m \frac{Z_p(s) Z_m(s)}{Z_p(s)\left[R_p(s) - k_p D^*(s)\right]} = W_m(s).$$

The Adaptive Law. As in the three special cases, the control parameter vector must be adjusted as a function of time using all the available data regarding the plant. The differential equation describing the plant together with the controller can be represented as

$$\begin{aligned} \dot{x}_p(t) &= A_p x_p(t) + b_p\left(\theta^T(t)\omega(t)\right) \\ \dot{\omega}_1(t) &= \Lambda\omega_1(t) + \ell\left(\theta^T(t)\omega(t)\right) \\ \dot{\omega}_2(t) &= \Lambda\omega_2(t) + \ell\left(h_p^T x_p(t)\right). \end{aligned} \tag{5.18}$$

As in the previous sections, we define the following parameter errors:

$$\psi(t) \stackrel{\Delta}{=} k(t) - k^*, \;\; \phi_0(t) \stackrel{\Delta}{=} \theta_0(t) - \theta_0^*, \;\; \phi_1(t) \stackrel{\Delta}{=} \theta_1(t) - \theta_1^*,$$

$$\phi_2(t) \stackrel{\Delta}{=} \theta_2(t) - \theta_2^*, \phi(t) \stackrel{\Delta}{=} \left[\psi(t),\; \phi_1^T(t),\; \phi_0(t),\; \phi_2^T(t)\right]^T.$$

Then Eq. (5.18) can be rewritten as

$$\dot{x} = A_{mn}x + b_{mn}\left[\phi^T\omega + k^*r\right]; \qquad y_p = h_{mn}^T x \tag{5.19}$$

where

$$A_{mn} = \begin{bmatrix} A_p + b_p\theta_0^* h_p^T & b_p\theta_1^{*T} & b_p\theta_2^{*T} \\ \ell\theta_0^* h_p^T & \Lambda + \ell\theta_1^{*T} & \ell\theta_2^{*T} \\ \ell h_p^T & 0 & \Lambda \end{bmatrix} \qquad b_{mn} = \begin{bmatrix} b_p \\ \ell \\ 0 \end{bmatrix};$$

$$h_{mn} = \left[h_p^T,\, 0,\, 0\right]^T, \text{ and } x \stackrel{\Delta}{=} [x_p^T,\, \omega_1^T,\, \omega_2^T]^T. \tag{5.20}$$

Since $W_o(s) \equiv W_m(s)$ when $\theta(t) \equiv \theta^*$, it follows that the reference model can be described by the $(3n-2)^{\text{th}}$ order differential equation

$$\dot{x}_{mn} = A_{mn}x_{mn} + b_{mn}k^*r; \qquad y_m = h_{mn}^T x_{mn}$$

where

$$x_{mn} = [x_p^{*T},\, \omega_1^{*T},\, \omega_2^{*T}]^T,$$

$$h_{mn}^T\left[sI - A_{mn}\right]^{-1} b_{mn} = \frac{k_p}{k_m}W_m(s).$$

$x_p^*(t)$, $\omega_1^*(t)$, and $\omega_2^*(t)$ can be considered as signals in the reference model corresponding to $x_p(t)$, $\omega_1(t)$, and $\omega_2(t)$ in the overall system. Therefore, the error equation for the overall system may be expressed as

$$\dot{e}(t) = A_{mn}e(t) + b_{mn}\left[\phi^T(t)\omega(t)\right]; \qquad e_1(t) = h_{mn}^T e(t) \tag{5.21}$$

where $e = x - x_{mn}$ and $e_1 = y_p - y_m$. Equation (5.21) is of dimension $3n-1$ while the corresponding transfer function $W_m(s)$ is of order n. The modes corresponding to the remaining $2n-1$ poles are uncontrollable and/or unobservable but asymptotically stable since $Z_p(s)$ is Hurwitz. From Lemma 5.1, the adaptive laws can be written by inspection as

$$\begin{aligned} \dot{k}(t) &= -sgn(k_p)e_1(t)r(t) \\ \dot{\theta}_0(t) &= -sgn(k_p)e_1(t)y_p(t) \\ \dot{\theta}_1(t) &= -sgn(k_p)e_1(t)\omega_1(t) \\ \dot{\theta}_2(t) &= -sgn(k_p)e_1(t)\omega_2(t). \end{aligned} \tag{5.22}$$

Using the same arguments as in the special cases, it follows that $\lim_{t\to\infty} e_1(t) = 0$.

Comment 5.6 The basic structure of the controller described here (Fig. 5.7) is used in many of the problems in all the following chapters. We shall refer to Eq. (5.22) as the standard adaptive law for plants with $n^* = 1$, in subsequent chapters.

Comment 5.7 In error equation (5.21), $\omega(t)$ is not assumed to be bounded. Since $e(t)$, the state error vector, is shown to be bounded, and the reference model is asymptotically stable, the boundedness of $x(t)$ follows. Since $\omega(t)$ constitutes a part of this vector, it is also bounded, assuring that $\lim_{t\to\infty} e_1(t) = 0$.

Comment 5.8 If the reference model has a transfer function $W_m(s)$, which is of relative degree unity but is not SPR, the following modification is required:

Let $W_m(s) = W(s)\overline{W}_m(s)$ where $\overline{W}_m(s)$ is strictly positive real and $W(s)$ is a transfer function of relative degree zero with poles and zeros in $\mathbb{C}^-$. If $r'(t) = W(s)r(t)$, then $r'(t)$ rather than $r(t)$ is used as the reference input to the plant.

A similar modification can also be adopted when $W_m(s)$ has a relative degree $n_m^* > 1$. If $W_m(s) = W(s)\overline{W}_m(s)$ where $\overline{W}_m(s)$ is SPR and $W(s)$ is asymptotically stable, the reference input into the plant is modified as $r'(t) = W(s)r(t)$.

Comment 5.9 In Section 5.2, it was stated that the instability of the plant transfer function can be explicitly assumed in the statement of the adaptive control problem. When the plant is asymptotically stable and the designer is aware of this fact, the nature of the problem changes significantly, since stability of the overall system ceases to be the primary issue. The designer may merely identify the parameters of the plant accurately before controlling it (a procedure suggested for the control of flexible space structures). Alternately, if an adaptive algorithm is used to improve performance but the behavior of the system tends to become unstable, the designer always has recourse to shutting off control, thereby keeping all the signals bounded. The same is also true if at least one specific value of the feedback gain vector θ is known for which the overall system is stable. In the problems discussed in the book, such assumptions are not made, and the main aim of adaptation is to determine a control parameter vector θ, which will stabilize the system before the system response becomes unbounded.

Simulation 5.1 The following adaptive control problem was simulated to study the behavior of solutions when the relative degree of the plant is unity. The transfer functions of the plant and the model were chosen to be

$$W_p(s) = \frac{s+1}{(s-2)(s-1)} \quad \text{and} \quad W_m(s) = \frac{1}{s+1} \quad \text{respectively.}$$

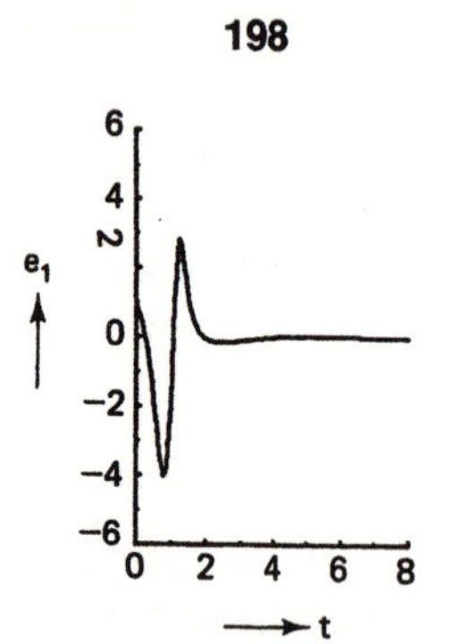

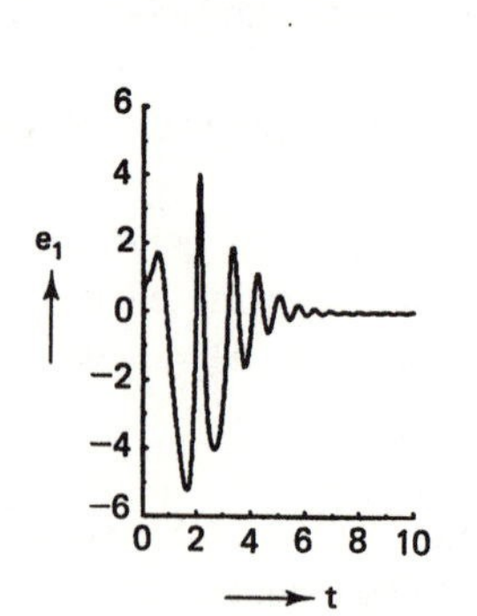

6
4
2
e1
0
−2
−4
−6
0 10 20 30 40
t

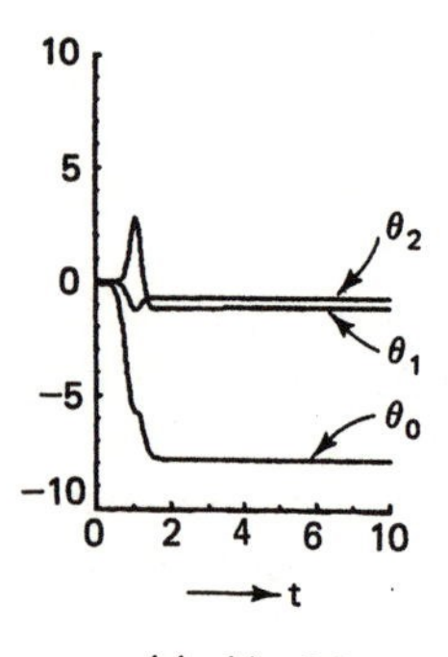

(a) r(t) = 0.0

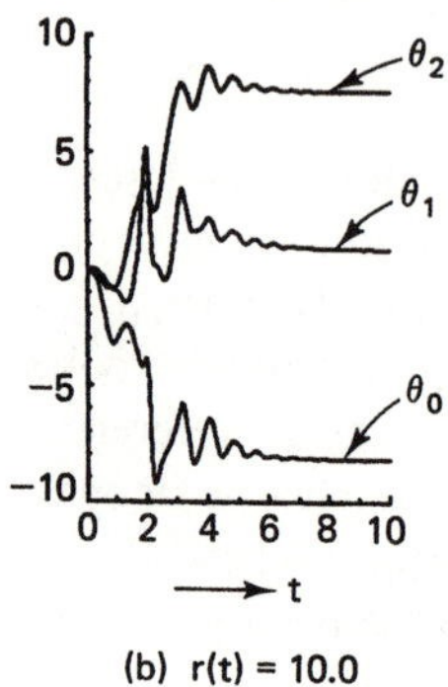

(b) r(t) = 10.0

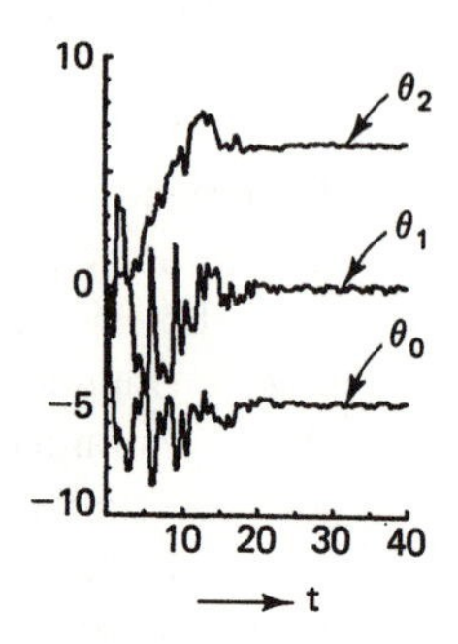

(c) r(t) = 5 cos t + 10 cos 5t

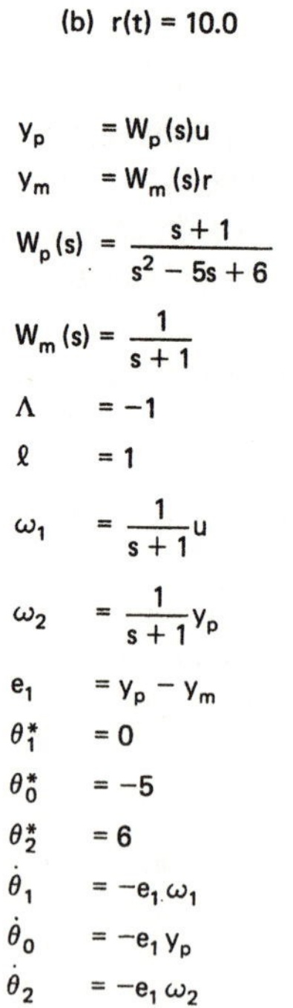

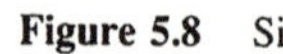

Figure 5.8 Simulation results when $n^* = 1$.

The fixed control parameters Λ and ℓ in Eq. (5.18) were chosen as $\Lambda = -1$ and $\ell = 1$. The high frequency gain k_p was assumed to be known and the control parameters $\theta_1(t)$, $\theta_0(t)$, and $\theta_2(t)$ were adjusted as shown in Section 5.3.3. All initial conditions were chosen to be zero. The resultant output error e_1 and the control parameters are shown in Figs. 5.8(a)-(c) for the cases (a) $r(t) \equiv 0$, (b) $r(t) \equiv 5$, and (c) $r(t) = 5 \cos t + 10 \cos 5t$. The true values of the control parameters are $\theta_1^* = 0$, $\theta_0^* = -5$, and $\theta_2^* = 6$. As in Chapters 3 and 4, we note that, in all cases, $e_1(t)$ tends to zero asymptotically, while the parameter errors do not converge to zero when $r(t)$ is zero or a constant. However, they converge to zero when $r(t)$ has two distinct frequencies (see Chapter 6).

5.4 ADAPTIVE CONTROL OF PLANTS WITH RELATIVE DEGREE ≥ 2

The algebraic as well as the analytic parts of the general adaptive control problem are considerably simplified when $n^* = 1$, as shown in the previous section. The algebraic part reduces to the simple problem of determining a polynomial $Z_2(s)$ of degree $n-1$, such that given a monic polynomial $Z_1(s)$ of degree n, $Z_1(s) + Z_2(s)$ can be made to have desired roots. The analytic part, on the other hand, is simplified by choosing the reference model to be strictly positive real so that adaptive laws can be determined using the Kalman-Yakubovich lemma (Lemmas 2.3 and 2.4). Both parts of the adaptive control problem are rendered difficult when the relative degree of the plant exceeds unity. Demonstrating that the controller structure has enough freedom to assure the existence of a solution, when plant parameters are known, involves the solution of more complex algebraic equations. The Bezout identity, discussed in Section 5.4.1, plays an important role here. Even assuming that such a controller structure exists, no simple lemma such as Lemma 5.1 is readily available for determining stable adaptive laws, since the reference model cannot be chosen to be strictly positive real. We deal with the algebraic part in Section 5.4.2 before addressing the analytic part, which is substantially more complex.

5.4.1 Bezout Identity

The existence of a constant vector θ^* for which the transfer function of the plant together with the controller equals that of the model can be established in a rather straightforward manner for the case $n^* = 1$. For the general problem where $n^* \geq 2$, however, this is more involved and requires the use of the Bezout identity given in Lemma 5.2.

Lemma 5.2 Let $Q(s)$ and $P(s)$ be polynomials of degrees n and m respectively which are relatively prime. Then polynomials $M(s)$ and $N(s)$ exist such that

$$M(s)Q(s) + N(s)P(s) = Q^*(s) \tag{5.23}$$

where $Q^*(s)$ is an arbitrary polynomial.

Before proceeding to prove Lemma 5.2, we consider some simple examples involving first- and second-degree polynomials.

Example 5.1

Let $P(s) = 1$ and $Q(s) = s^2 + 2s + 3$ and let $Q^*(s) = s^3 + q_1^* s^2 + q_2^* s + q_3^*$ be any arbitrary monic third-degree polynomial. By Lemma 5.2, polynomials $(s + a)$ and $(bs + c)$ where $a, b, c \in \mathbb{R}$ exist such that

$$Q^*(s) = (s+a)(s^2+2s+3) + (bs+c).$$

This can be demonstrated by equating the coefficients of the two polynomials leading to the equation

$$\begin{bmatrix} 1 & 0 & 0 \\ 2 & 1 & 0 \\ 3 & 0 & 1 \end{bmatrix} \begin{bmatrix} a \\ b \\ c \end{bmatrix} = \begin{bmatrix} q_1^* - 2 \\ q_2^* - 3 \\ q_3^* \end{bmatrix}. \tag{5.24}$$

Since the matrix in Eq. (5.24) is triangular, it is nonsingular, and hence, a solution (a, b, c) exists for arbitrary values of q_1^*, q_2^* and q_3^*.

Example 5.2

If $P(s) = s + 1$, proceeding in the same fashion as in Example 5.1, we obtain the vector equation

$$\begin{bmatrix} 1 & 1 & 0 \\ 2 & 1 & 1 \\ 3 & 0 & 1 \end{bmatrix} \begin{bmatrix} a \\ b \\ c \end{bmatrix} = \begin{bmatrix} q_1^* - 2 \\ q_2^* - 3 \\ q_3^* \end{bmatrix}. \tag{5.25}$$

In this case the matrix in Eq. (5.25) is nonsingular, since its rows are linearly independent, and hence polynomials $s + a$ and $bs + c$ exist such that

$$Q^*(s) = (s+a)(s^2+2s+3) + (bs+c)(s+1)$$

is satisfied. In more general cases, the difficulty lies in demonstrating the nonsingularity of the corresponding matrix.

Example 5.3

If $P(s) = s + 1$ and $Q(s) = s^2 + 3s + 2$ it is seen that $P(s)$ and $Q(s)$ are not relatively prime and the conditions of Lemma 5.2 are not met. The difficulty encountered is evident on writing the equations corresponding to Eqs. (5.24) and (5.25) of the previous examples, given by

$$\begin{bmatrix} 1 & 1 & 0 \\ 3 & 1 & 1 \\ 2 & 0 & 1 \end{bmatrix} \begin{bmatrix} a \\ b \\ c \end{bmatrix} = \begin{bmatrix} q_1^* - 3 \\ q_2^* - 2 \\ q_3^* \end{bmatrix}. \tag{5.26}$$

Since the third row of the matrix in Eq. (5.26) is a linear combination of the first two rows, it is singular, and hence a solution does not exist for all choices of the vector q^* where $q^{*T} = [q_1^*, q_2^*, q_3^*]$.

Proof of Lemma 5.2: Let

$$Q(s) = \sum_{i=0}^{n} q_i s^i \qquad \text{and} \qquad P(s) = \sum_{i=0}^{m} p_i s^i.$$

If $m = 0$ so that $P(s) = p_m$, then Eq. (5.23) can be solved by choosing $M(s) = R^*(s)$ and $p_m N(s) = Q^*(s) - R^*(s)Q(s)$ where $R^*(s)$ is the proper part of $Q^*(s)/Q(s)$.

Let $m \geq 1$. The proof of the lemma can be conveniently given in two stages. We first show that polynomials $A(s)$ and $B(s)$ of degrees $\leq m-1$ and $\leq n-1$ respectively exist satisfying the equation

$$A(s)Q(s) + B(s)P(s) \;=\; 1. \tag{5.27}$$

If

$$A(s) = [\alpha_1 s^{m-1} + \alpha_2 s^{m-2} + \cdots + \alpha_m], \qquad B(s) = [\beta_1 s^{n-1} + \beta_2 s^{n-2} + \cdots + \beta_n],$$

equating coefficients on both sides of Eq. (5.27) leads to the vector equation

$$\left[\begin{array}{ccccc|cccccccc}
q_n & 0 & \cdots & \cdots & 0 & p_m & 0 & & \cdots & \cdots & & 0 \\
q_{n-1} & q_n & 0 & \cdots & 0 & p_{m-1} & p_m & 0 & \cdots & \cdots & & 0 \\
\vdots & & & & \vdots & \vdots & & & & & & \vdots \\
q_{n-m} & & \cdots & & q_{n-1} & p_0 & p_1 & \cdots & p_m & 0 & \cdots & 0 \\
\vdots & & & & \vdots & 0 & p_0 & p_1 & \cdots & p_m & 0 & \cdots \\
q_0 & & \cdots & & q_{m-1} & \vdots & & & & & & \\
0 & q_0 & & \cdots & q_{m-2} & 0 & \cdots & & 0 & p_0 & \cdots & p_{m-2} \\
 & & \ddots & & \vdots & \vdots & & & \ddots & & \ddots & \\
0 & & \cdots & 0 & q_0 & 0 & & \cdots & & & 0 & p_0
\end{array}\right]
\left[\begin{array}{c} \alpha_1 \\ \alpha_2 \\ \vdots \\ \vdots \\ \alpha_m \\ \beta_1 \\ \beta_2 \\ \vdots \\ \vdots \\ \beta_n \end{array}\right]$$

$$= \left[\begin{array}{c} 0 \\ 0 \\ \vdots \\ \vdots \\ 0 \\ 0 \\ 1 \end{array}\right] \tag{5.28}$$

Since $P(s)$ and $Q(s)$ are relatively prime, the matrix in Eq. (5.28) is nonsingular and hence it follows that the $(m+n)$ parameters α_i and β_j $(i = 1, 2, \ldots, m,\ j = 1, 2 \ldots, n)$ exist such that Eq. (5.28) is satisfied. Equation (5.28) can be used in turn to prove the existence of $M(s)$ and $N(s)$ in Eq. (5.23).

Multiplying Eq. (5.27) on both sides by the polynomial $Q^*(s)$, we have

$$Q^*(s)A(s)Q(s) + Q^*(s)B(s)P(s) \;=\; Q^*(s). \tag{5.29}$$

Let $R^*(s)$ be the proper part of $Q^*(s)B(s)/Q(s)$ so that

$$\frac{Q^*(s)B(s)}{Q(s)} = R^*(s) + \frac{N(s)}{Q(s)}.$$

Equation (5.29) can be expressed in terms of $R^*(s)$ as

$$\left[Q^*(s)A(s) + R^*(s)P(s)\right] Q(s) + \left[Q^*(s)B(s) - R^*(s)Q(s)\right] P(s) = Q^*(s). \qquad (5.30)$$

Hence, defining

$$M(s) = \left[Q^*(s)A(s) + R^*(s)P(s)\right], \; N(s) = Q^*(s)B(s) - R^*(s)Q(s),$$

the result follows.

Comment 5.10 By the choice of $N(s)$, it follows that $N(s)$ is of degree $n-1$. If $Q^*(s)$ is a polynomial of degree $n_1 > n$, since $M(s) = [Q^*(s) - N(s)P(s)]/Q(s)$, the degree of $M(s) = \max(n_1 - n, m - 1)$. Hence, if $n_1 \geq 2n - 1$, then $N(s)/M(s)$ is a proper transfer function. If, in addition, $Q^*(s)$ and $Q(s)$ are monic polynomials, then $M(s)$ is monic.

5.4.2 $n^* \geq 2$: Algebraic Part

As in the simple cases discussed in Sections 5.3.1-5.3.3, the first question that has to be addressed is whether the controller has adequate freedom to satisfy the requirements of the MRAC problem, when the plant parameters are known. Using the Bezout identity, it is shown here that even for the general case, the basic controller structure used in Section 5.3 is adequate. This implies that if the plant transfer function is defined as in Eq. (5.3), then $\theta(t) \equiv \theta^*$ exists such that $W_o(s)$ in Section 5.3 is equal to $W_m(s)$.

Since $\lambda(s)$ is of degree $n-1$ while $Z_m(s)$ is of degree $m \leq n-2$, it is no longer possible to choose $\lambda(s) = Z_m(s)$ as in Section 5.3. Hence $\lambda(s)$ is chosen to contain $Z_m(s)$ as a factor. This is expressed as

$$\lambda(s) = Z_m(s)\lambda_1(s)$$

where $\lambda_1(s)$ is a Hurwitz polynomial of degree $(n - m - 1)$. The expression for the overall transfer function $W_o(s)$ may be derived as

$$W_o(s) = \frac{k_c k_p Z_p(s)\lambda_1(s)Z_m(s)}{R_p(s)\left[\lambda(s) - C(s)\right] - k_p Z_p(s)D(s)}. \qquad (5.31)$$

The existence of the control parameter vector θ^* is equivalent to the existence of the polynomials $C(s)$ and $D(s)$ [or equivalently $[\lambda(s) - C(s)]$ and $D(s)$] such that the denominator polynomial in Eq. (5.31) can be made equal to $Z_p(s)\lambda_1(s)R_m(s)$. Identifying $R_p(s)$ with $Q(s)$, $-k_pZ_p(s)$ with $P(s)$ and $Z_p(s)\lambda_1(s)R_m(s)$ with $Q^*(s)$ in Eq. (5.23), the problem is to determine $C^*(s)$ and $D^*(s)$ such that

$$\left[\lambda(s) - C^*(s)\right] Q(s) + D^*(s)P(s) = Q^*(s). \qquad (5.32)$$

Figure 5.9 Modification of the control input: $n^* = 2$.

Since all the conditions of Lemma 5.2 are satisfied, it follows that $C^*(s)$ and $D^*(s)$ exist satisfying Eq. (5.32), or equivalently, the desired control parameter vector θ^* exists.

5.4.3 $n^* = 2$: Analytic Part

Although the methods developed in Section 5.4.2 are applicable to plants with arbitrary relative degree, the case when the relative degree is equal to two may be considered to be unique since the methods developed in the previous section can also be used for its solution. We consider this special case first, since it provides the motivation for the changes in the adaptive laws necessary to achieve stable solutions for plants with arbitrary relative degree.

When $W_p(s)$ has relative degree two, the reference model has to be chosen so that $W_m(s)$ has the same relative degree. With the controller as in Section 5.3, the Bezout identity assures the existence of a controller parameter vector θ^* such that the plant, together with the fixed controller, has the same transfer function as the model. The error equation can therefore be expressed as

$$e_1(t) = \frac{k_p}{k_m} W_m(s)\phi^T(t)\omega(t).$$

The difficulty in determining stable adaptive laws stems from the fact that $W_m(s)$ is not SPR. If the controller can include differentiators, a transfer function $L(s) = s + a$ could be chosen such that $W_m(s)L(s)$ is SPR allowing the same procedure as in the previous sections to be used. This motivates a modified structure of the controller to achieve the same objective without explicit differentiation [14].

Consider the system shown in Fig. 5.9 whose input is $\omega(t)$ and output is $u(t)$, where the latter is the input into the plant. If $\overline{\omega}(t) = L^{-1}(s)\omega(t)$, $u(t)$ can be expressed as

$$\begin{aligned} u(t) &= \dot{\theta}^T(t)\overline{\omega}(t) + \theta^T(t)\dot{\overline{\omega}}(t) + a\theta^T(t)\overline{\omega}(t) \\ &= \dot{\theta}^T(t)\overline{\omega}(t) + \theta^T(t)\omega(t). \end{aligned} \tag{5.33}$$

If $\dot{\theta}(t)$ is specified by an adaptive law, $u(t)$ can be realized from Eq. (5.33) without explicit differentation.

In the controller structure used in Section 5.3, if the parameter vector $\theta(t)$ is replaced by an operator of the form $L(s)\theta(t)L^{-1}(s)$, it follows that

(i) the algebraic part is unaffected since $L(s)\theta L^{-1}(s) \equiv \theta$ when θ is a constant, and

(ii) the error equation is of the form

$$e_1(t) = \frac{k_p}{k_m} W_m(s)L(s)\phi^T(t)\overline{\omega}(t).$$

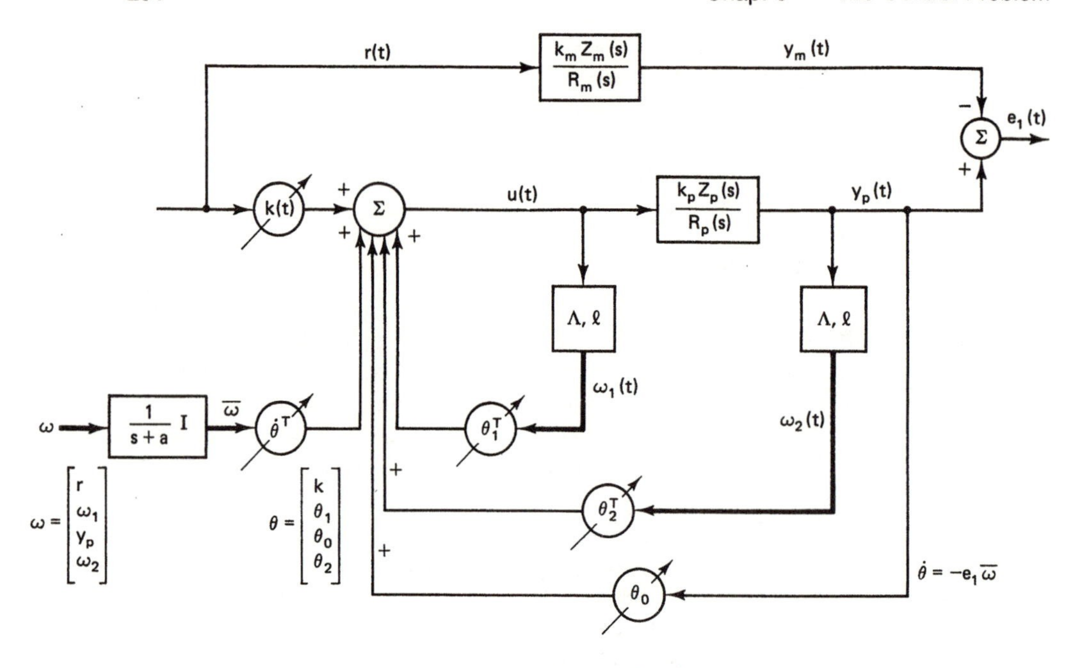

Figure 5.10 $n^* = 2$.

Hence, if an adaptive law of the form

$$\dot{\theta}(t) = \dot{\phi}(t) = -sgn(k_p)e_1(t)\overline{\omega}(t) \tag{5.34}$$

is used, the overall system is stable and $\lim_{t\to\infty} e_1(t) = 0$. By using the expression for $\dot{\theta}$ from Eq. (5.34) in Eq. (5.33), the control input becomes

$$u(t) = \theta^T(t)\omega(t) - sgn(k_p)e_1(t)\overline{\omega}^T(t)\overline{\omega}(t).$$

The controller structure therefore has the form shown in Fig. 5.10.

Comment 5.11 The importance of the case where the relative degree n^* of the plant is two is worth emphasizing. This is the only case where positive realness of the reference model is not required to generate the desired input using a differentiator free controller. In particular, by making $u = \theta^T\omega - sgn(k_p)e_1\overline{\omega}^T\overline{\omega}$, differentiation is avoided.

The procedure above cannot be extended to plants with relative degree $n^* \geq 3$ since second and higher derivatives of $\theta(t)$ with respect to time are not known.

5.4.4 $n^* \geq 2$: Analytic Part

In Fig. 5.11, a block diagram of the reference model and the plant are shown along with the control parameter $\theta(t)$, which determines the input $u(t)$ to the plant.

The transfer functions of the plant and the reference model are given by

$$W_p(s) = \frac{k_p Z_p(s)}{R_p(s)} \quad \text{and} \quad W_m(s) = \frac{k_m Z_m(s)}{R_m(s)}$$

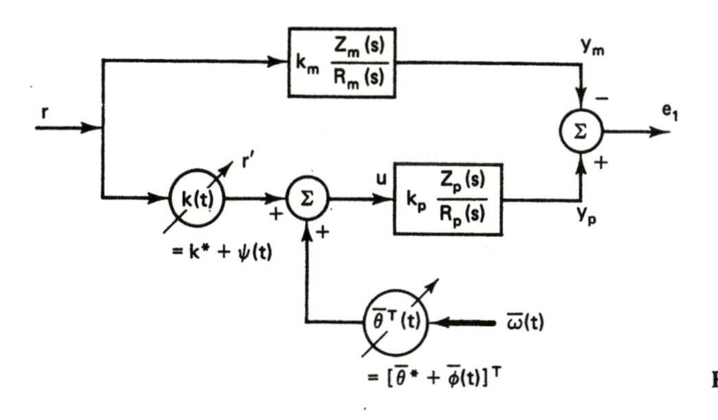

Figure 5.11 $n^* \geq 2$: The analytic part.

respectively. From Section 5.4.2, it is known that for $\theta(t) \equiv \theta^*$, the transfer function of the plant together with the controller is $W_m(s)$. Using the same procedure outlined in Section 5.3, the output error $e_1(t)$ can be expressed in terms of the parameter errors. The transfer function from the input r' in Fig. 5.11 to the output y_p of the plant is $(k_p/k_m)W_m(s)$, when the parameter values are constant and equal to θ^*. Hence, expressing the input $u(t)$ to the plant as

$$\begin{aligned} u(t) &= \theta^T(t)\omega(t) \\ &= \theta^{*T}\omega(t) + \phi^T(t)\omega(t), \end{aligned}$$

the output $y_p(t)$ is described by

$$y_p(t) = W_m(s)r(t) + \frac{k_p}{k_m}W_m(s)\phi^T(t)\omega(t).$$

Since the output of the model $y_m(t) = W_m(s)r(t)$, we obtain that

$$e_1(t) = \frac{k_p}{k_m}W_m(s)\phi^T(t)\omega(t). \tag{5.35}$$

Unlike the simpler cases described in Section 5.3, it is not immediately evident how the adaptive law is to be chosen, since $W_m(s)$ in Eq. (5.35) is not strictly positive real. For some time this difficulty was considered insurmountable by many researchers in the field until it was realized that the concept of the augmented error introduced by Monopoli [8] could be used judiciously to get around the difficulty. Monopoli's contribution represented a major breakthrough, and even at present, the solution to general model reference adaptive control problems involves the use of augmented errors, when the relevant transfer functions are not strictly positive real.

Around 1980, several solutions were given to the MRAC problem both for continuous and discrete time systems. While Narendra *et al.*[13] and Morse [9] gave solutions for the former, Goodwin *et al.* [6] and Narendra and Lin [12] gave solutions to the latter. Several other researchers including Fuchs [5], Landau [7], and Egardt [4] also gave similar solutions. These solutions of the adaptive control problem represented a major milestone in the development of the theory and set the stage for many of the advances in the following years.

In the sections following, we use the development suggested by Narendra and Lin in [11], which is a refinement of the treatment in [14]. Although this problem can also be treated in parts as in Section 5.3, we prefer to deal with the entire system here, making only the distinction between the two cases (i) k_p known, and (ii) k_p unknown.

5.4.5 k_p known

When the high frequency gain k_p of the plant is known, it is no longer necessary to adjust $k(t)$ adaptively, and the latter can be chosen to be a constant equal to k_m/k_p. With no loss of generality we assume that $k_p = k_m = 1$ so that the transfer functions of both plant and model have monic numerator and denominator polynomials. The controller consequently has only $2n - 1$ adjustable parameters. The plant and model transfer functions are respectively,

$$W_p(s) = \frac{Z_p(s)}{R_p(s)} \qquad \text{and} \qquad W_m(s) = \frac{Z_m(s)}{R_m(s)}.$$

The controller equations are therefore given by

$$\begin{aligned} \dot{\omega}_1 &= \Lambda\omega_1 + bu \\ \dot{\omega}_2 &= \Lambda\omega_2 + by_p \\ u &= r + \overline{\theta}^T\overline{\omega} \end{aligned} \tag{5.36}$$

where

$$\overline{\theta}^T \triangleq [\theta_1^T, \theta_0, \theta_2^T], \qquad \text{and} \qquad \overline{\omega}^T \triangleq [\omega_1^T, y_p, \omega_2^T].$$

The overall system equations can be described as

$$\dot{x} = A_{mn}x + b_{mn}\left[\overline{\phi}^T\overline{\omega} + r\right]; \qquad y_p = h_{mn}^T x \tag{5.37}$$

where A_{mn}, b_{mn}, h_{mn} and x are given by Eq. (5.20), $\overline{\theta}^{*T} = [\theta_1^{*T}, \theta_0^*, \theta_2^{*T}]$, $\overline{\phi} = \overline{\theta} - \overline{\theta}^*$, and

$$W_m(s) = h_{mn}^T(sI - A_{mn})^{-1}b_{mn}.$$

The error equations are obtained as

$$\dot{e} = A_{mn}e + b_{mn}\overline{\phi}^T\overline{\omega}; \qquad e_1 = h_{mn}^T e \tag{5.38}$$

where $e = x - x_{mn}$ and $e_1 = y_p - y_m$. Equation (5.38) can be equivalently stated as

$$e_1 = W_m(s)\overline{\phi}^T\overline{\omega}, \tag{5.39}$$

which is of the form described in Eq. (5.21). Since $W_m(s)$ is not strictly positive real, it does not follow that an adaptive law of the form $\dot{\overline{\phi}} = -e_1\overline{\omega}$ will stabilize the overall system.

Augmented Error. The viewpoint taken in the approach suggested by Monopoli [8] is that the error equation (5.39) can be suitably modified to result in a form amenable to analysis using earlier methods. In particular, an augmented error ϵ_1 is synthesized that permits stable adaptive laws to be implemented. Two methods of generating an augmented error are described in this section. Although method 2 suggested by Narendra and Valavani [14] was proposed earlier chronologically, method 1 suggested later by Narendra and Lin [11] is considered here first. The latter is currently more commonly used in the literature and is closer in spirit to the procedure adopted for adaptive control in discrete time systems. However, since the two procedures are quite distinct analytically, both are included.

Method 1. Since the transfer function $W_m(s)$ of the reference model is known, an auxiliary error signal e_2 can be generated as

$$e_2(t) \triangleq \overline{\theta}^T(t)W_m(s)I\overline{\omega}(t) - W_m(s)\overline{\theta}^T(t)\overline{\omega}(t) \tag{5.40}$$

using the signal $\overline{\omega}(t)$ in the adaptive loop, where I is a $(2n-1)\times(2n-1)$ identity matrix. Defining

$$\overline{\zeta}(t) \triangleq W_m(s)I\overline{\omega}(t)$$

Eq. (5.40) can be equivalently stated as

$$e_2(t) = \overline{\theta}^T(t)\overline{\zeta}(t) - W_m(s)\overline{\theta}^T(t)\overline{\omega}(t). \tag{5.41}$$

When $\overline{\theta}(t) \equiv \overline{\theta}^*$, where $\overline{\theta}^*$ is the desired parameter vector, $W_m(s)$ and $\overline{\theta}^*$ in Eq. (5.41) commute so that

$$\delta_1(t) = \overline{\theta}^{*T}\overline{\zeta}(t) - W_m(s)\overline{\theta}^{*T}\overline{\omega}(t)$$

is an exponentially decaying signal due to initial conditions. Further, since $\overline{\theta}(t) = \overline{\theta}^* + \overline{\phi}(t)$, the auxiliary signal $e_2(t)$ can be expressed in terms of the parameter error $\overline{\phi}(t)$ as

$$e_2(t) = \overline{\phi}^T(t)\overline{\zeta}(t) - W_m(s)\overline{\phi}^T(t)\overline{\omega}(t) + \delta_1(t). \tag{5.42}$$

Let $\epsilon_1(t)$ be defined as (Fig. 5.12)

$$\epsilon_1(t) = e_1(t) + e_2(t). \tag{5.43}$$

The signal $e_2(t)$ is referred to as the auxiliary error while $\epsilon_1(t)$, which is the sum of the true output error and the auxiliary error, is referred to as the augmented error. It therefore follows from Eqs. (5.39) and (5.42) that

$$\epsilon_1(t) = \overline{\phi}^T(t)\overline{\zeta}(t) + \delta_1(t). \tag{5.44}$$

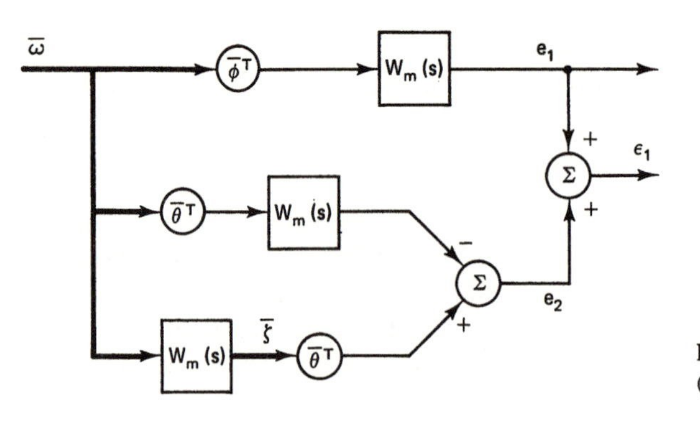

Figure 5.12 Augmented error: Method 1 (k_p known).

A few comments are in order here regarding the auxiliary error $e_2(t)$ and the augmented error $\epsilon_1(t)$. If the adaptive procedure were to result in the parameter vector converging to a constant, the auxiliary error $e_2(t)$ would tend to zero and the augmented error would asymptotically approach the true output error $e_1(t)$ of the system. Hence, the aim of the adaptation is to ensure that the augmented error tends to zero. Since the error equation (5.44) is in the standard form used in Chapter 4 [refer to Eq.(4.31)], this appears promising. This is in contrast to the error equation (5.39), which relates the true error $e_1(t)$ to the signal $\overline{\omega}(t)$, from which no stable adaptive laws can be directly obtained.

With the emphasis shifting to the augmented error and Eq. (5.44), it is normal to assume that an adaptive law similar to that used in Chapter 4 having the form

$$\dot{\overline{\phi}}(t) = -\epsilon_1(t)\overline{\zeta}(t)$$

would suffice to assure stability. However, this is not the case since the boundedness of $\overline{\omega}(t)$ and hence $\overline{\zeta}(t)$ has not been established. As shown in Section 5.5, where a detailed proof of stability is given, a normalizing factor $[1+\overline{\zeta}^T\overline{\zeta}]$ is required for this purpose. Hence, the adaptive law is modified to have the form[2]

$$\dot{\overline{\phi}}(t) = -\frac{\epsilon_1(t)\overline{\zeta}(t)}{1+\overline{\zeta}^T(t)\overline{\zeta}(t)}. \tag{5.45}$$

Method 2. In contrast to Method 1, the vector $\overline{\zeta}$ is defined here as

$$\overline{\zeta} \triangleq W(s)I\overline{\omega}$$

where $W_m(s)W^{-1}(s) = \overline{W}(s)$ is strictly positive real. Since $W_m(s)$ has a relative degree greater than one and $\overline{W}(s)$ is SPR, it follows that $W(s)$ is a realizable asymptotically stable transfer function. The auxiliary error e_2 and the corresponding augmented error ϵ_1 are indicated in Fig. 5.13 and described by the equations

[2]Equation (5.45) is referred to as the standard adaptive law for plants with $n^* \geq 2$.

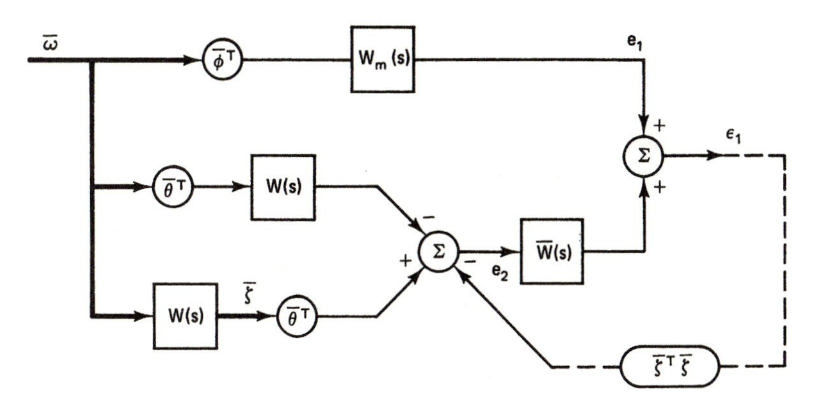

Figure 5.13 Augmented error: Method 2 (k_p known).

$$
\begin{aligned}
e_2 &\stackrel{\Delta}{=} \overline{\theta}^T\overline{\zeta} - W(s)\overline{\theta}^T\overline{\omega} \\
\epsilon_1 &\stackrel{\Delta}{=} e_1 + \overline{W}(s)e_2.
\end{aligned}
\tag{5.46}
$$

Since $e_1 = W_m(s)\overline{\phi}^T\overline{\omega}$ from Eq. (5.39), ϵ_1 can also be expressed as

$$
\epsilon_1 = \overline{W}(s)\overline{\phi}^T\overline{\zeta} + \delta_2(t) \tag{5.47}
$$

where $\delta_2(t)$ is an exponentially decaying signal due to initial conditions. Once again, since $\overline{W}(s)$ is SPR, Eq. (5.47) has the same form as the error equations obtained in Chapters 3 and 4 and in Section 5.3, and it is natural to attempt to use an adaptive law of the form

$$
\dot{\overline{\phi}} = -\epsilon_1\overline{\zeta}. \tag{5.48}
$$

The stability analysis in Section 5.5 reveals that this may not be adequate to assure stability. Equation (5.47) has to be modified to include the feedback signal $\epsilon_1\overline{\zeta}^T\overline{\zeta}$ shown in Fig. 5.13, to obtain a stable adaptive law. With this modification, the equation for the augmented error can be expressed as

$$
\begin{aligned}
\epsilon_1 &= e_1 + \overline{W}(s)\left[e_2 - \epsilon_1\overline{\zeta}^T\overline{\zeta}\right] \\
&= \overline{W}(s)\left[\overline{\phi}^T\overline{\zeta} - \epsilon_1\overline{\zeta}^T\overline{\zeta}\right] + \delta_2(t).
\end{aligned}
\tag{5.49}
$$

5.4.6 k_p unknown

When the high frequency gain k_p of the plant is unknown, an additional adjustable feedforward gain $k(t)$ has to be used in the controller. The desired steady state value of this gain is $k_m/k_p(= k^*)$ so that

$$
k(t) = k^* + \psi(t).
$$

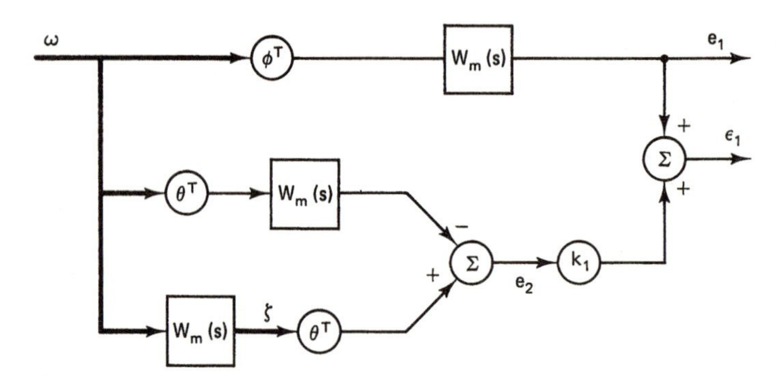

Figure 5.14 Augmented error: Method 1 (k_p unknown).

Hence, $k(t)$ along with the $(2n-1)$-vector $\overline{\theta}(t)$ shown in Fig. 5.11, need to be adjusted in a stable fashion. The equations describing the plant and the controller are given by Eq. (5.18) (see Section 5.3). While at first glance the error equations appear to be considerably more involved, they can be reduced to the same standard form as before by using a change of variables. For the sake of completeness we treat both methods described in Section 5.4.5.

Method 1. The auxiliary error was used in Method 1 when k_p is known to obtain an error equation of the form in Eq. (5.39). For this purpose, it was assumed that $k_m = k_p = 1$. When k_p is not known, this is no longer directly possible. This is due to the fact that in the error equation

$$e_1 = \frac{k_p}{k_m} W_m(s)\phi^T \omega \tag{5.35}$$

the gain k_p is unknown. Hence, the network generating the augmented error $\epsilon_1(t)$ must contain an additional gain as shown in Fig. 5.14.

Specifically, this is achieved by adding $k_1(t)e_2(t)$ (rather than $e_2(t)$) to the true output error $e_1(t)$ to form the augmented error, that is,

$$\epsilon_1 = e_1 + k_1(t)e_2 \tag{5.50}$$

where $e_2 = [\theta^T W_m(s)I - W_m(s)\theta^T]\omega$. If $k_1(t)$ is expressed as $k_1(t) = k_p/k_m + \psi_1(t)$, k_p/k_m can be considered as the desired value of $k_1(t)$, and $\psi_1(t)$ as the corresponding parameter error. The augmented error can now be expressed as

$$\epsilon_1 = \frac{k_p}{k_m}\phi^T \zeta + \psi_1 e_2 + \delta_3(t) \tag{5.51}$$

where $\zeta = W_m(s)I\omega$ and $\delta_3(t)$ is an exponentially decaying term due to initial conditions.

Since Eq. (5.51) is similar in form to Eq. (5.44) (except for the unknown constant k_p/k_m), the adaptive laws for adjusting $\phi(t)$ and $\psi_1(t)$ are given by

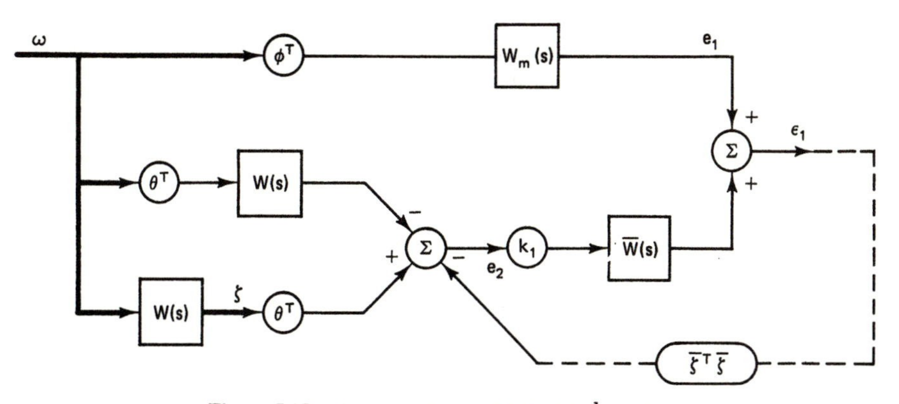

Figure 5.15 Augmented error: Method 2 (k_p unknown).

$$\begin{aligned} \dot{\phi} &= -sgn(k_p)\frac{\epsilon_1 \zeta}{1+\overline{\zeta}^T\overline{\zeta}} \\ \dot{\psi}_1 &= -\frac{\epsilon_1 e_2}{1+\overline{\zeta}^T\overline{\zeta}} \end{aligned} \tag{5.52}$$

where $\overline{\zeta} = W_m(s)I\overline{\omega}$. In the section following we show that these adaptive laws assure the global boundedness of all the signals in the overall adaptive system and that $\lim_{t\to\infty} e_1(t) = 0$.

Method 2. As in Method 1, the augmented error network has to be modified when k_p is unknown to obtain an error equation from which stable adaptive laws can be obtained. The modification once again takes the form of an additional gain $k_1(t)$, which is included in series with the transfer function $\overline{W}(s)$ as shown in Fig. 5.15. Since the analytical details are quite similar to those in Method 1, we merely include below the equations defining the various signals and the resultant augmented error equation.

$$\begin{aligned} \overline{W}(s) &= W_m(s)W^{-1}(s) \text{ is SPR} \\ \zeta &\stackrel{\Delta}{=} \overline{W}(s)\omega \\ e_2 &\stackrel{\Delta}{=} \left[\theta^T\zeta - W(s)\theta^T\omega\right] \\ \epsilon_1 &\stackrel{\Delta}{=} e_1 + \overline{W}(s)\left[k_1 e_2 - \epsilon_1\overline{\zeta}^T\overline{\zeta}\right] \end{aligned} \tag{5.53}$$

where $W(s)$ is asymptotically stable. Equation (5.53) can be further simplified as

$$\epsilon_1 = \overline{W}(s)\left[\frac{k_p}{k_m}\phi^T\zeta + \psi_1 e_2 - \epsilon_1\overline{\zeta}^T\overline{\zeta}\right] + \delta_4(t) \tag{5.54}$$

where $\psi_1(t) = k_1(t) - k_p/k_m$ and $\delta_4(t)$ is due to initial conditions and decays exponentially. Then adaptive laws of the form

$$\begin{aligned} \dot{\phi} &= -sgn(k_p)\epsilon_1\zeta \\ \dot{\psi}_1 &= -\epsilon_1 e_2 \end{aligned} \tag{5.55}$$

can be shown to assure the global stability of the adaptive system.

5.5 PROOF OF GLOBAL STABILITY OF THE ADAPTIVE SYSTEM

In contrast to all the methods used thus far, Lyapunov's method cannot be applied directly to demonstrate the global stability of the adaptive systems discussed in this section. The method merely serves to establish the boundedness of the parameter errors. Following this, using arguments based on the input-output relationships of various linear subsystems of the overall nonlinear system, it is shown that the state vector of the latter is in $\mathcal{L}^\infty$. The proof for this is by contradiction and is given at the end of the section. This marks a major departure in the approach used to prove the boundedness of all the signals. The same procedure also is needed in the chapters following, where the adaptive control of plants with relative degrees greater than 1 are discussed.

In view of its complexity, we present below, in qualitative terms, the major steps involved in the proof of global stability of the adaptive system:

Step 1: From the adaptive laws in Eqs. (5.45), (5.48), (5.52), or (5.55), it follows directly that the parameter vector θ is uniformly bounded and that $\dot{\theta} \in \mathcal{L}^2$.

Step 2: If $\theta(t)$ is considered as a bounded time-varying parameter in a linear system, we conclude that the state of the system can grow at most exponentially.

Step 3: Assuming that the signals grow in an unbounded fashion, the specific structure of the controller is used to relate the growth rates of the signals in the system. This is possible because most of the subsystems are linear and time invariant, and hence the methods outlined in Appendix B and Chapter 2 are directly applicable. In particular, the signals y_p, ω_2, ω and ζ are shown to grow at the same rate. This is achieved without reference to any specific adaptive law.

Step 4: From the fact that $\dot{\theta} \in \mathcal{L}^2$, it is shown that the signals ω_2 and ω do not grow at the same rate, which results in a contradiction. As a consequence, it follows that all signals are uniformly bounded and that $\lim_{t\to\infty} e_1(t) = 0$.

In the section that follows, we discuss the four steps above in detail when k_p is assumed to be known, using methods 1 and 2. The proof of stability when k_p is unknown is then given briefly.

5.5.1 k_p known

Before proceeding to consider the stability of the adaptive system, we shall briefly discuss the effect of the initial conditions in the auxiliary network resulting in the exponentially decaying term $\delta_1(t)$. Since the overall system is nonlinear, it is necessary to show that ignoring this term does not affect the stability analysis. We shall show below that the boundedness of $\overline{\theta}$ as well as $\dot{\overline{\theta}} \in \mathcal{L}^2$ stated in step 1 are unaffected by $\delta_1(t)$. Steps 2-4 involve arguments regarding unbounded signals and hence are unaltered by the presence of $\delta_1(t)$.

Method 1

Step 1. If the term $\delta_1(t)$ in Eq. (5.44) is neglected, the adaptive law has the form

$$\dot{\overline{\phi}} = -\frac{\overline{\zeta}\,\overline{\zeta}^T}{1+\overline{\zeta}^T\overline{\zeta}}\overline{\phi}. \tag{5.45}$$

Using a Lyapunov function candidate $V(\overline{\phi}) = \overline{\phi}^T\overline{\phi}/2$, we obtain the time derivative evaluated along the trajectories of Eq. (5.45) as

$$\dot{V}(\overline{\phi}) = -\frac{\epsilon_1^2}{1+\overline{\zeta}^T\overline{\zeta}} \leq 0.$$

Hence, $\overline{\phi}(t)$ is uniformly bounded for all $t \geq t_0$. Further, since

$$-\int_{t_0}^{\infty} \dot{V}\,dt = -V(\infty) + V(t_0) < \infty,$$

it follows that $\epsilon_1/(1+\overline{\zeta}^T\overline{\zeta})^{1/2} \in \mathcal{L}^2$ or equivalently

$$\epsilon_1 = \beta(t)\sqrt{1+\overline{\zeta}^T\overline{\zeta}} \qquad \beta \in \mathcal{L}^2.$$

From Eq. (5.45), it can be concluded that

$$\dot{\overline{\phi}} \in \mathcal{L}^2 \cap \mathcal{L}^\infty.$$

Effect of Initial Conditions. If $\delta_1(t)$ in Eq. (5.44) is not neglected, the adaptive law has the form

$$\dot{\overline{\phi}} = -\frac{\overline{\zeta}\,\overline{\zeta}^T}{1+\overline{\zeta}^T\overline{\zeta}}\overline{\phi} - \frac{\overline{\zeta}\delta_1}{1+\overline{\zeta}^T\overline{\zeta}}. \tag{5.56}$$

$\delta_1(t)$ can be expressed as the output of a system described by the equations

$$\dot{z}_1 = Az_1 \qquad \delta_1 = h^T z_1 \tag{5.57}$$

with $z_1(t_0) = z_{10}$, A is an asymptotically stable matrix and (h^T, A) is detectable. Choosing a Lyapunov function candidate

$$V(\overline{\phi}, z_1) = \frac{1}{2}\overline{\phi}^T\overline{\phi} + \frac{1}{4}z_1^T P z_1$$

where $P = P^T > 0$ is such that $A^T P + PA = -hh^T - Q$, where $Q > 0$, we obtain for the time derivative of V along the trajectories of Eqs. (5.56) and (5.57) as

$$\begin{aligned}\dot{V} &\leq -\frac{(\overline{\phi}^T\overline{\zeta})^2}{1+\overline{\zeta}^T\overline{\zeta}} - \frac{(\overline{\phi}^T\overline{\zeta})\delta_1}{1+\overline{\zeta}^T\overline{\zeta}} - \frac{\delta_1^2}{4} \\ &= -\left(\frac{\overline{\phi}^T\overline{\zeta} + \delta_1/2}{\sqrt{1+\overline{\zeta}^T\overline{\zeta}}}\right)^2 - \frac{\overline{\zeta}^T\overline{\zeta}\delta_1^2}{4(1+\overline{\zeta}^T\overline{\zeta})} \leq 0. \end{aligned} \quad (5.58)$$

From Eq. (5.58), it follows that $\overline{\phi} \in \mathcal{L}^\infty$ and $\dot{\overline{\phi}} \in \mathcal{L}^2$. Hence the result is identical to that obtained by omitting the term $\delta_1(t)$ in the error equation. In the following sections, we shall consequently ignore exponentially decaying terms due to initial conditions for which similar arguments can be given.

Step 2. The closed-loop system, consisting of plant and controller, is given by Eq. (5.37) and can be rewritten as

$$\dot{x}(t) = \left[A_{mn} + b_{mn}\overline{\phi}^T(t)C\right]x(t) + b_{mn}r(t) \quad (5.59)$$

where C is a $(2n-1) \times (3n-2)$ matrix of the form

$$C = \begin{bmatrix} 0 & I & 0 \\ h_p^T & 0 & 0 \\ 0 & 0 & I \end{bmatrix}. \quad (5.60)$$

Since $\overline{\phi}(t) \in \mathcal{L}^\infty$, Eq. (5.59) can be considered as a linear time-varying differential equation with bounded coefficients. It follows that $\|x(t)\|$ grows at most exponentially. Also, since r is a piecewise-continuous uniformly bounded function, all components of $x(t)$ and $\overline{\zeta}(t)$ belong to $\mathcal{PC}_{[0,\infty)}$.

Step 3. Let the signals of the system grow in an unbounded fashion, that is, $\lim_{t\to\infty}\sup_{\tau\leq t}\|x(\tau)\| = \infty$. In Eq. (5.36), Λ is an asymptotically stable matrix, and $\{u(\cdot), \omega_1(\cdot)\}$ and $\{y_p(\cdot), \omega_2(\cdot)\}$ are corresponding input-output pairs. Hence, from Lemma 2.6, we have

$$\|\omega_2(t)\| = O[\sup_{\tau\leq t}|y_p(\tau)|] \quad (5.61)$$

$$\|\omega_1(t)\| = O[\sup_{\tau\leq t}|u(\tau)|]. \quad (5.62)$$

Since $W_m(s)$ is an asymptotically stable transfer function, we conclude that

$$\|\overline{\zeta}(t)\| = O[\sup_{\tau\leq t}\|\overline{\omega}(\tau)\|]. \tag{5.63}$$

Also, since the output of the plant is given by

$$y_p(t) = W_m(s)\left[\overline{\phi}^T(t)\overline{\omega}(t) + r(t)\right]$$

where $\overline{\phi}$ is uniformly bounded, we have

$$|y_p(t)| = O\left[\sup_{\tau\leq t}\|\overline{\omega}(\tau)\|\right]. \tag{5.64}$$

Since y_p is a component of $\overline{\omega}$, from Eq. (5.64), we can conclude that

$$\|\underline{\omega}(t)\| \sim \sup_{\tau\leq t}\|\overline{\omega}(\tau)\| \qquad \text{where } \underline{\omega}(t) = [\omega_1^T(t), \omega_2^T(t)]^T. \tag{5.65}$$

From Appendix C,

$$\left|\frac{d}{dt}\|\overline{\omega}(t)\|\right| \leq \|\dot{\overline{\omega}}(t)\| \leq c_1\|\overline{\omega}(t)\| + c_2, \qquad c_1, c_2 \in \mathbb{R}^+. \tag{5.66}$$

that is, $\overline{\omega}(t)$ can be considered to be the effective state of the system. From Eqs. (5.65) and (5.66), it follows that

$$\|\dot{\omega}_1(t)\| = O[\sup_{\tau\leq t}\|\underline{\omega}(\tau)\|].$$

Since $W_p(s)\omega_1(t) = \omega_2(t)$ and $W_p(s)$ has its zeros in $\mathbb{C}^-$, it follows from Lemma 2.8 that

$$\|\omega_1(t)\| = O\left[\sup_{\tau\leq t}\|\omega_2(\tau)\|\right]. \tag{5.67}$$

Similarly, since $W_m(s)\overline{\omega}(t) = \overline{\zeta}(t)$, from Eq. (5.66) it follows that

$$\|\overline{\omega}(t)\| = O\left[\sup_{\tau\leq t}\|\overline{\zeta}(\tau)\|\right]. \tag{5.68}$$

Finally, since ω_2 is a sub-vector of $\underline{\omega}$, from Eq. (5.67), it follows that

$$\|\omega_2(t)\| \sim \sup_{\tau\leq t}\|\underline{\omega}(\tau)\|. \tag{5.69}$$

All the results in step 3 can therefore be summarized from Eqs. (5.61), (5.63), (5.64), (5.65), (5.68), and (5.69) as

$$\sup_{\tau\leq t}|y_p(\tau)| \sim \sup_{\tau\leq t}\|\omega_2(\tau)\| \sim \sup_{\tau\leq t}\|\overline{\omega}(\tau)\| \sim \sup_{\tau\leq t}\|\overline{\zeta}(\tau)\|. \tag{5.70}$$

From Eq. (5.67), and since $u = r + \overline{\theta}^T \overline{\omega}$, we also have

$$|u(t)|, \|\omega_1(t)\| \quad = \quad O\left[\sup_{\tau \leq t} \|\overline{\omega}(\tau)\|\right]. \tag{5.71}$$

Step 4. From Fig. 5.12, we have

$$e_2(t) \quad = \quad \left[\overline{\phi}^T(t) W_m(s) I - W_m(s) \overline{\phi}^T(t)\right] \overline{\omega}(t).$$

Since $\dot{\overline{\phi}} \in \mathcal{L}^2$, from Lemma 2.11, it follows that

$$e_2(t) \quad = \quad o\left[\sup_{\tau \leq t} \|\overline{\omega}(\tau)\|\right]. \tag{5.72}$$

Hence, the output of the plant can be expressed as

$$\begin{aligned} y_p(t) &= y_m(t) + e_1(t) = y_m(t) + \epsilon_1(t) - e_2(t) \\ &= y_m(t) + \beta(t)\sqrt{1 + \overline{\zeta}^T(t)\overline{\zeta}(t)} + o\left[\sup_{\tau \leq t} \|\overline{\omega}(\tau)\|\right]. \end{aligned}$$

Since $\omega_2(t) = (sI - \Lambda)^{-1} \ell y_p(t)$, we have

$$\|\omega_2(t)\| \quad \leq \quad \left\|(sI - \Lambda)^{-1} \ell y_m(t)\right\| + o\left[\sup_{\tau \leq t} \|\overline{\zeta}(\tau)\|\right] + o\left[\sup_{\tau \leq t} \|\overline{\omega}(\tau)\|\right]$$

from Lemma 2.9 and, hence,

$$\|\omega_2(t)\| \quad = \quad o[\sup_{\tau \leq t} \|\overline{\omega}(\tau)\|].$$

This contradicts Eq. (5.70) according to which $\omega_2(t)$ and $\overline{\omega}(t)$ grow at the same rate if they grow in an unbounded fashion. Hence, all the signals in the feedback loop are uniformly bounded.

Since $\overline{\zeta}$ is bounded, the augmented error ϵ_1 is bounded. Since $\dot{\overline{\zeta}}$ is also bounded, $\dot{\epsilon}_1$ is bounded. Finally, $\dot{\overline{\phi}} \in \mathcal{L}^2$ implies that $\epsilon_1 \in \mathcal{L}^2$. Hence, $\lim_{t \to \infty} \epsilon_1(t) = 0$. From Eq. (5.72), we obtain that $\lim_{t \to \infty} e_2(t) = 0$ and therefore the true error $e_1(t)$ tends to zero asymptotically.

Method 2

Step 1. Here, the equation that relates the parameter error to the output error has the form [refer to Eq. (5.49)]

$$\dot{\epsilon} \quad = \quad A_1 \epsilon + b_1 \left(\overline{\phi}^T \overline{\zeta} - \epsilon_1 \overline{\zeta}^T \overline{\zeta}\right); \qquad \epsilon_1 \quad = \quad h_1^T \epsilon \tag{5.73}$$

where

$$h_1^T(sI - A_1)^{-1}b_1 \quad \triangleq \quad \overline{W}(s)$$

is strictly positive real. As mentioned in the previous section, the adaptive law is chosen as

$$\dot{\overline{\phi}} = -\epsilon_1 \overline{\zeta}. \tag{5.48}$$

If $V(\epsilon, \overline{\phi}) = \epsilon^T P \epsilon + \overline{\phi}^T \overline{\phi}$ is a Lyapunov function candidate, its time derivative evaluated along the solutions of Eqs. (5.73) and (5.48) has the form

$$\dot{V}(\epsilon, \overline{\phi}) = \epsilon^T \left[A_1^T P + P A_1\right] \epsilon + 2\epsilon^T P b_1 \left(\overline{\phi}^T \overline{\zeta} - \epsilon_1 \overline{\zeta}^T \overline{\zeta}\right) - 2\epsilon_1 \overline{\phi}^T \overline{\zeta}.$$

Since $\overline{W}(s)$ is strictly positive real, there exist symmetric positive-definite matrices P and Q such that the equations $A_1^T P + P A_1 = -Q$ and $P b_1 = h_1$ are satisfied simultaneously. Hence,

$$\dot{V}(\epsilon, \overline{\phi}) = -\epsilon^T Q \epsilon - 2\epsilon_1^2 \overline{\zeta}^T \overline{\zeta} \leq 0,$$

so that ϵ and $\overline{\phi}$ are uniformly bounded. We can also conclude that $\epsilon \in \mathcal{L}^2$ and $\epsilon_1 \overline{\zeta} \in \mathcal{L}^2$, or equivalently from Eq. (5.48), $\dot{\overline{\phi}} \in \mathcal{L}^2$.

Step 2. Since the feedback loop is common to both methods, as in Method 1, we can conclude that the state x, in Eq. (5.59), can grow at most exponentially.

Step 3. The same conclusions as in Eqs. (5.70) and (5.71) can once again be drawn since $\overline{\phi} \in \mathcal{L}^\infty$.

Step 4. From Fig. 5.13,

$$e_2(t) = o\left[\sup_{\tau \leq t} \|\overline{\omega}(\tau)\|\right] \tag{5.72}$$

since $\dot{\overline{\phi}} \in \mathcal{L}^2$ and $W(s)$ is an asymptotically stable transfer function. Since $\overline{W}(s)$ is also asymptotically stable and $\epsilon_1 \overline{\zeta} \in \mathcal{L}^2$, the results of Lemma 2.9 imply that

$$\overline{W}(s)\left[\left(\epsilon_1(t)\overline{\zeta}(t)\right)^T \overline{\zeta}(t)\right] = o\left[\sup_{\tau \leq t} \|\overline{\zeta}(\tau)\|\right]. \tag{5.74}$$

Once again, from Eqs. (5.72) and (5.74), the output of the plant can be expressed as

$$\begin{aligned} y_p(t) &= y_m(t) + e_1(t) \\ &= y_m(t) + \epsilon_1(t) - \overline{W}(s)[e_2(t) - \epsilon_1(t)\overline{\zeta}^T(t)\overline{\zeta}(t)] \\ &= y_m(t) + \epsilon_1(t) + o\left[\sup_{\tau \leq t} \|\overline{\omega}(\tau)\|\right] + o\left[\sup_{\tau \leq t} \|\overline{\zeta}(\tau)\|\right]. \end{aligned}$$

Since y_m and ϵ_1 are uniformly bounded, we have

$$y_p(t) = o\left[\sup_{\tau \leq t} \|\overline{\omega}(\tau)\|\right]. \tag{5.75}$$

Equation (5.75) contradicts Eq. (5.70) according to which y_p and $\|\overline{\omega}\|$ grow at the same rate. Hence, we conclude that all the signals in the feedback loop are uniformly bounded.

In addition, from Eq. (5.73), $\dot{\epsilon}$ is bounded. Since $\epsilon \in \mathcal{L}^2$, $\lim_{t\to\infty} \epsilon(t) = 0$, and therefore $\lim_{t\to\infty} \epsilon_1(t) = 0$. From Eq. (5.72), $\lim_{t\to\infty} e_2(t) = 0$ and hence $e_1(t)$ tends to zero asymptotically.

Simulation 5.2 The following adaptive control problem was simulated to study the behavior of solutions when the relative degree of the plant is greater than 1. The transfer functions of the plant and the model were chosen to be

$$W_p(s) = \frac{1}{s(s-1)} \quad \text{and} \quad W_m(s) = \frac{1}{s^2+2s+1}$$

respectively. The fixed control parameters $\Lambda = -1$ and $\ell = 1$ in Eq. (5.18). Three parameters $\theta_1(t)$, $\theta_0(t)$, and $\theta_2(t)$ were adjusted using method 1. The true values of the control parameters are $\theta_1^* = -3$, $\theta_0^* = -7$, and $\theta_2^* = 6$. The augmented error ϵ_1 and the norm of the parameter error vector $\|\phi\|$ are shown in Figs. 5.16(a)-(c) for the cases (a) $r(t) \equiv 0$, (b) $r(t) \equiv 10$, and (c) $r(t) = 10 \cos t + 50 \cos 5t$ and initial conditions were chosen as indicated in Fig. 5.16. Once again, as in Simulation 5.1, in all cases, $e_1(t)$ tends to zero asymptotically, while the parameter errors do not converge to zero when $r(t)$ is zero or a constant. The latter converge to zero when $r(t)$ has two distinct frequencies.

5.5.2 k_p unknown

Method 1

Step 1. The error equation in this case is given by

$$\epsilon_1 = \frac{k_p}{k_m}\phi^T\zeta + \psi_1 e_2 \tag{5.51}$$

and the adaptive laws for adjusting ϕ and ψ_1 are

$$\begin{aligned} \dot{\phi} &= -sgn(k_p)\frac{\epsilon_1\zeta}{1+\overline{\zeta}^T\overline{\zeta}} \\ \dot{\psi}_1 &= -\frac{\epsilon_1 e_2}{1+\overline{\zeta}^T\overline{\zeta}} \end{aligned} \tag{5.52}$$

where $\overline{\zeta} = W_m(s)I\overline{\omega}$. A quadratic function V defined as

$$V = \frac{1}{2}\left[\frac{|k_p|}{k_m}\phi^T\phi + \psi_1^2\right]$$

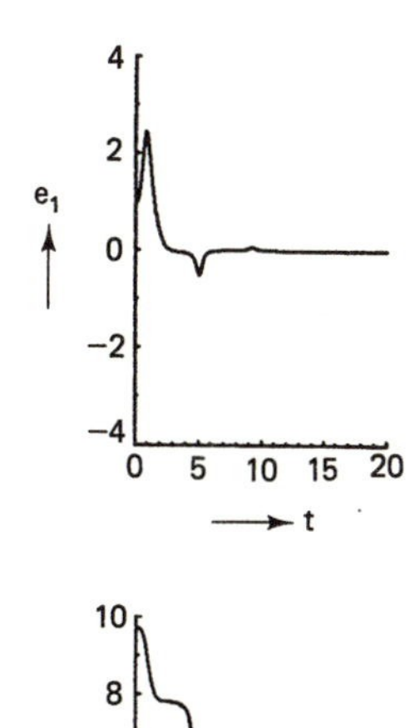

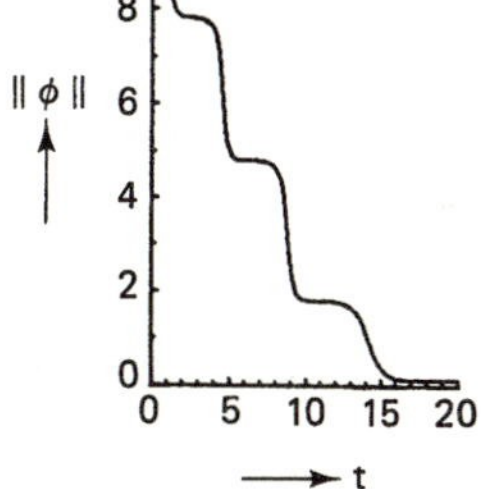

(a) $r(t) = 0.0$ (b) $r(t) = 10.0$ (c) $r(t) = 10\cos t + 50\cos 5t$

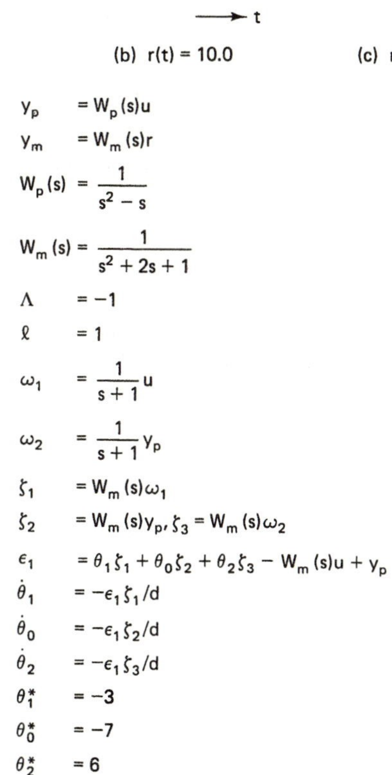

$$y_p = W_p(s)u$$
$$y_m = W_m(s)r$$
$$W_p(s) = \frac{1}{s^2 - s}$$
$$W_m(s) = \frac{1}{s^2 + 2s + 1}$$
$$\Lambda = -1$$
$$\ell = 1$$
$$\omega_1 = \frac{1}{s+1}u$$
$$\omega_2 = \frac{1}{s+1}y_p$$
$$\zeta_1 = W_m(s)\omega_1$$
$$\zeta_2 = W_m(s)y_p, \zeta_3 = W_m(s)\omega_2$$
$$\epsilon_1 = \theta_1\zeta_1 + \theta_0\zeta_2 + \theta_2\zeta_3 - W_m(s)u + y_p$$
$$\dot{\theta}_1 = -\epsilon_1\zeta_1/d$$
$$\dot{\theta}_0 = -\epsilon_1\zeta_2/d$$
$$\dot{\theta}_2 = -\epsilon_1\zeta_3/d$$
$$\theta_1^* = -3$$
$$\theta_0^* = -7$$
$$\theta_2^* = 6$$

Figure 5.16 Simulation results when $n^* = 2$.

yields a time derivative $\dot{V}$, and the latter can be evaluated along the trajectories of Eqs. (5.51) and (5.52) as

$$\dot{V} = -\frac{\epsilon_1^2}{1+\overline{\zeta}^T\overline{\zeta}} \leq 0.$$

Therefore, ϕ and ψ_1 are in $\mathcal{L}^\infty$. Also, since $\int_{t_0}^{\infty} \dot{V}dt < \infty$, it follows that

$$\frac{\epsilon_1}{\sqrt{1+\overline{\zeta}^T\overline{\zeta}}} \in \mathcal{L}^2.$$

Since $\zeta = [y_m, \overline{\zeta}^T]^T$, we also have $\zeta/(1+\overline{\zeta}^T\overline{\zeta})^{1/2} \in \mathcal{L}^\infty$. Therefore

$$\dot{\phi} \in \mathcal{L}^2.$$

Step 2. The closed-loop system, consisting of plant and controller, is given by Eq. (5.19) and can be rewritten as

$$\dot{x}(t) = \left[A_{mn} + b_{mn}\overline{\phi}^T(t)C\right]x(t) + b_{mn}(\psi(t)+k^*)r(t) \tag{5.76}$$

where $\phi = [\psi, \overline{\phi}^T]^T$, and C is given by Eq. (5.60). Since $\phi \in \mathcal{L}^\infty$, Eq. (5.76) can be considered as a linear time-varying differential equation with bounded coefficients. Hence, it follows that $\|x(t)\|$ grows at most exponentially.

Step 3. The only difference in the control loop between the cases when k_p is known and when it is unknown is the presence of two additional parameters $k(t)$ in the feedforward path and $k_1(t)$ in the generation of the augmented error. Since step 1 assures the boundedness of $k(t)$ and $k_1(t)$, the relation between the growth rates of various signals in the loop is unaffected; once again, assuming that all signals grow in an unbounded fashion, we can conclude that

$$\sup_{\tau\leq t}|y_p(\tau)| \sim \sup_{\tau\leq t}\|\omega_2(\tau)\| \sim \sup_{\tau\leq t}\|\overline{\omega}(\tau)\| \sim \sup_{\tau\leq t}\|\overline{\zeta}(\tau)\|.$$

Since $\omega = [r, \overline{\omega}^T]^T$ and $\zeta = [y_m, \overline{\zeta}^T]^T$, and r is uniformly bounded, we have

$$\sup_{\tau\leq t}\|\overline{\omega}(\tau)\| \sim \sup_{\tau\leq t}\|\overline{\zeta}(\tau)\| \sim \sup_{\tau\leq t}\|\omega(\tau)\| \sim \sup_{\tau\leq t}\|\zeta(\tau)\|$$

and

$$|u(t)|, \|\omega_1(t)\| = O\left[\sup_{\tau\leq t}\|\omega(\tau)\|\right].$$

Step 4. From Fig. 5.14, we have

$$e_2(t) = [\phi^T(t)W_m(s)I - W_m(s)\phi^T(t)]\omega(t).$$

Since $\dot{\phi} \in \mathcal{L}^2$, from Lemma 2.11, it follows that

$$e_2(t) = o\left[\sup_{\tau \leq t} \|\omega(\tau)\|\right].$$

The output of the plant can be expressed as

$$\begin{aligned} y_p(t) &= y_m(t) + e_1(t) = y_m(t) + \epsilon_1(t) - k_1(t)e_2(t) \\ &= y_m(t) + \beta(t)\sqrt{1 + \overline{\zeta}^T(t)\overline{\zeta}(t)} + o\left[\sup_{\tau \leq t} \|\omega(\tau)\|\right]. \end{aligned}$$

As in Section 5.5.1, we can conclude that

$$\|\omega_2(t)\| = o[\sup_{\tau \leq t} \|\omega(\tau)\|].$$

This contradicts the conclusion drawn from step 3 where it was shown that $\omega_2(t)$ and $\omega(t)$ grow at the same rate if they grow in an unbounded fashion. Hence, all the signals in the feedback loop are uniformly bounded.

It can once again be shown that $\lim_{t \to \infty} e_1(t) = 0$.

Method 2

Step 1. Equation (5.54) can be rewritten as

$$\dot{\epsilon} = A_1\epsilon + b_1\left((k_p/k_m)\phi^T\zeta + \psi_1 e_2 - \epsilon_1\overline{\zeta}^T\overline{\zeta}\right); \qquad \epsilon_1 = h_1^T\epsilon$$

where

$$h_1^T(sI - A_1)^{-1}b_1 \stackrel{\Delta}{=} \overline{W}(s)$$

is strictly positive real. The adaptive laws for adjusting ϕ and ψ_1 are given by Eq. (5.55). A Lyapunov function candidate V of the form

$$V = \epsilon^T P\epsilon + (|k_p|/k_m)\phi^T\phi + \psi_1^2$$

has a time derivative evaluated along the solutions of Eqs. (5.54) and (5.55) given by

$$\begin{aligned} \dot{V} = \; &\epsilon^T\left[A_1^T P + PA_1\right]\epsilon + 2\epsilon^T Pb_1\left((k_p/k_m)\phi^T\zeta + \psi_1 e_2 - \epsilon_1\overline{\zeta}^T\overline{\zeta}\right) \\ &- 2(k_p/k_m)\epsilon_1\phi^T\zeta - 2\epsilon_1\psi_1 e_2. \end{aligned}$$

As before, since $\overline{W}(s)$ is strictly positive real, symmetric positive-definite matrices P and Q, such that $A_1^T P + P A_1 = -Q$ and $P b_1 = h_1$, can be chosen. Hence,

$$\dot{V} \;=\; -\epsilon^T Q \epsilon - 2\epsilon_1^2 \overline{\zeta}^T \overline{\zeta} \leq 0,$$

so that ϵ, ϕ, and ψ_1 belong to $\mathcal{L}^\infty$. As before, $\epsilon \in \mathcal{L}^2$ and $\epsilon_1 \overline{\zeta} \in \mathcal{L}^2$. Since $\zeta = [y_m, \overline{\zeta}^T]^T$, it follows that $\dot{\phi} \in \mathcal{L}^2$.

Step 2. Since the feedback loop is the same as in the previous cases, we can conclude that the state x, in Eq. (5.59), can grow at most exponentially.

Step 3. The same conclusions regarding growth rates of the various signals can be drawn as in method 1, since $\phi \in \mathcal{L}^\infty$.

Step 4. From Fig. 5.15,

$$e_2(t) \;=\; o\left[\sup_{\tau \leq t} \|\omega(\tau)\|\right]$$

since $\dot{\phi} \in \mathcal{L}^2$ and $W(s)$ is an asymptotically stable transfer function. Also, since $\overline{W}(s)$ is asymptotically stable and $\epsilon_1 \overline{\zeta} \in \mathcal{L}^2$, Eq. (5.74) holds. The output of the plant can be expressed as

$$\begin{aligned} y_p(t) &= y_m(t) + e_1(t) \\ &= y_m(t) + \epsilon_1(t) - \overline{W}(s)[k_1(t) e_2(t) - \epsilon_1(t) \overline{\zeta}^T(t) \overline{\zeta}(t)] \\ &= y_m(t) + \epsilon_1(t) + o[\sup_{\tau \leq t} \|\omega(\tau)\|] + o[\sup_{\tau \leq t} \|\overline{\zeta}(\tau)\|] \end{aligned}$$

since $e_2(t) = o[\sup_{\tau \leq t} \|\omega(\tau)\|]$ and $\epsilon_1 \overline{\zeta} \in \mathcal{L}^2$. Since y_m and ϵ_1 are uniformly bounded, and ω and $\overline{\zeta}$ grow at the same rate, we have

$$y_p(t) \;=\; o[\sup_{\tau \leq t} \|\omega(\tau)\|].$$

Hence, as before, we have a contradiction and all signals are uniformly bounded. In a similar fashion, it can be shown that $\lim_{t \to \infty} e_1(t) = 0$.

5.6 COMMENTS ON THE CONTROL PROBLEM

In view of the importance of the control problem in the design of adaptive systems, a detailed stability analysis was presented in the preceding sections. The same analysis is also found to be applicable to more complex situations considered in the chapters following. In this section we briefly review the assumptions made in Sections 5.3 and

5.4, the methods used to generate the adaptive laws and the stability analysis used in Section 5.5 to prove the boundedness of the solutions.

The Relative Degree

From Sections 5.3 and 5.4, it is seen that the stability analysis of the adaptive system is very different for the two cases when $n^* = 1$ and $n^* \geq 2$. This is found to recur in all the problems considered in later chapters which use a direct approach. When $n^* = 1$, the existence of an explicit Lyapunov function assures the stability of the overall system. This is no longer the case for $n^* \geq 2$ (see Section 5.5) where the existence of a Lyapunov function merely assures the boundedness of all the parameters. This accounts for the fact that the solutions to more complex problems in which $n^* = 1$ cannot be directly extended to the case where the relative degree of the plant is greater than one. Examples of such systems are given in Chapter 8.

Assumptions (I)

The global stability of the adaptive control of LTI plants was shown in this chapter, provided the plant transfer function $W_p(s)$ satisfies assumptions (I) mentioned in Section 5.2. Exact knowledge of the relative degree n^* [assumption (ii)] was needed to determine the reference model $W_m(s)$. An upper bound on the order n of $W_p(s)$ [assumption (i)] was required to establish the existence of a controller that can match the closed-loop system transfer function to $W_m(s)$. Assumptions (ii) and (iii) therefore stem from the algebraic part of the adaptive control problem. The sign of the high frequency gain [assumption (i)] and the condition that the zeros of $W_p(s)$ lie in $\mathbb{C}^-$ [assumption (iv)] were needed for the generation of stable adaptive laws. The need for assumption (i) was evident even in Chapter 3 while dealing with simple adaptive systems. Assumption (iv), on the other hand, assures that there are no pole-zero cancellations in the right half of the complex plane. It was also explicitly used to derive Lemma 2.8 which was, in turn, needed to establish the relations in Eqs. (5.67) and (5.68) in the proof of stability. The various ways in which assumptions (I) can be relaxed are presented in Chapter 9.

Methods 1 and 2

Two methods were suggested for determining the adaptive laws for adjusting the control parameter vector θ. Using the notation introduced in Section 5.3.4, we can say that the two methods use the same controller but different tuners. In Method 1, the augmented error equation obtained has the same form as the error equation obtained in the simpler case when $n^* = 1$. However, an additional normalization factor is used in the adaptive law. In Method 2, the error equation is modified by using nonlinear feedback but the adaptive law remains the same as in the simple cases. The modifications suggested in methods 1 and 2 are aimed at assuring that $\theta \in \mathcal{L}^\infty$ and that $\dot{\theta} \in \mathcal{L}^2$.

Stability Analysis

Whether Method 1 or Method 2 is used, an explicit Lyapunov function assuring the stability of the overall system cannot be obtained for the case when $n^* \geq 2$. Instead,

Lyapunov functions are used in both methods to generate adaptive laws that merely assure the boundedness of $\theta(t)$. This, in turn, implies that the state of the system can grow at most exponentially.

Since the structure of the controller using the two methods is the same, the growth rates of various signals in the adaptive loop can be analyzed without explicit reference to the adaptive law. However, the fact that $\dot{\theta} \in \mathcal{L}^2$, which is a consequence of the adaptive law used, results in a contradiction if it is assumed that the signals grow in an unbounded fashion. This, in turn, assures the boundedness of all the signals in the system and it can also be shown that the output error $e_1(t)$ tends to zero asymptotically.

From the preceding discussion, it is clear that $\theta \in \mathcal{L}^\infty$ and $\dot{\theta} \in \mathcal{L}^2$ are the two properties of the controller that play an important role in the proof of stability.

5.7 SOME REMARKS ON THE CONTROLLER STRUCTURE

In the preceding sections it was shown that, under assumptions (I), a linear time-invariant plant with unknown parameters can be controlled in a stable fashion to follow a desired output $y_m(t)$ asymptotically. The emphasis throughout was on stability analysis rather than the details pertaining to the choice of the controller structure or its parametrization. We address some of these issues in this section.

(a) The adaptive controller in Fig. 5.7 is of order $2n-2$ [Eq. (5.16)] and has $2n$ adjustable parameters when $n^* = 1$. When k_p is known, the latter reduces to $2n-1$ parameters. When $n^* \geq 2$, the adaptive controller is of the same order, but an auxiliary signal ζ was generated to implement the tuner. To generate the signal ζ as $\zeta(t) = W_m(s)I\omega(t)$, $2n^2$ integrators are needed. However, ζ can also be realized using only $4n-2$ integrators as shown below, since the two transfer functions $W_m(s)$ and $(sI-\Lambda)^{-1}\ell$ commute:

$$u'(t) = W_m(s)u(t); \qquad y_p'(t) = W_m(s)y_p(t)$$

$$\begin{aligned} \dot{\zeta}_1 &= \Lambda\zeta_1 + \ell u' \\ \dot{\zeta}_2 &= \Lambda\zeta_2 + \ell y_p' \\ \zeta &= [y_m,\ \zeta_1^T,\ y_p', \zeta_2^T]^T. \end{aligned}$$

The total number of adjustable parameters in this case is $2n+1$.

(b) The parametrization used in Eq. (5.16) can be replaced by the following parametrization

$$\begin{aligned} \dot{\omega}_1 &= F\omega_1 + gu \\ \dot{\omega}_2 &= F\omega_2 + gy_p \\ u &= kr + \theta_1^T\omega_1 + \theta_2^T\omega_2 \end{aligned} \tag{5.77}$$

where $F \in \mathbb{R}^{n \times n}$ is an asymptotically stable matrix, and (F, g) is controllable. This is motivated by Representation 2 discussed in Chapter 4 and results in a system of order $2n$ containing $2n + 2$ adjustable parameters.

(c) Whether the representation in Eqs. (5.16) or (5.77) is used, the methods described thus far only assure the convergence of the output error to zero. The convergence of the parameter vector $\theta(t)$ to its desired value θ^* or the rate at which the output error converges to zero have not been discussed. These depend on the nature of the reference input and are considered in Chapter 6. In this context the choice of (Λ, ℓ) [or (F, g)] in the representation in Eq. (5.16) [or Eq. (5.77)] becomes relevant.

(d) The order of the controller as given in (a) is seen to depend on the order n of the plant. Whether the controller is parametrized as in Eq. (5.16) or Eq. (5.77), the desired parameter vector is unique since the corresponding Bezout identity is given by

$$\left[\lambda(s) - C^*(s)\right] Q(s) + D^*(s)P(s) = Q^*(s). \tag{5.32}$$

From the parametrization in Eq. (5.16), $Q^*(s)$ is a $(2n-1)^{\text{th}}$ degree monic polynomial, and $C^*(s)$ and $D^*(s)$ have $2n - 1$ coefficients that can be chosen uniquely. With the controller as in Eq. (5.77), $Q^*(s)$ becomes a $2n^{\text{th}}$ degree polynomial, while both $C^*(s)$ and $D^*(s)$ are polynomials of degree $n - 1$, and, hence, can be uniquely solved in Eq. (5.32). If higher order controllers are used, then the resultant parameter vector θ^* is not unique. This follows from Eq. (5.32), by choosing $\lambda(s)$ to be an $\overline{n}^{\text{th}}$ degree polynomial, $\overline{n} > n$, with $C^*(s)$ and $D^*(s)$ as polynomials of degree $\overline{n} - 1$. $Q^*(s)$ is then a monic polynomial of degree $\overline{n} + n$. As shown in Chapter 8, the additional $\overline{n} - n$ degrees of freedom can be used in certain cases to reject external disturbances.

(e) In all the previous sections it was assumed that n, the order of the plant, is known. When only $\overline{n}$, an upper bound on the order of the plant is specified, the controller is of dimension $2\overline{n}$ and, hence, is overparametrized. In such a case, the comments made in (d) apply.

(f) One of the important modifications made in the adaptive law when $n^* \geq 2$ is the use of a normalization factor. For example, in Eq. (5.52) (with $sgn(k_p) = 1$),

$$\dot{\theta} = \dot{\phi} = -\frac{\epsilon_1 \zeta}{1 + \overline{\zeta}^T \overline{\zeta}}$$

where the normalization factor is $1 + \overline{\zeta}^T \overline{\zeta}$. In the proof of stability, this is used to show that the augmented error is of the form

$$\epsilon_1 = \beta(t)\sqrt{1+\overline{\zeta}^T\overline{\zeta}}$$

where $\beta \in \mathcal{L}^2$. Although increasing the normalization factor does not affect the stability arguments used, it obviously decreases the speed of adaptation. In the chapters following, we use other normalization factors to assure robustness in the presence of perturbations.

(g) As in the case of $n^* = 1$, the analysis in Sections 5.4 and 5.5 can still be carried out if the relative degree n_m^* of $W_m(s)$ is greater than n^*, by choosing the reference input into the plant as $r'(t)$ where $r'(t) = W(s)r(t)$, $W_m(s) = \overline{W}_m(s)W(s)$, $\overline{W}_m(s)$ is asymptotically stable, minimum phase, and of relative degree n^*, and $W(s)$ is asymptotically stable.

5.8 STABLE ADAPTIVE CONTROL USING THE INDIRECT APPROACH

A direct approach was used in Sections 5.3-5.5 for controlling a linear time-invariant plant with unknown parameters. As mentioned in Chapter 1, an alternative way of controlling the plant is by the indirect approach, where the parameters of the plant are estimated and the estimates are, in turn, used to determine the control parameters. Several parametrizations of the identification model for estimating the plant parameters exist and in each case the manner in which the control parameters are to be computed must be determined. The question to be answered is whether the process of estimation followed by control is also globally stable and whether intrinsic differences exist between the direct and indirect control methods.

In Chapter 3, both direct and indirect control of a first order plant were discussed. It was seen that when the high frequency gain was known, the two approaches are equivalent. When the high frequency gain was unknown and the control parameters were computed by solving an algebraic equation using plant parameter estimates (that is, by the certainty equivalence principle), prior knowledge of the lower bound of $|k_p|$ was also needed. However, if the control parameters as well as plant parameter estimates are adjusted dynamically as indicated in Section 3.3.4, this condition could be dispensed with. Hence, both direct and indirect methods lead to stable solutions with the same prior information. In the paragraphs that follow, a similar comparison is made between the two methods for a general n^{th} order plant.

Let the transfer function $W_p(s)$ of the plant to be adaptively controlled satisfy assumptions (I). We discuss the relation between direct and indirect control by considering the two cases (i) when k_p is known, and (ii) when k_p is unknown. In both cases, the problems of relative degree $n^* = 1$ and $n^* \geq 2$ are treated separately.

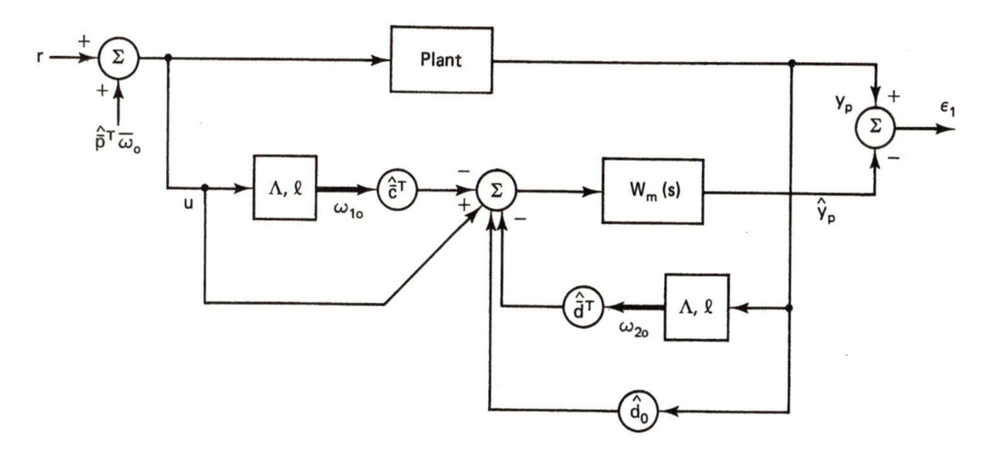

Figure 5.17 Indirect control: $n^* = 1$.

5.8.1 $n^* = 1$.

Case (i) k_p ***known*** [14]. In Chapter 4, an adaptive observer was suggested that generates the estimates of the plant output and plant parameters. We therefore choose an observer as described in Eq. (4.21) using representation 1. Since k_p is known, with no loss of generality, we can assume that $k_p = k_m = 1$, and that the parameter $\widehat{c}_0 \equiv 1$. The transfer function $1/(s+\lambda)$ is replaced by $W_m(s)$, the transfer function of the reference model. The resultant adaptive observer is described by the equations (Fig. 5.17)

$$\begin{aligned}
\dot{\omega}_{1o} &= \Lambda\omega_{1o} + \ell u \qquad \overline{\omega}_o = [\omega_{1o}^T, y_p, \omega_{2o}^T]^T \\
\dot{\omega}_{2o} &= \Lambda\omega_{2o} + \ell y_p \qquad\qquad\qquad\qquad (5.78)\\
\widehat{y}_p &= W_m(s)\left[u - \widehat{\overline{p}}^T\overline{\omega}_o\right] \qquad \widehat{\overline{p}} = [\widehat{\overline{c}}^T, \widehat{d}_0, \widehat{\overline{d}}^T]^T.
\end{aligned}$$

The adaptive law for adjusting $\widehat{\overline{p}}$ is given by

$$\dot{\widehat{\overline{p}}} = -\epsilon_1\overline{\omega}_o$$

where $\epsilon_1 = y_p - \widehat{y}_p$, and the adaptive controller is chosen as

$$u = r + \widehat{\overline{p}}^T(t)\overline{\omega}_o. \qquad (5.79)$$

Since

$$\widehat{y}_p(t) = W_m(s)\left[u(t) - \widehat{\overline{p}}^T(t)\overline{\omega}_o(t)\right] = W_m(s)r(t) = y_m(t)$$

it follows that the overall adaptive system is identical to that described in Section 5.3.4. Since it was shown in that section that all the signals are bounded and that $e_1(t) = y_p(t) - y_m(t)$ tends to zero, we can conclude that the indirect approach with an adaptive observer and controller given by Eqs. (5.78) and (5.79) leads to global stability as well.

Case (ii) k_p ***unknown.*** *(Algebraic Adjustment of Parameters)* Since k_p is unknown, $2n$ parameters including $\widehat{c}_0$ need to be adjusted in the adaptive observer, so that $\omega_o = [u, \omega_{1o}^T, y_p, \omega_{2o}^T]^T$ and $\widehat{p} = [\widehat{c}_0, \widehat{\overline{p}}^T]^T$, where $\widehat{\overline{p}}$ is as defined in Eq. (5.78). An adaptive controller of the form of Eq. (5.79) cannot be used here, since it leads to an algebraic loop. Hence, $u(t)$ is generated as

$$\begin{aligned} u(t) &= k(t)\left[r(t) + \widehat{\overline{p}}^T(t)\omega_o(t)\right] \\ k(t) &= \frac{1}{\widehat{c}_0(t)}. \end{aligned}$$

The desired value of $k(t)$ is k_m/k_p. It follows that if $\widehat{p}$ is adjusted using the same procedure as in (i), $k(t)$ can become unbounded. If a lower bound on $|k_p|$ is known, measures similar to those in Chapter 3 can be taken to adjust $\widehat{c}_0$ so that it is bounded away from zero. With this caveat, the same procedure as in case (i) can be used to design an indirect adaptive controller that is globally stable.

k_p ***unknown.*** *(Dynamic Adjustment of Parameters)* Once again, an apparent lack of symmetry between direct and indirect control becomes evident in the control of a general plant, since more prior information is needed for stable adaptive control using the indirect approach. However, this is found to be due to the method of adjustment of the controller parameters. If the unknown control parameters are adjusted dynamically, it has been shown [3] that stable adaptation can be achieved without a knowledge of the lower bound on $|k_p|$.

The parametrization of the plant, the observer, and the controller used in such a dynamic adjustment are as follows (see Fig. 5.18):

$$\begin{aligned} \dot{\omega}_1 &= \Lambda\omega_1 + \ell u \\ \dot{\omega}_2 &= \Lambda\omega_2 + \ell y_p \\ \dot{y}_p &= -\gamma y_p + k_p u + k_p\beta^T\omega_1 + \alpha_0 y_p + \alpha^T\omega_2 \end{aligned} \qquad \text{(Plant)}$$

$$\begin{aligned} W_m(s) &= k_m\frac{Z_m(s)}{R_m(s)} \\ Z_m(s) &= s^{n-1} + b_m^T S_{n-2}, \quad R_m(s) = s^n + a_m^T S_{n-1} \\ b_m &= [b_0, \ldots, b_{n-2}]^T, \; a_m = [a_0, \ldots, a_{n-1}]^T, \; S_i = [1, s, \ldots, s^i] \end{aligned} \qquad \text{(Model)}$$

$$\begin{aligned} \dot{\widehat{\omega}}_1 &= \Lambda\widehat{\omega}_1 + \ell u \\ \dot{\widehat{\omega}}_2 &= \Lambda\widehat{\omega}_2 + \ell y_p \\ \dot{\widehat{y}}_p &= -\gamma\widehat{y}_p + \widehat{k}_p u + \widehat{k}_p\widehat{\beta}^T\widehat{\omega}_1 + \widehat{\alpha}_0 y_p + \widehat{\alpha}^T\widehat{\omega}_2 \end{aligned} \qquad \text{(Observer)}$$

$$u = kr + \theta_1^T\widehat{\omega}_1 + \theta_0 y_p + \theta_2^T\widehat{\omega}_2 \qquad \text{(Controller)}$$

In the equation above, $\gamma > 0$, $\Lambda \in \mathbb{R}^{(n-1)\times(n-1)}$ is asymptotically stable, (Λ, ℓ) is controllable and is known. $k_p, \alpha_0 \in \mathbb{R}$ and $\alpha, \beta \in \mathbb{R}^{n-1}$ represent the unknown parameters of the plant. The following adaptive laws result in the global stability of the overall system.

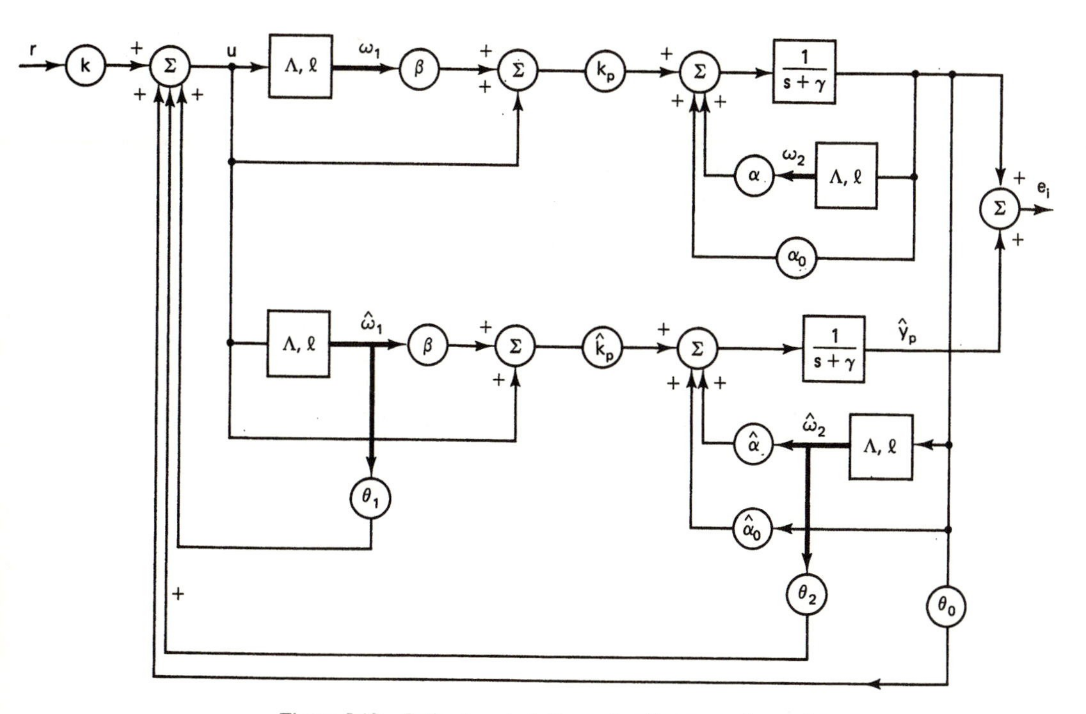

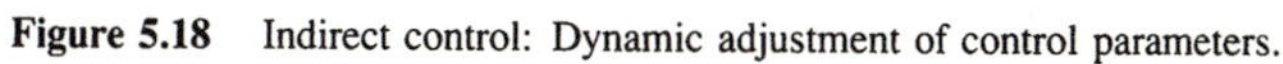
Figure 5.18 Indirect control: Dynamic adjustment of control parameters.

$$
\begin{aligned}
\dot{\widehat{k}}_p &= -e_i\widehat{\omega}_1^T\widehat{\beta} - e_i u + \theta_0\epsilon_{\theta_0} + \theta_2^T\epsilon_{\theta_2} - k\epsilon_k \\
\dot{\widehat{\beta}} &= -sgn(k_p)e_i\widehat{\omega}_1 - \epsilon_{\theta_1} \\
\dot{\widehat{\alpha}}_0 &= -e_i y_p + \epsilon_{\theta_0} \\
\dot{\widehat{\alpha}} &= -e_i\widehat{\omega}_2 + \epsilon_{\theta_2}
\end{aligned}
\qquad \text{(Estimator Parameters)}
$$

$$
\begin{aligned}
\dot{k} &= -sgn(k_p)\epsilon_k \\
\dot{\theta}_1 &= \epsilon_{\theta_1} \\
\dot{\theta}_0 &= sgn(k_p)\epsilon_{\theta_0} \\
\dot{\theta}_2 &= sgn(k_p)\epsilon_{\theta_2}
\end{aligned}
\qquad \text{(Controller Parameters)}
$$

and the closed-loop estimation errors ϵ_k, ϵ_{θ_1}, ϵ_{θ_0}, and ϵ_{θ_2} are defined as

$$
\begin{aligned}
\epsilon_k &= k\widehat{k}_p - k_m \\
\epsilon_{\theta_1} &= -\theta_1 + \widehat{\beta} \\
\epsilon_{\theta_0} &= -\widehat{k}_p\theta_0 + \gamma - \widehat{\alpha}_0 + b_{n-2} - a_{n-1} \\
\epsilon_{\theta_2} &= \overline{b}_m + (a_{n-1} - b_{n-2})b_m - \widehat{\alpha} - \widehat{k}_p\theta_2 - a_m
\end{aligned}
$$

where $\overline{b}_m = [0, b_0, \ldots, b_{n-3}]^T$.

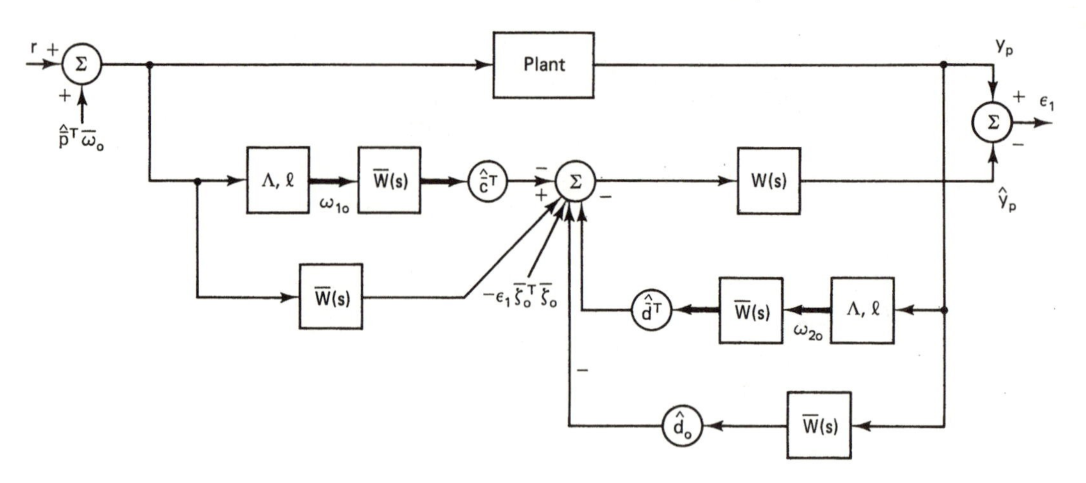

Figure 5.19 Indirect control: $n^* \geq 2$.

5.8.2 $n^* \geq 2$.

Although the problem of control is substantially more complex when $n^* \geq 2$, the equivalence between direct and indirect control can be established using the same procedure as in the previous sections. We demonstrate this only for the case where k_p is known. Similar results also can be achieved using the dynamic adjustment of control parameters when k_p is unknown.

k_p ***known.*** Since $W_m(s)$ is not strictly positive real in this case, the observer is modified as shown in Fig. 5.19.

$$\hat{y}_p = W(s)\left[\overline{W}(s)u - \hat{\bar{p}}^T\overline{W}(s)I\bar{\omega}_o - \epsilon_1\bar{\zeta}_o^T\bar{\zeta}_o\right]$$

where $W_m(s) = W(s)\overline{W}(s)$, $W(s)$ is strictly positive real, $\hat{\bar{p}}$ and $\bar{\omega}_o$ are defined as in Eq. (5.78), and $\bar{\zeta}_o = \overline{W}(s)\bar{\omega}_o$. If the controller is chosen as in Eq. (5.79), then the error $\epsilon_1 = y_p - \hat{y}_p$ is given by

$$\epsilon_1 = y_p - y_m + W(s)\left[\hat{\bar{p}}^T\overline{W}(s)I\omega_o - \overline{W}(s)\hat{\bar{p}}^T\bar{\omega}_o - \epsilon_1\bar{\zeta}_o^T\bar{\zeta}_o\right]$$

which is of the form of the augmented error generated using method 2, described in Section 5.6. Hence, the adaptive law

$$\dot{\hat{\bar{p}}} = -\epsilon_1\bar{\zeta}_o$$

assures that all the signals in the system are bounded. The latter follows since the equations describing the overall system are then identical to that obtained using method 2 with the direct control approach.

Comment 5.12 In the discussions above we have repeatedly stated that under a specified set of conditions, the direct and indirect control methods are equivalent. However, the term equivalence is used to imply that stable adaptation is achieved using the two methods with the same amount of prior information. The mathematical relationships between the two approaches and conditions under which one may be better to use than the other, require further investigation.

5.9 COMBINED DIRECT AND INDIRECT APPROACH

In the preceding sections we considered the direct and indirect methods as two distinct approaches to adaptive control. The principal aim is to reduce the error between the plant and model outputs asymptotically by adjusting the control parameters and the latter, in turn, is effected by using the estimates of the plant parameters in indirect control and the output error in direct control. However, improved response of the adaptive system in terms of speed and accuracy, as well as robustness in the presence of perturbations, may be possible if both methods were somehow combined. We can argue qualitatively that additional information regarding the plant cannot result in the deterioration of the response characteristics of the overall system. The difficulty arises since the information acquired using the two methods cannot, in general, be combined conveniently to determine the adaptive laws. However, recently, such a method of combining the two methods in a stable fashion was proposed by Duarte and Narendra [2]. It is our opinion that the approach will find wide application in the future. Since the scope of this combined approach is quite extensive, we cannot go into all the details in this chapter. Instead, we provide its main concepts in this section together with a simple example. For a more detailed treatment of this topic, the reader is referred to [2]. The following list represents the main features of the combined approach:

(i) At every instant, the plant parameter estimates and the controller parameter estimates are adjusted continuously and are described by differential equations. Solving algebraic equations such as the Bezout identity in conventional indirect control, is thereby eliminated.

(ii) At every instant, assuming that the plant and controller parameter estimates are constant, the closed loop transfer function can be determined. The deviations of the coefficients of this transfer function from those of the desired model transfer function are used to define a vector of closed-loop estimation errors. These errors, which depend on both identification as well as control parameter errors, provide the coupling between the direct and indirect methods.

(iii) In the indirect method, the identification parameters are adjusted using adaptive laws based on the identification error between plant and identification model outputs. In the direct method, the control parameters are adjusted using direct adaptive control laws. Both sets of adaptive laws are modified in the combined approach by using the closed-loop estimation errors in such a manner that the overall system has a quadratic Lyapunov function.

Example 5.4

Let the plant, the reference model, the identification model and the controller be described by the following equations:

$$
\begin{aligned}
\dot{x}_p &= a_p x_p + k_p u \qquad k_p \neq 0 && \text{(Plant)}\\
\dot{x}_m &= a_m x_m + k_m r \qquad a_m < 0 && \text{(Reference Model)}\\
\dot{\widehat{x}}_p &= a_m \widehat{x}_p + (\widehat{a}_p - a_m)x_p + \widehat{k}_p u && \text{(Identification Model)}\\
u &= kr + \theta x_p && \text{(Controller)}
\end{aligned}
\tag{5.80}
$$

Let $e_i = x_p - \widehat{x}_p$ and $e_c = x_p - x_m$. If $\phi = \theta - \theta^*$, $\psi = k - k^*$, $\widetilde{a}_p = \widehat{a}_p - a_p$, and $\widetilde{k}_p = \widehat{k}_p - k_p$ then error equations describing the overall system can be derived in terms of e_i, e_c, ϕ, ψ, $\widetilde{a}_p$ and $\widetilde{k}_p$ from Eq. (5.80) as follows:

$$
\begin{aligned}
\dot{e}_i &= a_m e_i - \widetilde{a}_p x_p - \widetilde{k}_p u\\
\dot{e}_c &= a_m e_c + k_p \phi x_p + k_p \psi r.
\end{aligned}
\tag{5.81}
$$

At every instant of time, independent of the identification and control procedure used, the values of θ, k, $\widehat{a}_p$, and $\widehat{k}_p$ are known. The aim is then to determine adaptive laws for updating them continuously so that the overall system has bounded solutions and the output errors e_i and e_c tend to zero asymptotically. This can be achieved using only direct control, or indirect control. To combine the two, we use the closed-loop estimation errors defined in Chapter 3:

$$
\begin{aligned}
\epsilon_\theta &= \widehat{a}_p + \widehat{k}_p \theta - a_m\\
\epsilon_k &= \widehat{k}_p k - k_m.
\end{aligned}
$$

We note that when $\widehat{a}_p$, $\widehat{k}_p$, θ, and k assume their true values a_p, k_p, θ^*, and k^* respectively, the closed-loop estimation errors are zero. The identification and control laws of the system are now expressed in terms of ϵ_θ and ϵ_k as follows:

$$
\begin{aligned}
\dot{\widehat{a}}_p &= e_i x_p - \epsilon_\theta & \dot{\widehat{k}}_p &= e_i u - \theta\epsilon_\theta - k\epsilon_k\\
\dot{\theta} &= -sgn(k_p)\left[e_c x_p + \epsilon_\theta\right] & \dot{k} &= -sgn(k_p)\left[e_c r + \epsilon_k\right]
\end{aligned}
\tag{5.82}
$$

The use of e_i in the adjustment of $\widehat{a}_p$ and $\widehat{k}_p$ indicates that the plant parameter estimates are updated partly on the basis of the identification error and partly on the basis of the closed-loop estimation errors. Similarly, the control parameter adjustment depends partly on the control error e_c and partly on the closed-loop estimation error. The relative weights to be given to the various signals have to be determined by the designer. From the definition of the closed-loop estimation errors, and Eqs. (5.81) and (5.82), it can be shown that a quadratic Lyapunov function exists and that $\lim_{t\to\infty} e_i(t) = \lim_{t\to\infty} e_c(t) = 0$. It can also be shown that the closed-loop estimation errors ϵ_θ and ϵ_k tend to zero.

Comment 5.13 Even the simple example just described indicates that there is considerable freedom in the choice of the constant gains in the adaptive laws [not included in Eq. (5.82)]. In [2] and [3], the same method is extended to general n^{th} order linear time-invariant plants with arbitrary relative degrees. How better performance can be achieved in practical systems by combining the two methods is currently being investigated.

5.10 SUMMARY

The adaptive control of a general finite dimensional linear time-invariant system with unknown parameters is discussed in this chapter. This problem, which was solved in 1980, was responsible for renewed interest in the field in the years that followed. In view of its historical importance, as well as the fact that it is the starting point of most of the problems discussed in the chapters following, it is discussed in detail in this chapter.

For purposes of analysis, the problem can be conveniently divided into two cases. In the first case, the plant to be controlled is assumed to have a transfer function of relative degree unity and the reference model is chosen to be strictly positive real. The second case deals with the control of plants with relative degree greater than one. To motivate the choice of the structure of the controller, three simple problems are discussed where the high frequency gain k_p, the zeros of the plant transfer function and the poles are successively assumed to be unknown. The solution to the general adaptive control problem, when all $2n$ parameters of the plant are unknown, follows directly from these three special cases.

When the plant transfer function has a relative degree greater than unity, the adaptive procedure is found to be significantly more complex. In contrast to the case where $n^* = 1$, Lyapunov's method is no longer sufficient to prove the boundedness of all the signals in the system, and additional arguments based on growth rates of signals must be used. The adjustments of the control parameters are carried out using an augmented error signal ϵ_1.

It is shown that equivalent results can also be obtained using an indirect approach. As in the direct control case, the problem is more complex when the relative degree of the plant is greater than one. Using different parametrizations of the plant, as well as dynamic adjustment of plant parameter estimates, it is shown that with the same prior information, either direct or indirect methods can be used to adaptively control the system.

At the end of the chapter, new results in adaptive control theory based on the idea of combining direct and indirect control methods are described briefly.

PROBLEMS

Part I

1. Given a dynamical system described by the differential equation

$$\begin{aligned} \dot{x} &= Ax + bu \\ y &= h^T x \end{aligned}$$

where $\{h^T, A, b\}$ is an observable and controllable triple, show that an observer and a controller exist so that the poles of the overall system can be located where desired. Indicate clearly where the controllability and observability conditions are needed in the proof.

2. **(a)** Show that the relative degree n^* of the transfer function $W_p(s)$ of a plant cannot be decreased by using either feedforward or feedback compensators whose transfer functions are proper.

   (b) A plant has a transfer function

$$W_p(s) = \frac{1}{s^2 - 3s + 2}.$$

   Determine stable feedforward and feedback compensators so that the overall transfer function of the system is

$$W_m(s) = \frac{2}{s^2 + 3s + 2}.$$

Part II

3. The transfer function of a linear time-invariant plant has the form

$$W_p(s) = \frac{k_p(s^2 + as + b)}{s^3 + cs^2 + ds + e}$$

where $k_p, a, b, c, d,$ and e are unknown constants, but the sign of k_p is known. The polynomial $s^2 + as + b$ is known to be Hurwitz. The transfer function of a reference model is given to be

$$W_m(s) = \frac{2}{s+1}.$$

   (a) Determine a controller structure and suitable adaptive laws for adjusting the control parameters so that the error between the output y_p of the plant and the output y_m of the model tends to zero asymptotically.

   (b) Indicate where the assumptions concerning the plant are used in proving the stability of the adaptive system.

4. **(a)** In Problem 3, if parameters k_p, a, and b are known with

$$k_p = 2, a = 3, \text{ and } b = 2,$$

   indicate the simplification that results in the controller structure.

   (b) If, in the same problem, k_p, a, and b are unknown, but c, d, and e are known with

$$c = 6,\ d = 11,\ \text{and } e = 6,$$

how is the controller structure simplified?

5. **(a)** If, in Problem 3, the dimension of the controller is increased, show that the values of the controller parameters, for which the plant and the reference model transfer functions are identical, are not unique.

(b) Show that the use of adaptive laws, similar to those in Problem 3, results in stable pole-zero cancellations.

6. The plant to be adaptively controlled has a transfer function

$$W_p(s) = \frac{k_p}{s+2}$$

while the transfer function of the reference model is

$$W_m(s) = \frac{1}{s+2}.$$

(a) Assuming that only the sign of k_p is known, determine a method of generating a control input u so that the output of the plant approaches the output of the model asymptotically. Assume that the reference input r to the model is uniformly bounded and can be measured.

(b) Can the control input in 6(a) be generated without knowledge of the sign of k_p?

7. If, in Problem 6, the model and plant transfer functions are, respectively,

$$\frac{1}{s^2+3s+2} \quad \text{and} \quad \frac{k_p}{s^2+3s+2}$$

indicate how you would have to modify your solution to assure that all signals in the system will remain bounded.

8. In an adaptive control problem, the plant and reference model transfer functions are, respectively,

$$W_p(s) = \frac{k_p(s+a)}{s^3+bs^2+cs+d} \quad \text{and} \quad W_m(s) = \frac{1}{s^2+s+1}.$$

(a) Determine the controller structure and the adaptive laws for adjusting the control parameters.

(b) Using the Bezout identity, show that constant values of the controller parameters exist for which the overall system transfer function is identical to that of the model.

9. In Lemma 5.2, as part of the proof it is shown that polynomials $A(s)$ and $B(s)$ exist such that

$$A(s)Q(s) + B(s)P(s) = 1.$$

This is based on the nonsingularity of a matrix whose elements include the coefficients of the polynomials $P(s)$ and $Q(s)$. Show that the condition that $P(s)$ and $Q(s)$ are relatively prime is both necessary and sufficient for the nonsingularity of the matrix.

10. Indicate the various steps in the proof of stability of the adaptive control problem for the case where the relative degree of the plant is greater than 2. (Assume that k_p is known.)

11. If in the adaptive control problem, k_p is unknown, show that the proof of stability can be given along the same lines as in Section 5.5.2 with $\overline{\zeta}$ replaced by ζ in the adaptive law (5.52).

12. In the indirect control of an LTI plant with unknown parameters, the plant parameters are first estimated and the control parameters are adjusted based on these estimates. Show that, in general, the relationship between the two sets of parameters is nonlinear.

Part III

13. In the adaptive control of any linear time-invariant plant, the concept of strict positive realness is found to play an important role. Do you consider this an indispensable part of adaptive control?

14. **(a)** Four assumptions were made regarding the plant transfer function that were sufficient to generate a stable adaptive control law. In your opinion, are these conditions also necessary for stable adaptive control?

(b) Discuss the question above in the context of the adaptive control of a plant whose transfer function is $W_p(s) = k_p Z_p(s)/R_p(s)$ where k_p is the only unknown parameter.

REFERENCES

1. Chen, C.T. *Introduction to linear system theory.* New York, NY:Holt, Rinehart and Winston, Inc., 1970.
2. Duarte, M.A., and Narendra, K.S. *"Combined direct and indirect adaptive control of plants with relative degree greater than one."* Technical Report No. 8715, Center for Systems Science, Yale University, New Haven, CT, Nov. 1987.
3. Duarte, M.A., and Narendra, K.S. *"Combined direct and indirect approach to adaptive control."* Technical Report No. 8711, Center for Systems Science, Yale University, New Haven, CT, Sept. 1987.
4. Egardt, B. *Stability of adaptive controllers.* Berlin:Springer-Verlag, 1979.
5. Fuchs, J.J. "Discrete adaptive control – A sufficient condition for stability and applications." *IEEE Transactions on Automatic Control* 25:940–946, 1980.
6. Goodwin, G.C., Ramadge, P.J., and Caines, P.E. "Discrete time multivariable adaptive control." *IEEE Transactions on Automatic Control* 25:449–456, 1980.
7. Landau, I.D. *Adaptive control.* New York:Marcel Dekker, Inc., 1979.
8. Monopoli, R.V. "Model reference adaptive control with an augmented error signal." *IEEE Transactions on Automatic Control* 19:474–484, 1974.
9. Morse, A.S. "Global stability of parameter adaptive control systems." *IEEE Transactions on Automatic Control* 25:433–439, June 1980.
10. Morse, A.S. "High-gain adaptive stabilization." In *Proceedings of the Carl Kranz Course*, Munich, W. Germany, 1987.
11. Narendra, K.S., and Lin, Y.H. "Design of stable model reference adaptive controllers." In *Applications of Adaptive Control*, pp. 69–130, New York:Academic Press, 1980.
12. Narendra, K.S., and Lin, Y.H. "Stable discrete adaptive control." *IEEE Transactions on Automatic Control* 25:456–461, June 1980.

13. Narendra, K.S., Lin, Y.H., and Valavani, L.S. "Stable adaptive controller design - Part II: proof of stability." *IEEE Transactions on Automatic Control* 25:440–448, June 1980.
14. Narendra, K.S., and Valavani, L.S. "Stable adaptive controller design-Direct control." *IEEE Transactions on Automatic Control* 23:570–583, Aug. 1978.
15. Valavani, L.S. *Stable adaptive control.* Ph.D. thesis, Yale University, 1979.

6

Persistent Excitation

6.1 INTRODUCTION

The adaptive laws described thus far involved the adjustment of the time derivative of the parameters as a function of some of the internal signals of the system. Hence, such parameters should also be regarded as state variables of the overall system, making the latter nonlinear. Hence, the state vector of the entire adaptive system is composed of the state variables of the plant and controller on the one hand and the adjustable parameters on the other. As mentioned in Chapters 2-5, the stability of the adaptive system is most conveniently described in the space of the state error vector e and the parameter error vector ϕ. In all cases it was established that $\lim_{t\to\infty} e(t) = 0$. However, the convergence of $\phi(t)$ to zero in simulation studies was shown to depend on the properties of certain signals in the system. This property, referred to as *persistent excitation*, is consequently a central one in adaptive systems and is discussed in detail in this chapter.

The concept of persistent excitation (PE) has been around since the 1960s, when it arose naturally in the context of system identification. The term PE was coined to express the qualitative notion that the input signal to the plant should be such that all the modes of the plant are excited if perfect identification is to be possible. It was only in 1966 that a formal statement, relating PE to the convergence of the identification model parameters to their true values, was given by Åström and Bohlin [2]. Several equivalent definitions in terms of the frequency content of the input signal, nonsingularity of an autocovariance matrix, and linear independence of the values of a time-varying vector over an interval followed [15,16,11]. All these are related in one form or another to the existence of a solution to a set of linear equations that arise in the identification problem.

In the late 1970s, when work on the adaptive control problem was in progress, it became clear that the concept of PE also played an important role in the convergence of the controller parameters to their desired values. However, questions unique to the control problem were encountered, requiring a closer examination of the concept. In particular, conditions under which some internal signals of the nonlinear adaptive control system are persistently exciting had to be established. This entailed a detailed study of both linear and nonlinear, algebraic as well as dynamic transformations, under which the property of PE remains invariant [18].

Recent work on the robustness of adaptive systems in the presence of bounded disturbances, time-varying parameters, and unmodeled dynamics of the plant has revealed that the concept of PE is also intimately related to the speed of convergence of the parameters to their final values, as well as the bounds on the magnitudes of the parameter errors. In view of the central role played by PE in all aspects of adaptive identification and control problems, we provide a general framework in this chapter within which the concept can be studied.

6.2 PERSISTENT EXCITATION IN ADAPTIVE SYSTEMS

The notion of PE is readily introduced by considering several simple identification and control problems.

Example 6.1

Let two functions denoted as the input $u : \mathbb{R}^+ \to \mathbb{R}$ and output $y : \mathbb{R}^+ \to \mathbb{R}$ be related by the equation

$$u(t)\theta \quad = \quad y(t)$$

where θ is an unknown constant. The problem is to estimate θ from input-output measurements. If an estimator is described by the equation $\widehat{\theta}(t)u(t) = \widehat{y}(t)$, where $\widehat{\theta}(t)$ denotes the estimate of θ at time t, the adaptive law

$$\dot{\widehat{\theta}}(t) \quad = \quad -(\widehat{y}(t) - y(t))u(t) \quad = \quad -e(t)u(t)$$

can be used to determine $\widehat{\theta}(t)$ as described in Chapter 4. This yields the error differential equation

$$\dot{\phi}(t) \quad = \quad -u^2(t)\phi(t) \tag{6.1}$$

where $\phi(t) = \widehat{\theta}(t) - \theta$ denotes the parameter error. The behavior of $\phi(t)$ in Eq. (6.1) for the choice of seven different simple input functions u, is shown in Fig. 6.1. For inputs (i) $u(t) \equiv 5$, (ii)$u(t) = 5 \cos t$, and (iii) $u(t) = 5s_1(t)$ where $s_1(t)$ is a unit pulse train of frequency 5, it is seen that $\phi(t) \to 0$ or $\widehat{\theta}(t) \to \theta$ exponentially. When the input tends to zero as in (iv) where $u(t) = 1/(1+t)^{1/2}$, the parameter error tends to zero but not exponentially. For exponentially decaying inputs (v) $u(t) = 5e^{-t}$, and (vi) $u(t) = 5\,e^{-t} \cos t$, $\phi(t)$ does not tend to zero. Finally, in (vii), $u(t) = s_2(t)$ where

$$s_2(t) \quad = \quad \begin{cases} 1 & t \in [t_i, t_i + 0.2] \\ 0 & t \in [t_i + 0.2, t_{i+1}] \end{cases}$$

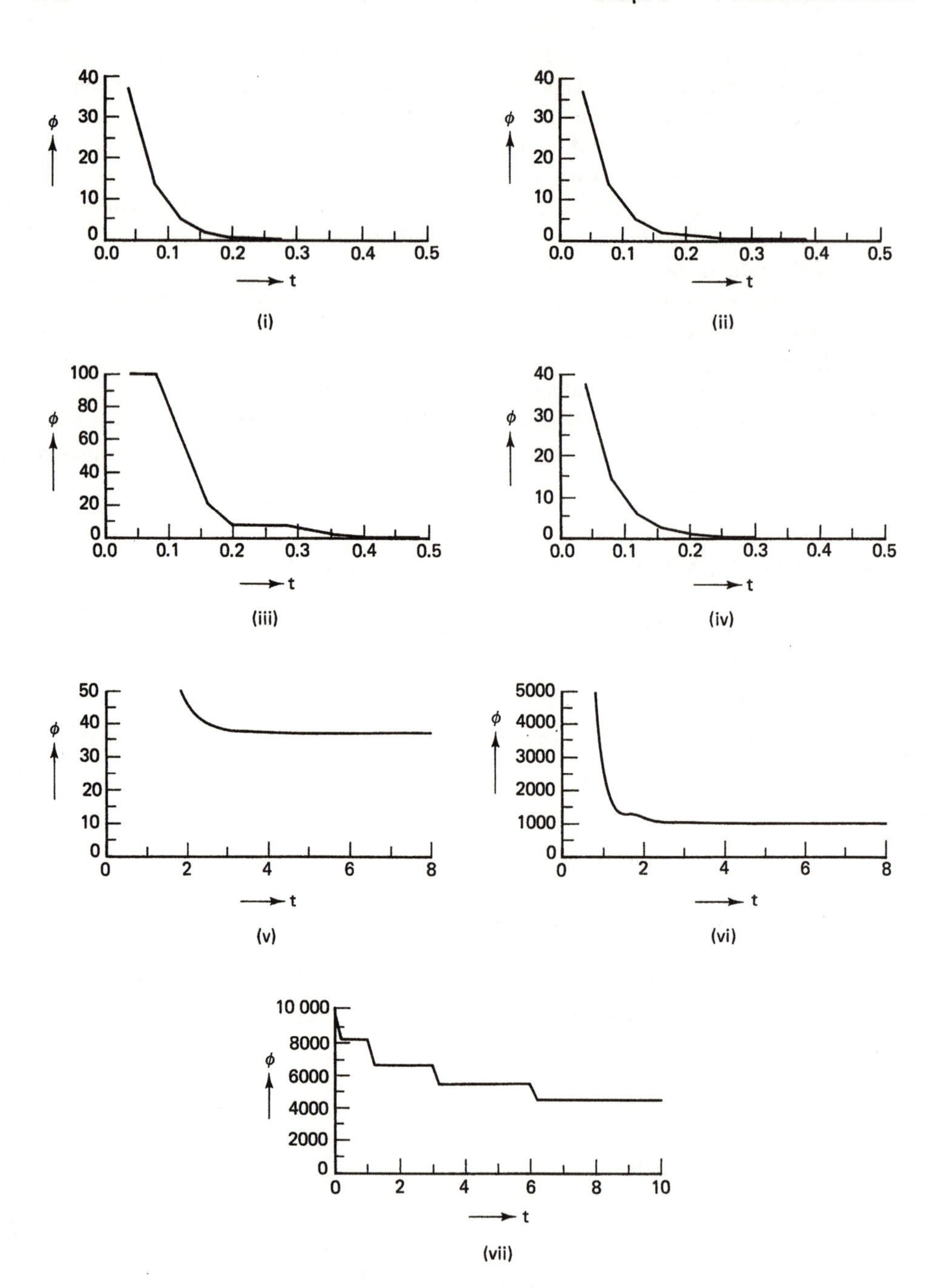

Figure 6.1 Example 6.1.

and $t_{i+1} = t_i + i$, $i = 1, 2, \ldots$, for which $\phi(t)$ tends to some constant value other than zero. The fact that $\widehat{\theta}(t)$ may not tend to θ even when $u(t)$ does not tend to zero is shown by (vii) where successive pulses are of constant width but the interval between them increases monotonically with time in an unbounded fashion.

Hence, a sufficient condition for the uniform convergence of the parameters to their true values is that u assume large values over finite intervals on the real line and that the supremum of the distance between such intervals be finite. These ideas are seen to carry over to the vector case in Example 6.2.

Example 6.2

Let the input $u : \mathbb{R}^+ \to \mathbb{R}^n$ and output $y : \mathbb{R}^+ \to \mathbb{R}$, be related by the equation

$$u^T(t)\theta = y(t)$$

where $\theta \in \mathbb{R}^n$ and is unknown. It was shown in Chapter 3 that an estimator of the form

$$u^T(t)\widehat{\theta}(t) = \widehat{y}(t)$$

can be used, where the estimate $\widehat{\theta}(t)$ is updated according to the adaptive law

$$\dot{\widehat{\theta}}(t) = -[\widehat{y}(t) - y(t)]u(t).$$

Defining the parameter error vector as $\phi(t) = \widehat{\theta}(t) - \theta$, the equivalent error equation can be written as

$$\dot{\phi}(t) = -u(t)u^T(t)\phi(t). \tag{6.2}$$

Hence, for the parameter estimate to converge to its true value, the origin of Eq. (6.2) must be asymptotically stable. The convergence behavior of $\phi = [\phi_1, \ \phi_2]^T$, with $n = 2$ and four different inputs, is shown in Fig. 6.2. These inputs are given by

$$\text{(i)} \quad u = \begin{bmatrix} 5 \cos t \\ 5 \sin t \end{bmatrix} \qquad \text{(ii)} u = \begin{bmatrix} 5 \\ 5(1 + \sin t) \end{bmatrix},$$

$$\text{(iii)} \quad u = \begin{bmatrix} 5 \\ 5 \end{bmatrix} \quad \text{and} \quad \text{(iv)} \ u = \begin{bmatrix} 5 \cos t \\ e^{-t} \end{bmatrix}.$$

For inputs (i) and (ii), $\phi(t)$ converges exponentially to zero. When the input vector is a constant as in (iii), or when one of its components tends to zero as in (iv), the parameter vector does not converge to its true value, as seen in Fig. 6.2(iii) and (iv). In (iii), the parameter error $\phi(t)$ converges to (2.4,−2.4) which is orthogonal to the constant vector u. In (iv), the first component $\phi_1(t)$ tends to zero while the second converges to a constant value −4.5.

Example 6.3

The identification of the parameters of a plant, described by a first-order equation, was discussed in Section 3.3. The differential equations describing the plant, and the estimator used to identify the unknown plant parameter a_p are respectively,

$$\begin{aligned} \dot{y}_p &= a_p y_p + u \\ \dot{\widehat{y}}_p &= a_m \widehat{y}_p + (\widehat{a}_p(t) - a_m) y_p + u \qquad a_m < 0 \end{aligned}$$

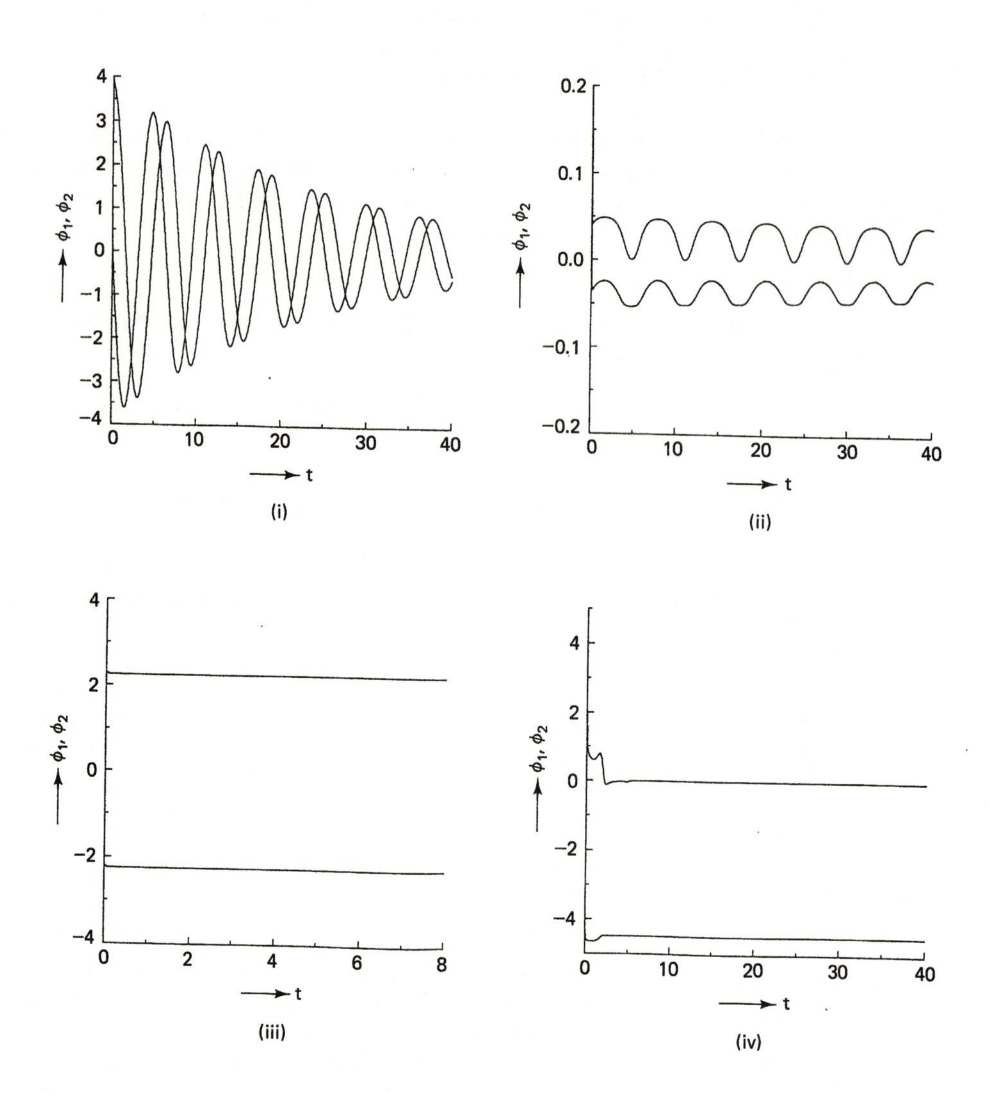

Figure 6.2 Example 6.2.

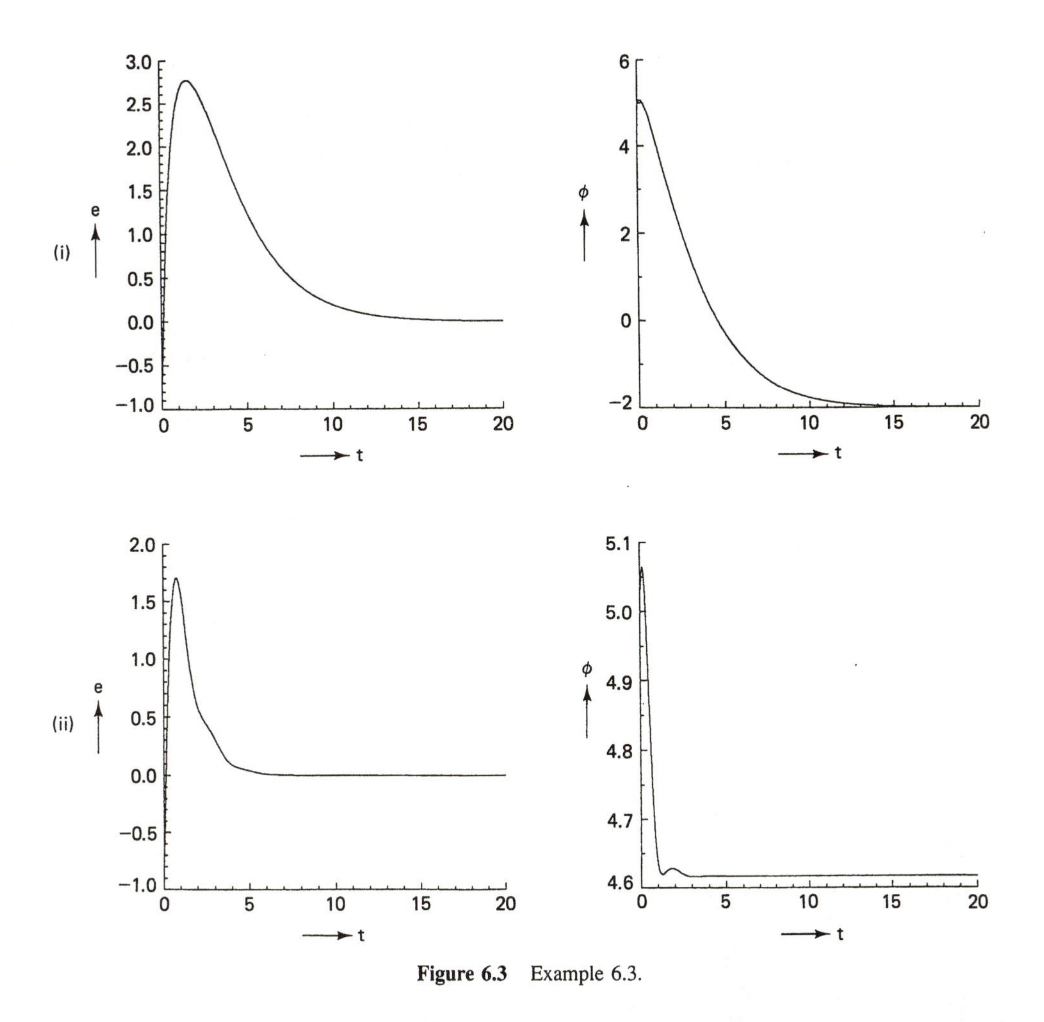

Figure 6.3 Example 6.3.

and the corresponding error equation is

$$\dot{e} \;=\; a_m e + \phi y_p \tag{6.3a}$$

where $e(t) = \widehat{y}_p(t) - y_p(t)$, and $\phi(t) = \widehat{a}_p(t) - a_p$. The adaptive law used to adjust ϕ has the form

$$\dot{\phi} \;=\; -e y_p. \tag{6.3b}$$

Equations (6.3a) and (6.3b) represent a set of two differential equations whose origin is stable. Figure 6.3 shows the behavior of the output e and parameter errors ϕ in Eqs. (6.3a) and (6.3b) for two different inputs (i) $u(t) \equiv 1$, and (ii) $u(t) = \exp(-t), t \geq 0$ and $a_m = -1$. As seen in the figure, the output error e tends to zero, but the parameter error ϕ does not always tend to zero asymptotically.

Example 6.4 (Vector Case)

The extension of the problem considered in example 6.3 to the vector case was described in Chapter 3 when the state variables of the plant were accessible, and in Chapter 4 in the context of adaptive observers. The error equations in these cases assume the form [refer to Eqs. (3.64) and (4.24)]

$$\dot{e}(t) = Ae(t) + b\phi^T(t)\omega(t)$$

where ϕ is a vector of parameter errors, and ω is a bounded vector of signals that can be measured. In Eq. (4.24), only a linear combination $e_1(= h^T e)$ of the state error e could be measured and the transfer function between $\phi^T\omega$ and e_1 was strictly positive real, that is, $h^T(sI - A)^{-1}b = 1/(s+\lambda)$, $\lambda > 0$. The adaptive law for adjusting ϕ has the form

$$\dot{\phi}(t) = -e^T(t)Pb\omega(t)$$

in Eq. (3.64) and

$$\dot{\phi}(t) = -e_1(t)\omega(t)$$

in Eq. (4.24). In both cases, we are interested in the asymptotic stability of the equation

$$\begin{bmatrix} \dot{e}(t) \\ \dot{\phi}(t) \end{bmatrix} = \begin{bmatrix} A & b\omega^T(t) \\ -\omega(t)h^T & 0 \end{bmatrix} \begin{bmatrix} e(t) \\ \phi(t) \end{bmatrix}. \tag{6.4}$$

Figure 6.4 illustrates the behavior of solutions of Eq. (6.4) for

$$\text{(i) } \omega(t) = [5, 5(1+\sin t)]^T, \quad \text{and} \quad \text{(ii) } \omega = [5, 5]^T,$$

with $A = -1$, $b = h = 1$, and $\phi = [\phi_1, \phi_2] \in \mathbb{R}^2$. In Fig. 6.4(ii), ϕ tends to the value $[1, -1]^T$, which is orthogonal to $\omega(t)$, so that $\phi^T\omega \to 0$ asymptotically.

Example 6.5 (Control Problem - Vector Case)

Example 6.4 deals with the identification problem of a stable plant so that $\omega(t)$ in Eq. (6.4) can be assumed to be uniformly bounded. However, as observed in Chapter 5, while the form of the error equation [refer to Eq. (5.21)] in the control problem is identical to Eq. (6.4), $\omega(t)$ cannot be assumed to be bounded a priori. In this case, $\omega(t)$ can be expressed as the sum of a bounded signal $\omega^*(t)$ and $Ce(t)$, where C is a rectangular matrix with constant elements and $e(t)$ is the state error (refer to Section 6.5). Hence, Eq. (6.4) can be rewritten as

$$\begin{bmatrix} \dot{e}(t) \\ \dot{\phi}(t) \end{bmatrix} = \begin{bmatrix} A & b\left(\omega^*(t) + Ce(t)\right)^T \\ -\left(\omega^*(t) + Ce(t)\right)h^T & 0 \end{bmatrix} \begin{bmatrix} e(t) \\ \phi(t) \end{bmatrix}. \tag{6.5}$$

The objective in this case is to determine conditions on the vector ω^* so that the origin of the nonlinear differential equation (6.5) is u.a.s. This problem is treated in Section 6.5.2.

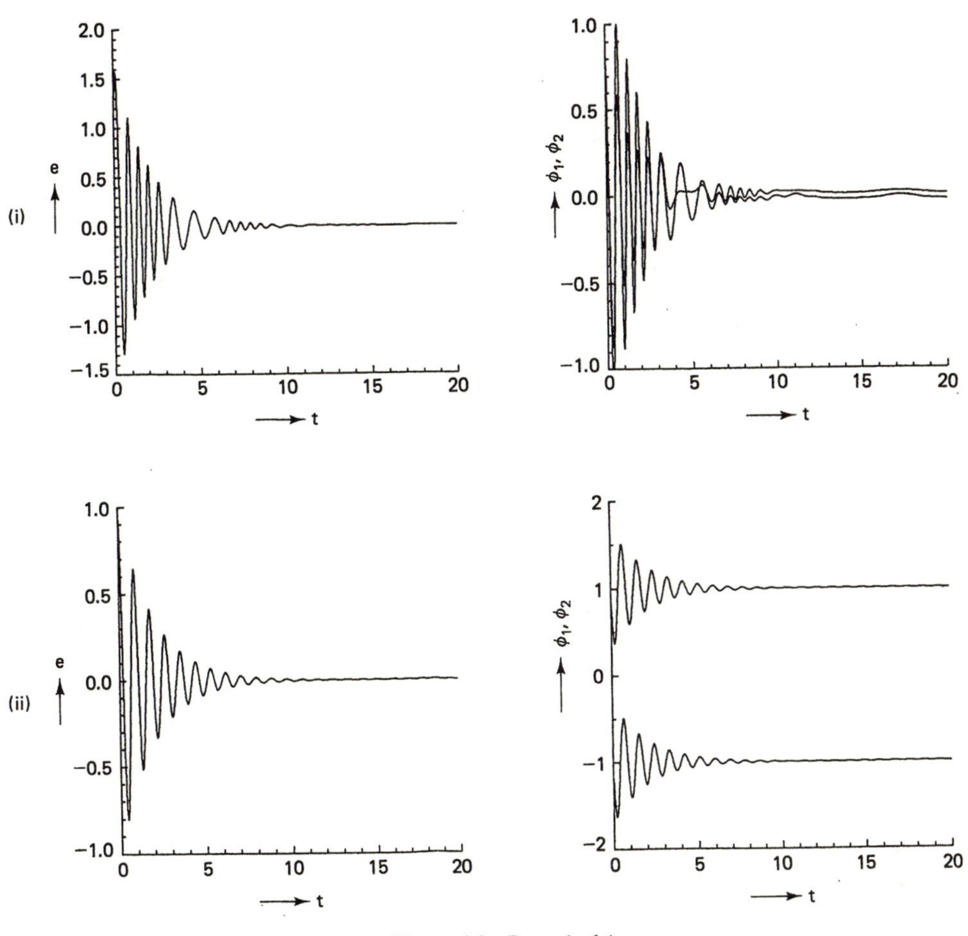

Figure 6.4 Example 6.4.

Example 6.6

When the relative degree of the plant to be controlled is greater than one, it was shown in Chapter 5 [refer to Eq. (5.44)] that the augmented error equation has the form

$$\phi^T(t)\zeta(t) = \epsilon_1(t)$$

where once again, $\zeta(t)$ is not known a priori to be bounded. By expressing $\zeta(t)$ as the sum of a bounded signal $\zeta^*(t)$ and a continuous function of the error $\epsilon_1(t)$, the problem can be stated as the determination of conditions on the signal $\zeta^*(t)$ so that $\phi(t)$ tends to zero asymptotically. This problem is also treated in Section 6.5.2.

In all the examples above, the problem of assuring parameter convergence is equivalent to demonstrating the asymptotic stability of a class of nonlinear time-varying differential equations. The conditions that are needed to ensure this result form the subject matter of this chapter.

6.3 DEFINITIONS

Since the parametrization of the plant in most identification problems takes the form $y_p(t) = \theta^T u(t)$ where $y_p(t)$ is the plant output and $u(t)$ is a known n-dimensional vector at time t, the error equation $e_1(t) = \phi^T(t)u(t)$ occurs frequently in such problems. The corresponding adaptive law, as shown in the previous chapters, has the form

$$\dot{\phi}(t) \quad = \quad -e_1(t)u(t) \;=\; -u(t)u^T(t)\phi(t). \tag{6.6}$$

Naturally, the conditions on the function u, which would assure that $\lim_{t\to\infty} \phi(t) = 0$, occupied the attention of many researchers in the 1960s and 1970s. Åström and Bohlin stated in 1966 [2] that if the autocorrelation function [1]

$$R_u(\tau) \quad \stackrel{\Delta}{=} \quad \lim_{T\to\infty} \frac{1}{T}\int_{t_0}^{t_0+T} u(t)u^T(t+\tau)dt \tag{6.7}$$

of u exists and is positive-definite, then $\phi(t) \to 0$ as $t \to \infty$. They referred to such a signal u as being *persistently exciting*. Sondhi and Mitra require that u be nondegenerate as defined in [19], and called it the *mixing condition*. According to Yuan and Wonham [21], the intuition behind the concept of persistent excitation is that the input should be *rich* enough to excite all the modes of the system that is being probed. They required that u be restricted to the class $\mathcal{P}_{[0,\infty)}$ of piecewise-differentiable functions with bounded discontinuities and that u be *persistently spanning*, that is, there exist positive constants K and L such that for every $t \geq 0$, there is a sequence of N numbers $T_i \in [t, t+L], i = 1, \ldots, N$ with

$$\left\| [u(T_1),\ u(T_2), \ldots,\ u(T_N)]^{-1} \right\| \leq K.$$

Anderson's conclusion in [1] was that the input u should have a minimum complexity to an extent determined by the dimension of ϕ and that it must persist for $\phi(t)$ to tend to zero asymptotically.

When $\phi(t) \in \mathbb{R}$, it follows directly from Eq. (6.6) that $\phi(t)$ will tend to zero asymptotically if $u(t) = sin\ \omega t$. Similarly, if $\phi(t) \in \mathbb{R}^2$, it can be shown that $u^T(t) = [sin\ \omega t,\ \cos\ \omega t]$ would result in $\lim_{t\to\infty} \phi(t) = 0$. This eventually led to the choice of a vector $u(t)$ with $[n/2]$ distinct frequencies to assure the convergence of $\phi(t)$ to zero when $\phi(t) \in \mathbb{R}^n$, where $[x]$ denotes the smallest integer greater than or equal to x. In many of the early papers [13], such a signal was commonly used to prove parameter convergence. This notion was formalized by Boyd and Sastry in [5], using the concept of spectral measures.

The uniform asymptotic stability of the equilibrium state of Eq. (6.6) was considered in detail in Chapter 2 and necessary and sufficient conditions were given in Theorem 2.16. One of the several equivalent conditions given there can be stated as follows: The piecewise-continuous uniformly bounded function $u : \mathbb{R}^+ \to \mathbb{R}^n$ satisfies the inequality

$$\int_t^{t+T_0} u(\tau)u^T(\tau)d\tau \geq \alpha I \qquad \forall t \geq t_0 \tag{6.8}$$

[1] In [2], a condition for discrete signals is given. Equation (6.7) can be considered as a continuous version of this condition.

for positive constants $t_0, T_0,$ and α if, and only if, $\phi = 0$ in Eq. (6.6) is u.a.s. Equivalently, for every unit vector w in $\mathbb{R}^n$,

$$\frac{1}{T_0}\int_t^{t+T_0} |u^T(\tau)\mathrm{w}|d\tau \geq \epsilon_0 \qquad \forall t \geq t_0 \tag{6.9}$$

for positive constants t_0, T_0, and ϵ_0. The condition in Eq. (6.8) implies that the integral of the semidefinite matrix $u(t)u^T(t)$ over a finite interval of length T_0 is a positive-definite matrix. According to the condition in Eq. (6.9), in the interval $[t, t+T_0]$, the vector $u(t)$ has a finite projection along any unit vector w over a finite measure of time.

An alternate class of differential equations that arises frequently in adaptive identification [Eqs. (4.24),(4.26)] and control [Eqs. (5.47),(5.48)] is given below:

$$\dot{x} = \begin{bmatrix} A & bu^T(t) \\ -u(t)b^T & 0 \end{bmatrix} x \tag{6.10}$$

where $x^T \triangleq [x_1^T, x_2^T], x_1 : \mathbb{R}^+ \to \mathbb{R}^m, x_2 : \mathbb{R}^+ \to \mathbb{R}^n, A$ is an $m \times m$ asymptotically stable matrix with $A + A^T = -Q < 0, (A, b)$ is controllable, and $u : \mathbb{R}^+ \to \mathbb{R}^n$ is piecewise-continuous and bounded. The stability of the equilibrium state of Eq. (6.10) was discussed in detail in Chapter 2, and Theorem 2.17 provides both necessary and sufficient conditions for its uniform asymptotic stability. As stated in this theorem, the equilibrium state $x = 0$ is uniformly asymptotically stable if, and only if, positive constants $T_0, \delta_0,$ and ϵ_1 exist so that $\exists\, t_2 \in [t, t+T_0]$ with

$$\left|\frac{1}{T_0}\int_{t_2}^{t_2+\delta_0} u^T(\tau)\mathrm{w}d\tau\right| \geq \epsilon_1 \qquad \forall t \geq t_0 \tag{6.11}$$

for every unit vector $\mathrm{w} \in \mathbb{R}^n$. Since both systems in Eqs. (6.6) and (6.10) are linear, the stability is exponential and global. The conditions in Eqs. (6.9) and (6.11), which assure, respectively, the exponential stability of systems in Eqs. (6.6) and (6.10), have been used in the literature to define persistent excitation of u in the corresponding cases.

One of the difficulties that arises in such definitions in continuous time systems is due to the fact that the two conditions in Eqs. (6.9) and (6.11) are not identical. In fact, Example 2.8 in Chapter 2 shows that a function that is persistently exciting for the system in Eq. (6.6) may not be persistently exciting with respect to the system in Eq. (6.10). Hence, if a general definition of persistent excitation is to be given, it should be based on Eq. (6.11) rather than Eq. (6.9). The discussion above reveals that PE can be defined either as an intrinsic property of a class of signals or in terms of the uniform asymptotic stability of a class of linear time-varying differential equations. This question is reexamined in Section 6.6, in the context of the stability of a class of nonlinear differential equations, using the Lyapunov theory.

Although any function u that satisfies Eq. (6.11) would qualify as a persistently exciting signal, the condition itself is complex and difficult to verify. In contrast to this condition, Eq. (6.9) is analytically simpler to use and has geometrically appealing features. Hence, we restrict the class of functions to which u belongs so that the two conditions in Eqs. (6.9) and (6.11) are equivalent. This is the class of piecewise-differentiable functions $\mathcal{P}_{[0,\infty)}$ introduced by Yuan and Wonham [21] and is defined below:

Definition 6.1 Let C_δ be a set of points in $[0,\infty)$ for which there exists a $\delta > 0$ such that for all $t_1, t_2 \in C_\delta$, $t_1 \neq t_2$ implies $|t_1 - t_2| \geq \delta$. Then $\mathcal{P}_{[0,\infty)}$ is defined as the class of real valued functions on $[0,\infty)$ such that for every $u \in \mathcal{P}_{[0,\infty)}$, there corresponds some δ and C_δ such that

(i) $u(t)$ and $\dot{u}(t)$ are continuous and bounded on $[0,\infty)/C_\delta$ and

(ii) for all $t_1 \in C_\delta$, $u(t)$ and $\dot{u}(t)$ have finite limits as $t \uparrow t_1$ and $t \downarrow t_1$.

A vector u is said to belong to $\mathcal{P}_{[0,\infty)}$ if every component of u belongs to $\mathcal{P}_{[0,\infty)}$.

By confining the function u to $\mathcal{P}_{[0,\infty)}$, it is seen that either Eq. (6.9) or (6.11) can be used to define the class of persistently exciting signals (see Problem 7). In the sections following we shall use the definition that is most appropriate in any given context. A point worth noting is that the distinctions above do not arise in discrete systems and hence a precise definition of PE is found to be relatively straightforward [4].

Definition 6.2 The set of all functions $u : \mathbb{R}^+ \to \mathbb{R}^n$ with $u \in \mathcal{P}_{[0,\infty)}$ that satisfies the condition in Eq. (6.9) [or equivalently Eq. (6.11)] over a period T_0 for all $t \geq t_0$ is denoted by $\Omega_{(n,t_0,T_0)}$. [18]

The subscripts n, t_0, and T_0 in the definition refer to the dimension of the space, the initial time, and the interval over which the function u is persistently exciting. In many cases, the initial time is clear from the context and, hence, can be omitted so that the set can be denoted simply as $\Omega_{(n,T_0)}$. It also immediately follows from Definition 6.2 that if u is persistently exciting at time t_0, it is also persistently exciting for any time $t_1 \geq t_0$. Also if $u \in \Omega_{(n,t_0,T_0)}$, then u is persistently exciting for any interval of length $T_1 \geq T_0$.

6.3.1 Examples

In this section, we present some simple examples of signals that are persistently exciting and some that are not.

Any nonzero constant scalar signal, as well as a sinusoid of any frequency, is persistently exciting in a one dimensional space, that is, $\in \Omega_{(1,T)}$. If a scalar function $u \in \mathcal{L}^1$, $\mathcal{L}^2$, or $\lim_{t\to\infty} u(t) = 0$, then $u \notin \Omega_{(1,T)}$. Using similar reasoning, we can also see that the signal $s_2 \notin \Omega_{(1,T)}$ where s_2 is defined as in Example 6.1. Another interesting example is the signal $u(t) = \sin\sqrt{t}$. With such a u, the differential equation (6.6) is only asymptotically stable and not u.a.s. and $u \notin \Omega_{(1,T)}$.

If $u(t) = [\sin\,\omega t,\ \cos\,\omega t]^T$, or if u has two spectral lines [20], then $u \in \Omega_{(2,T)}$. If however $u(t) = [\sin\,\sqrt{t},\ \cos\,\sqrt{t}]^T$, then $u \notin \Omega_{(2,T)}$.

Even when a vector has independent components over an interval T, it may not be persistently exciting. For example, if $u(t) = [\sin\,t,\ \sin\,t + e^{-t}]^T$, then the components are linearly independent over any finite interval, but $u \notin \Omega_{(2,T)}$ for any $T > 0$, since the projection of $u(t)$ along the direction = $[1, \quad -1]^T$ tends to zero as $t \to \infty$.

6.4 PROPERTIES OF PERSISTENTLY EXCITING FUNCTIONS

In this section, we present some results pertaining to algebraic and dynamic transformations of persistently exciting signals [18]. Unless stated otherwise, all signals will be assumed to belong to $\mathcal{P}_{[0,\infty)}$.

6.4.1 Algebraic Transformations

The projection of a persistently exciting signal on a subspace is also persistently exciting. Hence, all linear combinations of the elements of a persistently exciting n-dimensional vector are persistently exciting. The converse is not true, that is, an n-dimensional vector u need not be persistently exciting if its projections along n linearly independent directions are persistently exciting in 1-dimension. This is true *only if* the projection along any direction $\in \Omega_{(1,T)}$. These simple properties of PE functions are collected and stated concisely in the lemmas following.

Lemma 6.1 Let $u \in \Omega_{(n,T)}$ and M be an $(m \times n)$ constant matrix, $m \leq n$. Then $Mu \in \Omega_{(m,T)}$ if, and only if, M is of full rank.

Lemma 6.2 Let $u : \mathbb{R}^+ \to \mathbb{R}^n$. Then $u \in \Omega_{(n,T)}$ if, and only if, $\alpha^T u \in \Omega_{(1,T)}$ for every constant nonzero vector $\alpha \in \mathbb{R}^n$.

The proofs of Lemmas 6.1 and 6.2 are simple and follow directly from Definition 6.2.

The class of almost periodic functions can be used to illustrate some of the properties above.

Example 6.7

Assume in the following that $\omega_i \neq \omega_j$ if $i \neq j$.

a. $s_3(t) = a_1 \sin \omega_1 t + a_2 \sin \omega_2 t + \cdots + a_n \sin \omega_n t \in \Omega_{(1,T)}$, $a_i \neq 0$ for some $i \in \{1, 2, \ldots, n\}$.

b. $s_4(t) = [a_1 \sin \omega_1 t, \; a_2 \sin \omega_2 t, \; \ldots, a_n \sin \omega_n t]^T \in \Omega_{(n,T)}$, $a_i \neq 0$ for any $i \in \{1, 2, \ldots, n\}$.

c. For any nonsingular $n \times n$ matrix M, $Ms_4 \in \Omega_{(n,T)}$.

An n-dimensional vector u that does not belong to $\Omega_{(n,T)}$ may be persistently exciting in a subspace. Some simple examples are given below:

1. $u(t) = [\sin t, \; \sin t]^T \notin \Omega_{(2,T)}$ but $u_1(t) = [a, \; b]u(t) \in \Omega_{(1,T)}$ if $b \neq -a$.
2. $u(t) = [\sin t, \; \cos t, \; \sin t + \cos t]^T \notin \Omega_{(3,T)}$ but

$$u_2(t) = \begin{bmatrix} 0 & 1 & 0 \\ 0 & 0 & 1 \end{bmatrix} u(t) \in \Omega_{(2,T)}.$$

Definition 6.3 and Lemma 6.3 concern functions that are persistently exciting in a subspace of $\mathbb{R}^n$.

Definition 6.3 A bounded function $u : \mathbb{R}^+ \to \mathbb{R}^n$ is said to be persistently exciting in $\mathbb{R}^r$ for $r \leq n$, if there exists an $(r \times n)$ matrix P such that $Pu \in \Omega_{(r,T)}$ and r is the largest integer for which this holds. The set of all such functions is denoted by $\Omega^r_{(n,T)}$. We denote $\Omega^n_{(n,T)}$ as $\Omega_{(n,T)}$.

Lemma 6.3 Let M be a constant matrix. Then the following hold:

1. If $u \in \Omega^{r_1}_{(n,T)}$ then $Mu \in \Omega^{r_1}_{(n,T)}$ if M is square and nonsingular.
2. If $u \in \Omega_{(n,T)}$, then $Mu \in \Omega^{r_2}_{(m,T)}$ if M is an $(m \times n)$ matrix of rank r_2.
3. If $u \in \Omega^{r_1}_{(n,T)}$, then $Mu \in \Omega^{r_3}_{(m,T)}$ if M is an $(m \times n)$ matrix of rank r_2 where $r_3 = dim[\Re(M) \cap \Re(P)]$ and $\Re(A)$ denotes the range space of a matrix A.

An immediate application of Definition 6.3, in the study of adaptive systems, is related to partial convergence. This is given in Theorem 6.1. The following lemma is useful in proving the theorem.

Lemma 6.4 If $u : \mathbb{R}^+ \to \mathbb{R}^n$, and u is not persistently exciting in any subspace of $\mathbb{R}^n$, then $u \in \mathcal{L}^1$.

Proof. The proof follows directly from the fact that every component u_i of u is not persistently exciting in $\mathbb{R}$ and hence belongs to $\mathcal{L}^1$.

Theorem 6.1 If $\dot{x} = -uu^Tx$, and $u \in \Omega^r_{(n,T)}$, then there exists an $(r \times n)$ matrix T_1 such that $y_1(t) = T_1x(t) \to 0$ as $t \to \infty$.

Proof. Let T be an $(n \times n)$ orthogonal matrix. Premultiplying both sides of equation $\dot{x} = -uu^Tx$ by T and defining $Tx = y$, we obtain $\dot{y} = -[Tuu^TT^T]y$. If

$$T \triangleq \begin{bmatrix} T_1 \\ T_2 \end{bmatrix} \quad \text{and} \quad y \triangleq \begin{bmatrix} y_1 \\ y_2 \end{bmatrix}$$

where T_1 is an $(r \times n)$ matrix, and T_2 is an $[(n-r) \times n]$ matrix, then $T_ix = y_i$ $(i = 1, 2)$. Since $u \in \Omega^r_{(n,T)}$, T can be chosen so that $v_1 = T_1u \in \Omega_{(r,T)}$ and $v_2 = T_2u$ is such that $v_2^T\mathrm{w} \notin \Omega_{(1,T)}$ for any unit vector w in $\mathbb{R}^{n-r}$. From Lemma 6.4, it follows that $v_2 \in \mathcal{L}^1$. Since

$$\begin{aligned} \dot{y}_1 &= -v_1v_1^Ty_1 - v_1v_2^Ty_2 \\ \dot{y}_2 &= -v_2v_1^Ty_1 - v_2v_2^Ty_2 \end{aligned}$$

it follows that y_1 and y_2 are bounded. Since $v_1 \in \Omega_{(r,T)}$ and $v_2 \in \mathcal{L}^1$, it follows that $\lim_{t\to\infty} y_1(t) = 0$ [9].

Time-varying Transformations. Lemma 6.1 established that the PE of a vector is invariant under any nonsingular time-invariant transformation. However, this is not the case when the nonsingular transformation is a time-varying one, as shown below.

Example 6.8

Let

$$u(t) = \begin{bmatrix} 2+\sin t \\ -1 \end{bmatrix}, \qquad M(t) = \begin{bmatrix} 1 & 2+\sin t \\ 0 & 2+\sin t \end{bmatrix}.$$

Then $u \in \Omega_{(2,T)}$ and $M(t)$ is nonsingular for all $t \geq 0$. But $M(t)u(t) = [0, \ -2-\sin t]^T$ and $\notin \Omega_{(2,T)}$.

Often, in the study of adaptive systems, the PE of some signals must be deduced from the PE of others in a feedback loop. For example, one is frequently interested in determining if the sum of two persistently exciting signals is also persistently exciting. Results related to such properties of PE functions are stated below in Lemma 6.5. Their proofs follow directly from the definition of PE and hence are omitted.

Lemma 6.5

(i) If $u : \mathbb{R}^+ \to \mathbb{R}^n$ and any component of $u(t) \to 0$ as $t \to \infty$, $\in \mathcal{L}^1$, or $\in \mathcal{L}^2$, then $u \notin \Omega_{(n,T)}$ for any T.

(ii) If $u_1 \in \Omega_{(n,t_0,T)}$, $u_2 : \mathbb{R}^+ \to \mathbb{R}^n$ and $u_2(t) \to 0$ as $t \to \infty$, then $u_1 + u_2 \in \Omega_{(n,t_1,T)}$ for some $t_1 \geq t_0$. The same result also holds if $u_2 \in \mathcal{L}^1$ or $\mathcal{L}^2$.

(iii) If $u_1, u_2 \in \Omega_{(n,T)}$, then $u_1 + \epsilon u_2 \in \Omega_{(n,T)}$ for some sufficiently small $\epsilon \in \mathbb{R}$.

Lemma 6.5 indicates that if a signal is persistently exciting, the addition of another signal that is small in some sense, does not affect its persistent excitation. For example, if a signal $\overline{s}(t)$ is of the form $\overline{s}(t) = s_3(t) + u(t)$, where $s_3(t)$ is defined as in Example 6.7, and $u(t) \to 0$ as $t \to \infty$, then $\overline{s}(t)$ also belongs to $\Omega_{(1,T)}$. The same holds if $u \in \mathcal{L}^1$ or $\mathcal{L}^2$ or is sufficiently small.

6.4.2 Dynamic Transformations

From the previous sections, it is clear that if $u \in \Omega_{(n,T)}$, a linear algebraic transformation can only result in a function $y \in \Omega_{(m,T)}$, where $m \leq n$. Vector functions that are persistently exciting in dimensions greater than n can be generated from a function $u \in \Omega_{(n,T)}$ using dynamic transformations. In this section, we consider the effect of linear dynamic transformations on persistently exciting signals.

We first consider the following example to illustrate the relation between controllability and persistent excitation. This is generalized in Lemma 6.6.

Example 6.9

Let

$$\dot{x} = \begin{bmatrix} \dot{x}_1 \\ \dot{x}_2 \end{bmatrix} = \begin{bmatrix} -1 & 0 \\ -1 & -1 \end{bmatrix} \begin{bmatrix} x_1 \\ x_2 \end{bmatrix} + \begin{bmatrix} 0 \\ 1 \end{bmatrix} u$$

where $u \in \Omega_{(1,T)}$. We see that x_1 is not controllable, and $x_1(t) \to 0$ as $t \to \infty$. Hence, $x \notin \Omega_{(2,T)}$.

Consider the LTI dynamical system described by the vector differential equation:

$$\dot{x} = Ax + Bu \tag{6.12}$$

where $x : \mathbb{R}^+ \to \mathbb{R}^n$, $u : \mathbb{R}^+ \to \mathbb{R}^p$, A and B are constant matrices with appropriate dimensions, A is asymptotically stable, and $p \leq n$.

Lemma 6.6 A necessary condition for x to belong to $\Omega_{(n,T)}$ is that (A, B) be controllable.

Proof. By the canonical decomposition theorem it is known that if the rank of the controllability matrix is $n_c < n$, Eq. (6.12) can be transformed to the form

$$\begin{bmatrix} \dot{x}^c \\ \dot{x}^{\overline{c}} \end{bmatrix} = \begin{bmatrix} A_c & A_{12} \\ 0 & A_{\overline{c}} \end{bmatrix} \begin{bmatrix} x^c \\ x^{\overline{c}} \end{bmatrix} + \begin{bmatrix} B_c \\ 0 \end{bmatrix} u$$

by an equivalence transformation where $x^c : \mathbb{R}^+ \to \mathbb{R}^{n_c}$ and $x^{\overline{c}} : \mathbb{R}^+ \to \mathbb{R}^{n-n_c}$, x^c is controllable and $x^{\overline{c}}$ is uncontrollable. Since $A_{\overline{c}}$ is asymptotically stable, by Lemma 6.5(i), it follows that the vector $x \notin \Omega_{(n,T)}$. From Definition 6.3, if the input u is such that $x^c \in \Omega_{(n_c,T)}$, then it follows that $x \in \Omega^{n_c}_{(n,T)}$.

From Lemma 6.2, we see that any nonsingular linear time-invariant transformation of a persistently exciting signal is also persistently exciting. Hence, if the state of a dynamical system is persistently exciting, the states corresponding to the Jordan as well as the control canonical representations are also persistently exciting. This is stated in Lemma 6.7.

Lemma 6.7 Consider the single input system described by the equation

$$\dot{x} = Ax + bu, \qquad x : \mathbb{R}^+ \to \mathbb{R}^n, \tag{6.13}$$

A is asymptotically stable, and (A, b) is controllable. Let the corresponding Jordan and the controllable canonical representations be given by $\dot{z} = Jz + du$ and $\dot{\overline{x}} = \overline{A}\overline{x} + \overline{b}u$ respectively. The state vector x in Eq. (6.13) belongs to $\Omega_{(n,T)}$ if, and only if, $z \in \Omega_{(n,T)}$ or $\overline{x} \in \Omega_{(n,T)}$.

Proof. The proof follows directly from the fact that both z and $\overline{x}$ are nonsingular time-invariant transformations of x.

In a system of the type described in Eq. (6.13), we are interested primarily on conditions to be imposed on u so that x is persistently exciting. By Lemma 6.7, x is persistently exciting if, and only if, the state of any equivalent system is persistently exciting. However, the lemma does not provide any clue as to how any one of these equivalent states can be made persistently exciting by the proper choice of the input u. The following lemma attempts to address this question by providing sufficient conditions for the output of a single-input-single-output system to be persistently exciting [6].

Lemma 6.8 Let $y(t) = W(s)u(t)$ where $u, y : \mathbb{R}^+ \to \mathbb{R}$, the autocorrelation function R_u defined as in Eq. (6.7) exists and is uniform with respect to t_0, and $W(s)$ is an asymptotically stable transfer function with zeros in the open left-half plane. If $u \in \Omega_{(1,t_0,T_0)}$, then $y \in \Omega_{(1,t_1,T_1)}$ where $t_1 \geq t_0$ and $T_1 \geq T_0$.

The following sublemma is found to be useful [6] in proving Lemma 6.8.

Sublemma 6.1. Let a function $x : \mathbb{R}^+ \to \mathbb{R}^n$ be such that its autocovariance function R_x exists and is uniform with respect to t_0. Then $x \in \Omega_{(n,T)}$ if, and only if, $R_x(0)$ is positive-definite.

Proof of Sublemma 6.1. If $R_x(0) > 0$, then it immediately follows that $x \in \Omega_{(n,T)}$. If $x \in \Omega_{(n,T)}$, then from Eq. (6.9),

$$\frac{1}{T}\int_t^{t+T} |x^T(\tau)\mathrm{w}|^2 d\tau \ \geq \ \epsilon_0^2 \qquad \forall\, t \geq t_0$$

where ϵ_0 is a positive constant. Hence, for all unit vectors $\mathrm{w} \in \mathbb{R}^n$,

$$\mathrm{w}^T R_x(0)\mathrm{w} \ = \ \lim_{T\to\infty} \frac{1}{T}\int_t^{t+T} |x^T(\tau)\mathrm{w}|^2 d\tau \ \geq \ \epsilon_0^2 > 0.$$

Proof of Lemma 6.8. From Sublemma 6.1, it follows that a necessary and sufficient condition for u to belong to $\Omega_{(1,t_0,T_0)}$ is that $R_u(0) > 0$. Hence,

$$R_u(0) \ = \ \int S_u(d\omega) > 0 \tag{6.14}$$

where $S_u(\cdot)$ is the spectral measure of u. Neglecting the effect due to initial conditions that decay exponentially, the autocorrelation $R_y(0)$ can be expressed as

$$\begin{aligned} R_y(0) &= \int W(i\omega)S_u(d\omega)W^*(i\omega) \\ &= \int |W(i\omega)|^2 S_u(d\omega). \end{aligned}$$

Since $W(i\omega)$ has no zeros on the imaginary axis, from Eq. (6.14) it follows that $R_y(0) > 0$. Hence, $y \in \Omega_{(1,t_1,T_1)}$ for some $t_1 \geq t_0$ and $T_1 \geq T_0$.

Corollary 6.1 If in Lemma 6.8, $y, u : \mathbb{R}^+ \to \mathbb{R}^n$, then $u \in \Omega_{(n,t_0,T_0)} \Rightarrow y \in \Omega_{(n,t_1,T_1)}$ for some $t_1 \geq t_0, T_1 \geq T_0$.

Proof. The proof is based on Lemmas 6.2 and 6.8. Since the projection of u along a unit vector w belongs to $\Omega_{(1,t_0,T_0)}$, the projection of y along the same vector w is also in $\Omega_{(1,t_1,T_1)}$ by Lemma 6.8. Since this is true for any unit vector w, it follows from Lemma 6.2 that $y \in \Omega_{(n,t_1,T_1)}$.

Corollary 6.1 implies that the output vector of a transfer matrix $W(s)I$, where $W(s)$ is a transfer function as defined in Lemma 6.8, is persistently exciting if the vector input is persistently exciting. However, for a general $n \times n$ transfer matrix $\Lambda(s)$ with $y(t) = \Lambda(s)u(t)$, the same need not hold as shown in the following example.

Example 6.10

If $u(t) = [(-\sin t + \alpha \cos t), (-\sin t + \beta \cos t)]^T$, $\alpha, \beta > 0$, $\alpha \neq \beta$, then $u \in \Omega_{(2,T)}$. If

$$\Lambda(s) = \begin{bmatrix} \frac{1}{s+\alpha} & 0 \\ 0 & \frac{1}{s+\beta} \end{bmatrix}, \quad y(t) = \Lambda(s)u(t), \text{ and } y(0) = [1,1]^T,$$

then $y(t) = [\cos t + a_1 e^{-\alpha t}, \ \cos t + a_2 e^{-\beta t}]^T \notin \Omega_{(2,T)}$, where $a_1, a_2 \in \mathbb{R}$.

From the example above, we can also conclude that generally, in a dynamical system

$$\dot{x}(t) = Ax(t) + u(t)$$

where $A : \mathbb{R}^{n\times n}$ and is asymptotically stable, $u \in \Omega_{(n,T)}$ does not imply $x \in \Omega_{(n,T)}$.

Lemma 6.8, along with Corollary 6.1, forms the basis for many of the results that follow. An immediate consequence of Corollary 6.1 is that Lemma 6.8 can be further generalized as described in Theorem 6.2.

Theorem 6.2 Let $\dot{x}_1 = A_1x_1 + b_1r$ and $\dot{x}_2 = A_2x_2 + b_2r$ be two dynamical systems where A_i are asymptotically stable $(n \times n)$ matrices and (A_i, b_i) are controllable for $i = 1, 2$. Then $x_1 \in \Omega_{(n,t_1,T_1)}$ if, and only if, $x_2 \in \Omega_{(n,t_2,T_2)}$ for some t_1 and $t_2 \geq t_0$, where the input $r(t)$ is defined for all $t \geq t_0$.

Proof. Taking the transfer functions from r to $\overline{x}_1$ and r to $\overline{x}_2$ (where $\overline{x}_1$ and $\overline{x}_2$ correspond to x_1 and x_2, respectively, in the controllable canonical forms), it follows that $\overline{x}_1 = W(s)\overline{x}_2$ where $W(s)$ is a transfer function with zero relative degree whose poles and zeros are in the open left-half of the complex plane. Hence, $x_1 \in \Omega_{(n,t_1,T_1)}$ if, and only if, $x_2 \in \Omega_{(n,t_2,T_2)}$.

By Theorem 6.2, the state x_1 of an asymptotically stable and controllable dynamical system of nth order is persistently exciting if, and only if, the state of any other asymptotically stable and controllable dynamical system of the same order with the same input is persistently exciting. This has important implications since the PE of the state of an unknown system can be concluded from the PE of the state of a known system.

For example, if the n-dimensional state x_1 of a system with a bounded input r is given by

$$x_1(t) = \left[\frac{1}{s+a_1}, \frac{1}{s+a_2}, \cdots, \frac{1}{s+a_n}\right]^T r(t)$$

with $a_i > 0, a_i \neq a_j, i \neq j, i,j = 1,2,\ldots,n$, then, it can be checked using the condition in Eq. (6.8) whether $x_1 \in \Omega_{(n,T)}$ since such a state x_1 is easy to construct. If $x_1 \in \Omega_{(n,T)}$, it implies that states of all controllable asymptotically stable systems of dimension n, with the same input r, also belong to $\Omega_{(n,T)}$. This was demonstrated by Dasgupta *et al.* in [8]. Similar examples are given in (a)-(c) below:

(a) $x_2(t) = \left[\frac{1}{(s+a)}, \frac{1}{(s+a)^2}, \cdots, \frac{1}{(s+a)^n}\right]^T r(t)$ with $a > 0$.

(b) $x_3(t) = \left[\frac{1}{q(s)}, \frac{s}{q(s)}, \cdots, \frac{s^{n-1}}{q(s)}\right]^T r(t)$ where $q(s)$ is any Hurwitz polynomial of degree n.

(c) $x_4(t) = \left[r(t), \dot{r}(t), \ldots, \frac{d^{n-1}r}{dt^{n-1}}(t)\right]^T$ if $\frac{d^i r}{dt^i}$ exists and is bounded for $i = 1,2,\ldots,n-1$.

The examples above indicate that to ensure that the n-dimensional state of a controllable asymptotically stable system with a bounded input r is persistently exciting, the state of another known system with the same input has to be checked for persistent excitation. The question arises then as to whether this can be carried out by deriving conditions directly on the reference input rather than on the state of another system. By using the concept of spectral lines, necessary and sufficient conditions on r that ensure this can be obtained. This elegant method is due to Boyd and Sastry [6], who take a frequency domain approach to the problem of PE in a specific adaptive control system. We state their main result in Theorem 6.3 below.

Theorem 6.3 Let $r \in \mathcal{P}_{[0,\infty)}$ and let the autocorrelation function of r exist. If x is the state of the dynamical system defined as in Eq. (6.13), then the necessary and sufficient condition for the state $x \in \Omega_{(n,T)}$ is that the spectral measure of r be concentrated on at least $k \geq n$ points.

Proof. From Eq. (6.13), it follows that $x(t) = W(s)r(t)$, where $W(s)$ is the transfer vector given by $W(s) = (sI - A)^{-1}b$. By Sublemma 6.1, x is persistently exciting if, and only if, $R_x(0) > 0$. Since

$$\begin{aligned} R_x(0) &= \int S_x(d\omega) \\ &= \int W(i\omega)S_r(d\omega)W^*(i\omega) \end{aligned}$$

it follows that $x \in \Omega_{(n.T)}$ if, and only if,

$$\int W(i\omega)S_r(d\omega)W^*(i\omega) \quad > \quad 0.$$

Let S_r be concentrated at k points. Then the autocorrelation matrix $R_x(0)$ can be expressed as

$$R_x(0) \quad = \quad \sum_{i=1}^{k}\{W(i\omega_i)W^*(i\omega_i)\}S_r(\omega_i).$$

Since Eq. (6.13) is a controllable system, $W(s)$ has linearly independent elements and therefore $\sum_{i=1}^{k}\{W(i\omega_i)W^*(i\omega_i)\}$ is a positive-definite matrix, if $k \geq n, \omega_i \neq \omega_j, i \neq j, i, j = 1, 2, \ldots n$. Hence, $R_x(0) > 0$ if, and only if, the spectral measure $S_r(\omega_i)$ is concentrated on at least $k \geq n$ points.

Theorem 6.3 requires that the autocorrelation $R_r(t)$ exists in addition to the condition that $r \in \mathcal{P}_{[0,\infty)}$.

The following corollary finds special application in problems that arise in adaptive systems.

Corollary 6.2 Let r satisfy the assumptions given in Theorem 6.3. The condition that r has spectral measures at $k \geq n+1$ points is necessary and sufficient for the vector $z = [x^T, r]^T$ to belong to $\Omega_{(n,T)}$ where x is defined as in Theorem 6.3.

Proof. The proof follows along similar lines to that of Theorem 6.3 since $z(t) = \overline{W}(s)r(t)$ where $\overline{W}(s) = [\{(sI - A)^{-1}b\}^T, 1]^T$ has linearly independent elements.

From the discussions above, it follows that for a given reference input, either the state of every controllable and asymptotically stable system of order n belongs to $\Omega_{(n,t_0,T_0)}$ or none of them belongs to $\Omega_{(n,t_0,T_0)}$. Also, controllability was assumed in all the dynamical systems considered. If this condition is not satisfied, similar lemmas and theorems can be stated by replacing the entire state x by the controllable part x_c, and conditions for x to belong to $\Omega^{n_c}_{(n,T)}$ can be derived, where n_c is the dimension of x_c.

6.4.3 Almost Time-Invariant Systems

The last section dealt with conditions on the reference input that assured the PE of the state of an LTI system by appropriately choosing the reference input. As a next step, we study the effect of an almost time-invariant system on a persistently exciting signal. Many of the adaptive systems can be considered as slowly varying systems that asymptotically become linear time-invariant. We present the following lemma, which indicates that the property of PE is preserved when a signal is passed through a controllable almost time-invariant system.

Lemma 6.9 Let a time-invariant system be described by the equation

$$\dot{z}(t) \quad = \quad Az(t) + br(t) \qquad z : \mathbb{R}^+ \to \mathbb{R}^n$$

and an almost time-invariant system given by

$$\dot{x}(t) = [A + B(t)]x(t) + br(t) + w(t) \qquad x : \mathbb{R}^+ \to \mathbb{R}^n$$

where A is an asymptotically stable matrix, $B(t)$ is a bounded time-varying matrix, r is a bounded scalar input, and w is a bounded n−dimensional vector disturbance with $\lim_{t\to\infty} w(t) = 0$. If (i) $\|B(t)\| \leq c_1$ for all $t \geq 0$, where c_1 is a sufficiently small constant, or (ii) $B(\cdot) \in \mathcal{L}^1$ or $\mathcal{L}^2$, then

$$z \in \Omega_{(n,t_0,T)} \Rightarrow x \in \Omega_{(n,t_1,T)}$$

where $t_1 \geq t_0$.

Proof. (i) Using Theorem 2.4, it follows that the origin of $\dot{x} = [A + B(t)]x$ is exponentially stable when (i) is satisfied. Since r and w are bounded, x is also bounded. If $e = x - z$, we obtain the differential equation

$$\dot{e}(t) = Ae(t) + B(t)x + w(t).$$

Since A is asymptotically stable, x is uniformly bounded, $\lim_{t\to\infty} w(t) = 0$, and $\|B(t)\| \leq c_1$, it follows that $\|e(t)\| \leq kc_1$, where $k > 0$. Therefore there exists a c_1 for which

$$z \in \Omega_{(n,t_0,T)} \Rightarrow x \in \Omega_{(n,t_1,T)}, \qquad t_1 \geq t_0$$

since $x = z + e$.

(ii) Using Theorem 2.4 and Lemma 2.2, it can once again be shown that $x(t)$ is the output of an exponentially stable system with bounded input $br(t)+w(t)$. Since $w(t) \to 0$ as $t \to \infty$, and $B(\cdot) \in \mathcal{L}^1$ or $\mathcal{L}^2$, $e(t) \to 0$ as $t \to \infty$. Hence, the result follows.

6.4.4 Persistent Excitation of Unbounded Signals

In all the discussions in this section, we have assumed that the vector function, whose persistent excitation we are interested in, is uniformly bounded. In many adaptive control problems, however, the boundedness of the internal signals cannot be assumed a priori and hence a need exists to extend the concepts derived thus far to such signals as well. Ideas along these lines have been explored in the case of discrete systems in [10] and [12] and for continuous systems in [3,10,17]. Work is currently in progress in this area. In our opinion, the concept of PE of unbounded signals in the continuous case is not sufficiently well developed to warrant a detailed presentation in this chapter but represents an important area for future research. We have therefore confined our attention only to bounded signals. The following comments concerning the results in [3,10,17] are, however, useful.

If u is unbounded, the condition

$$\int_t^{t+T_0} u(\tau)u^T(\tau)d\tau \geq \alpha I \qquad \forall t \geq t_0 \tag{6.8}$$

is not sufficient to assure uniform asymptotic stability of the origin in Eq. (6.6) or Eq. (6.10) [17]. For example, if $u^T(t) = [\exp(\alpha t), \exp(\beta t)]$, $\alpha > \beta > 0$, it satisfies the condition in Eq. (6.8), but the origin of Eq. (6.6) is not u.a.s. In [3], on the other hand, it is shown that under the same condition in Eq. (6.8), if the differential equation (6.6) is modified to have the form

$$\dot{x} = -\Gamma(t)u(t)u^T(t)x$$

where $\Gamma(t)$ is a suitably chosen time-varying matrix, the origin is exponentially stable. This is discussed in Section 6.5.3(a) for the case when u is bounded. The reader is referred to [3] for the case when u is not assumed to be bounded.

6.5 APPLICATIONS TO ADAPTIVE SYSTEMS

The general concepts introduced as well as the specific results derived in Sections 6.3 and 6.4 are applied to adaptive identification and control problems in this section. The convergence of parameter estimates to their true values in adaptive observers is considered first. The same problem in adaptive control is then addressed and the importance of persistent excitation in the practical design of adaptive systems follows.

6.5.1 Parameter Convergence in Adaptive Observers

In Chapter 4, it was shown that a nonminimal representation of any single input single-output LTI system of dimension n is given by

$$\begin{aligned} \dot{\omega}_1 &= F\omega_1 + gu \\ \dot{\omega}_2 &= F\omega_2 + gy_p \\ y_p &= c^T\omega_1 + d^T\omega_2 \end{aligned} \tag{6.15}$$

where $F \in \mathbb{R}^{n\times n}$ is asymptotically stable, and (F, g) is controllable. The pair (F, g), and the n-vectors c and d represent the parameters of the plant. To estimate the latter when they are unknown, an adaptive observer structure described by

$$\begin{aligned} \dot{\widehat{\omega}}_1 &= F\widehat{\omega}_1 + gu \\ \dot{\widehat{\omega}}_2 &= F\widehat{\omega}_2 + gy_p \\ \widehat{y}_p &= \widehat{c}^T\widehat{\omega}_1 + \widehat{d}^T\widehat{\omega}_2 \end{aligned} \tag{6.16}$$

was suggested in Section 4.3. The error e_1 between $\widehat{y}_p$ and y_p satisfies the equation

$$e_1 = \phi^T\widehat{\omega} + \varepsilon \tag{6.17}$$

where $\omega = [\omega_1^T, \ \omega_2^T]^T$, $\widehat{\omega} = [\widehat{\omega}_1^T, \widehat{\omega}_2^T]^T$, $\theta = [c^T, \ d^T]^T$, $\widehat{\theta} = [\widehat{c}^T, \ \widehat{d}^T]^T$, $\phi = \widehat{\theta} - \theta$, $\widetilde{\omega} = \widehat{\omega} - \omega$, and $\varepsilon = \theta^T\widetilde{\omega}$. The observer parameter $\widehat{\theta}$ is adjusted as

$$\dot{\widehat{\theta}} = \dot{\phi} = -e_1\widehat{\omega}. \tag{6.18}$$

Stability Analysis. Following the approach used in the preceding chapters, it is first shown that $V = \phi^T\phi/2$ is a Lyapunov function for the system described by Eqs. (6.17) and (6.18). This ensures that $\phi \in \mathcal{L}^\infty$ (see Chapter 4) and that

$$\lim_{t\to\infty} e_1(t) = 0.$$

Uniform Asymptotic Stability. The convergence of the parameter vector $\widehat{\theta}(t)$ was not discussed in Chapter 4. The conditions on the external input u that are sufficient for $\widehat{\theta}(t)$ to converge to the true value θ, are given below. These can be conveniently expressed in two stages.

1. Neglecting the exponentially decaying term ε, we can express Eqs. (6.17) and (6.18) as

$$\dot{\phi} \;=\; -\widehat{\omega}\widehat{\omega}^T\phi. \tag{6.19}$$

Equation (6.19) is of the form of Eq. (6.6). Hence, if $\widehat{\omega} \in \Omega_{(2n,T)}$, that is, if positive constants T and ϵ_1 exist so that

$$\frac{1}{T}\int_t^{t+T} |\widehat{\omega}^T(\tau)\mathrm{w}d\tau| \;\geq\; \epsilon_1 \qquad \forall t \geq t_0$$

for every unit vector $\mathrm{w} \in \mathbb{R}^n$, the origin of Eq. (6.19) is uniformly asymptotically stable. Since $\widehat{\omega}$ is an internal signal of the observer, our aim is to determine the conditions in terms of the input u, which is an independent variable.

2. Equation (6.16) is of the form $\dot{\widehat{\omega}} = A\widehat{\omega} + bu$ where A is asymptotically stable, (A, b) is controllable and $\widehat{\omega}(t) \in \mathbb{R}^{2n}$. Hence, if $u \in \mathcal{P}_{[0,\infty)}$ and has $2n$ spectral lines (or is such that any $2n$-dimensional state of a controllable asymptotically stable system belongs to $\Omega_{(2n,T)}$), then $\widehat{\omega} \in \Omega_{(2n,T)}$. Hence, $\lim_{t\to\infty} \phi(t) = 0$.

Comment 6.1 When the plant to be identified is partially known, the condition on r for identification of the unknown parameters can be relaxed. The reader is referred to [7] for further details.

Comment 6.2 In Simulations 3.1, 3.2, 4.1-4.5, 6.1, and 6.2, parameter convergence in identification problems was studied. In all cases, it was seen that an external reference input with n frequencies is sufficient for the convergence of $2n$ parameters.

If representation 1 [refer to Eq. (4.24)] is used instead of Eq. (6.15) to design an adaptive observer, convergence of parameters to their true values can be shown in a similar manner. The plant can be represented as

$$\begin{bmatrix} \dot{\omega}_1 \\ \dot{\omega}_2 \\ \dot{x}_1 \end{bmatrix} = \begin{bmatrix} \Lambda & 0 & 0 \\ 0 & \Lambda & \ell \\ \overline{c}^T & \overline{d}^T & -\lambda + d_0 \end{bmatrix} \begin{bmatrix} \omega_1 \\ \omega_2 \\ x_1 \end{bmatrix} + \begin{bmatrix} \ell \\ 0 \\ c_0 \end{bmatrix} u \tag{6.20}$$

where $\lambda > 0$, the constant $(n-1) \times (n-1)$ matrix Λ, and the $(n-1)$-vector ℓ are known, and $c_0, \bar{c}$, d_0, and $\bar{d}$ are the unknown constant parameters of the plant. The error equation that relates the parameter error ϕ to the output error e_1 is of the form [refer to Eqs. (4.24),(4.26)]

$$\begin{bmatrix} \dot{e}_1 \\ \dot{\phi} \end{bmatrix} = \begin{bmatrix} -\lambda & \widehat{\omega}^T \\ -\widehat{\omega} & 0 \end{bmatrix} \begin{bmatrix} e_1 \\ \phi \end{bmatrix}$$

where $\widehat{\omega}$ is the estimate of $\omega = [u, \omega_1^T, y_p, \omega_2^T]^T$. A Lyapunov function of the form $V = (e_1^2 + \phi^T \phi)/2$ can be used to show that e_1 and ϕ are bounded (see Section 4.3). Since the matrix in Eq. (6.20) is asymptotically stable, and the system is controllable, from Corollary 6.2, an input $u \in \mathcal{P}_{[0,\infty)}$ with $2n$ spectral lines implies that $\omega \in \Omega_{(2n,T)}$. Since $\widetilde{\omega}(t) = \widehat{\omega}(t) - \omega(t)$ tends to zero asymptotically, $\widehat{\omega} \in \Omega_{(2n,t_1,T)}$ for some $t_1 \geq t_0$. Hence, $\lim_{t\to\infty} \phi(t) = 0$.

6.5.2 Parameter Convergence in Adaptive Control

In Chapter 5, the stability of the MRAC system was discussed in detail. It was shown that when the controller parameters are adjusted according to an adaptive law, the output error $e_1(t)$ would tend to zero. In this section, we discuss the conditions under which the parameter vector θ also converges to its desired value θ^*. For the benefit of the reader, we recapitulate the conditions under which the adaptive law was obtained.

The plant P to be controlled has an input-output pair $\{u(\cdot), y_p(\cdot)\}$ and a transfer function

$$W_p(s) = k_p \frac{Z_p(s)}{R_p(s)}$$

where the coefficients of the numerator and denominator polynomials are unknown. A reference model with an input-output pair $\{r(\cdot), y_m(\cdot)\}$ is chosen with $r \in \mathcal{P}_{[0,\infty)}$ and an asymptotically stable transfer function $W_m(s)$. Both $W_p(s)$ and $W_m(s)$ satisfy the same assumptions as in Chapter 5. The adaptive controller structure is given by Eqs. (5.16) for the case when $n^* = 1$ and by Eqs. (5.16) and (5.50) for $n^* \geq 2$ when method 1 is used. It was shown in Sections 5.3 and 5.5 that, in both cases, the signals in the system remain bounded and that $\lim_{t\to\infty} e_1(t) = 0$, where $e_1 = y_p - y_m$. Since the two cases are analytically different, we treat them separately in the following paragraphs:

Case (i) $n^* = 1$. In this case, $W_m(s)$ is strictly positive real. The overall system equations become

$$\dot{x} = Ax + b(\phi^T \omega + k^* r); \qquad y_p = h^T x$$

where $h^T(sI - A)^{-1} b k^* = W_m(s)$, $x = [x_p^T, \omega_1^T, \omega_2^T]^T$, and $\omega = [r, \omega_1^T, y_p, \omega_2^T]^T$. Since the transfer function of the plant together with the controller is equal to that of the model when $\phi(t) \equiv 0$, a nonminimal representation of the model is given by

$$\dot{x}_{mn} = Ax_{mn} + bk^* r; \qquad y_m = h^T x_{mn}$$

where $x_{mn} = [x_p^{*T}, \omega_1^{*T}, \omega_2^{*T}]^T$. The error equations between the system and the model are therefore given by

$$\dot{e} = Ae + b\phi^T \omega; \qquad e_1 = h^T e. \tag{6.21a}$$

The adaptive law is chosen in the standard way to be

$$\dot{\phi} = -sgn(k_p) e_1 \omega. \tag{6.21b}$$

If $\omega^* = [r, \omega_1^{*T}, y_m, \omega_2^{*T}]^T$, it follows that $\omega = \omega^* + Ce$ where C is a $2n \times (3n-2)$ matrix given by

$$C = \begin{bmatrix} 0 & 0 & 0 \\ 0 & I & 0 \\ h_p^T & 0 & 0 \\ 0 & 0 & I \end{bmatrix}$$

and I is an $(n-1) \times (n-1)$ identity matrix.

Stability Analysis.

(i) From a quadratic Lyapunov function $V = [e^T Pe + \phi^T \phi]/2$, it follows that e and ϕ are uniformly bounded and $e \in \mathcal{L}^2$.

(ii) Since the state of the model is bounded, it follows that ω is bounded.

(iii) Since $\dot{e}$ is bounded, it follows from (i) that $\lim_{t \to \infty} e(t) = 0$.

Parameter Convergence. From Theorem 2.17, and Eqs. (6.21a) and (6.21b), we conclude that the origin is u.a.s. if $\omega \in \Omega_{(2n.T)}$. Since

$$\omega = \omega^* + Ce, \qquad \text{and} \qquad \lim_{t \to \infty} e(t) = 0,$$

it follows that $\omega \in \Omega_{(2n,t_1,T)}$ if, and only if, $\omega^* \in \Omega_{(2n,t_0,T)}$ where $t_1 \geq t_0$. Since ω^* corresponds to the controllable part of the state x_{mn}, parameter convergence follows, if, and only if, r has at least $2n$ spectral lines.

Case (ii) $n^* \geq 2$. Here, $W_m(s)$ cannot be chosen to be strictly positive real and an augmented error ϵ_1 must be generated to implement a stable adaptive law, as discussed in Chapter 5. If Method 1 is used, from Eqs. (5.50) and (5.52), the augmented error and the adaptive laws can be derived as

$$\begin{aligned} \epsilon_1 &= \frac{k_p}{k_m} \phi^T \zeta + \psi_1 e_2 \\ \dot{\phi} &= -sgn(k_p) \frac{\epsilon_1 \zeta}{1 + \overline{\zeta}^T \overline{\zeta}} \\ \dot{\psi}_1 &= -\frac{\epsilon_1 e_2}{1 + \overline{\zeta}^T \overline{\zeta}} \end{aligned}$$

where $W_m(s)I\omega = \zeta$, $\omega = [r, \overline{\omega}^T]^T$, $\overline{\zeta} = W_m(s)I\overline{\omega}$, and $e_2 = \phi^T\zeta - W_m(s)I\phi^T\omega$.

Stability Analysis. Using a quadratic Lyapunov function, it is shown that ϕ and therefore ϵ_1 are bounded. Using arguments involving growth rates of unbounded signals, it can be established that all the signals in the loop are bounded (see Section 5.5) and that $\lim_{t\to\infty} e(t) = 0$. Hence, $\omega(t) \to \omega^*(t)$ and $\zeta(t) \to \zeta^*(t)$, as $t \to \infty$, where ω^* is defined in case (i), and $\zeta^* = W_m(s)I\omega^*$.

Parameter Convergence. Since $e_2(t)$ tends to zero as $t \to \infty$, the vector $[\zeta^T, e_2]^T \notin \Omega_{(2n+1,T)}$. However, if $\zeta \in \Omega_{(2n.T)}$, from Theorem 6.1, it follows that $\lim_{t\to\infty} \phi(t) = 0$. Since $\zeta(t) \to \zeta^*(t)$ as $t \to \infty$, $\zeta \in \Omega_{(2n,t_2.T_1)}$ if, and only if, $\zeta^* \in \Omega_{(2n,t_1,T_1)}$ where $t_2 \geq t_1$. By imposing the same conditions on $r(t)$ as in case (i), we obtain that $\phi(t)$ tends to zero asymptotically.

Comment 6.3 In Simulation 5.1, the control of a plant with relative degree unity was considered. It was seen that when the reference input has two frequencies, the control parameters θ_1, θ_0, and θ_2 converge to their true values. In Simulation 5.2, $n^* = 2$, and the three control parameters once again converge to their desired values when the reference input has two distinct frequencies.

Comment 6.4 The equilibrium states of the error differential equations in the two cases discussed in this section are uniformly asymptotically stable in the large, but exponentially stable only in a finite ball around the origin.

Comment 6.5 The importance of Theorem 6.2 in the adaptive control problem is evident while attempting to establish the PE of ω^*. While ω^* is the controllable part of the adaptive system, we cannot determine directly whether or not it is persistently exciting since we do not have prior knowledge of θ^*. However, the PE of the state of any known $2n$-dimensional controllable and asymptotically stable system with r as the input, also assures the PE of ω^* (Theorem 6.2).

Comment 6.6 Since $e_2(t) \to 0$ as $t \to \infty$, $\psi_1(t)$ does not tend to zero, but $e(t)$ and $\phi(t)$ tend to zero asymptotically. This is sufficient for the control parameter $\theta(t) = [\overline{\theta}^T(t)\ k(t)]^T$ to converge to the true value $\theta^* = [\overline{\theta}^{*T},\ k_m/k_p]^T$. This was pointed out in [8].

Comment 6.7 If the augmented error ϵ_1 is generated alternately using method 2, it can be shown in a similar fashion that

(i) $\omega^* \in \Omega_{(2n.t_0.T_0)}$ and hence $\zeta^* = \overline{W}(s)\omega^* \in \Omega_{(2n.t_1.T_1)}$,

(ii) $\tilde{\zeta}(t) \triangleq \zeta(t) - \zeta^*(t) \to 0$ as $t \to \infty$, and

(iii) $\zeta \in \Omega_{(2n.t_2.T_1)}$ and, hence, $\lim_{t\to\infty} \phi(t) = 0$.

6.5.3 Design Considerations

In the design of practical adaptive observers and controllers, we are interested in accuracy as well as the speed of response. The results of the previous section addressed the problem of parameter convergence in adaptive identification and control. Once stability is assured by the adaptive law, the persistent excitation of the reference input ensures that the parameter error ϕ converges to zero. Hence, the origin in the (e, ϕ) space is uniformly asymptotically stable, and the criterion of accuracy is automatically met. However, nothing has been said about the speed with which the errors tend to zero. We outline some design considerations in this section that affect the speed of convergence of the overall adaptive system.

(a) Convergence Rate. The differential equations (6.6) and (6.10), which were used to define the persistent excitation of a signal u, are also prototypes of the error differential equations that arise in adaptive systems in the ideal case. We shall first discuss the convergence rate of Eq. (6.6).

In the differential equation

$$\dot{\phi} = -uu^T\phi \qquad u, \phi : \mathbb{R}^+ \to \mathbb{R}^n \tag{6.6}$$

let $u \in \Omega_{(n,T_0)}$, so that positive constants t_0, T_0, and ϵ_0 exist such that

$$\frac{1}{T_0}\int_t^{t+T_0} |u^T(\tau)\mathrm{w}|d\tau \geq \epsilon_0 \qquad \forall\ t \geq t_0 \tag{6.9}$$

for every unit vector $\mathrm{w} \in \mathbb{R}^n$. Choosing a quadratic function

$$V(\phi(t)) = \frac{1}{2}\phi^T(t)\phi(t),$$

it was shown in Chapter 2 that

$$V(t_0 + T_0) \leq \gamma V(t_0), \qquad \gamma = 1 - \frac{2T_0\epsilon_0^2}{(1 + u_{\max}^2 T_0)^2}, \; \|u(t)\| < u_{\max}.$$

The value of γ determines the rate at which $V(t)$ decreases over an interval of time T_0. The closer γ is to zero [or equivalently $2T_0\epsilon_0^2/(1 + u_{\max}^2 T_0)^2$ is to unity], the faster is the rate of convergence. The expression for γ, however, indicates that the relationship between the choice of the reference input and the rate of convergence is not a simple one. For example, if the amplitude of the reference input is small, ϵ_0 is small, and, hence, the convergence is slow; if however the amplitude of u is made very large, $u_{\max}^4$ is large compared with ϵ_0^2 so that once again the conveigence rate is slow. Hence, for a prescribed value of ϵ_0 and T_0, an optimal value of $u_{\max}$ exists. While ϵ_0, the degree of PE of u, can be increased by increasing the amplitude of the reference input, the rate of convergence cannot be increased beyond a certain maximum value. The introduction of a fixed adaptive gain has a similar effect on the rate of convergence.

A similar analysis is also possible for the rate of convergence of Eq. (6.10). In this case, the degree of PE is defined by ϵ_1 in the inequality in Eq. (6.11) and can be increased by increasing the amplitude of the reference input. Once again, a simple relation does not exist between the rate of convergence and the degree of PE of u, and increasing the amplitude of u does not improve the speed of convergence. This is particularly evident in Simulations 4.1(b), 4.2(b), and 5.2(c) where the gains chosen correspond to those values that yield the fastest response. As mentioned in Chapter 5, in Simulation 5.2(c), the speed of response of the control parameters was quite sensitive to variations in the adaptive gain. Further, different adaptive gains were found to be optimal for different initial conditions.

The fact that arbitrarily fast speed of convergence cannot be obtained by increasing the amplitude of u in both the above cases can be traced back to the fact that u is a vector function. If $u : \mathbb{R}^+ \to \mathbb{R}^n$, $u(t)u^T(t)$ is a matrix of rank 1 for all $t \geq t_0$. If rapid convergence to zero is desired, either u has to be replaced by U, where $U : \mathbb{R}^+ \to \mathbb{R}^{n \times n}$ and $U(t)$ is of full rank for all $t \geq t_0$, or the adaptive law has to be modified suitably. Several methods are currently known that assure arbitrarily fast convergence of the parameter vector and were discussed in the context of adaptive observers in Chapter 4. These include

(i) the hybrid algorithm,

(ii) the integral algorithm,

(iii) the algorithm with time-varying adaptive gains, and

(iv) the algorithm using multiple models.

In (i), (ii), and (iv), the vector u is replaced by a matrix U while in (iii) a time-varying gain matrix $\Gamma(t)$ is included in the adaptive law. In Simulations 4.3-4.5, these algorithms were implemented for parameter identification. It was seen that in all cases, the rate of convergence was faster than that resulting from Eq. (6.6). In the paragraphs that follow, we provide an analytical justification for this observation. Method (i) is described in detail in Chapter 8 and we shall confine our attention to cases (ii)-(iv) here.

(ii) The Integral Algorithm. As described in Chapter 4 (Section 4.4.5), the integral algorithm suggested by Kreisselmeier uses all past data, collected over the interval $[t_0, t]$ and suitably weighted, to adjust the parameter vector $\theta(t)$ at time $t(\geq t_0)$. This results in the adaptive law

$$\dot{\theta}(t) = \dot{\phi}(t) = -\gamma[\Gamma(t)\theta(t) + \delta(t)], \qquad \gamma > 0$$

which is equivalent to

$$\dot{\phi}(t) = -\gamma\Gamma(t)\phi(t) \tag{6.22}$$

where the matrix $\Gamma(t)$ is positive-definite for all $t \geq t_0 + T$. The equations describing the evolution of $\Gamma(t)$ and $\delta(t)$ are given by

$$\begin{aligned}\dot{\Gamma}(t) &= -q\Gamma(t) + \omega(t)\omega^T(t) & \Gamma(t_0) &= 0\\ \dot{\delta}(t) &= -q\delta(t) + \omega(t)y_p(t) & \delta(t_0) &= 0\end{aligned}$$

where $q > 0$.

From Eq. (6.22), it is clear that if $\omega \in \Omega_{(n,T)}$, $\lim_{t\to\infty} \phi(t) = 0$, and the speed of convergence of the parameter error depends on $\gamma\lambda_{\min}(t)$, where $\lambda_{\min}(t)$ is the minimum eigenvalue of $\Gamma(t)$ at time t. Hence, by increasing the value of γ, the speed of convergence can be increased.

(iii) Time-varying gains. An alternate method for adjusting the parameters is to use time-varying gains. This results in the adaptive law

$$\dot{\phi} = -\gamma\Gamma(t)e_1\omega \qquad \gamma > 1/2 \tag{4.38}$$

$$\dot{\Gamma} = -\Gamma\omega\omega^T\Gamma \tag{4.37}$$

where $\Gamma(t_0) = \Gamma^T(t_0) = \Gamma_0 > 0$. As mentioned in Chapter 4, a Lyapunov function candidate $V(\phi) = \phi^T\Gamma^{-1}(t)\phi$ establishes the stability in the large of the origin $\phi = 0$ in Eq. (4.38). In [3], it is shown that $\phi = 0$ in (4.38) is exponentially stable if the following conditions are satisfied:

(1) an unbounded sequence $\{t_i\}$ exists such that $T_{\min} \leq t_{i+1} - t_i \leq T_{\max}$ and the value of $\Gamma(t)$ is reset at $t = t_i$ as $\Gamma(t_i) = (1/\beta)I$,

(2) $\Gamma(t)$ is adjusted as in Eq. (4.37) for $t \in [t_0, \infty)/\{t_i\}$, and

(3) $\omega \in \Omega_{(n,T_{\min})}$, that is, $\displaystyle\int_t^{t+T_{\min}} \omega(\tau)\omega^T(\tau)d\tau \geq \alpha I.$

The method above is referred to as *covariance resetting* in the literature.

(iv) The Algorithm Based on Multiple Models. In Chapter 4 (Section 4.4.6), it was shown that if the equation relating the unknown constant parameter vector θ and the functions $\omega : \mathbb{R}^+ \to \mathbb{R}^{2n}$ and $y : \mathbb{R}^+ \to \mathbb{R}$ is of the form

$$\omega^T(t)\theta = y(t)$$

$2n-1$ asymptotically stable transfer functions $W_1(s), W_2(s), \ldots, W_{2n-1}(s)$, can be used to process both $\omega(t)$ and $y(t)$ to obtain the equation

$$\Omega^T(t)\theta = Y(t)$$

where $\Omega(t)$ is a $2n \times 2n$ matrix, and $Y(t) \in \mathbb{R}^{2n}$ for all $t \geq t_0$. This, in turn, yields the adaptive law

$$\dot{\hat{\theta}}(t) = -\gamma \left[\widehat{\Omega}(t)\widehat{\Omega}^T(t)\right] \hat{\theta}(t) + \gamma \left[\widehat{\Omega}(t) Y(t)\right],$$

which corresponds to the error equation [Eq. (4.45)]

$$\dot{\phi}(t) = -\gamma \widehat{\Omega}(t)\widehat{\Omega}^T(t)\phi(t), \qquad \gamma > 0.$$

If $u(t)$ is bounded and persistently exciting, and the transfer functions $W_i(s)$ are so chosen that $\Omega(t)$ is positive-definite for every $t \geq t_0$, arbitrarily fast convergence rates can be realized in theory by making the adaptive gain γ sufficiently large.

Comment 6.8 In the adaptive control literature as well as in the preceding sections, the phrase *arbitrarily fast convergence* has been used. Strictly speaking, this is not true since arbitrarily fast convergence is achieved only with zero initial conditions in the plant and when the matrix $\Gamma(t)$ in (i) and $\widehat{\Omega}(t)$ in (ii) are positive-definite and the initial conditions are close to zero. If the system is started at time t_0, the condition above is met only at the end of one period T, that is, at time $t_0 + T$. Hence, arbitrarily fast convergence implies that the convergence takes place in a time arbitrarily close to $t_0 + T$ and not to t_0!

(b) Choice of the Controller Structure. In all adaptive systems that use an adaptive law of the form in Eq. (6.6) or (6.10), it has been observed over the years that while the output error converges to zero in a relatively short time, the parameter error converges extremely slowly to its true value, even when the reference input is persistently exciting and contains n distinct frequencies. This phenomenon can be observed in Simulations 4.1, 4.2, 5.1, and 5.2. This has been attributed to the stiffness of the corresponding error differential equations. It is precisely in such cases that the alternate methods suggested in (a) have to be used. However, when the poles or the zeros of the plant transfer function are known, the convergence of the control parameters is found to be rapid [refer to Simulation 4.1(c)-(d)] and adaptive laws of the form of Eq. (6.6) or (6.10) may be adequate. The slow convergence of the parameters when neither the poles nor the zeros of the plant transfer function are known may be due to the fact that the signal ω^* (and therefore ω) may not be sufficiently persistently exciting in $\mathbb{R}^{2n}$. One way of improving the rate of convergence may be by choosing the fixed parameters of the matrix Λ and the vector ℓ in the controller appropriately. Precisely how this can be accomplished is not known at present.

6.6 PERSISTENT EXCITATION AND UNIFORM ASYMPTOTIC STABILITY

The discussions in the preceding sections have revealed the close connection that exists between PE of reference inputs and the stability of adaptive systems. By the definition given in Section 6.3, PE determines a class of signals for which the origins of two linear differential equations are u.a.s. These are adequate to prove the stability of most of the systems discussed in this book. However, as adaptive systems become more complex, conditions under which they are u.a.s.l. will also become more difficult to determine and may no longer coincide with those needed to assure the PE of the relevant signal, as defined in Section 6.3. The following examples indicate that the conditions in Eqs. (6.9) and (6.11) may be neither necessary nor sufficient for the uniform asymptotic stability in the large of nonlinear differential equations whose linear parts correspond to Eqs. (6.6) and (6.10).

Example 6.11

a. Consider the second order differential equation

$$\dot{x} = -u(t)u^T(t)x + f(x) \tag{6.23}$$

where

$$x^T = [x_1, x_2], f(x)^T = [x_2(x_1^2 + x_2^2), \quad -x_1(x_1^2 + x_2^2)]$$

and $u : \mathbb{R}^+ \to \mathbb{R}^2$. It is seen that as for Eq. (6.13), $V(x) = x^T x/2$ yields $\dot{V} = -(u^T x)^2 \leq 0$ along any trajectory of Eq. (6.23), since $x^T f(x) = 0$. When $u(t) \equiv 0$, the differential equation (6.23) has solutions of the form

$$x(t) = [\sin(\|x(t_0)\|t), \cos(\|x(t_0)\|t)]^T$$

and $x \in \Omega_{(2,T)}$. If

$$u(t) = [\cos(\|x(t_0)\|t), -\sin(\|x(t_0)\|t)]^T \quad \forall \ t \geq t_0,$$

then $u \in \Omega_{(2,T)}$ and the origin of the linear part $\dot{x} = -uu^T x$ is exponentially stable. However, since $u^T(t)x(t) = 0 \ \ \forall t \geq t_0$, the origin of the nonlinear differential equation is only stable but not u.a.s.l, and, hence, PE of u is not sufficient for the uniform asymptotic stability in the large of the origin of Eq. (6.23).

b. If in Eq. (6.23), $u(t) \equiv [1, 0]^T$, then the vector u is not persistently exciting and, hence, the origin of the linear part is not uniformly asymptotically stable. However, by the application of LaSalle's theorem [14], it follows that the origin of the nonlinear system is u.a.s.l. Hence the condition in Eq. (6.9) is not necessary for the uniform asymptotic stability in the large of the nonlinear system in Eq. (6.23). The following example indicates that a similar statement can also be made concerning Eq. (6.10).

Example 6.12

Consider the following nonlinear differential equation:

$$\dot{x} = \begin{bmatrix} -a & bu^T(t) \\ -u(t)b & 0 \end{bmatrix} x + f(x) \tag{6.24}$$

where

$$\begin{aligned} x &= [x_1, x_2, x_3]^T, \ x_i : \mathbb{R}^+ \to \mathbb{R}, \ i = 1, 2, 3, \ a, b \in \mathbb{R}, a > 0, \ b \neq 0, \ u : \mathbb{R}^+ \to \mathbb{R}^2 \\ f^T(x) &= [0, \ \ x_3(x_2^2 + x_3^2), \ \ -x_2(x_2^2 + x_3^2)]. \end{aligned}$$

Once again, the time derivative of $V(x) = x^T x/2$ along a trajectory of Eq. (6.24) is given by $\dot{V}(x,t) = -ax_1^2(t) \leq 0$ since $x^T f(x) = 0$. Choosing

$$u(t) = [\sin ct, \cos ct]^T,$$

and the initial condition

$$x(t_0) = [0, \sqrt{c} \cos ct_0, -\sqrt{c} \sin ct_0]^T,$$

the solution of (6.24) can be expressed as

$$x(t) = [0, \sqrt{c} \cos ct, -\sqrt{c} \sin ct]^T.$$

Hence, $\dot{V}(x,t)$ is identically zero and the origin of Eq. (6.24) is not uniformly asymptotically stable.

The examples above show that conventional definitions of persistent excitation (given in Section 6.3) may be neither necessary nor sufficient to assure uniform asymptotic stability of nonlinear systems. Hence, as adaptive systems become more complex and the definitions of persistent excitation given in Section 6.3 continue to be used, the asymptotic stability of such systems must be established on a system-by-system basis.

The stability of all the systems described in Chapters 3-6 was based on the existence of a Lyapunov function $V(x)$ with $\dot{V}(x,t) \leq 0$ along the trajectories of the system. Even though the nonautonomous field vector $f(x,t)$ is such that $\dot{V}(x,t)$ is only negative-semidefinite, since $V(x)$ decreases by a finite amount over a finite interval of time, uniform asymptotic stability of the origin could be concluded.

The conditions of Eqs. (6.9) and (6.11) used in the definition of persistent excitation assure precisely this for Eqs. (6.6) and (6.10). Since these two equations feature in all the adaptive systems and error models (refer to Chapter 7) discussed thus far, the property of PE became synonymous with the uniform asymptotic stability of adaptive systems. In more general adaptive systems of the form $\dot{x} = f(x, u(t))$, persistent excitation of a function u must be defined with respect to the system; if $V(x,t)$ is a Lyapunov function for such a system, PE of u must be defined as that property of u which implies that $V(x,t+T) - V(x,t) \leq \gamma(\|x\|) < 0$ for some $T > 0$, where $\gamma(\cdot)$ is a monotonic function.

6.7 SUMMARY

Persistent excitation features prominently in most of the current problems that arise in adaptive systems. It was originally defined as a condition on a signal that assured the convergence of a parameter vector in identification problems and later it was extended to control problems as well. In both cases, the condition is expressed in terms of an integral of a vector function. Since the input to the plant in the identification problem and the reference input in the control problem are scalar signals that can be chosen by the designer, there is a need to relate the conditions of persistent excitation of internal signals of the system to equivalent conditions on these scalar signals. The study of dynamic transformations of persistently exciting signals reveals that the frequency content of a scalar signal determines the persistent excitation of the vector signal obtained by a dynamic transformation.

In the identification problems, the vector input u is known to be bounded and can be suitably chosen by the designer. In the control problem, this is no longer the case and u depends on the adaptation process. Hence, while conditions for boundedness and stability can be stated in terms of the input to the plant in the former, it has to be stated in terms of signals that are independent of the adaptive process in the latter. Generally, these are the input and states of the reference model that is linear, time-invariant and asymptotically stable.

Definitions of persistent excitation given in Sections 6.3-6.5 are related to the uniform asymptotic stability of linear time-varying systems that frequently arise in adaptive control. In Section 6.6, it is shown that the properties attributed to persistently exciting signals may be neither necessary nor sufficient to assure uniform asymptotic stability of nonlinear systems.

PROBLEMS

Part I

1. If θ is a constant vector in $\mathbb{R}^n$, and the functions $u : \mathbb{R}^+ \to \mathbb{R}^n$ and $y : \mathbb{R}^+ \to \mathbb{R}$ are related by

$$u^T(t)\theta = y(t)$$

determine the conditions on u so that θ can be estimated from input-output data. Can θ be estimated exactly in a finite time?

2. **(a)** If in the equation

$$\dot{x}(t) = -u(t)u^T(t)x(t) \qquad x, u : \mathbb{R}^+ \to \mathbb{R}^n \tag{P6.1}$$

$u \notin \Omega_{(n,T)}$, can you characterize the asymptotic behavior of $x(t)$?

(b) In part (a), if $n = 2$ and

$$u(t) = [c_1 \sin t + e^{-t}, \; c_2 \sin t + e^{-t}]^T,$$

where c_1 and c_2 are constants, show that $c_2x_1 - c_1x_2$ does not tend to zero asymptotically, where $x = [x_1, x_2]^T$.

3. If $u : \mathbb{R}^+ \to \mathbb{R}^2$, show that $u = [\sin t, \; \cos t]^T \in \Omega_{(2,T)}$ while $u = [\sin \sqrt{t}, \; \cos \sqrt{t}]^T \notin \Omega_{(2,T)}$. Provide a qualitative justification for your answer.

4. If $u_1 \in \Omega_{(n,t_0,T)}$, and $u_2 \in \mathcal{L}^2$, show that $u_1 + u_2 \in \Omega_{(n,t_1,T)}$ for some $t_1 \geq t_0$.

Part II

5. Let a vector function $r : \mathbb{R}^+ \to \mathbb{R}^n$ be such that its ith component is $r_i(t) = a_i \sin \omega_i t$, where $a_i \neq 0$, $\omega_i \neq \omega_j$, for $i \neq j$, and $i, j = 1, \ldots, n$. Show that $r \in \Omega_{(n.T)}$.

6. If $r(t) = \sum_{i=1}^{n} a_i \sin \omega_i t$, where ω_i and a_i are as in Problem 5, show that the state x of a $2n^{\text{th}}$ order system $\dot{x} = Ax + br$ where A is asymptotically stable and (A, b) is controllable belongs to $\Omega_{(2n,T)}$.

7. If $u : \mathbb{R}^+ \to \mathbb{R}^n$ is a piecewise-continuous and piecewise-differentiable function in $\mathcal{P}_{[0,\infty)}$, show that the definitions of persistent excitation given in Eqs. (6.8) and (6.10) are equivalent.

8. Given two dynamical systems

$$\begin{aligned} \dot{x} &= A_1 x + b_1 u \qquad x : \mathbb{R}^+ \to \mathbb{R}^3 \\ \dot{z} &= A_2 z + b_2 u \qquad z : \mathbb{R}^+ \to \mathbb{R}^2 \end{aligned}$$

where A_1 and A_2 are asymptotically stable matrices, (A_i, b_i), $i = 1, 2$ are controllable

(a) show that $x \in \Omega_{(3,T)} \Rightarrow z \in \Omega_{(2,T_1)}$ for some $T_1 \geq T$.

(b) If $x \in \Omega^2_{(3,T)}$, does $z \in \Omega_{(2,T_1)}$ for some $T_1 \geq T$?

9. In Problem 1, if $u(t) \equiv 0$ for $t > t_0 + T$ and

$$\int_{t_0}^{t_0+T} u(\tau)u^T(\tau)d\tau = Q > 0$$

can you suggest an adaptive algorithm for determining the estimate $\widehat{\theta}$ of θ such that $\lim_{t\to\infty}[\widehat{\theta}(t) - \theta] = 0$?

10. Let the reference input r in an MRAC problem be as defined in Problem 6. The order of the plant transfer function is n and its relative degree $n^* = 1$. State the arguments used to prove that the controller parameters converge to their true values. Why is the corresponding problem with $n^* \geq 2$ more difficult?

Part III

11. **(a)** If

$$\dot{x} = \left[A - u(t)u^T(t)\right] x \qquad x, u : \mathbb{R}^+ \to \mathbb{R}^n \tag{P6.2}$$

where A is a constant matrix whose symmetric part is negative definite, show that the origin is u.a.s. even if $u \notin \Omega_{(n,T)}$.

(b) Determine sufficient conditions on u so that the origin of the system

$$\dot{x} = \left[A + u(t)u^T(t)\right] x \qquad x, u : \mathbb{R}^+ \to \mathbb{R}^n$$

is uniformly asymptotically stable under the same conditions on A as in (a).

12. **(a)** Given the differential equation (P6.1) where $u \in \Omega_{(n,T)}$, why can't the rate of convergence be increased without bound by increasing the degree of persistent excitation?

(b) What modification in the algorithm would you suggest to obtain arbitrarily fast convergence by increasing the degree of PE?

13. If $u \in \Omega_{(1,T)}$, can an algebraic nonlinear transformation $y = f(u)$ result in $y \in \Omega_{(m,T)}$, where $y : \mathbb{R}^+ \rightarrow \mathbb{R}^m$, and $m > 1$? Give a simple example.

14. Given that

$$\frac{1}{T}\int_t^{t+T} u(\tau)u^T(\tau)d\tau = P > 0$$

and the minimum eigenvalue of $P = \Lambda_{\text{min}}$ with $\text{Re}[\Lambda_{\text{min}}] > \alpha$, determine the stability properties of the equilibrium state of the differential equation

$$\dot{z} = -u(t)\,[u(t) + c]^T z \qquad z, u : \mathbb{R}^+ \rightarrow \mathbb{R}^n$$

where $||c|| \leq \alpha$.

REFERENCES

1. Anderson, B. D. O. "Exponential stability of linear equations arising in adaptive identification." *IEEE Transactions on Automatic Control* 22:83–88, 1977.
2. Åström, K.J., and Bohlin, T. "Numerical identification of linear dynamic systems from normal operating records." In *Theory of Self-Adaptive Control Systems*, pp. 96–111, New York:Plenum Press, 1966.
3. Bai, E.W., and Sastry, S.S. "Global stability proofs for continuous-time indirect adaptive control schemes." *IEEE Transactions on Automatic Control* 32:537–543, June 1987.
4. Bitmead, R.R. "Persistence of excitation conditions and the convergence of adaptive schemes." *IEEE Transactions on Information Theory* IT-30:183– 191, March 1984.
5. Boyd, S., and Sastry, S. "Necessary and sufficient conditions for parameter convergence in adaptive control." *Automatica* 22:629–639, Nov. 1986.
6. Boyd, S., and Sastry, S.S. "On parameter convergence in adaptive control." *Systems & Control Letters* 3:311–319, 1983.
7. Dasgupta, S., and Anderson, B.D.O. "Prior information and persistent excitation." Technical Report, Australian National University, 1987.
8. Dasgupta, S., Anderson, B.D.O., and Tsoi, A.C. "Input conditions for continuous time adaptive systems problems." Technical Report, Australian National University, 1983.
9. C.A. Desoer and M. Vidyasagar. *Feedback Systems: Input-Output Properties.* Academic Press, New York, 1975.
10. Elliott, H., Cristi, R., and Das, M. "Global stability of adaptive pole placement algorithms." *IEEE Transactions on Automatic Control* AC-30:348–356, April 1985.

11. Eykhoff, P., ed. "Identification and system parameter estimation part 2." In *Proceedings of the 3rd IFAC Symposium*, Amsterdam:North Holland Publishing Company, 1973.
12. Goodwin, G.C., and Teoh, E.K. "Persistency of excitation in the presence of possibly unbounded signals." *IEEE Transactions on Automatic Control* AC-30:595–597, June 1985.
13. Kudva, P., and Narendra, K.S. "An identification procedure for discrete multivariable systems." *IEEE Transactions on Automatic Control* 19:549–552, 1974.
14. LaSalle, J.P. "Some extensions of Lyapunov's second method." *IRE Transactions on Circuit Theory* 7:520–527, Dec. 1960.
15. Lion, P.M. "Rapid identification of linear and nonlinear systems." *AIAA Journal* 5:1835–1842, 1967.
16. Mendel, J.M. *Discrete techniques of parameter estimation: The equation error formulation.* New York:Marcel Dekker, Inc., 1973.
17. Morgan, A.P., and Narendra, K.S. "On the uniform asymptotic stability of certain linear time-varying differential equations with unbounded coefficients.' Technical Report No. 7807, Yale University, New Haven, CT, Nov. 1978.
18. Narendra, K.S., and Annaswamy, A.M. "Persistent Excitation of Adaptive Systems" *International Journal of Control* 45:127–160, 1987
19. Sondhi, M.M., and Mitra, D. "New results on the performance of a well-known class of adaptive filters." *Proceedings of the IEEE* 64:1583–1597, Nov. 1976.
20. Wiener, N. "Generalized harmonic analysis." *Acta Mathematica* 55:117–258, 1930.
21. Yuan, J. S-C., and Wonham, W. M. "Probing signals for model reference identification." *IEEE Transactions on Automatic Control* 22:530–538, August 1977.

7

Error Models

7.1 INTRODUCTION

In Chapters 3-6 several methods were described for the identification and control of linear time-invariant dynamical systems with unknown parameters. In all cases, the parameters of an estimator or a controller were adjusted in such a manner that the error between plant output and the output of an identification model or a reference model tended to zero. For example, in Chapter 3, the state variables of the given plant were assumed to be accessible so that parameter adjustment was based on the state error vector e between the plant and model. In Chapters 4 and 5, these results were extended to identification and control problems respectively, where only the outputs of the plant were assumed to be known. In both problems, the parameters are adjusted using the output error and therefore the resultant systems were invariably nonlinear. Since an external reference input r was generally present, the systems were also nonautonomous.

From the point of view of stability and performance, our concern in the problems above is mainly with the behavior of the output and parameter errors as functions of time. We shall refer to the mathematical model describing the evolution of these errors as the *error model*. It is obvious that different problem formulations and the corresponding choices of adaptive laws lead to different error models. The stability analyses of all the adaptive systems considered thus far have implicitly dealt with four classes of error models. By analyzing these error models, which are independent of the specific adaptive system investigated, it is possible to obtain insights into the behavior of a large class of adaptive systems. This makes the study of error models particularly attractive.

Error models, whose main aim is to study the stability properties of adaptive systems, occupy a central place in this book. A designer, familiar with such models, can develop an intuitive understanding of the stability questions that may arise, without a detailed analysis of the overall system dynamics. This is invaluable both in the choice of the structure of the adaptive controller as well as in the selection of an adaptive law.

The general description of an error model can be given as follows: The evolution of the state error vector e is described by the differential equation

$$\dot{e} = f(e, \phi, t) \qquad \text{(error equation)} \tag{7.1}$$

where $\phi(t)$ is the parameter error at time t. Equation (7.1) is referred to as the error equation. Generally, this can be obtained in any specific adaptive situation directly from the equations decribing the plant. The aim of adaptation as discussed in the preceding chapters is to determine a function g so that the differential equation

$$\dot{\phi} = g(e, t) \qquad \text{(adaptive law)} \tag{7.2}$$

together with Eq. (7.1) has $e = 0, \phi = 0$ for an equilibrium state, which has desired stability properties. Equation (7.2) is referred to as the adaptive law and together with the error equation (7.1) constitutes the error model. Obviously, for the implementation of the adaptive law in Eq. (7.2), it must be possible to determine the value of $g(.,.)$ at every instant of time from the measured signals of the overall system. Since the parameter error ϕ is unknown in any nontrivial adaptive problem, $g(.,.)$ cannot depend explicitly on ϕ. In terms of the error model described thus far, an adaptive identification or control problem may be stated as one of determining a set of m differential equations (the adaptive law) given a set of n differential equations (the error equation) so that the combined set of $m + n$ equations have desirable stability properties. In the following sections, we discuss four prototypes of error models that capture the essential features of the adaptive process.

In each section, the properties of the error model are listed under three headings. These are based on the nature of the external input u into the error model given by (i) $u \in \mathcal{L}^{\infty}$, (ii) $u \in \Omega_{(n,T)}$, and (iii) $u \in \mathcal{PC}_{[0,\infty)}$. Case (i) pertains to identification problems and control problems where the boundedness of all the signals has been established. Case (ii) is a special case of case (i) where the input is also persistently exciting. Case (iii) includes problems where it is not known a priori that the signals are bounded, but that they are defined over the interval $[0, \infty)$.

7.2 ERROR MODEL 1

In the parametrizations used for the representation of a plant for identification or control purposes, one which has occurred very frequently in the preceding chapters, has the form $\theta^T u(t) = y(t)$ where $u(t) \in \mathbb{R}^n$ and $y(t) \in \mathbb{R}$ can be measured for all $t \geq t_0$, and θ is a constant vector in $\mathbb{R}^n$. In identification problems, θ corresponds to the unknown parameters of the plant while in control problems it represents the desired control parameter vector. In both cases, an estimate $\widehat{\theta}(t)$ of θ is used to generate an output signal $\widehat{y}(t)$. Defining $\widehat{\theta}(t) - \theta = \phi(t)$ and $\widehat{y}(t) - y(t) = e_1(t)$, the error equation is obtained as (Fig. 7.1)

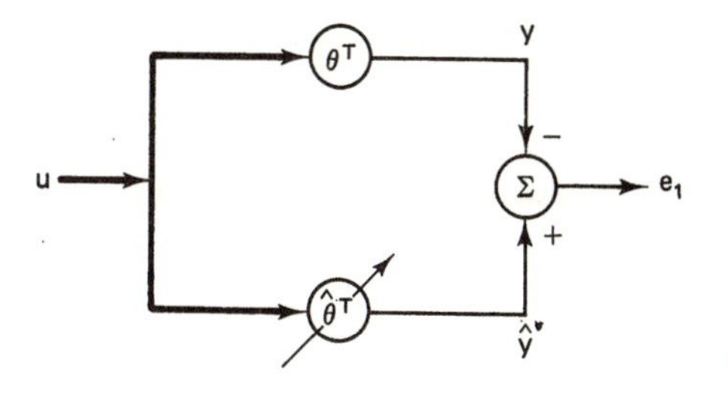

Figure 7.1 Error model 1.

$$\phi^T(t)u(t) = e_1(t). \tag{7.3}$$

Our aim is to determine a rule for adjusting $\phi(t)$ [or equivalently $\theta(t)$] from a knowledge of u and e_1 so that $\phi(t)$ tends to zero asymptotically.

For the simplest case where u, θ, and $\widehat{\theta}$ are scalar quantities, it follows that $e_1 u (= \phi u^2)$ contains information about the sign of the error ϕ and the law

$$\dot{\phi}(t) = -e_1(t)u(t) \tag{7.4a}$$

naturally suggests itself for the adjustment of ϕ. This results in a monotonic nonincreasing function $|\phi|$. The same adaptive law in Eq. (7.4a) can be employed even in the more complex case where u is a vector. While such a law may be adequate if the input u is known to be bounded, in problems where this cannot be assumed a priori, the following law

$$\dot{\phi}(t) = -\frac{e_1(t)u(t)}{1 + u^T(t)u(t)} \tag{7.4b}$$

is used, which includes a normalization term $(1 + u^T u)$. We shall refer to the error equation (7.3) together with either the adaptive law in Eq. (7.4a) or (7.4b) as error model 1.

With the adaptive law in Eq. (7.4a) the resultant error model has the form

$$\dot{\phi}(t) = -u(t)u^T(t)\phi(t). \tag{7.5}$$

Then, a scalar function $V(\phi) = \phi^T\phi/2$ has a time derivative $\dot{V}(\phi)$ evaluated along the trajectories of Eq. (7.5) given by $\dot{V}(\phi) = -e_1^2(t) \le 0$. From this, the following conclusions can be drawn:

(a) Since $V(\phi)$ is a Lyapunov function, ϕ is bounded[1].

(b) Integrating $\dot{V}$ from t_0 to ∞, we obtain

$$\int_{t_0}^{\infty} \dot{V}(\phi(t))dt = -\int_{t_0}^{\infty} e_1^2(t)dt = V(\phi(\infty)) - V(\phi(t_0)) < \infty.$$

Hence, $e_1 \in \mathcal{L}^2$.

We can draw additional conclusions about the behavior of error model 1 by making further assumptions regarding the input u:

[1] All the differential equations considered in this chapter are assumed to have initial conditions that are finite.

Case (i) $u \in \mathcal{L}^\infty$. If we assume that the vector u is uniformly bounded, we can conclude from equation (7.5) that $\dot{\phi}$ is also bounded. Also, since $e_1 \in \mathcal{L}^2$ and $\dot{\phi} = -e_1 u$, it follows that $\dot{\phi} \in \mathcal{L}^2$. Therefore $\dot{\phi} \in \mathcal{L}^2 \cap \mathcal{L}^\infty$. If we also assume that $\dot{u} \in \mathcal{L}^\infty$ then $\dot{e}_1 = \dot{\phi}^T u + \phi^T \dot{u}$, and therefore, $\dot{e}_1$ is bounded. From Lemma 2.12, this implies that

$$\lim_{t\to\infty} e_1(t) = 0 \qquad \text{and} \qquad \lim_{t\to\infty} \dot{\phi}(t) = 0.$$

Case (ii) $u \in \Omega_{(n,t_0,T_0)}$. Equation (7.5) is identical to the differential equation $\dot{x} = -uu^T x$ discussed in Chapter 2. Hence, from Theorem 2.16 it follows that the equilibrium state $\phi \equiv 0$ of Eq. (7.5) is u.a.s. or

$$\lim_{t\to\infty} \phi(t) = 0.$$

Case (iii) $u \in \mathcal{PC}_{[0,\infty)}$. In this case we consider the situation where $\lim_{t\to\infty} \sup_{\tau\leq t} [\|u(\tau)\|] = \infty$. To derive similar conditions on $\dot{\phi}$ as in case (i), we modify the adaptive law to have the form of Eq. (7.4b). With $V(\phi) = \phi^T\phi/2$, $\dot{V}(\phi)$ is evaluated along Eq. (7.4b) as

$$\dot{V} = -\frac{e_1^2}{1+u^T u} \leq 0.$$

Therefore $\phi(t)$ is bounded for all $t \geq t_0$. Using arguments similar to those in case (i), we also obtain that

$$\frac{e_1}{\sqrt{1+u^T u}} \in \mathcal{L}^2. \tag{7.6}$$

Hence,

$$\dot{\phi} = -\frac{e_1}{\sqrt{1+u^T u}} \cdot \frac{u}{\sqrt{1+u^T u}} \in \mathcal{L}^2. \tag{7.7}$$

Since $\dot{\phi}$ can also be expressed as

$$\dot{\phi} = -\frac{u}{\sqrt{1+u^T u}} \cdot \frac{u^T}{\sqrt{1+u^T u}} \cdot \phi,$$

and ϕ is bounded, it follows that $\dot{\phi}$ is also bounded. Also

$$\frac{d}{dt}\left[\frac{e_1}{\sqrt{1+u^T u}}\right] = \frac{(1+u^T u)\left[\phi^T \dot{u} + \dot{\phi}^T u\right] - e_1 u^T \dot{u}}{(1+u^T u)^{\frac{3}{2}}}.$$

If we impose an additional restriction on $\|u\|$ as

$$\|\dot{u}\| \leq M_1\|u\| + M_2 \qquad M_1, M_2 \in \mathbb{R}^+, \tag{7.8}$$

TABLE 7.1 PROPERTIES OF ERROR MODEL 1

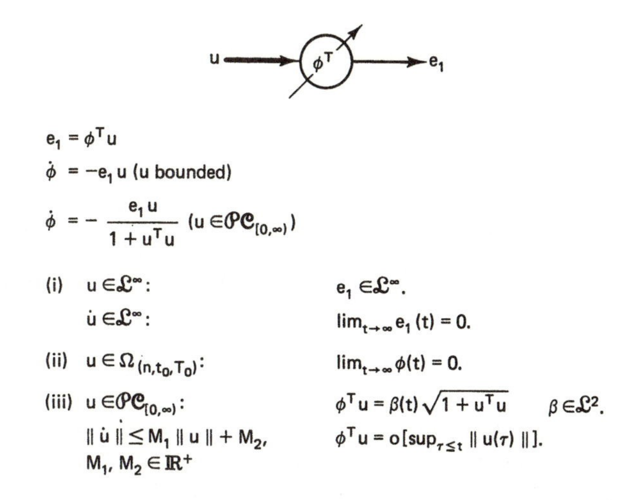

$e_1 = \phi^T u$

$\dot{\phi} = -e_1 u$ (u bounded)

$\dot{\phi} = -\dfrac{e_1 u}{1 + u^T u}$ $(u \in \mathcal{PC}_{[0,\infty)})$

(i) $u \in \mathcal{L}^\infty$:	$e_1 \in \mathcal{L}^\infty$.
$\dot{u} \in \mathcal{L}^\infty$:	$\lim_{t\to\infty} e_1(t) = 0$.
(ii) $u \in \Omega_{(n,t_0,T_0)}$:	$\lim_{t\to\infty} \phi(t) = 0$.
(iii) $u \in \mathcal{PC}_{[0,\infty)}$:	$\phi^T u = \beta(t)\sqrt{1 + u^T u}$ $\quad \beta \in \mathcal{L}^2$.
$\lVert \dot{u} \rVert \le M_1 \lVert u \rVert + M_2$, $M_1, M_2 \in \mathbb{R}^+$	$\phi^T u = o[\sup_{\tau \le t} \lVert u(\tau) \rVert]$.

it follows that

$$\left| \frac{d}{dt} \left[\frac{e_1}{\sqrt{1 + u^T u}} \right] \right| \le c,$$

where c is a constant, since ϕ and $\dot{\phi}$ are bounded. From Eq. (7.6), we can then conclude that

$$\lim_{t\to\infty} \frac{e_1(t)}{\sqrt{1 + u^T(t)u(t)}} = 0. \tag{7.9}$$

From the analysis above, it is clear that Eqs. (7.6) and (7.7) play an important role in the derivation of the condition in Eq. (7.9). The modification of the adaptive law from Eqs. (7.4a) to (7.4b) is therefore an important step. As shown in Chapter 5, this change eventually led to the resolution of the control problem.

In view of the frequent occurrence of the first error model in the analysis of adaptive identification and control problems, we summarize its properties below for easy reference. The same information is also contained in Table 7.1.

When the input u is bounded, the adaptive law in Eq. (7.4a) is used in error model 1. This assures the boundedness of ϕ and that e_1 and $\dot{\phi}$ belong to $\mathcal{L}^2$. If $\dot{u}$ is also known to be bounded, we conclude that $e_1(t)$ and $\dot{\phi}(t)$ will tend to zero. If it is further assumed that u is persistently exciting in $\mathbb{R}^n$, the parameter error $\phi(t)$ tends to zero. The assumption that u is bounded is valid in all the identification problems discussed in Chapters 3 and 4 and the conditions under which it is persistently exciting can be determined using the results of Chapter 6.

When u cannot be assumed to be bounded a priori, as in the control problem in Chapter 5, the adaptive law in Eq. (7.4b) has to be used. In general we obtain that $e_1 = \beta(1 + u^T u)^{1/2}$ where $\beta \in \mathcal{L}^2$. Further if it is assumed that the input u satisfies the

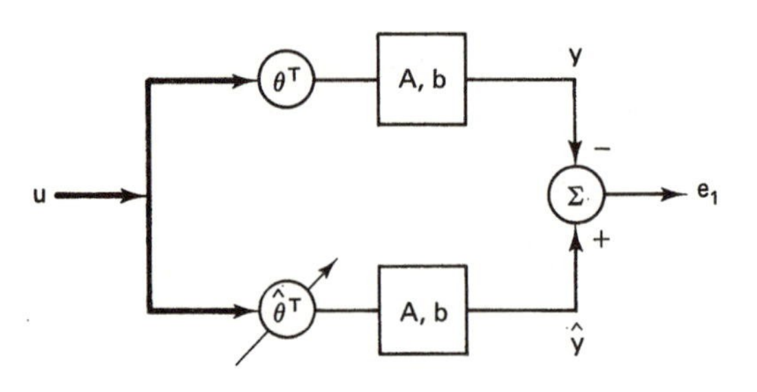

Figure 7.2 Error model 2.

condition in Eq. (7.8), it follows that $\beta(t) \to 0$ as $t \to \infty$ so that e_1 and $\|u\|$ grow at different rates, that is, $e_1(t) = o[\sup_{\tau \leq t} \|u(\tau)\|]$.

7.3 ERROR MODEL 2

In the mathematical description of the sytem that led to error model 1 in the previous section, the measured output is a linear combination of the elements u_i of the input u. Quite often this signal is not directly accessible and only the response of a dynamical system with $\theta^T u$ as the input can be measured (Fig. 7.2). Error model 2 arises in situations where the entire state of this dynamical system is accessible. Such a model is described by the differential equation

$$\dot{e} = Ae + b\phi^T u \tag{7.10a}$$

where A is an $n \times n$ asymptotically stable matrix, (A, b) is controllable, the parameter error $\phi : \mathbb{R}^+ \to \mathbb{R}^m$ and the state error $e : \mathbb{R}^+ \to \mathbb{R}^n$. Let the rule for adjusting ϕ be given by the differential equation

$$\dot{\phi} = -e^T Pbu \tag{7.10b}$$

where P is a symmetric positive-definite matrix such that $A^T P + PA = -Q < 0$. Equations (7.10a) and (7.10b) together constitute the error model 2. Defining the function $V(e, \phi) = e^T Pe + \phi^T \phi$, it follows that its time derivative along Eq. (7.10) is given by $\dot{V}(e, \phi) = -e^T Qe \leq 0$. Hence, $V(e, \phi)$ is a Lyapunov function and we can conclude the following:

(a) $e \in \mathcal{L}^\infty$, and $\phi \in \mathcal{L}^\infty$.

(b) Since $\int_{t_0}^{\infty} \dot{V} dt < \infty$, we have $e \in \mathcal{L}^2$.

As in the case of error model 1, further conclusions can be drawn based on the assumptions regarding the input vector u.

Case (i) $u \in \mathcal{L}^\infty$. By Eq. (7.10a) $\dot{e}$ is bounded and hence

$$\lim_{t \to \infty} e(t) = 0.$$

TABLE 7.2 PROPERTIES OF ERROR MODEL 2

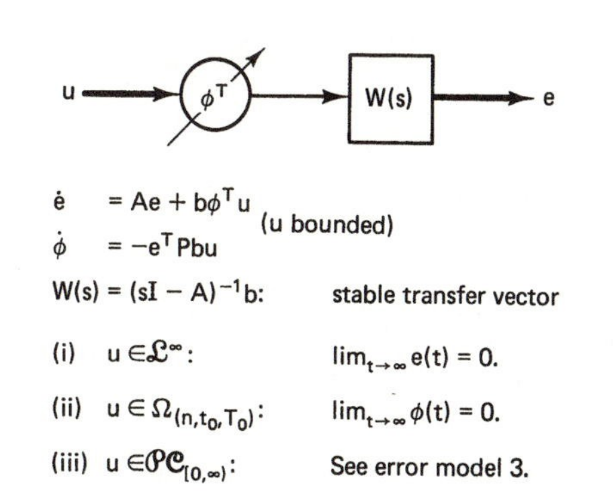

$$\dot{e} = Ae + b\phi^T u \qquad \text{(u bounded)}$$
$$\dot{\phi} = -e^T Pbu$$

$W(s) = (sI - A)^{-1}b$: stable transfer vector

(i) $u \in \mathcal{L}^\infty$:	$\lim_{t\to\infty} e(t) = 0.$
(ii) $u \in \Omega_{(n,t_0,T_0)}$:	$\lim_{t\to\infty} \phi(t) = 0.$
(iii) $u \in \mathcal{PC}_{[0,\infty)}$:	See error model 3.

Case (ii) $u \in \Omega_{(n,t_0,T_0)}$. Equation (7.10) has the same form as Eq. (2.32) and, hence, from Theorem 2.17 it follows that the equilibrium state $(e = 0, \phi = 0)$ of Eq. (7.10) is u.a.s. Hence,

$$\lim_{t\to\infty} \phi(t) = 0.$$

Case (iii) $u \in \mathcal{PC}_{[0,\infty)}$. Since this closely resembles the corresponding case in error model 3, we relegate the discussion to the following section.

The properties of error model 2 are given in Table 7.2. The adaptive law depends on the product of a linear combination of the state errors and the input u. Since the matrix P is used in the implementation of the adaptive law, it follows that the asymptotically stable matrix A defined by the system must be known. In this model, $\lim_{t\to\infty} e(t) = 0$ when u is bounded without requiring that $\dot{u}$ be bounded as in error model 1. The limited use of this error model in adaptive systems is due to the fact that it arises only in those cases where one has access to the entire state error vector e (or equivalently the state) of the given system.

7.4 ERROR MODEL 3

Error model 3 has the same structure as error model 2 with the exception that only one of the state variables y of the plant can be measured. The plant equations have the form

$$\dot{x} = Ax + b\theta^T u; \qquad y = h^T x$$

where $\{h^T, A, b\}$ is controllable and observable and A is asymptotically stable. The corresponding identification model has the form (Fig. 7.3)

$$\dot{\widehat{x}} = A\widehat{x} + b\widehat{\theta}^T u; \qquad \widehat{y} = h^T \widehat{x}.$$

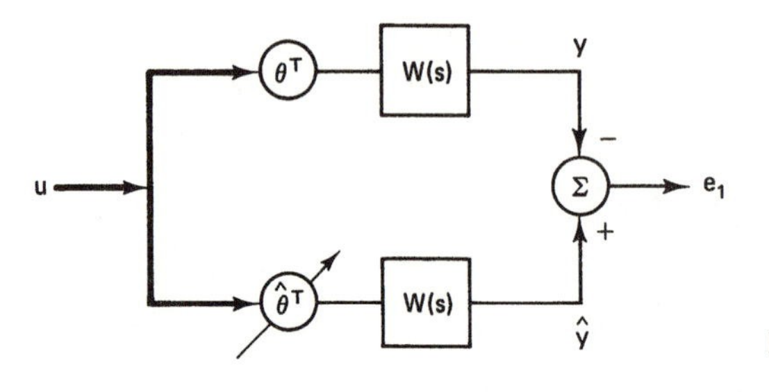

Figure 7.3 Error model 3.

Hence, the error differential equation has the form

$$\dot{e} = Ae + b\phi^T u; \qquad e_1 = h^T e. \tag{7.11a}$$

The fact that only e_1 rather than the entire state error vector e is assumed to be known at time t makes this model applicable to a much wider class of problems than error model 2. However, as might be expected, more stringent conditions must be imposed on the transfer function between $\phi^T u$ and e_1, if stable adaptive laws are to be synthesized. Although only asymptotic stability of the transfer vector $(sI - A)^{-1}b$ was assumed in error model 2, we assume that the transfer function $h^T(sI - A)^{-1}b$ is SPR in this case.

As in the error model 1, we consider separately the cases where $u \in \mathcal{L}^\infty$ and where it belongs to $\mathcal{PC}_{[0,\infty)}$. In the former case, the adaptive law

$$\dot{\phi} = -e_1 u \tag{7.11b}$$

is used. Equations (7.11a) and (7.11b) together constitute error model 3. Using the Lyapunov function candidate

$$V(e, \phi) = e^T P e + \phi^T \phi \tag{7.12}$$

the time derivative of V, evaluated along the solutions of Eq. (7.11), is given by

$$\dot{V}(e, \phi) = e^T \left[A^T P + PA\right] e + 2e^T P b \phi^T u - 2\phi^T e_1 u.$$

Since $h^T(sI - A)^{-1}b = W(s)$ is strictly positive real, it follows from Lemma 2.3 that there exists a positive definite matrix P and $Q = Q^T > 0$ such that

$$\begin{aligned} A^T P + PA &= -Q \\ Pb &= h. \end{aligned} \tag{7.13}$$

Hence, we have $\dot{V} = -e^T Q e \leq 0$. We therefore conclude that

(a) $e \in \mathcal{L}^\infty$, and $\phi \in \mathcal{L}^\infty$.

(b) $\int_{t_0}^{\infty} \dot{V} dt < \infty$ and hence e and $e_1 \in \mathcal{L}^2$.

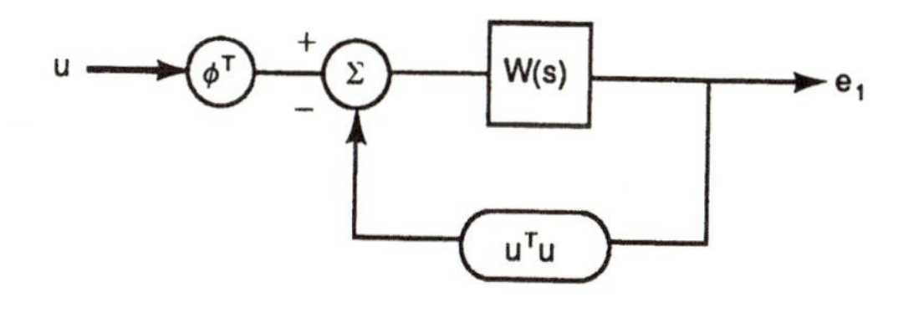

W(s) = $h^T(sI-A)^{-1}b$
SPR transfer function

Figure 7.4 Error model 3: $u \in \mathcal{PC}_{[0,\infty)}$.

Case (i) $u \in \mathcal{L}^\infty$. Since $e_1 \in \mathcal{L}^2$, we have $e_1 u \in \mathcal{L}^2$ and, hence, $\dot{\phi} \in \mathcal{L}^2$. From Eq. (7.11), $\dot{e}_1$ is bounded and, hence, $\lim_{t\to\infty} e_1(t) = 0$.

Case (ii) $u \in \Omega_{(n,t_0,T_0)}$. Using Eq. (7.13), Eq. (7.11) can be written as

$$\begin{bmatrix} \dot{e} \\ \dot{\phi} \end{bmatrix} = \begin{bmatrix} A & bu^T \\ -ub^T P & 0 \end{bmatrix} \begin{bmatrix} e \\ \phi \end{bmatrix}.$$

From Theorem 2.17 it follows that the equilibrium state of Eq. (7.11) is exponentially stable. Therefore $\lim_{t\to\infty} \phi(t) = 0$.

Case (iii) $u \in \mathcal{PC}_{[0,\infty)}$. If $\lim_{t\to\infty} \sup_{\tau\le t} \left[\|u(\tau)\|\right] = \infty$, the analysis above can no longer be used and the error differential equations are modified as given in Eq. (7.14). This is achieved by introducing a feedback signal $-e_1 u^T u$ as shown in Fig. 7.4. If the same adaptive law as given in Eq. (7.11b) is used, we obtain the modified error model 3 given by

$$\begin{aligned} \dot{e} &= Ae + b\left[\phi^T u - e_1 u^T u\right]; \qquad e_1 = h^T e \\ \dot{\phi} &= -e_1 u. \end{aligned} \tag{7.14}$$

With $V(e,\phi)$ defined as before in Eq. (7.12), we have $\dot{V} = -e^T Q e - 2e_1^2 u^T u$. Hence, in addition to e, $e_1 \in \mathcal{L}^2$, we also have $e_1 u \in \mathcal{L}^2$. This, in turn, implies that $\dot{\phi} \in \mathcal{L}^2$. The input to the strictly positive real transfer function $W(s)$ is $\phi^T u - e_1 u^T u$. Since $e_1 u \in \mathcal{L}^2$, from Lemma 2.9 it follows that the response due to the second term is

$$W(s)\left[e_1(t)u^T(t)u(t)\right] = o\left[\sup_{\tau\le t} \|u(\tau)\|\right].$$

Further, since $W(s)(\phi^T u - e_1 u^T u) = e_1$ and e_1 is bounded, it follows that the response due to the first term

$$W(s)[\phi^T(t)u(t)] = o\left[\sup_{\tau\le t} \|u(\tau)\|\right]. \tag{7.15}$$

TABLE 7.3 PROPERTIES OF ERROR MODEL 3

$\dot{e} = Ae + b\phi^T u;\ e_1 = h^T u$

$\dot{\phi} = -e_1 u$ (u bounded)

$\dot{e} = Ae + b\phi^T u - e_1 u^T u;\ e_1 = h^T u$

$\dot{\phi} = -e_1 u\ (u \in \mathcal{PC}_{[0,\infty)})$

(i) u bounded:	$\lim_{t\to\infty} e_1(t) = 0.$
(ii) $u \in \Omega_{(n,t_0,T_0)}$:	$\lim_{t\to\infty} \phi(t) = 0.$
(iii) $u \in \mathcal{PC}_{[0,\infty)}$:	$W(s)\phi^T u = o[\sup_{\tau\le t} \|u(\tau)\|].$

Table 7.3 contains the essential details of error model 3 for all the cases discussed.

By the Kalman-Yakubovich lemma, it is shown in the error model 3 that the linear combination of state errors, $e^T Pb$, used in the adaptive law in error model 2 can be made identical to the output error e_1, by the proper choice of the matrix P. Hence, the same adaptive law can also be implemented with this error model, but without explicit knowledge of P. When the input u is bounded, the same results as in error model 2 are obtained. In particular, $\phi \in \mathcal{L}^\infty$, $\lim_{t\to\infty} e_1(t) = 0$, and $\lim_{t\to\infty} \dot{\phi}(t) = 0$. When $u \in \mathcal{PC}_{[0,\infty)}$, the feedback term shown in Fig. 7.4 is needed to assure that $W(s)[\phi^T(t)u(t)] = o[\sup_{\tau\le t} \|u(\tau)\|]$.

As mentioned earlier, the rationale for choosing error models 2 and 3 lies in the fact that $\phi^T u$ cannot be measured directly. Hence, the reader may wonder how error model 3 as given in Table 7.3 can be realized with $e_1 u^T u$ as a feedback signal into $W(s)$. This is possible by feeding back the signal into the overall adaptive system as shown in Chapter 5, Fig. 5.13.

7.5 ERROR MODEL 4

Error models 1-3 form the basis for all the adaptive systems discussed in the preceding chapters. However, in many problems, the identification model or the controller cannot be parametrized to yield these error models. This implies that stable adaptive laws can no longer be chosen by inspection. Then, one attempts to process the accessible signals to yield error models 1-3 expressed in terms of the derived signals. Error model 4 represents one such attempt and was used in the solution of the control problem in Chapter 5. Consider the error equation

$$e_1(t) = W(s)\phi^T(t)u(t)$$

represented in block diagram form in Fig. 7.5 where $\phi, u : \mathbb{R}^+ \to \mathbb{R}^n$ and $e_1 : \mathbb{R}^+ \to \mathbb{R}$. If $W(s)$ is a known asymptotically stable transfer function that is not strictly positive real and e_1 is a scalar output, it does not conform to any of the error equations that we have discussed thus far. Therefore, none of the three adaptive laws discussed in Sections

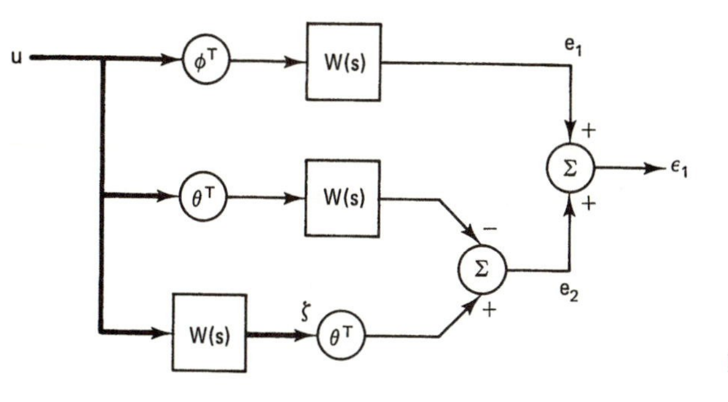

Figure 7.5 Error model 4 - Method 1.

7.2-7.4 can be applied directly. The following approach is used to modify the error equation to have the same form as that of the first error model.

Consider an auxiliary error e_2 defined by

$$e_2 \triangleq \left[\theta^T W(s)I - W(s)\theta^T\right] u \tag{7.16}$$

where I is an $n \times n$ unit matrix. Since $W(s)$ is an asymptotically stable transfer function, if the time-varying parameter vector $\theta(t)$ assumes the constant value θ^*, $e_2(t)$ will tend to zero exponentially. Hence, Eq. (7.16) can also be expressed in terms of a parameter error vector $\phi(t) = \theta(t) - \theta^*$ as

$$e_2 = \left[\phi^T W(s)I - W(s)\phi^T\right] u + \delta(t)$$

where $\delta(t)$ decays exponentially with time. Since $W(s)$ and θ are known, Eq. (7.16) can be implemented practically and an augmented error ϵ_1 can be generated as shown in Fig. 7.4 by adding the auxiliary error e_2 to the true error e_1, that is,

$$\epsilon_1 = e_1 + e_2.$$

This, in turn, yields

$$\begin{aligned} \epsilon_1 &= e_1 + \phi^T W(s)Iu - W(s)\phi^T u + \delta(t) \\ &= \phi^T \zeta + \delta(t) \end{aligned} \tag{7.17}$$

where the n-dimensional vector ζ is defined as

$$\zeta \triangleq W(s)Iu.$$

Neglecting the term $\delta(t)$, the error equation

$$\phi^T \zeta = \epsilon_1$$

has a form identical to that of the error equation (7.3). Depending on the prior information available regarding ζ, an adaptive law for adjusting ϕ can be chosen using the properties of error model 1. If u is bounded, ζ is also bounded since $W(s)$ is asymptotically stable. The adaptive law can be chosen as

$$\dot{\phi} = -\epsilon_1 \zeta.$$

If u and, hence, $\zeta \in \mathcal{PC}_{[0,\infty)}$, the adaptive law

$$\dot{\phi} = -\frac{\epsilon_1 \zeta}{1+\zeta^T \zeta}$$

is used. From the results of Section 7.2 we conclude the following:

(i) $\phi \in \mathcal{L}^\infty$ and $\dot{\phi} \in \mathcal{L}^2 \cap \mathcal{L}^\infty$.

(ii) $u \in \mathcal{L}^\infty \Rightarrow \zeta,\ \dot{\zeta} \in \mathcal{L}^\infty$, and, hence, $\lim_{t\to\infty} \epsilon_1(t) = \lim_{t\to\infty} \dot{\phi}(t) = 0$.

(iii) If $u \in \Omega_{(n,T)}$, then $\zeta \in \Omega_{(n,t_0,T_0)}$ and, hence, $\phi(t)$ and $\epsilon_1(t)$ tend to zero asymptotically.

(iv) If $u \in \mathcal{PC}_{[0,\infty)}$, then $\epsilon_1 = \beta[1+\zeta^T\zeta]^{1/2}$, $\beta \in \mathcal{L}^2$; if in addition $\|\dot{\zeta}\| \leq M_1\|\zeta\| + M_2$, $M_1, M_2 \in \mathbb{R}^+$, then $\epsilon_1(t) = o[\sup_{\tau\leq t} \|\zeta(\tau)\|]$.

Properties (i)-(iv) are stated in terms of the derived signals ζ and ϵ_1. Our interest in using error model 4, however, is to derive the relations between the signal u and the error e_1. Lemma 2.11, which was discussed in Chapter 2, is needed to establish these relations. We first state the lemma and derive the properties of the error model 4 from it.

Lemma 2.11 Let $W(s)$ be a proper transfer function whose poles and zeros are in $\mathbb{C}^-$, $u, \phi : \mathbb{R}^+ \to \mathbb{R}^n$, $y : \mathbb{R}^+ \to \mathbb{R}$ and

$$y(t) = \phi^T(t) W(s) I u(t) - W(s) \phi^T(t) u(t).$$

If $\dot{\phi} \in \mathcal{L}^2$, then

$$y(t) = o\left[\sup_{\tau\leq t} \|u(\tau)\|\right].$$

From Lemma 2.11, we conclude the following:

(i) If $u \in \mathcal{L}^\infty$, then $e_2(t) \to 0$ as $t \to \infty$. Hence, $\lim_{t\to\infty} e_1(t) = \lim_{t\to\infty}[\epsilon_1(t) - e_2(t)] = 0$.

(ii) If $u \in \Omega_{(n,t_0,T_0)}$, then $\phi(t)$ and $e_1(t)$ tend to zero asymptotically. This conclusion can be drawn even without using Lemma 2.11.

(iii) If $u \in \mathcal{PC}_{[0,\infty)}$, then

$$e_1(t) = \epsilon_1(t) - e_2(t) = \beta(t)\left(1+\zeta^T(t)\zeta(t)\right)^{\frac{1}{2}} + \gamma(t) \sup_{\tau\leq t} \|u(\tau)\| \tag{7.18}$$

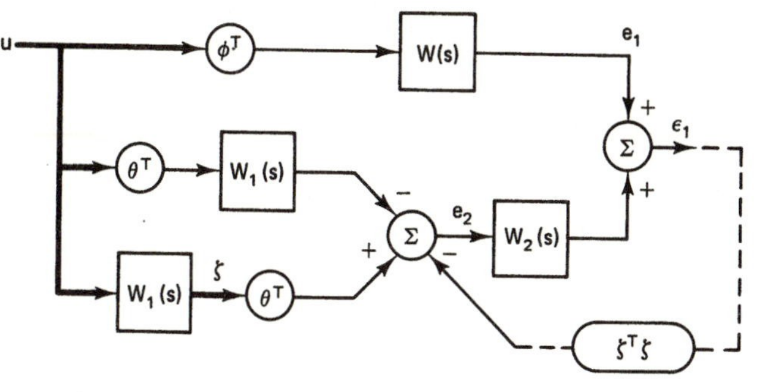

Figure 7.6 Error model 4 - Method 2.

where $\beta \in \mathcal{L}^2$ and $\lim_{t\to\infty}\gamma(t) = 0$. Equation (7.18) was used to establish the proof of stability of the adaptive system in Chapter 5.

Another version of error model 4 has the form shown in Fig. 7.5 (referred to as Method 2 in Chapter 5). The transfer functions $W_1(s)$ and $W_2(s)$ are chosen such that $W_1(s)W_2(s) = W(s)$ and $W_2(s)$ is SPR. The error model 4 discussed earlier can be considered as a special case of this version. The augmented error ϵ_1 is then obtained as

$$\begin{aligned}\epsilon_1 &= e_1 + W_2(s)e_2\\ e_2 &= \left[\theta^T W_1(s)I - W_1(s)\theta^T\right]u.\end{aligned}$$

This leads to the expression

$$\epsilon_1 = W_2(s)\phi^T\zeta \tag{7.19}$$

where $\zeta = W_1(s)Iu$, neglecting the exponentially decaying term due to initial conditions. Equation (7.19) is similar to Eq. (7.11a) and, hence, the results of the error model 3 can be applied as shown below:

(i) If $u \in \mathcal{L}^\infty$, then $\zeta, \dot{\zeta} \in \mathcal{L}^\infty$. Choosing the adaptive law

$$\dot{\phi} = -\epsilon_1\zeta$$

it can be shown that ϕ and ϵ_1 belong to $\mathcal{L}^\infty$. Also, $\lim_{t\to\infty}\epsilon_1(t) = 0$.

(ii) If $u \in \Omega_{(n,t_0,T_0)}$, then $\zeta \in \Omega_{(n,t_0,T_0)}$ and $\lim_{t\to\infty}\phi(t) = 0$.

(iii) If u and therefore $\zeta \in \mathcal{PC}_{[0,\infty)}$, then the structure of the error model is modified, as shown in Section 7.4, by including an additional feedback signal, which is shown in dotted lines in Fig. 7.5. Then the augmented error obeys the differential equation

$$\epsilon_1 = W_2(s)\left[\phi^T\zeta - \epsilon_1\zeta^T\zeta\right].$$

Since ϵ_1 is bounded, this assures that $W_2(s)\epsilon_1\zeta^T\zeta = o[\sup_{\tau\leq t}\|\zeta(\tau)\|]$ and $W_2(s)\phi^T\zeta = o[\sup_{\tau\leq t}\|\zeta(\tau)\|]$. Lemma 2.11 can once again be used to show that $e_2 = o[\sup_{\tau\leq t}\|u(\tau)\|]$. Hence, the true error e_1 satisfies the equation

$$\begin{aligned} e_1(t) &= \epsilon_1(t) - W_2(s)\left[e_2(t) - \epsilon_1(t)\zeta^T(t)\zeta(t)\right] \\ &= o[\sup_{\tau\leq t}\|u(\tau)\|] + o[\sup_{\tau\leq t}\|\zeta(\tau)\|]. \end{aligned} \tag{7.20}$$

Equation (7.20) can be used to prove global stability of the adaptive system, as shown in Chapter 5.

7.6 FURTHER COMMENTS ON ERROR MODELS 1-4

In all the error models described in Sections 7.2-7.5, it is assumed that the input vector $u(t)$ can be measured at time t. In the first error model, the scalar error e_1 has the form $\phi^T u$. In the second and third error models, $\phi^T u$ cannot be measured directly but it is assumed that we have access to the output of a dynamical system whose input is $\phi^T u$. In the second error model, the dynamical system is asymptotically stable with a state $e(t)$ at time t, which can be measured. In the third error model, where only a scalar error is accessible, a further assumption is made that the dynamical system is SPR.

Whether or not $u \in \mathcal{L}^\infty$, in every case, it follows directly that $\phi \in \mathcal{L}^\infty$. Further, in error models 2 and 3, but not in 1, it can be concluded that $e_1 \in \mathcal{L}^\infty$. When $u \in \mathcal{PC}_{[0,\infty)}$, a normalization factor $1/(1 + u^T u)$ is needed in the adaptive law in error model 1 to ensure that $\dot{\phi} \in \mathcal{L}^2 \cap \mathcal{L}^\infty$. The same result is achieved in the third error model by using a nonlinear feedback term. In both cases, we can conclude that e_1 grows at a slower rate than u.

In all the adaptive laws, the adjustment of the parameter vector depends on the output error e_1 and the input vector u. Hence, the parametrization is always chosen to be such that the unknown parameter occurs in a product term, in which the second term is a known signal.

To assure the global stability of the error models 1-3, the simplest adaptive laws were chosen. In the context of the adaptive observer, various modifications of these adaptive laws were discussed in Chapter 4 to improve parameter convergence rates. These included the use of constant or time-varying adaptive gains in the adaptive laws, integral algorithms, hybrid algorithms, and multiple models. The same modifications are also possible in the first error model. Although a time-invariant matrix can be used in the adaptive law with the third error model, extensions of the other ideas to this model have not appeared in the literature thus far.

The idea behind the fourth error model is to convert an error equation, which is not of the form described by any one of the error models 1-3, into one of these forms by the addition of suitable signals. Two versions of such modifications were presented, resulting in the first and third error models respectively. It was only by using such modifications that the adaptive control problem was finally solved in 1980.

The use of the fourth error model also reveals that considerable signal processing may be needed in some cases to determine a stable adaptive law. The augmented signals are not signals that exist in the system to be controlled but are derived from them. In this sense, the fourth error model succeeded in separating the generation of the adaptive laws from the control process.

7.7 USE OF ERROR MODELS IN ADAPTIVE SYSTEMS

As indicated earlier, the error models have all been used implicitly in the stability analysis of adaptive systems discussed in Chapters 3, 4, 5, and 6. The reader may have observed that all the problems discussed involve similar arguments, though some essential differences exist among the analyses presented in the various chapters. The former may be attributed to the common philosophy underlying all the error models, while the latter is due to the differences in the assumptions made regarding the plant to be controlled. These are outlined briefly in this section.

In Chapter 3, the identification and control of plants described by first-order differential equations were discussed. These give rise to second or third error models that are equivalent in this case. When the plant is of higher order and the entire state vector is accessible, the second error model is the only one that is applicable. In the identification problems, the error model has a bounded input and, hence, the parameter error vector ϕ and the state error vector e are bounded and the output error tends to zero. When the input is persistently exciting, it corresponds to case (ii) and, hence, $\lim_{t\to\infty} \phi(t) = 0$. In the control problem, when state variables are accessible, even though it is not known a priori that all the signals in the feedback loop are bounded, it follows from error model 2 that $||e||$ and $||\phi||$ are bounded. Since the signals of the model are bounded, the same is true of the signals in the plant as well and the arguments used earlier can be applied.

Error models 1 and 3 were used in the design of stable adaptive observers in Chapter 4. Since, in all the cases discussed, the relevant signals of the system can be assumed to be bounded, the analysis is considerably simplified. From the discussion of the error models it follows that $\lim_{t\to\infty} e_1(t) = 0$. Parameter convergence can be demonstrated using arguments based on persistent excitation and applying the results of case (ii).

In the control problem discussed in Chapter 5, the error equation is of the form

$$e_1 = W_m(s)\phi^T\omega$$

where $W_m(s)$ is the transfer function of the reference model. When the relative degree of the plant is unity, $W_m(s)$ can be chosen to be strictly positive real so that error model 3 is directly applicable. An immediate consequence of this is that ω can be shown to be bounded using an explicit Lyapunov function. Hence, a normalization term is not needed in the adaptive law, which consequently has the form

$$\dot{\phi} = -e_1\omega.$$

When the relative degree is greater than or equal to two, $W_m(s)$ is no longer strictly positive real and, hence, an augmented error has to be generated as described in error model 4. This leads to error models 1 or 3 expressed in terms of derived signals. Since these signals cannot be assumed to be bounded, either a normalization term in the adaptive law (with error model 1)or a nonlinear feedback term (with error model 3) has to be used.

Error models were first introduced in the adaptive control literature by Narendra and Kudva in [4]. Later these error models were introduced to present adaptive identification methods in a unified fashion in [2]. The use of error models in the control problem was treated by Lin and Narendra [1] who also introduced the fourth error model described in Section 7.5. All these models are for continuous time systems in which no external disturbance is present, and are used to characterize the stability properties of the equilibrium state.

The various error models can be extended directly to discrete systems as well. Recently several hybrid error models were developed in [3] for systems in which the signals are continuous but the parameters are adjusted at discrete-time instants. These are discussed in Chapter 8. It also has been shown that the error models are also useful for the analysis of systems when disturbances are present. In Chapter 8, a detailed analysis of such models is carried out in the context of robust adaptive systems.

PROBLEMS

Part I

1. **(a)** If in error model 1, with the error equation $\phi^T u = e_1$, the adaptive law is modified as

$$\dot{\phi} = -\Gamma e_1 u \qquad \Gamma = \Gamma^T > 0$$

show that (i) if $u \in \mathcal{L}^\infty$, then $\phi \in \mathcal{L}^\infty$, and (ii) if, in addition, $\dot{u} \in \mathcal{L}^\infty$, then $\lim_{t\to\infty} e_1(t) = 0$.

(b) If in (a), $\dot{u}$ is not bounded, give an example to show that $e_1(t)$ does not tend to zero as $t \to \infty$.

2. In error model 3, the errors e_1 and ϕ are related by the differential equation $\dot{e}_1 = -\alpha e_1 + \phi^T u$ where $\alpha > 0$. If the adaptive law is modified as

$$\dot{\phi} = -\Gamma e_1 u \qquad \Gamma = \Gamma^T > 0$$

show that (i) if $u \in \mathcal{L}^\infty$, then $\phi \in \mathcal{L}^\infty$, and $\lim_{t\to\infty} e_1(t) = 0$.

Part II

3. If in the multivariable version of the error model 1, $\Phi : \mathbb{R}^+ \to \mathbb{R}^{m\times n}$, $u : \mathbb{R}^+ \to \mathbb{R}^n$, $e_1 : \mathbb{R}^+ \to \mathbb{R}^m$, and

$$\Phi(t)u(t) = e_1(t)$$

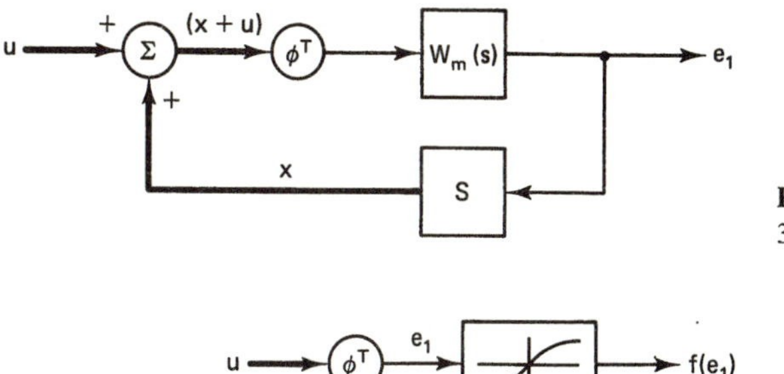

Figure 7.7 Modification of error model 3.

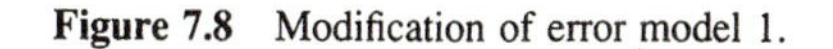

Figure 7.8 Modification of error model 1.

determine the adaptive law for adjusting Φ so that $\lim_{t\to\infty} e_1(t) = 0$. In particular, what signals are used in the adjustment of Φ_{ij}, the $i - j$th element of the matrix Φ? Derive conditions on the input u so that $\lim_{t\to\infty} \Phi(t) = 0$.

4. Error model 3 is modified as shown in Fig. 7.7 to include a linear time-invariant asymptotically stable feedback system S in the feedback path where $u, x : \mathbb{R}^+ \to \mathbb{R}^n$ and u is a piecewise-continuous bounded function.

(a) If the parameter error vector ϕ is adjusted as

$$\dot{\phi} = -e_1(x + u),$$

show that

$$\lim_{t\to\infty} e_1(t) = 0 \text{ and } \lim_{t\to\infty} x(t) = 0.$$

(b) How would you use this result to modify the adaptive system treated in Section 5.3 in Chapter 5 when the relative degree of the plant transfer function is unity?

5. In Problem 1, if u is the output of an LTI system, which is unstable, how would you modify the adaptive law? Determine the nature of e_1 in this case.

6. Repeat the problem for error model 3 assuming that u is the output of an LTI unstable system.

Part III

7. A number of methods for increasing the speed of response of adaptive systems were described in Chapters 3 and 6, such as the introduction of constant and time-varying adaptive gains, use of integral algorithms, hybrid algorithms, and multiple-inputs. All of these pertain to error model 1. Can the same methods also be applied to error model 3?

8. **(a)** Consider a modification of the first error model as shown in Fig. 7.8. In this case, only a nonlinear function of e_1 can be observed. If f is a no-memory first and third quadrant monotonic nonlinearity, indicate how the adaptive law for adjusting $\phi(t)$ can be modified so that $\lim_{t\to\infty} e_1(t) = 0$.

(b) Can a similar result be obtained using the third error model? Justify your answer.

REFERENCES

1. Lin, Y.H., and Narendra, K.S. "A new error model for adaptive systems." *IEEE Transactions on Automatic Control* 25:585–587, June 1980.

2. Narendra, K.S. "Stable identification schemes." In *System Identification: Advances and Case Studies*, R.K. Mehra and D.G. Lainiotis, eds., New York:Academic Press, 1976.

3. Narendra, K.S., Khalifa, I.H., and Annaswamy, A.M. "Error models for stable hybrid adaptive systems." *IEEE Transactions on Automatic Control* 30:339–347, April 1985.

4. Narendra, K.S., and Kudva, P. "Stable adaptive schemes for system identification and control - Parts I & II." *IEEE Transactions on Systems, Man and Cybernetics* SMC-4:542–560, Nov. 1974.

8

Robust Adaptive Control

8.1 INTRODUCTION

The adaptive control systems considered in Chapter 5 portrayed an idealized situation. A finite-dimensional LTI plant with unknown parameters was considered, whose input and output could be measured exactly. The transfer function $W_p(s)$ of the plant was required to satisfy assumptions (I). It was shown that an adaptive controller can be implemented so that all the signals in the system remain bounded and the output of the plant tends to the output of a reference model asymptotically. However, in practical applications, no plant is truly linear or finite dimensional. Plant parameters tend to vary with time, and measurements of system variables are invariably contaminated with noise. The plant model used for analysis is almost always approximate. It is precisely in these cases that adaptive control is most needed.

In spite of the idealized assumptions, the results in Chapter 5 had a great impact on workers in the field. It provided the impetus to attempt more realistic problems where there is uncertainty regarding either the plant or the disturbances. In particular, two major directions of research evolved during this period. The first of these deals with situations in which the plant satisfies assumptions (I) but adaptation is carried out under different types of perturbations mentioned earlier. In the perturbation-free plant, the objective was to design a stable adaptive controller that would ensure asymptotic model-following. In the presence of perturbations, exact model-following is no longer possible, and the objective is to design a controller that will ensure boundedness of all the signals in the system and satisfy performance specifications. We discuss this problem in this chapter. The second major research effort was directed toward relaxing assumptions (I) needed to solve the problem in Chapter 5. This is considered in Chapter 9.

Since the underlying differential equations that describe adaptive systems are nonlinear and time-varying, the addition of exogenous inputs in the form of external disturbances makes the analysis of the resultant adaptive systems considerably more difficult and constitutes a marked departure from that carried out in the ideal case. The simple reason for this is that not much can be inferred about the input-output properties of a nonlinear system from the behavior of the corresponding unforced system. For example, when external disturbances are present, it is no longer possible to ensure that the error between plant and model outputs will remain bounded if the standard adaptive law [Eq. (5.45)] is used. Even about 1980 it was realized that, when bounded external disturbances are present, the adaptive law in Eq. (5.45) could result in the parameter error ϕ growing in an unbounded fashion [5], [29]. It was also shown, using simulations, that other perturbations such as time-varying parameters and unmodeled dynamics [33] could result in unbounded solutions. All these clearly indicated that new approaches were needed to assure the boundedness of all the signals in the system and led to a body of work referred to as *robust adaptive control theory*. Robustness in this context implies that the adaptive systems perform in essentially the same manner, even when exogenous or endogenous perturbations are present.

In Section 8.2, we study adaptive identifiers and observers in the presence of disturbances. With the use of error models, problems of identification in the presence of noise, over-parametrization, and reduced-order identification can be treated in a unified fashion. The role of persistent excitation in this context is emphasized.

In Section 8.3, adaptive control in the presence of external disturbances that are bounded is considered, and methods to ensure that the signals in the closed-loop system remain bounded are discussed. This problem is referred to as robust adaptive control in the presence of bounded disturbances. Obviously, the controller structure as well as the specific adaptive law used will depend to a large extent on the prior information available regarding the disturbance. For a specific class of disturbances, it is shown that an adaptive controller can be designed so that its effect is cancelled at the output of the plant.

When more general disturbances are present, two different methods can be adopted to realize the desired objective. In the first method, the adaptive law is modified suitably. In the second method, on the other hand, the persistent excitation property of the reference input is used to assure the boundedness of the signals in the adaptive system, without any modifications in the standard adaptive law. It is found that the two methods can also be used for other perturbations such as time variations (Section 8.6) and unmodeled dynamics (Section 8.7) as well. Wherever possible, we first consider first-order systems to illustrate the idea on which each method is based, and present simulation studies that indicate the performance of the resultant system. Following this, we discuss higher order systems and derive conditions under which all the signals are bounded. The improvement in performance achieved using an adaptive controller is emphasized wherever possible.

8.2 ADAPTIVE OBSERVERS IN THE PRESENCE OF DISTURBANCES

8.2.1 Identifier

Let $W_p(s)$ be an asymptotically stable transfer function of a plant that has to be identified. In Chapter 4, we considered the case where $W_p(s)$ can be parametrized as $W_p(s) = \sum_{i=1}^{n} W_i(s)k_i$, $W_i(s)$ are known asymptotically stable transfer functions, and the identification problem is to estimate the unknown constants k_i. This is carried out by constructing a model whose transfer function is $\sum_{i=1}^{n} \hat{k}_i W_i(s)$ and adjusting $\hat{k}_i$ so that the error between model and plant outputs tends to zero. Let r be the reference input to the plant and model, and y_p, $\hat{y}_p$ be the outputs of the plant and model respectively. Defining

$$u_i(t) \stackrel{\triangle}{=} W_i(s)r(t), \phi_i(t) \stackrel{\triangle}{=} \hat{k}_i(t) - k_i, \text{ and } e_1(t) \stackrel{\triangle}{=} \hat{y}_p(t) - y_p(t), i = 1, 2, \ldots, n,$$

the error equation may be expressed as

$$\sum_{i=1}^{n} \phi_i(t)u_i(t) \stackrel{\triangle}{=} \phi^T(t)u(t) = e_1(t),$$

which corresponds to the first error model discussed in Chapter 7.

Case (i). If there is a bounded external disturbance $\nu(t)$ present at the output of the plant, the error equation has the form

$$e_1(t) = \phi^T(t)u(t) + \nu(t). \tag{8.1}$$

If the parameter error is adjusted as in the ideal case we have

$$\begin{aligned} \dot{\phi}(t) &= -\gamma e_1(t)u(t) \qquad \gamma > 0 \\ &= -\gamma u(t)u^T(t)\phi(t) - \gamma u(t)\nu(t). \end{aligned} \tag{8.2}$$

Case (ii). If the plant parameters vary with time so that $k(t)$ is a function of time, and the estimated plant output is generated as in the disturbance-free case, the adaptive law for adjusting the parameter estimates is given by

$$\dot{\hat{k}}(t) = -\gamma u(t)e_1(t) = -\gamma u(t)\left[u^T(t)\hat{k}(t) - y(t)\right] \qquad \gamma > 0.$$

However, $\hat{k}(t) = k(t) + \phi(t)$ so that the parameter error satisfies the equation

$$\dot{\phi}(t) = -\gamma u(t)u^T(t)\phi(t) - \dot{k}(t),$$

which is similar to Eq. (8.2). If it is assumed that the plant parameters vary slowly with time so that $\dot{k}$ is uniformly bounded, then $\dot{k}$ can be considered as a bounded disturbance.

Case (iii). Let a model of the form

$$\sum_{i=1}^{\overline{n}} k_i W_i(s) \qquad \overline{n} < n$$

be used to represent the plant. Since $\overline{n} < n$, this represents a reduced-order model. If

$$\sum_{i=\overline{n}+1}^{n} W_i(s) k_i \cdot r(t) \stackrel{\triangle}{=} \nu_1(t),$$

then the signal $\nu_1(t)$ due to the unmodeled part represents the residual error. Hence, the true output $\overline{y}_p$ of the plant can be expressed as

$$\overline{y}_p(t) = y_p(t) + \nu_1(t),$$

and the estimated output is given by

$$\hat{y}_p(t) = \sum_{i=1}^{\overline{n}} \hat{k}_i W_i(s) r(t).$$

The corresponding error equation is given by

$$e_1(t) = \phi^T(t) u(t) - \nu_1(t),$$

which is of the form of Eq. (8.1). However, $\phi(t)$ and $u(t) \in \mathbb{R}^{\overline{n}}$ rather than $\mathbb{R}^n$.

Hence, whether the plant output is explicitly corrupted by noise as in case (i), contains a time-varying parameter as in case (ii), or is approximated by a lower order model as in case (iii), if the ideal adaptive law is used, the error equations can be expressed in the form of Eq. (8.2). The latter equation needs to be analyzed to study the various problems in a unified fashion. As seen in the following sections, even in the adaptive observer and adaptive control problems, the error equation has the same form as Eq. (8.1), but u and ν are not always known to be bounded a priori.

8.2.2 Adaptive Observers

We now consider an adaptive observer discussed in Chapter 4 when observation noise is present. It is assumed that the unknown plant has an asymptotically stable rational transfer function $Z_p(s)/R_p(s)$ and the output of the plant $y_p(t)$ can be observed only in the presence of additive noise $\nu(t)$ as $\overline{y}_p(t) = y_p(t) + \nu(t)$. The input and noisy output are processed, as described in Chapter 4, through filters to generate the vectors $u_1(t)$ and $u_2(t)$. If

$$\theta \stackrel{\triangle}{=} [c^T,\, d^T]^T, \;\; \widehat{\theta} \stackrel{\triangle}{=} \left[\widehat{c}^T(t), \widehat{d}^T(t)\right]^T \;\; \text{and} \;\; u(t) \stackrel{\triangle}{=} \left[u_1^T(t),\, u_2^T(t)\right]^T,$$

then the output of the observer $\hat{y}_p$ can be expressed as

$$\hat{y}_p(t) = \widehat{\theta}^T(t) u(t).$$

It is evident that the output error between observer and plant outputs satisfies the equation

$$e_1(t) \;=\; \phi^T(t)u(t) + \overline{\nu}(t)$$

where $e_1 = \widehat{y}_p - y_p$, $\phi = \widehat{\theta} - \theta$, $\overline{\nu}(t) = [d^T(sI - F)^{-1}g - 1]\nu(t)$, and F, g and d are defined as in Chapter 4. If $\phi(t)$ is adjusted using the same adaptive law as before, the analysis of the behavior of the observer in the presence of noise reduces once again to the analysis of Eq. (8.2).

8.2.3 Error Model

In Chapter 7, the first error model was analyzed extensively where the parameter error and the output error were related by

$$e_1(t) \;=\; \phi^T(t)u(t).$$

If the parameter error is updated as $\dot{\phi} = -\gamma u u^T \phi$, it was shown that ϕ is bounded, $\lim_{t\to\infty} e_1(t) = 0$ if u and $\dot{u}$ are bounded, and that $\lim_{t\to\infty} \phi(t) = 0$ if $u \in \Omega_{(n,T)}$.

In the presence of bounded disturbances, error model 1 is modified as shown in Section 8.2.1 as

$$e_1(t) \;=\; \phi^T(t)u(t) + \nu(t).$$

With the same adaptive law as in the ideal case, the resultant nonhomogeneous differential equation is given by

$$\dot{\phi} \;=\; -\gamma u u^T \phi - \gamma u \nu. \tag{8.3}$$

Various assumptions can be made regarding $\nu(t)$ in Eq. (8.3). For example, if $\nu(t)$ is stochastic, uncorrelated with $u(t)$, $u \in \Omega_{(n,T)}$, and γ is a time-varying gain that tends to zero in a specific manner, it has been shown that $\|\phi(t)\|$ tends to zero in a mean square sense [21]. However, our interest lies in cases where only a bound on $\nu(t)$ can be assumed and further where $\nu(t)$ and $u(t)$ are correlated.

If $u \in \Omega_{(n,T)}$, the homogeneous part of Eq. (8.3) is u.a.s. and, hence, the bounded forcing term $-u(t)\nu(t)$ in Eq. (8.3) produces a bounded $\phi(t)$. Therefore questions concerning the boundedness of the parameters will arise only when u is not persistently exciting, that is, $u \notin \Omega_{(n,T)}$. In the following case, we study the behavior of Eq. (8.3) for different kinds of inputs.

Case (i) $u \in \Omega_{(n,T)}$. In Eq. (8.3), the following assumptions are made: For all $t \in [0, \infty)$, (i) $|\nu(t)| \le \nu_0$, (ii) $\int_t^{t+T} |u^T(\tau)\mathrm{w}|d\tau \ge \epsilon_0$ for all constant unit vectors w in $\mathbb{R}^n$ and (iii) a constant $u_0 > 0$ exists so that $\int_t^{t+T} \|u(\tau)\|d\tau \le u_0$. It can then be shown that [29]

$$\|\phi(t)\| \le \frac{T + \gamma u_0^2}{\epsilon_0} \cdot \nu_0.$$

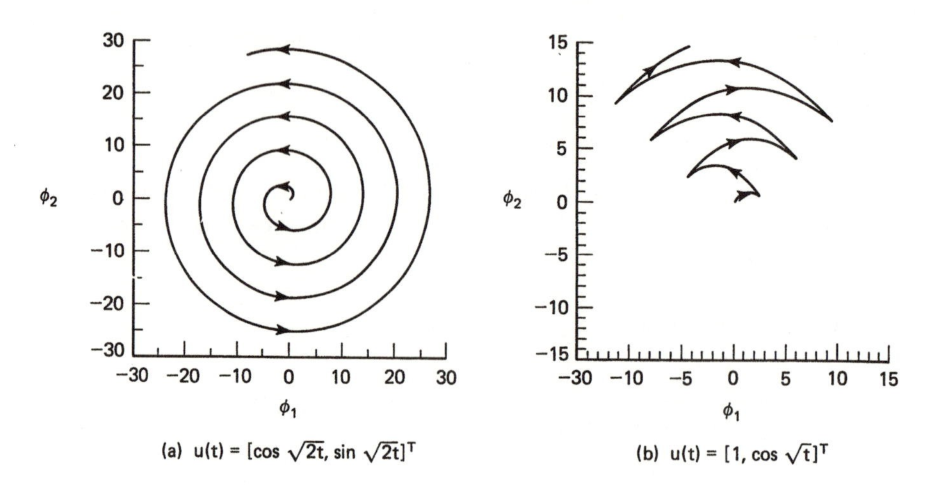

Figure 8.1 Two examples of unbounded behavior of $\phi(t)$.

Case (ii) $u \notin \Omega_{(n.T)}$. When u is not persistently exciting and $\nu(t) \equiv 0$, it is known that the equilibrium state of Eq. (8.3) is uniformly stable. We consider some simple examples below where Eq. (8.3) can have unbounded solutions for bounded ν.

(1) Scalar Equation: In the scalar differential equation

$$\dot{\phi}(t) = -u^2(t)\phi(t) - u(t)\nu(t) \tag{8.4}$$

let u and ν be bounded scalar functions. If $u(t) \equiv 0$ or $u(t) = 1/t^2$, then the output $\phi(t)$ is bounded for any disturbance $\nu(t)$. If $u(t) = 1/t^{1/2}$, the null solution of the homogeneous part of Eq. (8.4) is asymptotically stable but not u.a.s. If we choose $\nu(t) = sgn[u(t)]$, then $u(t)\nu(t) = |u(t)|$ and $\lim_{t\to\infty} \|\phi(t)\| = \infty$. These examples indicate that if u is persistently exciting, or if $u(t)$ tends to zero sufficiently fast, then $\|\phi(t)\|$ is bounded; in cases where it is neither (for example, $u(t) = 1/t^{1/2}$), $\phi(t)$ can become unbounded.

(2) Vector Equation. Following are two interesting examples that result in an unbounded parameter error vector $\phi : \mathbb{R}^+ \to \mathbb{R}^2$ in Eq. (8.3) (Fig. 8.1). Their importance lies in the fact that the relevant signals are not persistently exciting; unlike the previous example, in each of these two cases, the magnitude of u does not tend to zero but its angular velocity in $\mathbb{R}^2$ tends to zero.

1. $u(t) = \begin{bmatrix} \cos\sqrt{2t} \\ \sin\sqrt{2t} \end{bmatrix}$, $\nu(t) = -2$; then $\phi(t) = \begin{bmatrix} \cos\sqrt{2t} + \sqrt{2t}\sin\sqrt{2t} \\ \sin\sqrt{2t} - \sqrt{2t}\cos\sqrt{2t} \end{bmatrix}$

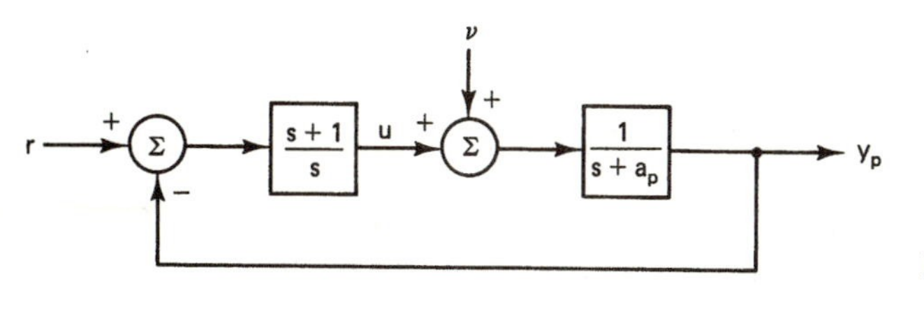

Figure 8.2 Exact noise cancellation.

2. $u(t) = \begin{bmatrix} 1 \\ \cos\sqrt{t} \end{bmatrix}$, $\nu(t) = -2\, sgn[\sin\sqrt{t}]$

and once again $\|\phi(t)\|$ grows as $\sqrt{t}$. In both cases, the homogeneous equation is asymptotically stable but not u.a.s.

8.3 ADAPTIVE CONTROL OF A FIRST-ORDER PLANT IN THE PRESENCE OF BOUNDED DISTURBANCES

The stability analysis of the control problem is more complex than that of the identification problem when an external disturbance is present. In this section, we consider this problem in three stages. In the first stage, we discuss situations where a suitable control input can be produced to cancel the effect of the disturbance exactly. In the second stage, where such perfect cancellation is not possible, methods are discussed that ensure that all the signals and parameters remain bounded. These methods involve different modifications of the adaptive law used in the ideal case in Chapter 5. The discussion of the methods in this section is confined to plants described by first-order differential equations and is for the most part qualitative in nature. Detailed proofs for general plants described by n^{th}-order differential equations ($n > 1$) are found in Section 8.4 and Appendix D. In the third and final stage, it is shown that boundedness of all the signals in the system can also be assured by increasing the degree of persistent excitation of the reference input. Once again, a detailed proof of this result for a general plant is included in Section 8.4.

8.3.1 Exact Cancellation of Disturbance

Consider a linear time-invariant plant described by the equation (Fig. 8.2)

$$y_p(t) = \frac{1}{s+a_p}[u(t) + \nu(t)] \tag{8.5}$$

where u is a control input, $\nu(t)$ is a constant disturbance with $\nu(t) \equiv \nu_0$, and a_p is a known positive constant. It is well known that the effect of such a disturbance can be cancelled at the output by having an integrator in a feedback loop. For example, if a proportional-integral (PI) controller is used so that the control input u is of the form

$$u(t) = \frac{s+1}{s}\left[r(t) - y_p(t)\right],$$

the output is given by

$$y_p(t) = \frac{s+1}{s^2+(a_p+1)s+1} r(t) + \frac{s}{s^2+(a_p+1)s+1} \nu(t). \tag{8.6}$$

Since a_p is positive, the transfer function from r to y_p is asymptotically stable. Since ν is a constant disturbance, the second term in Eq. (8.6) decays exponentially to zero.

If a_p is unknown and the disturbance $\nu(t)$ is identically zero, the methods outlined in Chapter 3 can be used to adaptively control the plant in Eq. (8.5) so that its output follows the output of a reference model asymptotically. In this case, a feedforward gain $k(t)$ and a feedback gain $\theta(t)$ have to be adjusted using the output error.

In the following paragraphs, we consider the problem of exact model following even when a disturbance is present. This involves prior information regarding the disturbance and over-parametrization of the adaptive controller. The output of the controller (or input to the plant) can be considered to consist of two components: one to cancel the effect of the disturbance ν and one to produce the desired output $y_p(t)$.

(a) Constant Disturbance. Consider the problem of exact model following when the reference model has a transfer function

$$W_m(s) = \frac{k_m}{s+a_m}$$

and the output of the plant is $W_p(s)(u(t)+\nu(t))$, where

$$W_p(s) = \frac{k_p}{s+a_p}.$$

The parameters a_p and k_p as well as the disturbance ν are assumed to be unknown constants. The controller structure has the form shown in Fig. 8.3 and is described by the differential equations

$$\begin{aligned}
\dot{\omega}_1 &= -\omega_1 + u \\
\dot{\omega}_2 &= -\omega_2 + y_p \\
u &= kr + c\omega_1 + d\omega_2 + \theta y_p.
\end{aligned}$$

If the four parameters k, c, d, and θ are adjusted adaptively using the methods described in Chapter 3-5, the output error $e_1(t)$ will tend to zero asymptotically. Since it is known a priori that the feedforward transfer function should contain an integrator, the number of adjustable parameters may be reduced to three by choosing $c = 1$. For constant values d^*, θ^*, and k^* of these parameters, the output $y_p(t)$ can be expressed as

$$\begin{aligned}
y_p(t) &= \left[\frac{k^* k_p(s+1)}{s^2+(a_p-k_p\theta^*)s-k_p(d^*+\theta^*)}\right] r(t) \\
&+ \left[\frac{k_p s}{s^2+(a_p-k_p\theta^*)s-k_p(d^*+\theta^*)}\right] \nu(t).
\end{aligned}$$

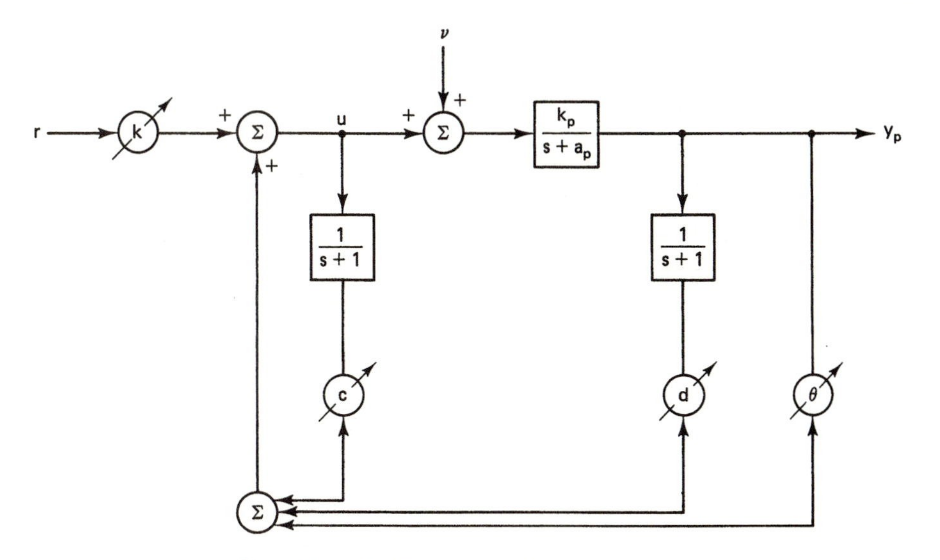

Figure 8.3 Model-matching and noise cancellation.

It is obvious that adequate freedom exists to match the closed-loop transfer function to the model transfer function $W_m(s)$, since the latter has a relative degree unity.

The responses of the system to different reference inputs in the presence of a constant disturbance are shown in Fig. 8.4. The plant is assumed to have a transfer function $2/(s-1)$, and the reference model is chosen to be $1/(s+2)$. If e_1 is the error between the plant output y_p and model output y_m, the parameters k, d, and θ are adjusted using the following adaptive laws:

$$\dot{k} = -e_1 r; \qquad \dot{d} = -e_1\omega_2; \qquad \dot{\theta} = -e_1 y_p.$$

In Fig. 8.4, y_p and y_m are shown for the cases (a) $r(t) = 0$, (b) $r(t) = 5$, and (c) $r(t) = 15 \cos t$, with a disturbance $\nu(t) = 2$ in each case. The initial value of y_m was chosen as $y_m(0) = 1$. All other initial conditions were assumed to be zero.

It is clear from the discussion above that the effect of the constant disturbance can be nulled out at the output by introducing a pole in the feedforward transfer function at $s = 0$. If the disturbance is a sinusoid of frequency ω, the poles in the feedforward path must be at $\pm i\omega$. If this frequency is unknown, then the poles must be adjusted adaptively. This is described briefly below.

(b) Sinusoidal Disturbance. In this case, five parameters must be adjusted to achieve perfect model-following while cancelling the undesired disturbance. Two of the parameters are needed for perfect model-following as described in Chapter 3 and four others are needed to locate the poles at the required values on the imaginary axis. However, since the characteristic polynomials of the filters are assumed to be known, one of the parameters can be chosen a priori as in the previous case and need not be adjusted on-line. The structure of the adaptive system is shown in Fig. 8.5. The outputs

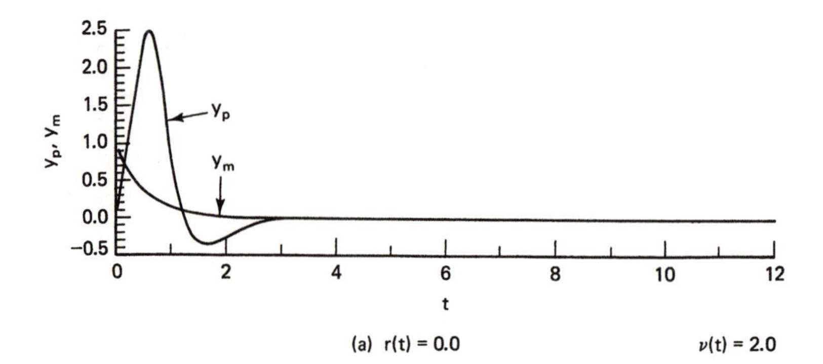

(a) $r(t) = 0.0$ $\nu(t) = 2.0$

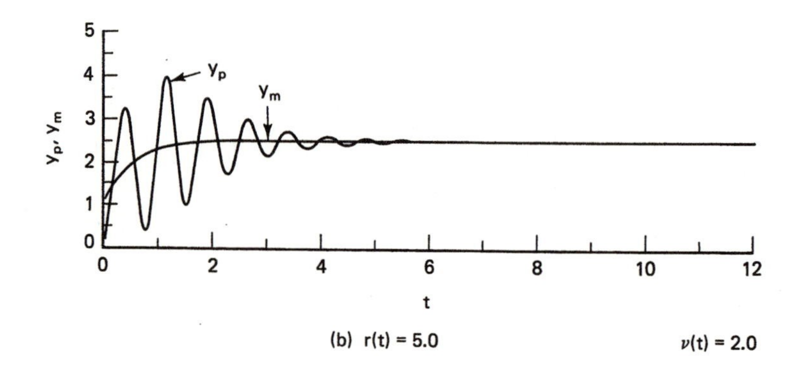

(b) $r(t) = 5.0$ $\nu(t) = 2.0$

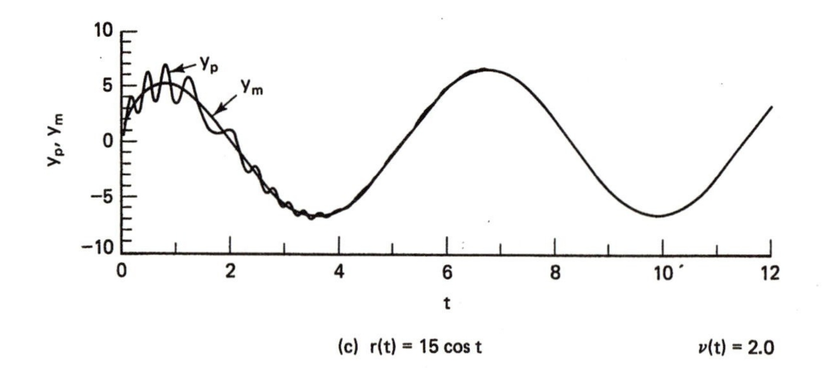

(c) $r(t) = 15 \cos t$ $\nu(t) = 2.0$

Figure 8.4 Model matching with constant disturbance.

of the model and the plant for different reference inputs and sinusoidal disturbances are also shown in Fig. 8.5. All initial conditions except that of the model output were set to zero and $y_m(0) = 1$.

The discussions above reveal that over-parametrization of the controller enables exact cancellation of disturbance signals, when the latter are expressed as the solutions of homogeneous stable linear time-invariant differential equations whose coefficients and initial conditions are unknown.

(c) Disturbance as the Output of a Differential Equation. In general, if the disturbance ν is such that $D(s)\nu(t) = 0$, where $D(s)$ is a monic stable polynomial of degree n_ν, the results of this section can be used to establish exact cancellation of the disturbance as well as asymptotic model-following. The adaptive controller in this case would consist of $2n_\nu + 2$ parameters. This problem has been treated in detail in [8].

In practice, the disturbance may have a continuous spectrum that is concentrated around a finite number of values. In such cases, perfect disturbance cancellation may not be possible but the method described in this section may still be adequate in many situations to meet performance specifications.

8.3.2 Bounded Disturbances: Statement of the Problem

Exact cancellation of the disturbance is possible, as described in the previous section, only when the precise nature of the disturbance is known a priori. In particular, if the order of the LTI differential equation generating the disturbance is known, the signal needed to cancel the disturbance can be computed internally by over-parametrization of the adaptive controller. Quite often, it may not be possible to describe the characteristics of the disturbance apart from the fact that it is bounded. Even the bound on the disturbance may not be known. The objective in such cases may be only to ensure the boundedness of all the signals in the system while the output error satisfies performance specifications.

The plant to be controlled adaptively, the corresponding controller, and the reference model are described by the equations

$$\begin{aligned}
\dot{y}_p(t) &= a_p y_p(t) + u(t) + \nu(t) && \text{(Plant)} \\
u(t) &= \theta(t) y_p(t) + r(t) && \text{(Controller)} \\
\dot{y}_m(t) &= -a_m y_m(t) + r(t) \qquad a_m > 0 && \text{(Reference model)}
\end{aligned} \tag{8.7}$$

where y_p and y_m are the outputs of the plant and model respectively, a_p is an unknown constant, r is a bounded reference input, ν is a bounded disturbance with $|\nu(t)| \leq \nu_0$, and θ is a control parameter.

The output error e_1 and the parameter error ϕ are defined as in the previous chapters as $e_1 = y_p - y_m$ and $\phi = \theta - \theta^*$. $\theta^*(= -a_p - a_m)$ is the constant value of the control parameter θ for which the plant transfer function is equal to that of the model. The error equation can then be written as

$$\dot{e}_1(t) = -a_m e_1(t) + \phi(t) y_p(t) + \nu(t). \tag{8.8a}$$

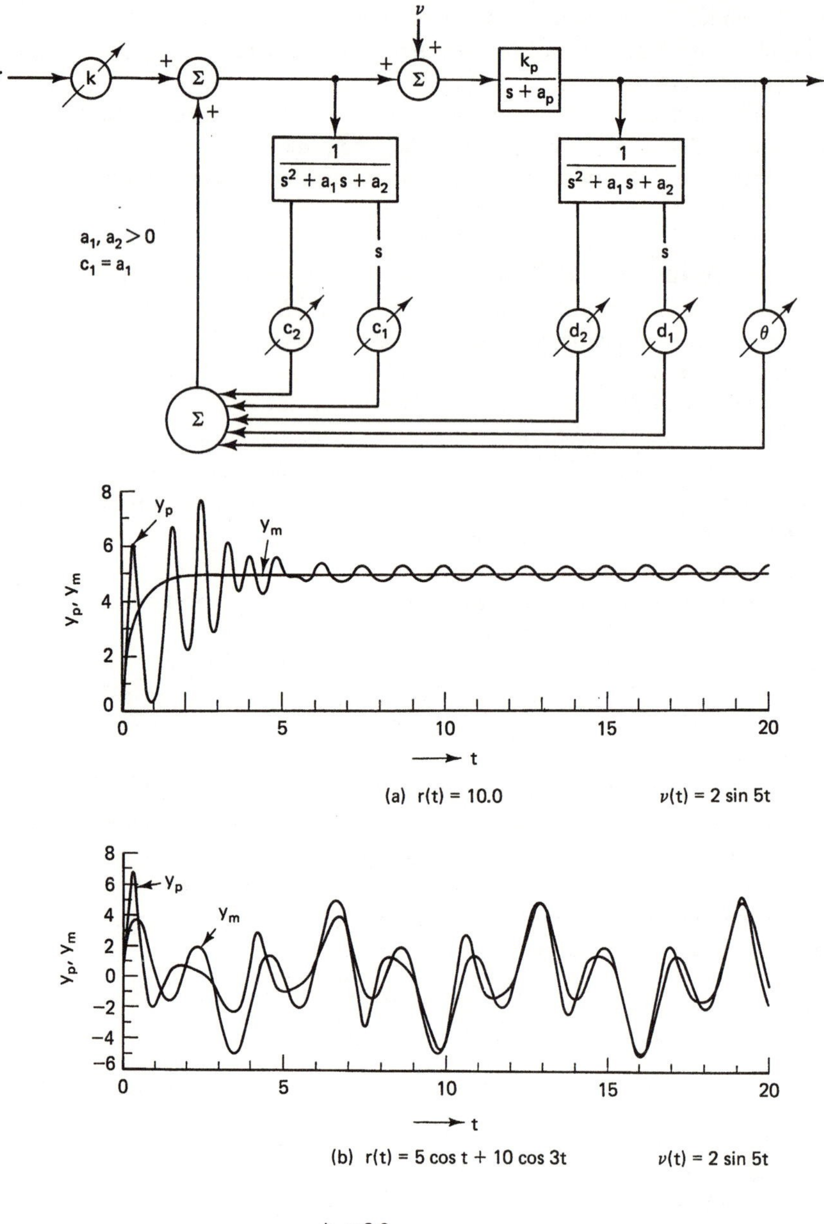

$k_p = 2.0$
$a_p = -1.0$
$a_1 = 3.0$
$a_2 = 2.0$
$k = 0.5$

$W_m(s) = \frac{1}{s+2}$

$y_m(0) = 1,\ y_p(0) = c_2(0) = \theta(0) = d_1(0) = d_2(0) = 0$

Figure 8.5 Cancellation of a sinusoidal disturbance.

The adaptive law for updating the control parameter $\theta(t)$ in the ideal case (refer to Chapter 5) is given by

$$\dot{\phi}(t) = -e_1(t)y_p(t). \tag{8.8b}$$

Our aim is to determine whether the same adaptive law of Eq. (8.8b) would also stabilize the system in the presence of bounded disturbances. We will briefly recapitulate the properties of Eq. (8.8) in the absence of disturbances.

The ideal system. In the absence of external disturbances, the nonlinear differential equation (8.8) reduces to

$$\begin{aligned} \dot{e}_1(t) &= -a_m e_1(t) + \phi(t)y_p(t) \\ \dot{\phi}(t) &= -e_1(t)y_p(t). \end{aligned} \tag{8.9}$$

The origin of the error equation (8.9) was shown to be (i) uniformly stable in Chapter 3, and (ii) u.a.s.l. in Chapter 6, if y_m is persistently exciting.

Perturbed system. To provide some insight into the behavior of the nonlinear system when a bounded disturbance ν with $|\nu(t)| \le \nu_0$ is present, we shall discuss two cases where the perturbed nonlinear system in Eq. (8.8) is autonomous.

Case (i) $y_m(t) \equiv 0$. The error equation (8.8) becomes

$$\begin{aligned} \dot{y}_p(t) &= -(a_m - \phi(t))\,y_p(t) + \nu(t) \\ \dot{\phi}(t) &= -y_p^2(t). \end{aligned} \tag{8.10}$$

Hence, $\phi(t)$ is a nonincreasing function of time and will be unbounded if $y_p \notin \mathcal{L}^2$. It can be shown that a necessary and sufficient condition for this to happen is $\nu \notin \mathcal{L}^2$ (for example, $\nu(t) \equiv \nu_0 \neq 0$). Figure 8.6 shows that $|\phi(t)|$ increases with time when $a_m = 1$ and $\nu(t) \equiv 2$ in Eq. (8.10), for initial conditions $e_1(0) = 1$ and $\phi(0) = 5$.

Case (ii) $y_m(t) \equiv y_0$. The ideal system in this case is autonomous, and by Theorem 2.7, its origin is u.a.s. since the largest invariant set in $E = \{x | e_1^2 = 0\}$ is the origin, where $x = [e_1, \phi]^T$. However, since the system is nonlinear, it no longer follows that a bounded input will result in a bounded output. If, for example, $\nu(t) \equiv -\nu_0$, and $\nu_0 > a_m y_0 > 0$, $\lim_{t\to\infty} e_1(t) = -y_0$ and $\lim_{t\to\infty} \phi(t) = -\infty$, which can be proved as follows:

Proof. Let D_1 be the open region enclosed by the line $e_1 = \nu_0/a_m$ and the curve $\phi = (a_m e_1 + \nu_0)/(e_1 + y_0)$ with $\phi \le 0$ (Fig. 8.7). From Eq. (8.10), it then follows that when $y_m(t) \equiv y_0$ and $\nu(t) \equiv -\nu_0$, all solutions that start on the boundary ∂D_1 enter D_1, as shown in Fig. 8.7. Since the system is autonomous and contains no singular points in D_1, all solutions originating in D_1 become unbounded and $\lim_{t\to\infty} \phi(t) = -\infty$ and $\lim_{t\to\infty} e_1(t) = -y_0$. Note that $y_p(t)(= e_1(t) + y_m(t)) \to 0$ as $t \to \infty$ so that $y_p(t)$ is not persistently exciting when the solutions are unbounded.

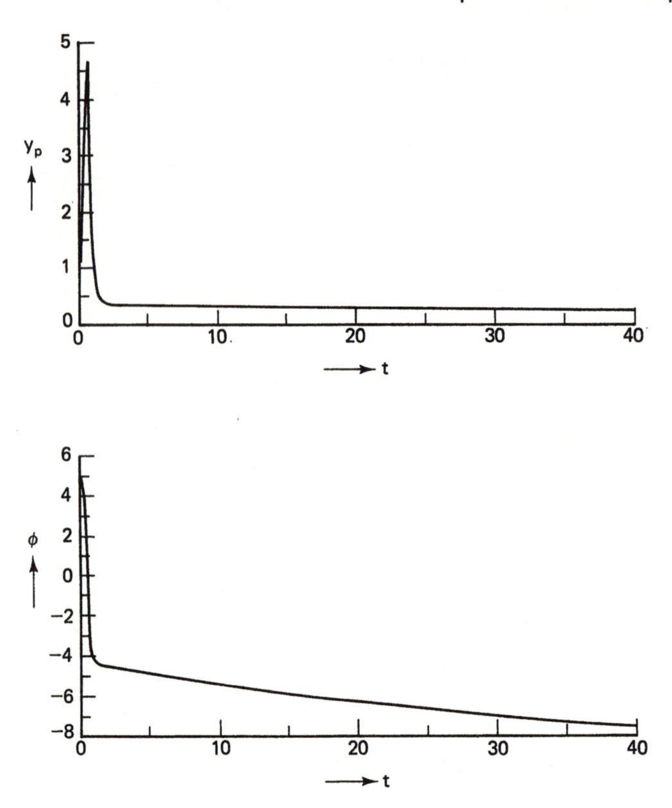

Figure 8.6 Unbounded behavior with adaptive control.

In Fig. 8.8, the solutions of Eq. (8.8) are depicted for $a_m = 1$, $y_m(t) \equiv 2$, and $\nu(t) \equiv -5$, and initial conditions $e_1(0) = 1$, $\phi(0) = 5$. Although $e_1(t)$ is bounded, $\lim_{t \to \infty} \phi(t) = -\infty$.

Cases (i) and (ii) show that the same adaptive law, which results in stability in the ideal case, can lead to unboundedness of the parameters when there are bounded disturbances. Such a drift in the parameter values is undesirable for a variety of reasons, not the least of which are the large transient excursions in the output error, which result when the characteristics of the reference input vary significantly after an interval of time. Hence, it is essential that steps be taken to assure that all parameter values remain bounded. As shown in the following section, this can be accomplished by

1. modifying the adaptive law, or

2. increasing the degree of persistent excitation of the reference input.

The first case is treated in Section 8.3.3 and the latter is dealt with in Section 8.3.4.

In both sections, we consider only plants described by first-order differential equations to acquaint the reader with the principal ideas involved. The motivations for the changes suggested, as well as the mathematical difficulties that such changes lead to,

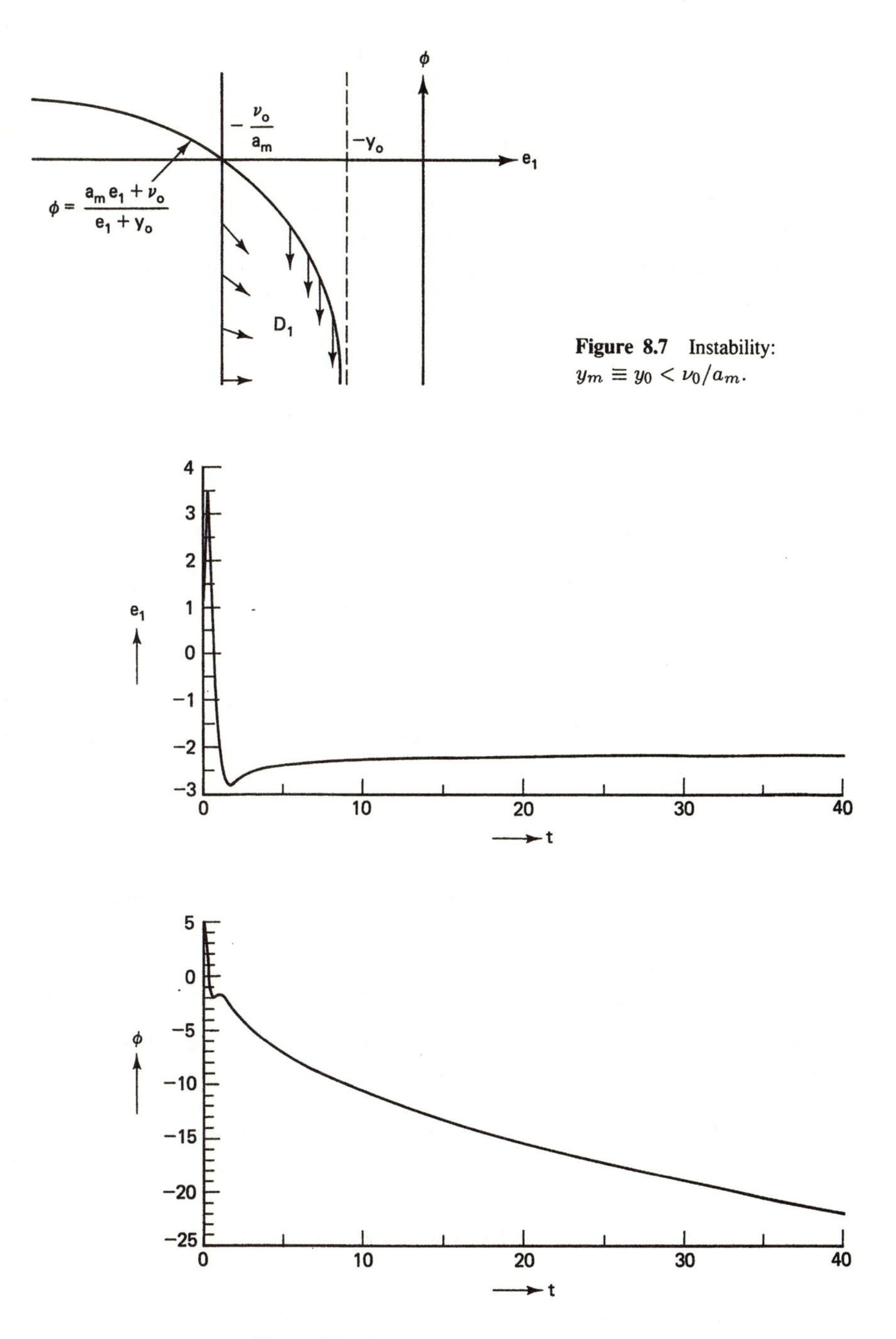

Figure 8.7 Instability: $y_m \equiv y_0 < \nu_0/a_m$.

Figure 8.8 Unbounded behavior with adaptive control.

are treated qualitatively at this stage. The proofs of boundedness are dealt with in detail in Section 8.4 and Appendix D for general plants described by n^{th}-order differential equations.

8.3.3 Category 1: Modifications of the Adaptive Law

a) Dead Zone. Consider the error equation (8.8a) corresponding to the adaptive system described by Eq. (8.7). Let the adaptive law in Eq. (8.8b) be used to update the parameter $\theta(t)$ in the feedback path. If $V(e_1, \phi) = (e_1^2 + \phi^2)/2$, the time derivative of V evaluated along the trajectories of Eq. (8.8) yields

$$\dot{V} = -a_m e_1^2 + e_1 \nu,$$

which assures that $V(e_1, \phi)$ is nonincreasing when

$$|e_1| > \frac{\nu_0}{a_m}. \tag{8.11}$$

Since V may increase when the condition in Eq. (8.11) is not satisfied, it follows that the signals can grow in an unbounded fashion in

$$D_1 \triangleq \left\{(e_1, \phi) \middle| \; |e_1| \leq \frac{\nu_0}{a_m}\right\}.$$

The scheme containing a dead zone suggested by Peterson and Narendra [31] overcomes this difficulty by stopping the adaptation process in D_1 by using the following adaptive law:

$$\dot{\phi} = \begin{cases} -e_1 y_p & \text{if } (e_1, \phi) \in D_1^c \\ 0 & \text{if } (e_1, \phi) \in D_1 \end{cases} \tag{8.12}$$

where D^c denotes the complement of D. This implies that the adaptation is stopped when the output error $e_1(t)$ becomes smaller than a prescribed value. This assures the boundedness of all the signals in the system. Similar algorithms were also proposed by Egardt [5] and Samson [34]. Some modifications have been suggested in [3,30].

Simulation 8.1 In all the simulations carried out in this section, the transfer function of the plant and the reference model are chosen to be $1/(s-4)$ and $1/(s+1)$ respectively. Hence, $a_p = 4$, $a_m = 1$, and $\theta^* = -5$. Figure 8.9 shows the solutions of Eqs. (8.7) and (8.12) when (i) $r(t) = 0$, $\nu(t) = 2$, (ii) $r(t) = 15 \cos t$, and $\nu(t) = 2 \cos 5t$. The size of the dead zone was chosen to be equal to 2. The initial value $y_p(0)$ was set to 1 and all other initial conditions were equal to zero. The simulations indicate that adaptation ceases completely after a finite time when the error amplitude is less than the magnitude of the dead zone.

Since the adaptive law contains a dead zone, this method cannot assure asymptotic convergence of the plant output to that of the model even when a disturbance is not present.

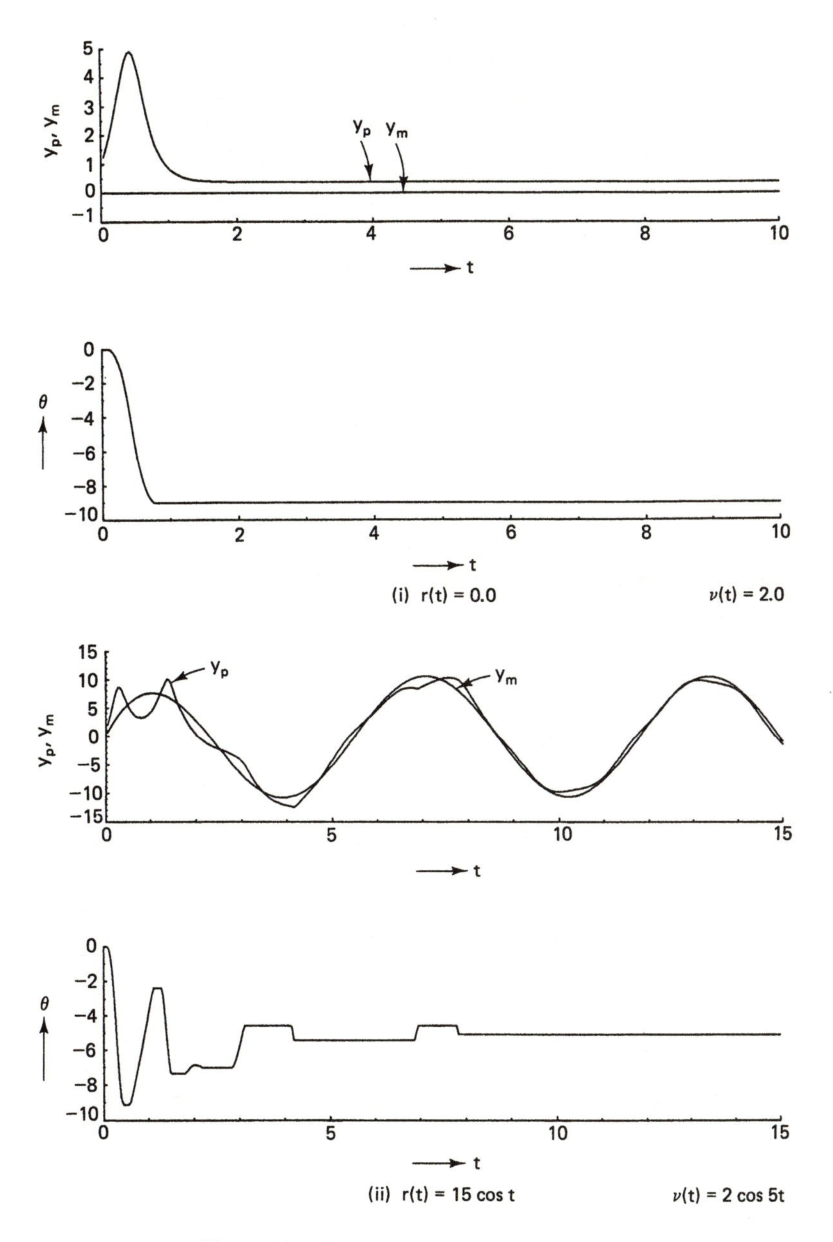

Figure 8.9 Adaptive control using dead zone.

b) Bound on $|\theta|$. The instability in the system described in (a) is due to the parameter error and not the output error e_1, which remains bounded. In many practical problems, bounds on the components of the desired parameter vector θ^* are generally known so that unboundedness of the vector $\theta(t)$ can be avoided by suitably modifying the adaptive law. Such a scheme was proposed by Kreisselmeier and Narendra [16] in which it is assumed that $|\theta(t)| \leq \theta^*_{\max}$, where $\theta^*_{\max}$ is a known constant. The adaptive law has the form

$$\dot{\phi} \quad = \quad -e_1 y_p - \theta \left(1 - \frac{|\theta|}{\theta^*_{\max}}\right)^2 f(\theta) \tag{8.13}$$

where

$$f(\theta) \quad = \quad \begin{cases} 1 & \text{if } |\theta| > \theta^*_{\max} \\ 0 & \text{otherwise.} \end{cases} \tag{8.14}$$

Equations (8.13) and (8.14) imply that when the control parameter θ lies in the interior of the constraint set $S = \{\theta |\, |\theta| \leq \theta^*_{\max}\}$, the same adaptive law as that given in Chapter 5 is used. A correction term is used to force the parameter vector back into S when it lies on the boundary ∂S or outside S. A similar procedure can be adopted even when the lower and upper bounds of θ are not symmetric with respect to the origin. For example, in the scalar case discussed earlier, if $\theta \in [\theta_1, \theta_2]$, the parameter adjustment is stopped when θ reaches the boundary and the sign of its derivative is such as to make θ leave the set S. Simulations of two simple examples are given below in which θ_1 and θ_2 are specified. The nature of the adaptation is seen to depend on both the interval $[\theta_1, \theta_2]$ as well as the reference input to the system.

Simulation 8.2 With the same plant and model transfer functions used in Simulation 8.1, the response of the adaptive system was observed using the adaptive algorithm in Eq. (8.13). The initial value $y_p(0)$ was set to 1 and all other initial conditions were equal to zero. The constraint set S, the reference input as well as the disturbance were varied for each simulation. In each case, the output error, as well as the parameter θ, are displayed.
(i) In this case, $\theta^*_{\max} = 6$ so that $S = [-6, 6]$. When $r(t) \equiv 0$, and $\nu(t) = 2$ [Fig. 8.10(a)], the steady-state value of the parameter lies close to the boundary of S. When $r(t) = 15 \cos t$ and $\nu(t) = 2$, θ almost never reaches the boundary of the set S (Fig. 8.10-(b)).
(ii) When the constraint set S is given by $S = [-6, -4]$, the trajectories are similar to those in case (i), since θ almost always lies in the interior of S.

In contrast to the method outlined in (a), the adjustment of the control parameter θ never ceases in this case. Hence, θ does not converge to a constant value but continues to vary with time. Also, if the disturbance were to tend to zero and the reference input is persistently exciting, the parameter θ will tend to its true value θ^* asymptotically.

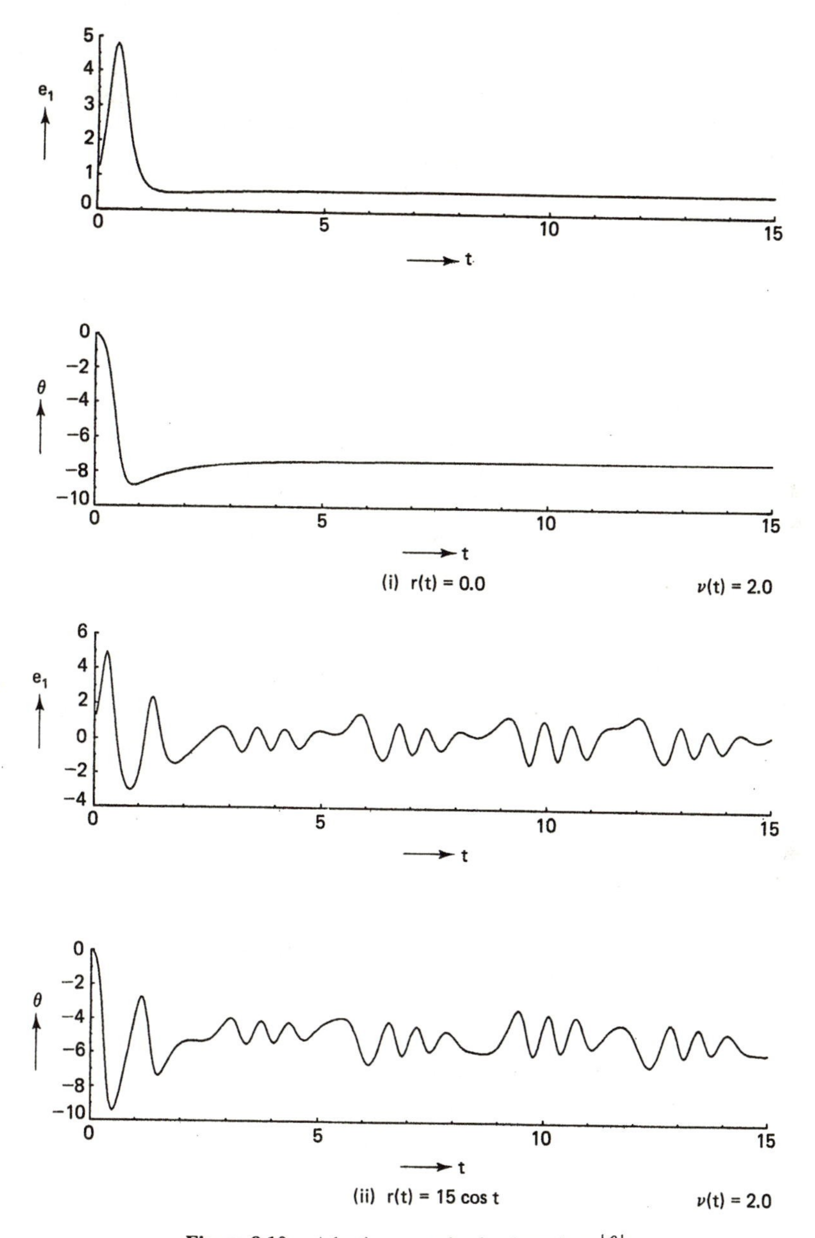

Figure 8.10 Adaptive control using bound on $|\theta|$.

c) The σ-modification scheme. Both schemes in (a) and (b) assumed prior information in the form of an upper bound ν_0 on the disturbance, and an upper bound $\theta^*_{\max}$ on the control parameter, respectively. A third scheme proposed by Ioannou and Kokotovic [10] to study the robust adaptive control problem in the presence of bounded disturbances, is the σ-modification scheme, which does not require any additional prior information about the plant beyond that required in the ideal case. An additional term $-\sigma\theta$ used in the adaptive law ensures that ϕ does not become unbounded:

$$\dot{\phi} = -e_1 y_p - \sigma\theta \qquad \sigma > 0. \tag{8.15}$$

If $V(e_1, \phi) = (e_1^2 + \phi^2)/2$, its time derivative along the trajectories of Eqs. (8.8a) and (8.15) is given by

$$\dot{V} = -a_m e_1^2 + e_1\nu - \sigma\phi^2 - \sigma\phi\theta^*.$$

Hence, $\dot{V} < 0$ outside a compact region $D_3 = \{(e_1, \phi) | \ |e_1| \leq k_3, |\phi| \leq k_4\}$, where k_3 and k_4 are positive constants. Therefore, all signals are bounded according to Theorem 2.24. The use of the additional term $-\sigma\theta$ in Eq. (8.15) makes the adaptive system robust under external disturbances. The drawback of the adaptive law is that the origin is no longer an equilibrium point of Eqs. (8.8a) and (8.15). This implies that even when the disturbance is removed and the reference input is persistently exciting, the errors do not converge to zero.

Simulation 8.3 Figure 8.11(a) shows the solutions of Eqs. (8.8a) and (8.15) with the plant and model described as in Simulation 8.1 for different initial conditions when $y_m(t) \equiv 0$ and there is no disturbance present. There are three stable equilibrium states at $(\pm\sqrt{-\sigma\,(a_m + \theta^*)},\ a_m)$, and $(0,\ -\theta^*)$. In Fig. 8.11(b), the same plant and model are considered with $y_m(t) \equiv y_0 \neq 0$ where y_0 is a constant, and $\nu(t) \equiv 0$. In this case, the equations have a single stable equilibrium state whose distance from the origin decreases as the amplitude of y_m is increased or the value of σ is decreased.

The magnitude of σ determines the magnitude of the errors so that prior information regarding the nature of the disturbance has to be used in the choice of this parameter.

d) The e_1-modification scheme. The fourth scheme considered in this category was suggested by Narendra and Annaswamy [24] and is motivated by the scheme in (c). The gain σ in Eq. (8.15) is replaced by a term proportional to $|e_1|$. The rationale for using such a term is that it tends to zero with the output error. Hence, when there is no disturbance present, if the mismatch between the plant and the model vanishes, the correction term tends to zero as well. The adaptive law is chosen as

$$\dot{\phi} = -e_1 y_p - \gamma|e_1|\theta \qquad \gamma > 0. \tag{8.16}$$

Here, the time derivative $\dot{V}$ along the trajectories of Eqs. (8.8a) and (8.16) is given by

$$\begin{aligned}\dot{V} &= -a_m e_1^2 + e_1\nu - \gamma|e_1|\phi^2 - \gamma|e_1|\phi\theta^* \\ &\leq -|e_1|\left[a_m|e_1| - \nu_0 + \gamma\phi\theta^* + \gamma\phi^2\right].\end{aligned}$$

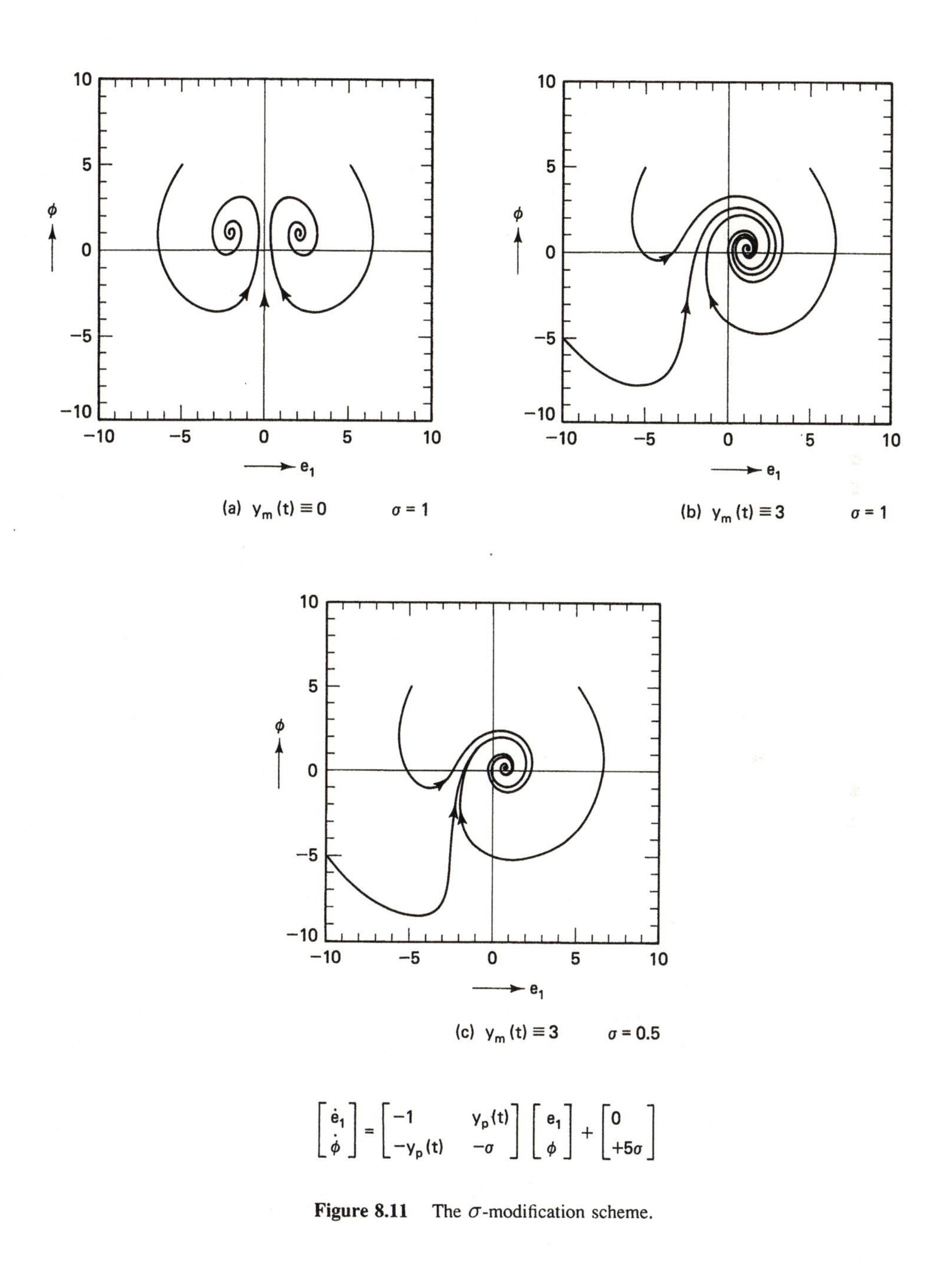

$$\begin{bmatrix} \dot{e}_1 \\ \dot{\phi} \end{bmatrix} = \begin{bmatrix} -1 & y_p(t) \\ -y_p(t) & -\sigma \end{bmatrix} \begin{bmatrix} e_1 \\ \phi \end{bmatrix} + \begin{bmatrix} 0 \\ +5\sigma \end{bmatrix}$$

Figure 8.11 The σ-modification scheme.

Hence, $\dot{V} \leq 0$ in D_4^c, where the set D_4 is defined as

$$D_4 \triangleq \{(e_1, \phi) |\ \phi^2 + \phi\theta^* + \frac{a_m}{\gamma}|e_1| - \frac{\nu_0}{\gamma} \leq 0\}.$$

By Theorem 2.24, it follows that all the solutions are bounded.

When $y_m(t) \equiv y_0$ and $\nu(t) \equiv 0$, the behavior of the solutions of Eqs. (8.8a) and (8.16) can be divided into the four cases below. In all these cases, the equilibrium points are determined by the intersection of the curve $\dot{V} = 0$ with the curve $-a_m e_1 + \phi(y_m + e_1) = 0$ (that is, $\dot{e}_1 = 0$).

(i) $|y_0| < \min(\gamma|\theta^*|, \frac{\gamma a_p^2}{4a_m})$: Eqs. (8.8a) and (8.16) have four equilibrium points including the origin [Fig. 8.12(a)]. Two of these four equilibrium points are stable foci (p_2, p_4), the third is a saddle (p_3), and the origin is a saddle focus (p_1).

(ii) $\gamma|\theta^*| < |y_0| < \frac{\gamma a_p^2}{4a_m}$: In this case, there are three equilibrium points, which include a stable focus (p_1), a stable node (p_2), and a saddle (p_3) [Fig. 8.12(b)].

(iii) $\frac{\gamma a_p^2}{4a_m} < |y_0| < \gamma|\theta^*|$: There are two equilibrium points in this case. The origin is a saddle focus (p_1) and the second equilibrium point is a stable focus (p_2) [Fig. 8.12(c)].

(iv) $|y_0| > \max(\gamma|\theta^*|, \frac{\gamma a_p^2}{4a_m})$: Here we have only one equilibrium point at the origin, which is a stable focus (p_1)[Fig. 8.12(d)].

From (iv), it follows that if the reference input is sufficiently persistently exciting so that $|y_0| > \max(\gamma|\theta^*|, \gamma a_p^2/4a_m)$, the only trajectory of Eqs. (8.8a) and (8.16) that lies on the curve $\dot{V} = 0$, is the origin. Hence, from Theorem 2.7, we can conclude that the origin of Eqs. (8.8a) and (8.16) is u.a.s.l. The same result can also be established when $y_m(t)$ is a general time-varying signal, which is persistently exciting and γ is sufficiently small.

Comment 8.1 We have considered the modifications to the adaptive law suggested in (c) and (d) in some detail to emphasize the fact that the solutions obtained in the case of a seemingly simple plant can lead to complex questions involving nonlinear differential equations. This is only aggravated in the problem of control of plants described by higher order differential equations as shown in Section 8.4. Also, since the adaptive laws are only sufficient to assure the boundedness of all signals, they are invariably found to be very conservative in many practical situations.

8.3.4 Category 2: Persistent Excitation of the Reference Input

For the ideal system shown in Eq. (8.9), uniform stability follows immediately from the existence of a Lyapunov function $V(e_1, \phi) = (e_1^2 + \phi^2)/2$. In addition, $\lim_{t\to\infty} e_1(t) = 0$. Further, if $y_p(t)$ is persistently exciting, the system is exponentially stable [23] and $\lim_{t\to\infty} \phi(t) = 0$ so that the overall transfer function of the plant together with the controller matches that of the model as $t \to \infty$. When a disturbance $\nu(t)$ is

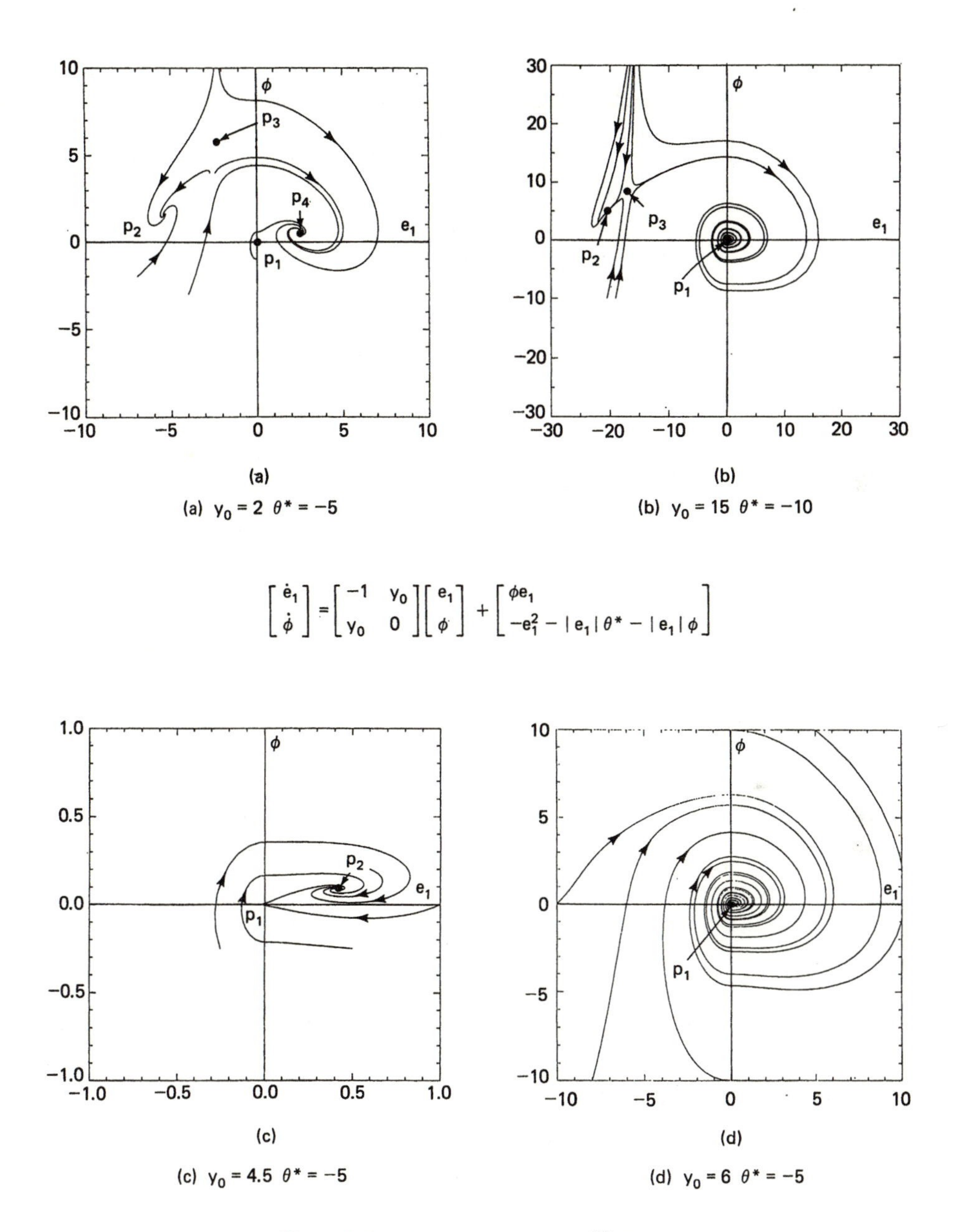

Figure 8.12 Trajectories when $y_m(t) \equiv y_0$.

present, it is tempting to proceed as in the ideal case and require $y_p(t)$ in Eq. (8.8) to be persistently exciting. This would ensure the exponential stability of the unperturbed system and, hence, result in a bounded error vector for bounded disturbances. However, such arguments, which are based on linear systems theory, are no longer valid since the system being considered is nonlinear and $y_p(t)$ is a dependent variable. Since stability of the overall system has not been established, $y_p(t)$ cannot be assumed to be bounded. Thus, proving that it is persistently exciting becomes specious. If $y_p(t)$ is expressed in terms of the model output $y_m(t)$ and the output error $e_1(t)$, the resultant nonlinear differential equations are of the form

$$\begin{aligned} \dot{e}_1 &= -a_m e_1 + \phi y_m + \phi e_1 + \nu \\ \dot{\phi} &= -e_1 y_m - e_1^2. \end{aligned} \tag{8.17}$$

This equation must be analyzed to determine the conditions under which the errors e_1 and ϕ will be bounded. The importance of Eq. (8.17) lies in the fact that the evolution of $e_1(t)$ and $\phi(t)$ are expressed in terms of the independent variable $y_m(t)$. In this section, we show that when the persistent excitation of the model output y_m is larger in some sense than the magnitude of the disturbance $\nu(t)$, the overall system in Eq. (8.17) will have bounded solutions. We shall first consider the case where both ν and y_m are constants. This simple case, together with the discussion in Section 8.3.2, provides the necessary insight to solve the general problem when the reference input and disturbance are time-varying. The latter is treated in Theorem 8.1.

Case (iii) $y_m(t) \equiv y_0$, $\nu(t) \equiv \nu_0$, $y_0 > \nu_0/a_m$. It was shown in Section 8.3.2 that Eq. (8.17) has unbounded solutions when $\nu(t) \equiv \nu_0$, where ν_0 is a constant, with $y_m(t) \equiv 0$ [case (i)], or $y_m(t) \equiv y_0 < \nu_0/a_m$ [case (ii)]. The case when $y_m(t) \equiv y_0$, $\nu(t) \equiv \nu_0$, with $y_0 > \nu_0/a_m$ is considered here. The equilibrium state of Eq. (8.17) is at $(0, -\nu_0/y_0)$, which can be shown to be u.a.s.l. This suggests the possibility that global boundedness of solutions of Eq. (8.17) can be achieved by increasing the amplitude of y_m relative to the amplitude of the disturbance. Such results can also be derived when both $y_m(t)$ and $\nu(t)$ are time-varying and are given in Theorem 8.1. Since a detailed proof of the corresponding theorem for the case of an nth order plant is treated in Section 8.4, we provide an outline of the proof here.

Theorem 8.1 Let $|y_m(t)| \leq y_0, |\nu(t)| \leq \nu_0$ and $y_m(t)$ be a persistently exciting signal, that is, positive numbers $T_0, \delta_0, \epsilon_0$ exist such that given $t \geq t_0$, there exists $t_2 \in [t, t+T_0]$ such that $[t_2, t_2 + \delta_0] \subset [t, t+T_0]$ and

$$\left| \frac{1}{T_0} \int_{t_2}^{t_2+\delta_0} y_m(\tau)\, d\tau \right| \geq \epsilon_0 \qquad \forall t \geq t_0. \tag{8.18}$$

Then

(a) if $y_0 < \nu_0/a_m$, there exists an input $\nu(t)$ and initial conditions for which $\lim_{t\to\infty} \phi(t) = -\infty$ and $e_1(t)$ approaches asymptotically the region $|e_1| \leq y_0 + \varepsilon$, where ε is a positive constant.

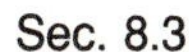

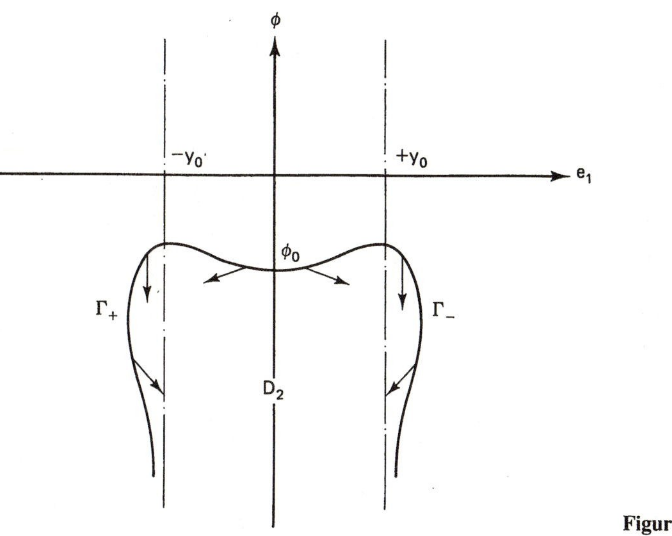

Figure 8.13 Proof of part (a)

(b) If

$$\epsilon_0 \geq \nu_0/a_m + \delta, \tag{8.19}$$

where δ is a positive constant, all the solutions of the differential equation (8.17) are bounded.

Proof.

(a) We provide a constructive proof for the first part of the theorem by choosing $\nu(t)$ as follows:

$$\nu(t) = \begin{cases} -sgn(y_m(t))\nu_0 & \text{if } |e_1(t)| \geq y_0, \text{ or } e_1(t) = 0 \\ +sgn(e_1(t))\nu_0 & \text{if } |e_1(t)| < y_0, e_1 \neq 0. \end{cases} \tag{8.20}$$

Consider the solution of the differential equation (8.17) with initial condition $(0, \phi_0)$ with $y_m(t) \equiv y_0$, $\nu(t) \equiv -\nu_0$, and $\phi_0 < 0$. Let Γ_+ denote the open curve along which the trajectory lies for all $t \geq 0$. Similarly, let Γ_- denote the curve along which the solution lies for all $t \geq 0$ when $y_m(t) \equiv -y_0$, and $\nu(t) \equiv \nu_0$ with the same initial condition. Let $\Gamma(\phi_0) = \Gamma_+ \cup \Gamma_-$. $\Gamma(\phi_0)$ divides $\mathbb{R}^2$ into two open regions D_5 and D_5^c, where $(0, \phi) \in D_5$ if $\phi < \phi_0$. Then all solutions of Eq. (8.17) with $|y_m(t)| \leq y_0$ and $\nu(t)$ chosen as in Eq. (8.20) with initial conditions on $\Gamma(\phi_0)$ lie either on $\Gamma(\phi_0)$ or enter D_5. Since this is true for every $\phi_0 < 0$, the solutions become unbounded and $\lim_{t\to\infty} \phi(t) = -\infty$ [see Fig. 8.13].

(b) We provide a qualitative proof of boundedness below. If $V(x) = x^T x/2$, where $x = [e_1, \phi]^T$, when the degree of PE of $y_m(t)$ satisfies Eq. (8.19), the change in $V(x(t))$ over a finite interval $[t_0, t_0 + T_0]$ is negative if $\|x(t_0)\|$ is sufficiently large. This follows by considering the two cases given in (i) and (ii) below, which are mutually exclusive and collectively exhaustive:

(i) $y_p(t)$ is persistently exciting over the interval $[t_0, t_0 + T_0]$.

(ii) $y_p(t)$ is not persistently exciting over the interval $[t_0, t_0 + T_0]$.

In case (i), the unperturbed system is exponentially stable, which implies that $V(x(t))$ decreases over an interval $[t_0, t_0+T_0]$ if $\|x(t_0)\|$ is large. In case (ii), $e_1(t)$ is persistently exciting since it is the difference between two signals $y_m(t)$ and $y_p(t)$, where the former is persistently exciting and satisfies the condition in Eq. (8.19), while the latter is not. Hence, it can be concluded once again that $V(x(t))$ decreases over the interval $[t_0, t_0+T_0]$ when $\|x(t_0)\|$ is large.

Simulation 8.4 The differential equation (8.17) was simulated with $a_m = 1$ for different initial conditions. Figure 8.14 shows that when $y_0 < \nu_0$, e_1 remains bounded while $\phi(t) \to -\infty$, but when $\epsilon_0 > \nu_0$, all the trajectories are bounded. In Fig. 8.15, the dependence of system response on disturbance frequencies is illustrated. A sinusoidal reference signal $y_m(t) = 20 \cos 1.2\,t$ was chosen and the responses were studied for different $\nu(t)$, with $\nu_0 = 5$. In all cases, the initial conditions were chosen as $e_1(0) = 1$, $\phi(0) = 5$. The limit set decreases with increasing frequency of $\nu(t)$. In Fig. 8.16, for a constant disturbance $\nu(t) \equiv 5$, the effect of different types of reference signals $y_m(t)$ was studied with $e_1(0) = 1$, $\phi(0) = -5$. It is found that lower input frequencies result in better performance.

8.3.5 Performance of the Adaptive System

Sections 8.3.2-8.3.4 emphasized the need to take precautionary measures while adaptively controlling a plant when external disturbances are present. Even though the stability of the equilibrium state and boundedness of the signals are important considerations, keep in mind that the adaptive system will be ultimately judged on the basis of its performance. In this section, we present some simulation results obtained from the adaptive control of a first-order plant with a transfer function $k_p/(s - a_p)$. k_p and a_p are assumed to be unknown but belong to a known compact region S. The plant is subjected to an external bounded input disturbance $d(t)$. The objective is to determine a control input u to the plant which will result in a small error $|y_p(T) - r(T)|$ between a command input $r(t)$ and the output $y_p(t)$ at a terminal time $t = T$, for all values of (k_p, a_p) in the specified region. Both the command input r and the disturbance d are piecewise constants and are shown in Fig. 8.17. The control input is generated as

$$u(t) = k(t)r(t) + \theta(t)y_p(t).$$

In the simulations carried out, the compact region S is defined by

$$S \triangleq \{(k_p, a_p) | \; -1 \leq a_p \leq 3, \qquad 1 \leq k_p \leq 3\}.$$

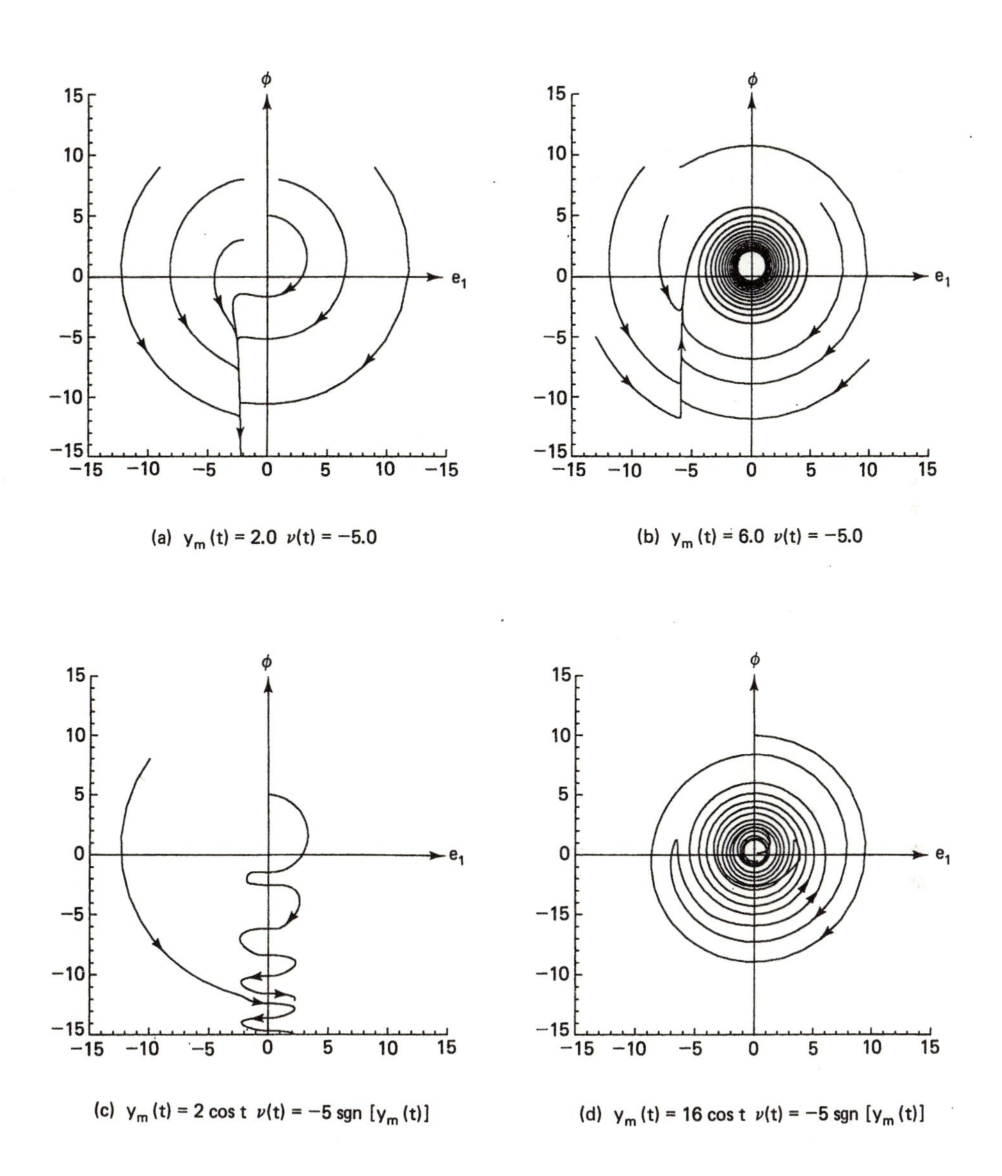

(a) $y_m(t) = 2.0 \quad \nu(t) = -5.0$

(b) $y_m(t) = 6.0 \quad \nu(t) = -5.0$

(c) $y_m(t) = 2 \cos t \quad \nu(t) = -5 \operatorname{sgn}[y_m(t)]$

(d) $y_m(t) = 16 \cos t \quad \nu(t) = -5 \operatorname{sgn}[y_m(t)]$

Figure 8.14 Robustness with persistent excitation.

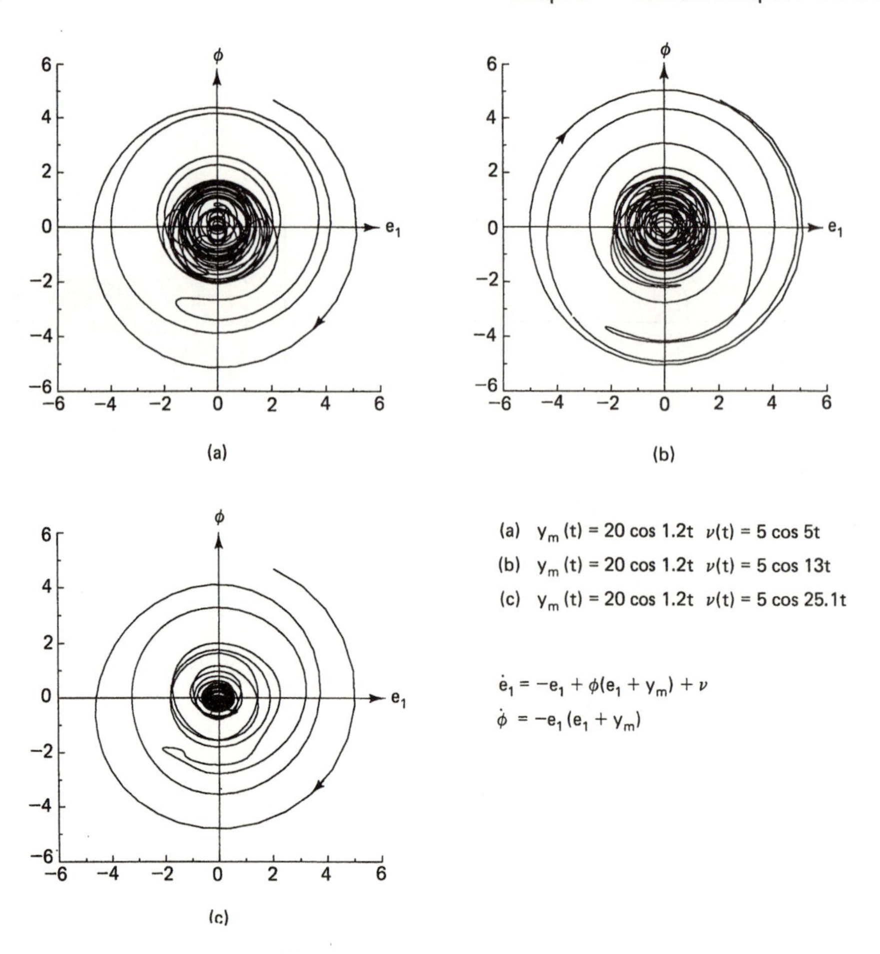

Figure 8.15 Dependence on disturbance frequencies.

and the terminal time $T = 20$. The control parameters $k(t)$ and $\theta(t)$ were adjusted using the standard adaptive law described in Chapter 3 with an adaptive gain $\gamma = 10$. Different reference models were chosen to improve the performance. Results are shown in Fig. 8.17 for three sets of values for the pair (k_p, a_p) and a reference model with a transfer function $5/(s+5)$. In each case, the command input r, the disturbance d, the plant output y_p, as well as the control input u are indicated. Although improved response can be achieved using adaptive control, the designer should be aware that instabilities of the type discussed in the previous section may occur.

8.4 ADAPTIVE CONTROL OF AN nth ORDER PLANT

The methods described in Section 8.3 to achieve robustness can be directly extended to general linear time-invariant plants when bounded external disturbances are present.

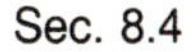

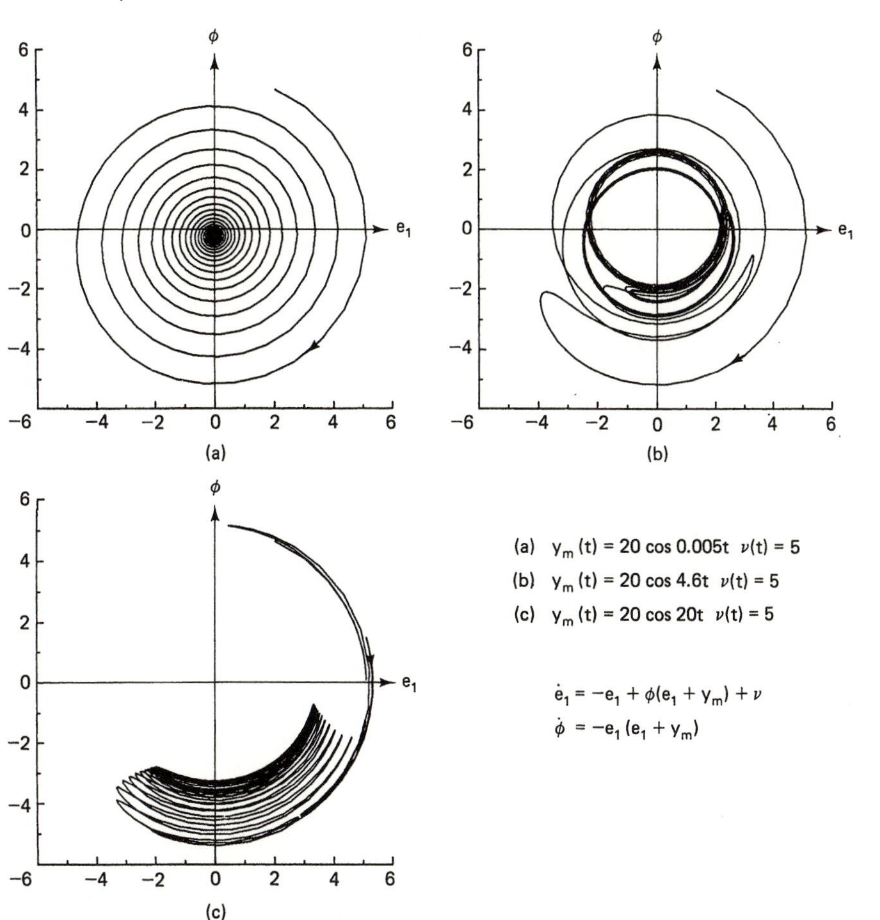

Figure 8.16 Dependence on input frequencies.

Although the concepts are the same, the analytical arguments used to assure the boundedness of all the signals, when the order n of the plant transfer function is greater than one, are more complex and are discussed in this section. The reader who is not interested in these details may proceed directly to Section 8.5 without any loss of continuity.

The plant to be adaptively controlled can be described by the differential equations (Fig. 8.18)

$$\begin{aligned} \dot{x}_p &= A_p x_p + b_p u + d_p \nu_1(t) \\ y_p &= h_p^T x_p + \nu_2(t) \end{aligned} \tag{8.21}$$

where $\nu_1 : \mathbb{R}^+ \to \mathbb{R}$ is a bounded input disturbance and $\nu_2 : \mathbb{R}^+ \to \mathbb{R}$ is an output disturbance that is assumed to be bounded and differentiable. The assumptions regarding the plant transfer function and the choice of the reference model are identical to those in the ideal case. The adaptive controller is described by the equations

$$\begin{aligned}\dot{\omega}_1 &= \Lambda\omega_1 + \ell u \\ \dot{\omega}_2 &= \Lambda\omega_2 + \ell y_p \\ u &= \theta^T(t)\omega\end{aligned} \tag{5.16}$$

where $\omega^T \triangleq [r, \omega_1^T, y_p, \omega_2^T]$, $\theta^T = [k, \theta_1^T, \theta_0, \theta_2^T]$, ω_1 and ω_2 are $(n-1)$-dimensional vectors, Λ is an asymptotically stable matrix, (Λ, ℓ) is controllable, and θ is the control parameter vector. Hence, a constant vector θ^* exists such that when $\theta(t) \equiv \theta^*$, the transfer function of the plant together with the controller matches that of the model exactly. The problem therefore can be restated as the determination of the adaptive law for adjusting $\theta(t)$ so that $e_1(t)$ and all other signals remain bounded.

The disturbances ν_1 and ν_2 in Eq. (8.21) can be replaced by either an equivalent output disturbance ν or a vector input disturbance $\overline{\nu}$. The equations describing the overall system are similar to Eq. (5.19), except for the presence of the term $\overline{\nu}(t) \in \mathbb{R}^{3n-2}$ and are as follows.

$$\dot{x} = Ax + b(\phi^T\omega + k^*r) + \overline{\nu}(t); \qquad y_p = h^T x \tag{8.22}$$

where $x^T = [x_p^T, \omega_1^T, \omega_2^T]^T$, $h^T(sI - A)^{-1}bk^* = W_m(s)$, and $\phi = \theta - \theta^*$. As shown in Comment 8.2, $\overline{\nu}(t)$ is uniformly bounded. Since the nonminimal representation of the model is given by

$$\dot{x}_{mn} = Ax_{mn} + br; \qquad y_m = h^T x_{mn},$$

the state error $e = x - x_{mn}$ satisfies the differential equation

$$\dot{e} = Ae + b\phi^T\omega + \overline{\nu}(t); \qquad e_1 = h^T e \tag{8.23a}$$

where e_1 is the output error.

Comment 8.2 (a) Assuming that the transfer functions from the input u and the output y_p to the summer, when $\theta(t) \equiv \theta^*$, are $W_1(s)$ and $W_2(s)$ respectively, the output y_p is given by

$$y_p = W_m(s)r + \nu(t)$$

where $\nu(t) \in \mathbb{R}$ is the equivalent output disturbance and is given by

$$\nu(t) = [1 + W_1(s)]\left[W_m(s)\nu_1(t) + W_m(s)W_p^{-1}(s)\nu_2(t)\right].$$

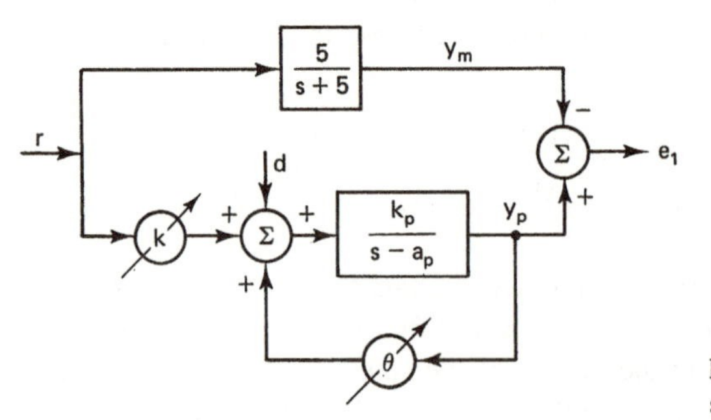

Figure 8.17 Performance of the adaptive system.

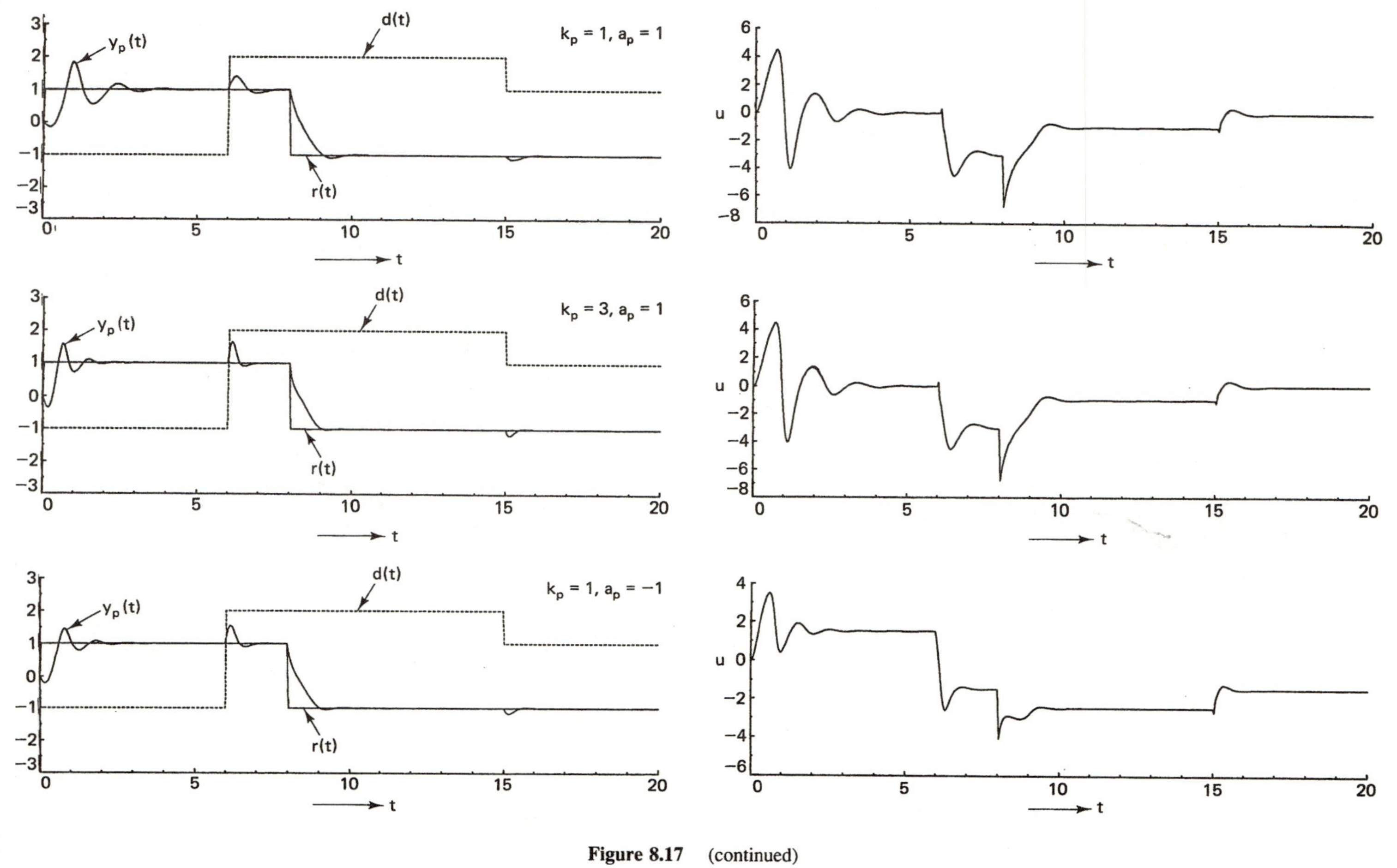

Figure 8.17 (continued)

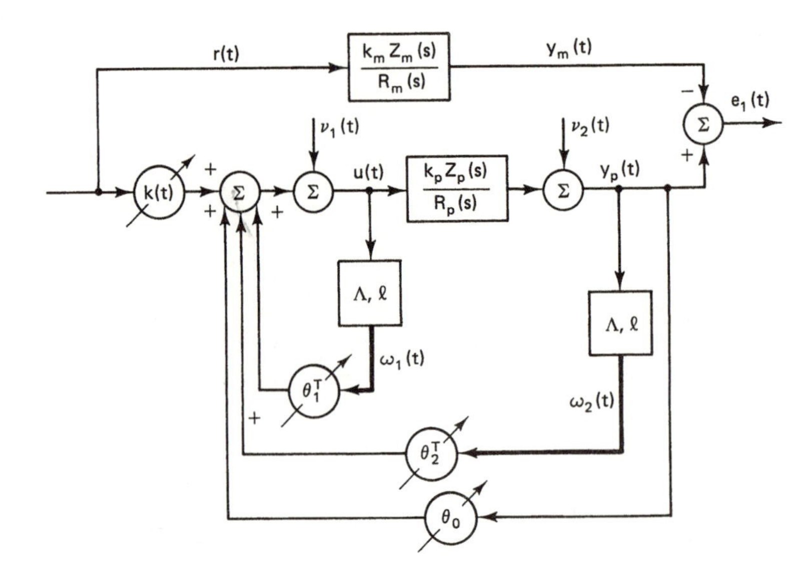

Figure 8.18 Robust adaptive control in the presence of bounded disturbances.

From the assumptions made regarding $W_p(s)$, $W_m(s)$, ν_1, and ν_2, $\nu(t)$ is bounded and differentiable.

(b) The equivalent input disturbance $\overline{\nu}$ can also be shown to be a bounded function whose elements are determined by ν and $\dot{\nu}$. For example, if $h = [1, 0, \ldots, 0]^T$,

$$\overline{\nu}(t) = [sI - A]\,[\nu(t), 0, \ldots, 0]^T.$$

Since ν is differentiable, it follows that $\overline{\nu}$ is bounded.

8.4.1 Robustness Without Persistent Excitation

As in Section 8.3, the robustness of the adaptive system can be achieved either by modifying the adaptive law or by making the reference input sufficiently persistently exciting. In this section, we discuss the former and treat directly the general case where the relative degree n^* of the plant transfer function is greater than one. We further assume, for ease of exposition, that k_p is known and $k_p = k_m = 1$. The results that follow can, however, be extended to the case when only $sgn(k_p)$ is known. For ease of notation, we define θ and ω as

$$\theta \triangleq [\theta_1^T, \theta_0, \theta_2^T]^T, \qquad \omega \triangleq [\omega_1^T, y_p, \omega_2^T]^T.$$

The adaptive law is based on the augmented error $\epsilon_1(t)$ as mentioned in Chapter 5, which is generated as follows:

$$e_2(t) = \left[\theta^T(t) W_m(s) I - W_m(s)\theta^T(t)\right] \omega(t)$$

$$\epsilon_1(t) = e_1(t) + e_2(t).$$

The augmented error satisfies the equation

$$\epsilon_1(t) = \phi^T(t)\zeta(t) + \nu(t) \tag{8.24}$$

where $\zeta(t) \triangleq W_m(s)I\omega(t)$ and $|\nu(t)| \leq \nu_0$.

Equation (8.24) forms the starting point of all the results presented in this section. The manner in which $\phi(t)$ is to be adjusted in the presence of the bounded disturbance $\nu(t)$, while assuring the boundedness of all the signals in the system, is the robust adaptive control problem in the present case. The four schemes described in Section 8.3 represent four different methods by which such control can be achieved successfully by suitably modifying the adaptive law.

In the ideal case, the adaptive law

$$\dot{\phi} = -\frac{\epsilon_1\zeta}{1+\overline{\zeta}^T\overline{\zeta}} \tag{5.52}$$

directly implies the boundedness of the parameter error vector ϕ. This, in turn, played an important role in the proof of global stability. In contrast to this, when a disturbance is present, the boundedness of ϕ can no longer be concluded from the adaptive law in Eq. (5.52). The first step in all four methods is therefore to demonstrate that ϕ is bounded.

Once the boundedness of the parameter vector is assured, the next step is to show that all the signals in the system also remain bounded. This is significantly more complex and is based on arguments involving the growth rates of signals in the system. For the first method, in which a dead zone is used in the adaptive law, the arguments follow along the same lines as in Chapter 5. For the other three methods, however, the proof is different.

(a) Use of a Dead Zone [31]. The adaptive law

$$\dot{\phi} = \begin{cases} -\dfrac{\epsilon_1\zeta}{1+\xi^T\xi} & |\epsilon_1| \geq \nu_0 + \delta \\ 0 & |\epsilon_1| < \nu_0 + \delta \end{cases} \tag{8.25}$$

where δ is an arbitrary positive constant and $\xi = [\zeta^T, \omega^T]^T$ ensures that the parameter is adjusted only when the magnitude of the augmented error is larger than a predetermined value. Note that in Eq. (8.25) as well as in the schemes that follow, the normalization factor is chosen to be different from that in Section 5.5 to facilitate a simpler proof.

The following properties of the adaptive system can be derived:

(i) The boundedness of ϕ follows directly from the adaptive law since $V(\phi)$ given by

$$V(\phi) = \frac{1}{2}\phi^T\phi,$$

has a time derivative

$$\dot{V}(\phi) = \begin{cases} -\dfrac{\phi^T\zeta(\phi^T\zeta+\nu)}{1+\xi^T\xi} & \text{if } |\epsilon_1| \geq \nu_0 + \delta \\ 0 & \text{if } |\epsilon_1| < \nu_0 + \delta \end{cases} \tag{8.26}$$

and hence is negative-semidefinite.

(ii) Since ϕ is bounded from (i), the same arguments as in the ideal case (Section 5.5) can be used to show that

$$\sup_{\tau \leq t} |y_p(\tau)| \sim \sup_{\tau \leq t} \|\omega_2(\tau)\| \sim \sup_{\tau \leq t} \|\omega(\tau)\| \sim \sup_{\tau \leq t} \|\zeta(\tau)\|. \tag{8.27}$$

if the signals in the adaptive loop grow in an unbounded fashion.

(iii) When $|\phi^T \zeta + \nu| > \nu_0 + \delta$, it follows that

$$\frac{\delta}{\nu_0 + \delta} |\phi^T \zeta + \nu| \leq |\phi^T \zeta| \leq \frac{2\nu_0 + \delta}{\nu_0 + \delta} |\phi^T \zeta + \nu|.$$

Hence, from Eq. (8.25), we have

$$\begin{aligned} \int_{t_0}^{\infty} \dot{\phi}^T \dot{\phi}\, dt &= \int_{\Omega_1} \frac{(\phi^T \zeta + \nu)^2 (\zeta^T \zeta)}{(1 + \xi^T \xi)^2} dt \\ &\leq \frac{\nu_0 + \delta}{\delta} \int_{\Omega_1} \frac{|\phi^T \zeta|\, |\phi^T \zeta + \nu|}{1 + \xi^T \xi} dt \end{aligned} \tag{8.28}$$

where $\Omega_1 = \{t |\; |\epsilon_1| > \nu_0 + \delta\}$. Since $\int_{t_0}^{\infty} \dot{V}\, dt < \infty$, it follows from Eqs. (8.26) and (8.28) that $\dot{\phi} \in \mathcal{L}^2$. From the same arguments used in Chapter 5, it follows that

$$\|\omega_2(t)\| = o \left[\sup_{\tau \leq t} \|\omega(t)\| \right] + \overline{y}(t) \tag{8.29}$$

where $\overline{y}(t)$ is a bounded signal. Hence $\omega_2(t)$, and consequently by Eq. (8.27), all other signals in the system are bounded.

As in the ideal case, the boundedness of ϕ and the fact that $\dot{\phi} \in \mathcal{L}^2$ are sufficient to establish the robustness of the adaptive system.

(b) Bound on $\|\theta^*\|$ [16]. In the previous approach, adaptation was terminated when the augmented error was small. In contrast to that, as stated in Section 8.3 for the scalar case, the search region in the parameter space is constrained in this approach. It is assumed that prior information is available regarding the domain S in which θ^* can lie. This is quantitatively stated as

$$S \triangleq \{\theta |\; \|\theta\| \leq \theta^*_{\max}\}.$$

where $\theta^*_{\max}$ is a positive constant. The adaptive law now takes the form

$$\dot{\phi} = -\frac{\epsilon_1 \zeta}{1 + \xi^T \xi} - \theta f(\theta) \tag{8.30}$$

where

$$f(\theta) = \begin{cases} \left(1 - \dfrac{\|\theta\|}{\theta^*_{\max}}\right)^2 & \text{if } \|\theta\| > \|\theta^*\|_{\max} \\ 0 & \text{elsewhere.} \end{cases} \tag{8.31}$$

The important feature of Eq. (8.30) is that when $\theta(t)$ is in the interior of S, the adaptive law is identical to that used in the ideal case (refer to Chapter 5). Hence, in the presence of the disturbance $\nu(t)$, adaptation never ceases and $\theta(t)$ does not converge to a constant value.

The following relations derived using the adaptive law in Eq. (8.30) are sufficient to prove the boundedness of all the signals in the system. In Sections 8.4.1 (c) and (d), it will be seen that similar relations between the augmented error ϵ_1 and the (equivalent) state ξ of the system are obtained when the σ-modification and e_1-modification schemes are used. Since the proofs for demonstrating the boundedness of the signals in the system, based on these relations, are similar, we shall comment on all of them toward the end of this section.

(i) With a scalar function $V = \phi^T\phi/2$, the time derivative $\dot{V}$ along Eq. (8.30) is obtained as

$$\dot{V} = -\frac{\epsilon_1(\epsilon_1 - \nu)}{1 + \xi^T\xi} - \phi^T\theta f(\theta).$$

from which the boundedness of $\|\phi\|$ follows directly.

(ii) Integrating $\dot{V}$ over an interval $[t_1, t_2]$, we have

$$\int_{t_1}^{t_2} \left[\frac{\epsilon_1^2}{1 + \xi^T\xi} + \phi^T\theta f(\theta)\right] dt = V(t_1) - V(t_2) + \int_{t_1}^{t_2} \frac{\epsilon_1 \nu}{1 + \xi^T\xi} dt$$

Since

$$\frac{|\epsilon_1 \nu|}{\sqrt{1 + \xi^T\xi}} \leq \nu_0 \left(\|\phi\| + \nu_0\right),$$

it follows that

$$\int_{t_1}^{t_2} \left[\frac{\epsilon_1^2}{1 + \xi^T\xi} + \phi^T\theta f(\theta)\right] dt \leq c_0 + c_1 \int_{t_1}^{t_2} \frac{1}{\sqrt{1 + \xi^T\xi}} dt$$

where c_0 and c_1 are constants.

(iii) From Eq. (8.30), it follows that

$$\|\dot{\phi}\| < \frac{c_2|\epsilon_1|}{\sqrt{1 + \xi^T\xi}} + c_3 f(\theta)$$

where c_2 and c_3 are positive constants.

(c) The σ-modification scheme [10]. When $n^* \geq 2$, the adaptive law using the σ-modification has the form [1]

$$\dot{\phi} = -\frac{\epsilon_1 \zeta}{1+\xi^T\xi} - \frac{\sigma\theta}{1+\xi^T\xi} \qquad \sigma > 0 \tag{8.32}$$

for the general problem when $n^* \geq 2$. The following three relations, similar to (i)-(iii) in scheme (b) can be derived:

(i) With $V(\phi) = \phi^T\phi/2$, we have

$$\begin{aligned}
\dot{V}(\phi) &= -\frac{\epsilon_1(\epsilon_1 - \nu)}{1+\xi^T\xi} - \frac{\sigma\phi^T\theta}{1+\xi^T\xi} \\
&\leq -\frac{1}{1+\xi^T\xi}\left[\sigma\phi^T\phi - \sigma\|\phi\|\,\|\theta^*\| + \epsilon_1^2 - \nu_0|\epsilon_1|\right] \\
&< 0 \qquad \forall\ \|\phi\| > \|\theta^*\| + \frac{\nu_0}{2\sqrt{\sigma}},\ |\epsilon_1| > \nu_0 + \sqrt{\sigma}\|\theta^*\|/2.
\end{aligned}$$

This assures the boundedness of ϕ.

(ii) Integrating $\dot{V}$ over $[t_1, t_2]$, it follows that

$$\int_{t_1}^{t_2}\left[\frac{\epsilon_1^2}{1+\xi^T\xi}\right]dt \leq c_0 + c_1\int_{t_1}^{t_2}\frac{1}{\sqrt{1+\xi^T\xi}}dt$$

for some constants c_0 and c_1 in $\mathbb{R}$.

(iii) As in (b), we also have

$$\|\dot{\phi}\| < \frac{c_2|\epsilon_1|}{\sqrt{1+\xi^T\xi}} + \frac{c_3}{1+\xi^T\xi}$$

where c_2 and c_3 are constants in $\mathbb{R}^+$.

(d) The e_1-modification scheme [24]. As in (c), the adaptive law using the e_1-modification scheme discussed for the scalar case can be extended to the case where ϕ is a vector. The adaptive law has the form

$$\dot{\phi} = -\frac{\epsilon_1\zeta}{1+\xi^T\xi} - \gamma\frac{|\epsilon_1|\theta}{1+\xi^T\xi} \qquad \gamma > 0. \tag{8.33}$$

---

[1] Equation (8.32) differs from that in [10], which has a somewhat different structure. However, our discussions apply to the latter scheme as well.

The relations (i)-(iii) derived in (c) also hold when Eq. (8.33) is used.

As mentioned earlier in this section, the relations (i)-(iii) stated in (b) and (c), which are very similar, are adequate to establish the boundedness of all the signals in the adaptive system. The important point is that the error equation (8.24), the error models resulting from the use of adaptive laws in Eqs. (8.25), (8.30), (8.32), or (8.33) as well as the stability analysis used, find application in more general adaptive control problems than that considered here. The robust control problem in the presence of time-varying parameters discussed in Section 8.6 and the reduced order controller problem in Section 8.7 are cases in point. It is therefore desirable to deal with them in a unified fashion. In Appendix D, a detailed proof of the boundedness of all the signals in the adaptive system is given when general relations of the form (i)-(iii) are valid. These can be modified suitably to apply to all the cases considered in Sections 8.4, 8.6 and 8.7. We merely comment here on the main features of the proof and indicate how it differs from that given in Chapter 5 for the ideal case.

When external perturbations are not present, the choice of the adaptive law is based on the existence of a Lyapunov function for the overall system. This assures that $\phi \in \mathcal{L}^\infty$ and $\dot{\phi} \in \mathcal{L}^2$ which, in turn, are used to prove the boundedness of all the signals. When the adaptive law is modified to have a dead zone to cope with external disturbances, the same arguments as in the ideal case can be used to show that $\phi \in \mathcal{L}^\infty$, $\dot{\phi} \in \mathcal{L}^2$ and that all the signals remain bounded. In sections (b), (c), and (d), the boundedness of ϕ can be established [refer to (i)] but $\dot{\phi}$ cannot be shown to belong to $\mathcal{L}^2$. Conditions (ii) and (iii) however can be considered as statements equivalent to the latter. Qualitatively, the boundedness of signals can be proved as follows: Since ϕ is bounded, if $||\xi(t)||$ is assumed to grow without bound, it can grow at most exponentially. Condition (ii) indicates that if $||\xi(t)||$ is sufficiently large over an interval $[t_1, t_2]$, then $\int_{t_1}^{t_2} ||\dot{\phi}||dt < \infty$, and therefore as the length of the interval increases, $||\dot{\phi}||$ must be small over most of the interval. When $||\phi||$ is small, the error signal $\phi^T\omega$ fed back into the plant is small and, hence, the solutions of the adaptive system, or equivalently $||\xi(t)||$, decrease exponentially. This enables us to conclude that during an interval of length T, $||\xi(t)||$ can grow exponentially over a set of finite measure T_1 and decay exponentially over a set of measure $T - T_1$. Therefore, by choosing T and, hence, $T - T_1$ to be arbitrarily large, $||\xi(t)||$ can be shown to violate the condition that it is bounded below over the entire interval $[t, t+T]$. The boundedness of all the signals in the system follows.

8.4.2 Robustness with Persistent Excitation

When the degree of persistent excitation of the input is greater than the amplitude of the disturbance, it was shown in Section 8.3 to result in the global boundedness of all the signals for a first-order plant. In this section, we show that a similar result can be derived for an nth order plant ($n > 1$) as well.

We recall that with the controller as in Chapter 5, the error equations resulting from the overall system can be derived as

$$\dot{e} = Ae + b\phi^T\omega + \overline{\nu}; \qquad e_1 = h^T e. \tag{8.23a}$$

If ω^* is the signal corresponding to ω when $\theta(t) \equiv \theta^*$, then we have $\omega = \omega^* + Ce$ where C is a constant matrix defined as

$$C = \begin{bmatrix} 0 & 0 & 0 \\ 0 & I & 0 \\ h_p^T & 0 & 0 \\ 0 & 0 & I \end{bmatrix}$$

and I is an $(n-1) \times (n-1)$ identity matrix. This results in the nonlinear differential equation

$$\begin{aligned} \dot{e} &= Ae + b\phi^T(\omega^* + Ce) + \overline{\nu}; \qquad e_1 = h^T e \\ \dot{\phi} &= -sgn(k_p)e_1(\omega^* + Ce) \end{aligned} \tag{8.34}$$

where $h^T(sI - A)^{-1}bk^* = W_m(s)$, $k^* = k_m/k_p$, and $W_m(s)$ is the transfer function of the reference model. Hence, the problem is to determine sufficient conditions under which the output of a nonlinear time-varying system is bounded for a given bounded input $\overline{\nu}(t)$ with $|\overline{\nu}(t)| \leq \overline{\nu}_0$. In what follows, the results are derived for the case when the relative degree of the plant is unity. The principal result is expressed in Theorem 8.2 [25].

Since $W_m(s)$ can be chosen to be strictly positive real when $n^* = 1$, given a symmetric positive-definite matrix Q, a matrix $P = P^T > 0$ exists such that $A^TP + PA = -Q$, $Pb = h$. Let λ_P and λ_Q be the maximum eigenvalue of P and minimum eigenvalue of Q, respectively. For ease of exposition, we assume that $sgn(k_p) = 1$.

Theorem 8.2 Let the $2n$–dimensional vector ω^* be persistently exciting and satisfy the condition in Eq. (6.11) for constants T_0, δ_0, and ϵ_0. Then all solutions of Eq. (8.34) are bounded if

$$\epsilon_0 \geq \gamma\overline{\nu}_0 + \delta, \qquad \gamma = \frac{2\lambda_P}{\lambda_Q} \tag{8.35}$$

and δ is an arbitrary positive constant.

Proof. Let $z^T \triangleq [e^T\sqrt{P},\ \phi^T]$. Then a quadratic function $W(z) = z^Tz + \phi^T\phi$ has a time derivative $\dot{W}(z) = -e^TQe + 2e^TP\overline{\nu}$. Defining the region $D = \{z|\ \ \|e\| \leq \gamma\overline{\nu}_0\}$, it follows that $W(z)$ can increase only in D and the maximum increase in $W(z)$ over any period T_1 is given by

$$W(z(t_0 + T_1)) - W(z(t_0)) \leq \frac{\gamma^2\overline{\nu}_0^2}{4}\lambda_Q T_1. \tag{8.36}$$

We consider the following two cases:

$$\text{Case (i)} \qquad \left|\frac{1}{T_0}\int_{t_2}^{t_2+\delta_0}(\omega^*(\tau) + Ce(\tau))^T w\, d\tau\right| \geq k\epsilon_0,\ k = \frac{\delta}{\gamma\overline{\nu}_0 + \delta} \tag{8.37}$$

If the condition in Eq. (8.37) is satisfied for any unit vector $\mathrm{w} \in \mathbb{R}^{2n}$, then $\omega = \omega^* + Ce$ is persistently exciting and Eq. (8.34) represents an exponentially stable system [23] with a bounded input $\overline{\nu}$. Hence, the system

$$\dot{z} = J(t)z + \mu(t) \tag{8.38}$$

is also exponentially stable with a bounded input μ where

$$J(t) \triangleq \begin{bmatrix} \sqrt{P}A\sqrt{P}^{-1} & \sqrt{P}b\omega^T(t) \\ -\omega(t)h^T\sqrt{P}^{-1} & 0 \end{bmatrix}, \quad \mu(t) \triangleq \begin{bmatrix} \sqrt{P} \\ 0 \end{bmatrix} \overline{\nu}(t).$$

From the results of [23], or equivalently, Theorem 2.17, it follows that for the unforced system $\dot{z} = J(t)z$, $W(z(t_0 + T_0)) \leq c^2 W(z(t_0))$ for all t_0, where $c < 1$. Equivalently, if $\Phi(t_0 + T_0, t_0)$ is the transition matrix of $\dot{z} = J(t)z$, $\|\Phi(t_0 + T_0, t_0)\| \leq c$ for all t_0. Hence, for the forced system in Eq. (8.38),

$$\|z(t_0 + T_0)\| \leq c\|z(t_0)\| + k_1\overline{\nu}_0$$

where k_1 is a constant. If $c_1 > k_1\overline{\nu}_0/(1-c)$, then

$$\|z(t_0)\| \geq c_1 \quad \Rightarrow \quad \|z(t_0 + T_0)\| < \|z(t_0)\|$$

or

$$W(z(t_0 + T_0)) < W(z(t_0)) \text{ if } W(z(t_0)) \geq c_1^2/2.$$

Case (ii)
$$\left| \frac{1}{T_0} \int_{t_2}^{t_2+\delta_0} (\omega^*(\tau) + Ce(\tau))^T \mathrm{w}\, d\tau \right| < k\epsilon_0 \tag{8.39}$$

If the condition in Eq. (8.37) is not satisfied, then Eq. (8.39) is satisfied for some constant unit vector $\mathrm{w} \in \mathbb{R}^{2n}$, that is, ω is not persistently exciting over the interval $[t_0, t_0 + T_0]$ in some direction w. As shown below, this once again assures that $\|z(t_0 + T_0)\| < \|z(t_0)\|$ for any initial condition.

From the inequality in Eq. (8.35) and the choice of k, it follows that $(1-k)\epsilon_0 \geq \gamma\overline{\nu}_0$. Since ω^* satisfies the inequality in Eq. (6.11), we have

$$\left| \int_{t_2}^{t_2+\delta_0} [Ce(\tau)]^T \mathrm{w}\, d\tau \right| \geq (1-k)\epsilon_0 T_0.$$

Since $\dot{W} = -e^T Qe + 2e^T P\overline{\nu}$, the decrease in $W(z)$ over the interval $[t_2, t_2 + \delta_0]$ satisfies the inequality

$$\begin{aligned} W(z(t_2)) - W(z(t_2 + \delta_0)) &\geq \lambda_Q \textstyle\int_{t_2}^{t_2+\delta_0} \|e(\tau)\|^2 d\tau - 2\lambda_P \overline{\nu}_0 \textstyle\int_{t_2}^{t_2+\delta_0} \|e(\tau)\| d\tau \\ &\geq \frac{\lambda_Q}{\delta_0} \int_{t_2}^{t_2+\delta_0} \|e(\tau)\| d\tau \left\{ \int_{t_2}^{t_2+\delta_0} \|e(\tau)\| d\tau - \gamma\overline{\nu}_0\delta_0 \right\} \\ &\geq \gamma^2\overline{\nu}_0^2 (T_0 - \delta_0)\lambda_Q \end{aligned}$$

by the choice of k. Since the minimum decrease in $W(z(t))$ over the interval $[t_2, t_2+\delta_0]$ is greater than the maximum increase in $W(z(t))$ over a period $(T_0 - \delta_0)$, it follows that $W(z(t_0+T_0)) < W(z(t_0))$ for any $W(z(t_0))$.

Since cases (i) and (ii) represent the only two possibilities over any interval $[t_0, t_0+T_0]$, we obtain that

$$W(z(t_0)) \geq c_1^2/2 \Rightarrow W(z(t_0+T_0)) < W(z(t_0)),$$

which implies that all solutions of Eq. (8.34) are uniformly bounded.

Sufficient conditions for boundedness of solutions are stated in the theorem above in terms of the degree of persistent excitation of ω^*. For design purposes, it is more desirable to express them in terms of the reference input r. From the results of Chapter 6, it follows that by choosing a reference input r with n distinct frequencies, ω^* can be made persistently exciting. The theorem above implies that a reference input r must be chosen so that ω^* is persistently exciting with a degree of PE large compared with the amplitude of the disturbance. Very little is currently known regarding the relationships between the degree of persistent excitation of inputs and outputs of linear systems. Such relationships must be better understood before the theorem above can be used in design.

Theorem 8.2 applies to systems with relative degree unity. Similar results have been established for discrete-time plants with arbitrary relative degree in [19].

8.4.3 Comments on Adaptive Algorithms for Bounded Disturbances

1. The schemes (a) and (c) in Section 8.4.1 are not u.a.s. when no external disturbances are present. Scheme (b) assures uniform asymptotic stability if the reference input is persistently exciting and scheme (d) assures uniform asymptotic stability if the reference input has a sufficiently large degree of PE. Additional prior information regarding bounds on the external disturbance and the parameter vector θ^* is needed to implement schemes (a) and (b) respectively. In contrast to this, such information is not needed for the implementation of schemes (c) and (d).

2. The scheme (a) using a dead zone is a conservative approach in which adaptation is stopped when the error becomes small. In addition, with such a dead zone, the output and parameter errors may not tend to zero even when the disturbance is removed. If a time-varying bound can be specified for the disturbance, a corresponding time-varying dead zone may be incorporated in the adaptive law, which will lead to smaller errors in the adaptive system.

3. The sufficient condition given in Section 8.4.2 for the boundedness of solutions in the presence of bounded disturbances is expressed in terms of the degree of PE of ω^*. The latter is determined by the reference input r, the frequency responses of the reference model, and the fixed control parameters (Λ, ℓ) in the adaptive loop. Hence, all the conditions above are important factors in the design of efficient adaptive controllers.

8.5 HYBRID ADAPTIVE CONTROL

Adaptive algorithms of the form described in the preceding chapters have been developed for the stable identification and control of continuous time as well as discrete time plants with unknown parameters. In continuous adaptive systems, the plant operates in continuous time and the controller parameters are adjusted continuously. Similarly, in discrete adaptive systems, the plant is modeled in discrete time and the various signals, as well as the control parameter vector, are defined at discrete instants. However, practical systems, for a variety of reasons, may contain both discrete and continuous elements. It may be desirable to make infrequent adjustments of control parameters at discrete instants, even as continuous signals are being processed in real time, making the overall system *hybrid*. The term was first introduced in the adaptive literature by Gawthrop in [7]. Even in purely discrete time systems, the adjustment of the adaptive control parameters may be carried out at rates significantly slower than the rate at which the system operates. The latter is referred to in the literature as *block processing*. Whether a discrete updating algorithm is used in a continuous or discrete plant, an important consideration is whether the overall system will have bounded solutions. In this section, we shall discuss systems of both types that are stable [27]. One attractive feature of such algorithms is their robustness. When there are no disturbances present in the system, the convergence of the parameter vector to its true value can be made arbitrarily fast by increasing the gain or the persistent excitation of the reference input. In the presence of bounded disturbances, it is seen that for a specific degree of persistent excitation, better robustness properties are achieved. From a practical viewpoint, the use of digital controllers for both continuous and discrete systems is inevitable, and the analysis of hybrid systems provides the theoretical basis for the use of such controllers.

In Section 8.5.1, we consider the three error models discussed in Chapter 7 and develop discrete adaptive laws in each case that assure the boundedness of the parameter error vector. Conditions under which the output and parameter error vectors tend to zero asymptotically are also derived. In Section 8.5.2, these error models are applied to the adaptive control problem. We show that the overall system is globally stable in the absence of disturbances. When there are external disturbances, the hybrid adaptive system is shown to be robust in the presence of persistently exciting signals.

8.5.1 Hybrid Error Models

The dynamical systems discussed in this section are continuous-time systems in which $t \in \mathbb{R}^+$, the set of all positive numbers. The quantities $u(t)$ and $e_1(t)$ are continuous time signals defined as $u : \mathbb{R}^+ \to \mathbb{R}^n$, and $e_1 : \mathbb{R}^+ \to \mathbb{R}$. The signal $\phi(t)$ is defined as

$$\phi(t) \stackrel{\triangle}{=} \phi_k \qquad t \in [t_k, t_{k+1}], \;\; k \in N$$

where N is the set of nonnegative integers, t_k is a monotonic unbounded sequence and ϕ_k is a constant over the interval $[t_k, t_{k+1}]$. The period $T_k = t_{k+1} - t_k$ with $0 < T_{\min} \leq T_k \leq T_{\max} < \infty$. Since $\phi(t)$ is a constant over any interval $[t_k, t_{k+1}]$, it implies that the parameter vector $\theta(t)$ is held constant with a value θ_k over the same interval and updated to a constant value θ_{k+1} at time t_{k+1}. The rules (adaptive laws) for adjusting $\phi(t)$, which are defined in discrete time, make the overall error model hybrid in nature.

The error models discussed here are the hybrid counterparts of the error models discussed in Chapter 7 and the behavior of each error model is analyzed for the three specific cases: (i) when $u \in \mathcal{L}^\infty$, (ii) when $u \in \Omega_{(n,T)}$, and (iii) when u is unbounded and $\in \mathcal{PC}_{[0,\infty)}$.

Error Model 1. As in the continuous-time case, the error model is given by

$$e_1(t) \quad = \quad \phi_k^T u(t)$$

where ϕ, e_1, and u represent, respectively, the parameter error, the output error, and a vector of inputs that can be measured at each instant. If

$$\Delta\phi_k \quad \stackrel{\Delta}{=} \quad \phi_{k+1} - \phi_k \;=\; \theta_{k+1} - \theta_k$$

can be adjusted at $t = t_{k+1}$, the objective is to determine an adaptive law for choosing the sequence $\{\Delta\phi_k\}$ using all available input-output data so that $\lim_{t\to\infty} e_1(t) = 0$. Two different approaches are presented below based on the manner in which ϕ_k is adjusted.

Approach 1. Let the adaptive law be given by the equation

$$\Delta\phi_k \quad = \quad -\frac{1}{T_k}\int_{t_k}^{t_{k+1}} \frac{e_1(t)u(t)}{1+u^T(t)u(t)}dt.$$

Consider the Lyapunov function candidate $V(k) = \phi_k^T\phi_k/2$. Then $\Delta V(k) = V(k+1) - V(k)$ is given by

$$\begin{aligned} \Delta V(k) \quad &= \quad \left[\phi_k + \frac{\Delta\phi_k}{2}\right]^T \Delta\phi_k \\ &= \quad -\frac{1}{2}\phi_k^T\left[2I - R_k\right]R_k\phi_k \end{aligned}$$

where R_k is the symmetric positive-semidefinite matrix

$$R_k \quad = \quad \frac{1}{T_k}\int_{t_k}^{t_{k+1}} \frac{u(\tau)u^T(\tau)}{1+u^T(\tau)u(\tau)}d\tau$$

with all its eigenvalues within the unit circle. Since $[2I - R_k] > \gamma I$ for some constant $\gamma > 0$,

$$\Delta V(k) \quad < \quad -\frac{1}{2}\gamma\phi_k^T R_k\phi_k \leq 0.$$

Hence, $V(k)$ is a Lyapunov function and assures the boundedness of ϕ_k. Further, since $\{\Delta V(k)\}$ is a nonnegative sequence with

$$\sum_{k=0}^{\infty} \Delta V(k) \quad < \quad \infty,$$

it follows that $\Delta V(k) \to 0$ as $k \to \infty$ or alternately, $\phi_k^T R_k \phi_k \to 0$ as $k \to \infty$. This can also be expressed as

$$\frac{1}{T_k}\int_{t_k}^{t_{k+1}} \frac{e_1^2(\tau)}{1+u^T(\tau)u(\tau)} d\tau \to 0 \qquad \text{as } k \to \infty.$$

The following arguments indicate that the adjustment of the control parameter vector along an average gradient will also result in perfect model following. As in Chapters 5 and 7, it is shown in [26] that

Case (i): If $u \in \mathcal{L}^\infty$, then $e_1 \in \mathcal{L}^2$. If in addition, $\dot{u}$ is also bounded, then $\lim_{t\to\infty} e_1(t) = 0$.

Case (ii): If $u \in \Omega_{(n,T_{\min})}$, then the matrix R_k is positive definite and, hence, the parameter error vector tends to zero as $k \to \infty$.

Case (iii): If $u \in \mathcal{PC}_{[0,\infty)}$, we can only conclude that

$$\frac{e_1(t)}{1+u^T(t)u(t)} = \beta(t) \qquad \beta \in \mathcal{L}^2$$

If in addition $\|\dot{u}\| \leq M_1\|u\| + M_2$ where $M_1, M_2 \in \mathbb{R}^+$, we can also conclude that

$$e_1(t) = o[\sup_{\tau \leq t} \|u(\tau)\|]. \tag{8.40}$$

This fact is used in the hybrid adaptive control problem to show that the error tends to zero asymptotically.

Approach 2. In the second approach, the error equations are integrated over a finite inverval $[t_k, t_{k+1}]$. Since the parameter error is assumed to be a constant over this interval, we obtain

$$\phi_k^T \int_{t_k}^{t_{k+1}} e_1(\tau)u(\tau)d\tau = \int_{t_k}^{t_{k+1}} e_1^2(\tau)d\tau$$

or equivalently the discrete error model

$$\phi_k^T \omega_k = \epsilon_k \tag{8.41}$$

where

$$\epsilon_k \triangleq \int_{t_k}^{t_{k+1}} e_1^2(t)dt \qquad \text{and} \qquad \omega_k \triangleq \int_{t_k}^{t_{k+1}} e_1(t)u(t)dt.$$

For the error model in Eq. (8.41), the adaptive law for updating ϕ_k can be derived as [28]

$$\Delta\phi_k = -\frac{\epsilon_k \omega_k}{1+\omega_k^T \omega_k}.$$

The error model in Eq. (8.41) is the discrete counterpart of error model 1 discussed in Chapter 7 and has been well analyzed in the adaptive literature. The results are found to be very similar to those obtained using error model 1 and can be stated as follows: (i) ϕ_k is bounded, (ii)$\Delta\phi_k \to 0$ as $k \to \infty$, and (iii) $|\epsilon_k| = o[\{1+\omega_k^T\omega_k\}^{1/2}]$ or ϵ_k grows more slowly than ω_k.

Once again, using the same methods as described in earlier chapters, it is shown in [26] that

Case (i): If $u, \dot{u} \in \mathcal{L}^\infty$, then $\lim_{t\to\infty} e_1(t) = 0$.

Case (ii): If $u \in \Omega_{(n,T_{\min})}$, then

$$P_k \triangleq \int_{t_k}^{t_{k+1}} u(\tau)u^T(\tau)d\tau$$

is uniformly positive-definite so that $\phi_k \to 0$ as $k \to \infty$.

Case (iii): If $u \in \mathcal{PC}_{[0,\infty)}$ and $\|\dot{u}\| \leq M_1\|u\| + M_2$, $M_1, M_2 \in \mathbb{R}^+$, then

$$e_1(t) = o[\sup_{\tau \leq t} \|u(\tau)\|]. \tag{8.42}$$

The relation in Eq. (8.42) is used in adaptive control problems to demonstrate the boundedness of all the signals in the controlled system. In [26] it is further shown that similar results can be established for error model 3 discussed in Chapter 7. These results indicate that hybrid adaptation (or block processing if the original system is discrete) can be used in all the adaptive schemes without fear of instability. Although they can be used to increase the rate of convergence in disturbance free situations, their practical importance arises from their robustness in the presence of disturbances.

8.5.2 Hybrid Adaptive Control

In Chapter 5, continuous adaptive laws were derived for the adjustment of the control parameter vector $\theta(t)$ to assure global stability in the ideal case. Using the same controller structure and the error models described in Section 8.5.1, it can be shown that the parameter vector $\theta(t)$ can also be adjusted using a hybrid algorithm while assuring global stability of the overall system. In such a case $\theta(t)$ is held constant over intervals of length T and updated at the ends of the intervals. In the following paragraphs, an outline of the proof of stability for the case when k_p is known is given. When k_p is not known, $sgn(k_p)$ and an upper bound on $|k_p|$ are needed to generate a stable adaptive law.

With the plant, reference model, controller structure, and augmented error chosen as in Section 8.4, we obtain the error model

$$\epsilon_1(t) = \phi^T(t)\zeta(t).$$

From the discussion of hybrid error models in Section 8.5.1, we note that using Approach 1, $\phi(t)$ can be adjusted at $t = t_{k+1}$ as

$$\Delta\phi_k = -\frac{1}{T_k}\int_{t_k}^{t_{k+1}} \frac{\epsilon_1(\tau)\zeta(\tau)}{1+\xi^T(\tau)\xi(\tau)}\,d\tau \tag{8.43}$$

where ξ is defined as in Section 8.4. Alternately, using Approach 2, the adaptive law is obtained as

$$\Delta\phi_k = -\frac{\int_{t_k}^{t_{k+1}} \epsilon_1^2(\tau)\, d\tau \int_{t_k}^{t_{k+1}} \epsilon_1(\tau)\zeta(\tau)\, d\tau}{1 + \int_{t_k}^{t_{k+1}} \epsilon_1(\tau)\zeta^T(\tau)\, d\tau \int_{t_k}^{t_{k+1}} \epsilon_1(\tau)\zeta(\tau)\, d\tau}. \tag{8.44}$$

Proof of stability using Approach 1. The proof of boundedness of all signals follows along the same lines as in Chapter 5 and can be summarized as follows: Let the signals grow in an unbounded fashion. Since $\epsilon_1(t) = \phi_k^T \zeta(t)$ over the interval $[t_k, t_{k+1}]$ and the adaptive law is given by Eq. (8.43), it follows that ϕ_k is bounded (from Section 8.5.1). Hence the state of the system can grow at most exponentially and as in Chapter 5, we can conclude that

$$\sup_{\tau\leq t} |y_p(\tau)| \sim \sup_{\tau\leq t} \|\omega_2(\tau)\| \sim \sup_{\tau\leq t} \|\omega(\tau)\| \sim \sup_{\tau\leq t} \|\zeta(\tau)\|. \tag{8.45}$$

From Eq. (8.40), we obtain that

$$\epsilon_1(t) = o[\sup_{\tau\leq t} \|\xi(\tau)\|].$$

Since $\lim_{k\to\infty} \Delta\phi_k = 0$, we can show that

$$e_2(t) = o[\sup_{\tau\leq t} \|\omega(\tau)\|].$$

Hence,

$$\begin{aligned} y_p &= y_m + e_1 = y_m + o[\sup_{\tau\leq t} \|\xi(\tau)\|] + o[\sup_{\tau\leq t} \|\omega(\tau)\|] \\ &= y_m + o[\sup_{\tau\leq t} \|\omega(\tau)\|]. \end{aligned}$$

This contradicts Eq. (8.45) and the assumption that all the signals grow in an unbounded fashion.

Proof of stability using Approach 2. With the adaptive law in Eq. (8.44), it once again follows that

(i) ϕ_k is bounded, and

(ii) $\epsilon_1(t) = o[\sup_{\tau\leq t} \|\zeta(\tau)\|] = o[\sup_{\tau\leq t} \|\omega(\tau)\|]$.

From (i), we obtain that y_p and ω grow at the same rate, while from (ii), we have the result that the same signals grow at different rates. This leads to a contradiction if we assume that the signals grow in an unbounded fashion.

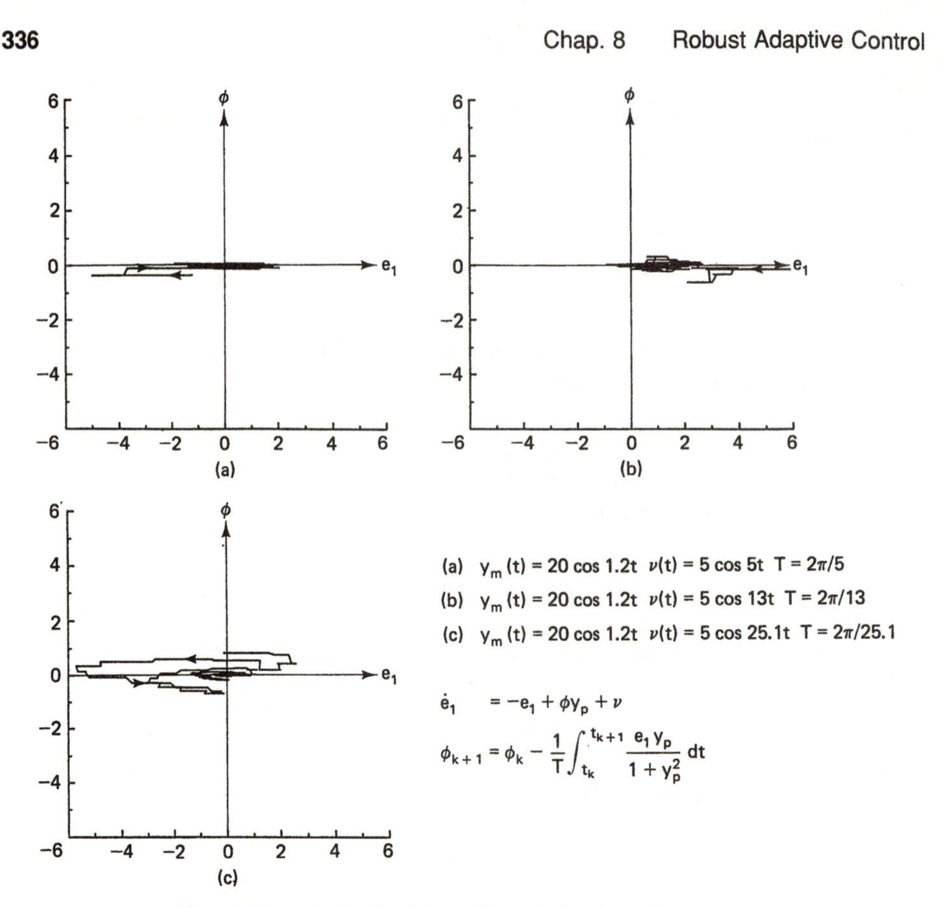

Figure 8.19 Hybrid algorithms: Effect of disturbance frequencies.

8.5.3 Hybrid Adaptive Control in the Presence of Bounded Disturbances

The importance of PE of the reference input in the context of robustness was discussed in detail in Section 8.4. The simulation results included in that section clearly revealed that the magnitude of the limit set depends on a number of factors such as the frequency content of both the reference input as well as the disturbance. If, in the same system, the control parameters are adjusted using a hybrid algorithm, the limit set is found to be substantially smaller in general, indicating improved performance. These are indicated in Figs. 8.19 and 8.20.

For continuous-time systems the result contained in Theorem 8.2 has not been extended to systems with relative degree greater than unity. However, it has been proved for discrete systems in [19]. This, in turn, makes the validity of Theorem 8.2 to general continuous-time systems of arbitrary order and relative degree more plausible.

It is our view that hybrid adaptation is always preferable to pointwise adaptation as far as robustness is concerned. The length of the interval over which the parameters are held constant should be chosen judiciously to trade off speed of adaptation against

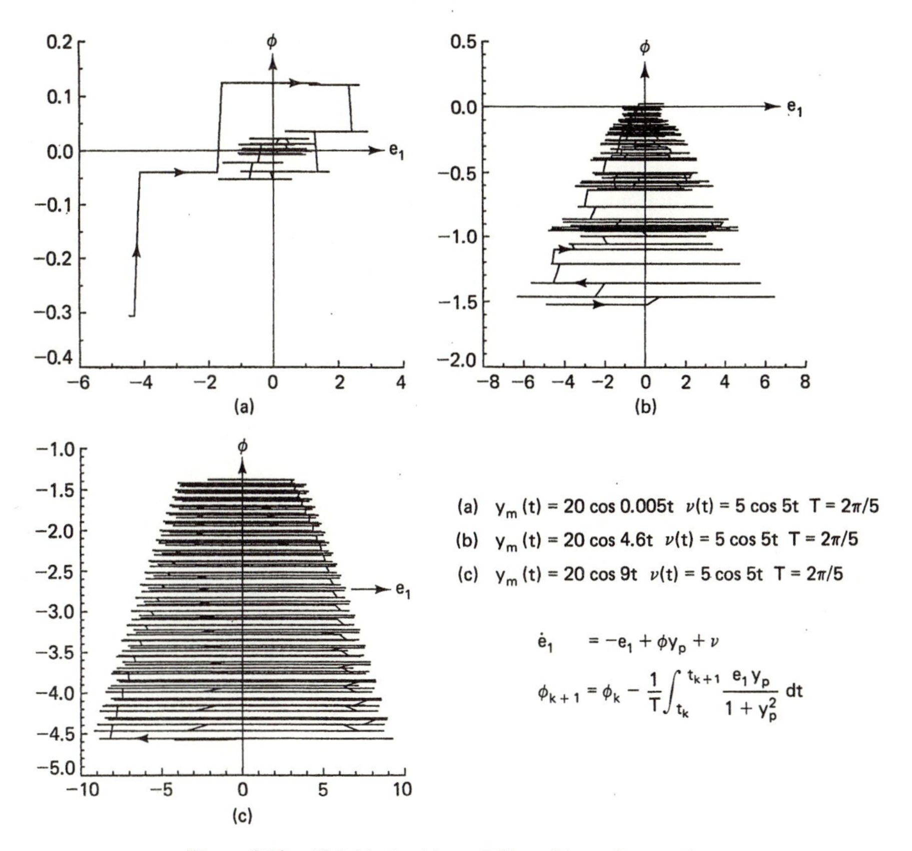

Figure 8.20 Hybrid algorithms: Effect of input frequencies.

the accuracy desired. It has been shown in [4] that even the length of the interval T_k can be adjusted adaptively to improve performance.

8.6 ADAPTIVE CONTROL OF TIME-VARYING PLANTS

One of the compelling reasons for considering adaptive methods in practical applications, as mentioned in Chapter 1, is to compensate for large variations in plant parameter values. Experience with adaptive systems has shown that in many cases where plant parameters vary slowly, the control parameters adjust themselves, using measured data, to meet performance specifications. Much of the current research is aimed at quantifying this empirical observation. Although considerable attention has been given to this problem since the early days of adaptive control, it is only recently that it has been studied on a rigorous basis [14,20,22,35]. It is therefore not surprising that, unlike the results in Sections 8.3-8.5, those presented here do not constitute a coherent theory but merely provide glimpses of developments in an area that is in a state of flux.

The effectiveness of adaptive control for plants with time-varying parameters, as in all other problems presented in this chapter, depends to a great extent on the prior information that is available. Such prior information includes the nature of the variations of the plant parameters, that is, whether they vary slowly, whether the variations are bounded, and whether bounds on the values of the parameters as well as their time derivatives are known. For example, it is generally assumed that the plant parameters belong to a bounded set but the bounds on the set may not always be known. When the parameters vary discontinuously, it is commonly assumed that such discontinuities occur over large intervals. Other common assumptions include slow but large variations or rapid but small variations of parameters.

If $\theta : \mathbb{R}^+ \to \mathbb{R}^m$ is a control parameter vector that is used to compensate for variations in plant parameters, it must first be shown that a vector function $\theta^* : \mathbb{R}^+ \to \mathbb{R}^m$ exists such that when $\theta(t) \equiv \theta^*(t)$, the performance specifications of the system are satisfied. However, this is not easy to accomplish in practice since the analysis of the system involves time-varying operators that do not commute. Assuming that a function θ^* can be shown to exist, the prior information given in terms of the plant parameters can be expressed in terms of $\|\theta^*(t)\|$ and $\|\dot{\theta}^*(t)\|$. In Section 8.6.1, we treat first-order plants with time-varying parameters to introduce some of these ideas as well as to indicate the nature of the analytical difficulties. In Section 8.6.2, recent results presented in four papers [14,20,22,35] dealing with the adaptive control of higher order plants with time-varying parameters are discussed.

8.6.1 Adaptive Control of a First-Order Plant

Let a plant be described by a first-order differential equation

$$\dot{x}_p(t) = a_p(t)x_p(t) + u(t) \tag{8.46}$$

and a reference model by the equation

$$\dot{x}_m(t) = a_m x_m(t) + r(t)$$

where $a_p(t)$ is an unknown time-varying parameter, a_m is a negative constant, and $r(t)$ is the reference input. With a control input $u(t) = \theta(t)x_p(t) + r(t)$, the output error e_1 defined as $e_1 = x_p - x_m$ satisfies the equation

$$\dot{e}_1(t) = a_m e_1(t) + (a_p(t) + \theta(t) - a_m)x_p(t).$$

If $a_p(t)$ is known exactly for all $t \geq t_0$, then the control parameter $\theta(t)$ can be chosen as $\theta(t) = \theta^*(t)$, where $\theta^*(t) = a_m - a_p(t)$, which would result in $\lim_{t\to\infty} e_1(t) = 0$. $\theta^*(t)$ consequently represents the function of time that $\theta(t)$ has to approach asymptotically for the output error to tend to zero. Since $a_p(t)$ and hence $\theta^*(t)$ are unknown in an adaptive situation, the adaptive law has to be chosen using the available input-output data to achieve the same objective.

Case (i) $\dot{a}_p(t) = -a_p(t) + c$. We first consider the case where $a_p(t)$ is the output of an asymptotically stable linear time-invariant system with a constant input. Hence, $\lim_{t\to\infty} a_p(t) = c$. As shown below, the output error tends to zero asymptotically with the standard adaptive law. The difference between the analysis here and the previous one is in the definition of the parameter error. Instead of the definition $\phi = \theta - \theta^*$, we define ϕ in the present case as $\phi(t) = \theta(t) - a_m + c$, that is, the parameter error is not measured with respect to the instantaneous value of $a_p(t)$, but with respect to its steady-state value. The error equations can therefore be expressed as

$$\begin{aligned} \dot{e}_1 &= a_m e_1 + \phi x_p + \widetilde{a}_p(t) x_p \\ \dot{\phi} &= -e_1 x_p \\ \dot{\widetilde{a}}_p &= -\widetilde{a}_p \end{aligned} \tag{8.47}$$

where $\widetilde{a}_p(t) = a_p(t) - c$. A quadratic function V and its time derivative along the trajectories of Eq. (8.47) are given as follows:

$$\begin{aligned} V &= \frac{1}{2}\left(e_1^2 + \phi^2 + \beta \widetilde{a}_p^2\right) \qquad \beta > 0 \\ \dot{V} &= \left(a_m + \widetilde{a}_p\right) e_1^2 + \widetilde{a}_p x_m e_1 - \beta \widetilde{a}_p^2. \end{aligned}$$

Since $\widetilde{a}_p(t)$ decays exponentially to zero, it follows that by choosing β sufficiently large, $\dot{V}(t) \leq 0$ for all t greater than some time t_1. Hence, the origin of the error equation (8.47) is uniformly stable in the large. When the reference input r is persistently exciting, it can also be shown to be u.a.s.l.

Comment 8.3 The important point to note here is that the boundedness of signals has been established without any modifications in the standard adaptive law. A definition of the parameter error that is different from that used earlier makes the problem analytically tractable. The same procedure may also be used in higher order systems where $a_p(t)$ is either a vector or a matrix and is the output of an asymptotically stable system with a constant input. This condition can be further relaxed by not requiring that $a_p(t)$ tend to a constant value a_p^* exponentially.

Comment 8.4 If $a_p(t)$ can assume only constant values and the switching from one value to another occurs at low frequencies, the same approach can be used to show that the output error will be bounded provided the switching frequency is sufficiently small. Figure 8.21 shows the response of an adaptive system when $a_p(t)$ switches periodically between the values 1 and 3.

Case (ii) $\ddot{a}_p + \omega^2 a_p = \alpha\omega^2$. In this case, $a_p(t) = \alpha + a_0 \sin(\omega t + \phi_0)$ where α, ω, a_0, and ϕ_0 are constants. Our interest is in determining how the adaptive system responds as the frequency ω of $a_p(t)$ increases. From a theoretical point of view, this represents a case where $|a_p|$ and $|\dot{a}_p|$ are bounded. The results of simulation studies are

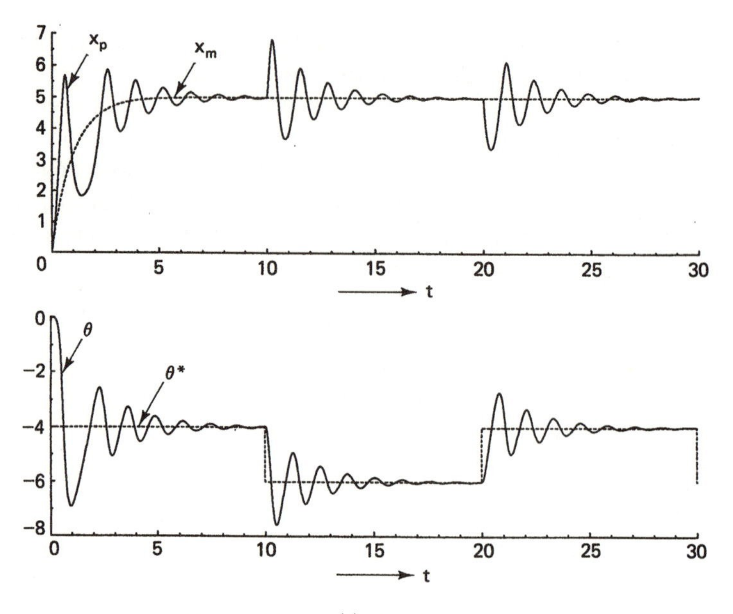

Figure 8.21 $a_p(t)$ piecewise constant.

shown in Fig. 8.22 and indicate that the response of the adaptive system is bounded for all the frequencies considered. When the frequency of the parameter variation is low, the standard adaptive law is seen to result in a small output error.

Despite the apparent simplicity of the problem above, it is surprisingly difficult to analyze quantitatively due to its nonlinear and nonautonomous nature. This is typical of many of the situations encountered with time-varying plant parameters and accounts for the sparsity of quantitative results in the area.

Case (iii) $|a_p(t)| \leq a_0$, $|\dot{a}_p(t)| \leq d_0$. The simple examples considered in cases (i) and (ii) reveal some of the difficulties encountered in the adaptive control of time-varying systems while the corresponding simulations indicate that successful adaptation is possible when plant parameters vary slowly. The general problem of adaptively controlling a first-order plant of the form of Eq. (8.46) may therefore be stated as one in which $a_p(t)$ varies in an unknown fashion with $a_p(t)$ and $\dot{a}_p(t)$ being bounded. The aim of the analysis then is to show that the control parameter θ as well as the other signals in the adaptive loop remain bounded.

If the parameter error is defined as $\phi(t) = \theta(t) - \theta^*(t)$ and the standard adaptive law

$$\dot{\theta} = -e_1 x_p$$

is used to adjust θ, the error equations can be expressed as

$$\begin{aligned} \dot{e}_1 &= a_m e_1 + \phi x_p \\ \dot{\phi} &= -e_1 x_p - \dot{\theta}^*. \end{aligned} \tag{8.48}$$

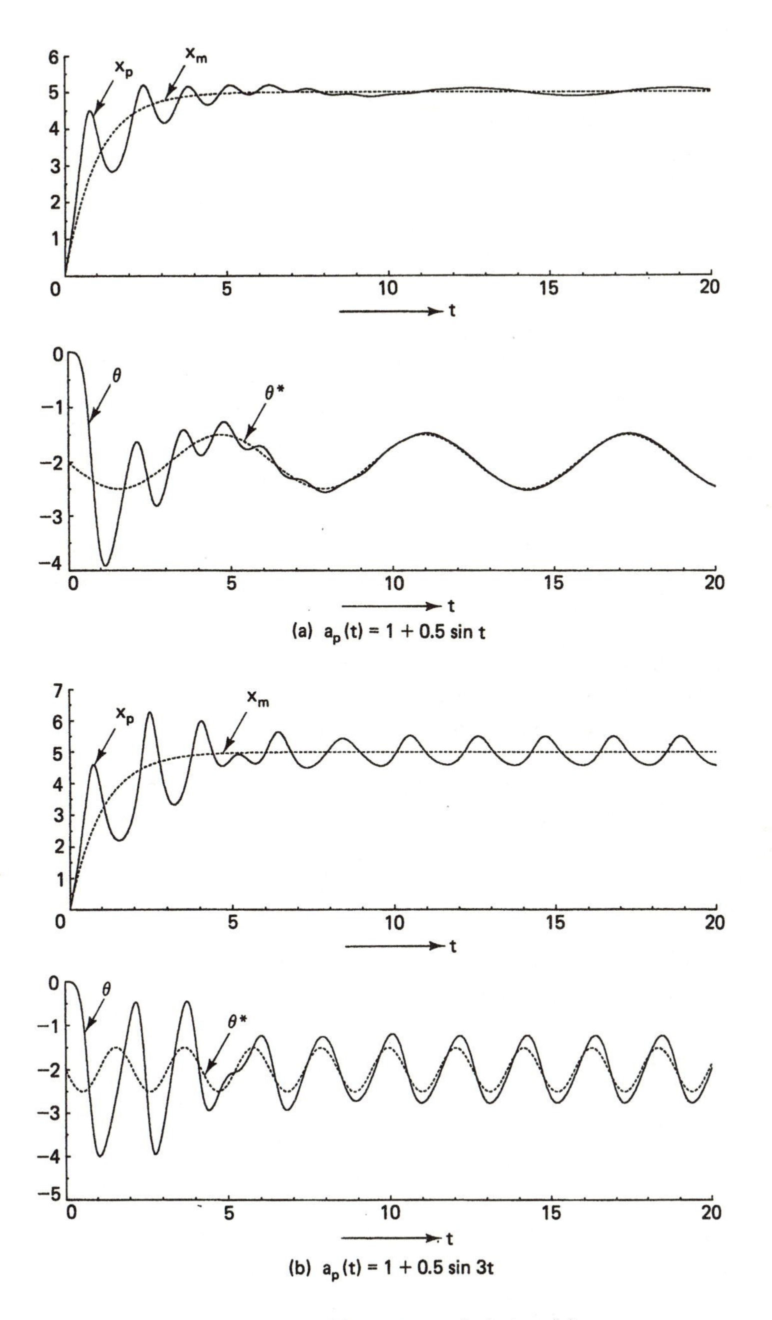

Figure 8.22 $a_p(t) = \alpha + a_0 sin(\omega t + \phi_0)$.

Since Eq. (8.48) is nonlinear and nonautonomous with a bounded input $\dot{\theta}^*$, it is not possible to assure the boundedness of e_1 and ϕ in a straightforward manner. However, some of the methods suggested in the previous section can be used to modify the adaptive laws to achieve this. In particular, the scheme (b), which leads to the equation

$$\dot{\phi} = -e_1 x_p - \theta f(\theta) - \dot{\theta}^* \tag{8.49}$$

where

$$f(\theta) = \begin{cases} \left(1 - \dfrac{|\theta|}{\theta^*_{max}}\right)^2 & \text{if } |\theta| > \theta^*_{max} \\ 0 & \text{otherwise} \end{cases}$$

can be shown to result in bounded solutions. The implementation of Eq. (8.49) requires the knowledge of a bound on θ^* (that is, $a_m + a_0$). Alternately, if the σ-modification is used [scheme (c)], the adaptive law is given by

$$\dot{\phi} = -e_1 x_p - \sigma\theta - \dot{\theta}^*$$

and assures the same result. This result can be proved using a quadratic function $V(e_1, \phi) = (e_1^2 + \phi^2)/2$, which yields a time derivative $\dot{V}(e_1, \phi)$ that is negative-definite outside a compact region containing the origin in the (e_1, ϕ) space.

Comment 8.5 The dead zone scheme (a) or the e_1-modification scheme (d) cannot be applied directly here, since in both cases, it does not readily follow that $\dot{V}(e_1, \phi)$ is negative-definite outside a compact region.

Comment 8.6 Extensive computer simulations have shown that a sufficiently persistently exciting reference input without any changes in the adaptive law would also produce bounded outputs in the case discussed. At present, no theoretical justification for this is available.

Comment 8.7 We have dealt with a first-order plant in some detail, since the procedures used in more general cases, discussed in Section 8.6.2, are qualitatively very similar. However, some of the analytical difficulties that occur are peculiar to higher order systems.

8.6.2 Time-varying Plants of Higher Order

As mentioned in the introduction, although work on adaptive control of plants with time-varying parameters has been in progress for some time, it is only in the last few years that conditions for their bounded response have been established in a rigorous fashion. In particular, while much of the early work was concerned with restricted classes of time variations, preliminary theoretical results for a general problem in which parameters drift slowly over a bounded set have been given recently for continuous time systems in [22] and [35] and for discrete time systems in [14] and [20]. Our aim in this section is to point out the problems that are peculiar to higher order systems as well as the features that are common to all the solutions given in the references above.

The principal steps in the analysis of adaptive systems in the ideal case include the derivation of the output error equation and the corresponding adaptive law described by

$$\begin{aligned} e_1 &= W_m(s)\phi^T\omega \qquad &\text{(error equation)} \\ \dot{\phi} &= -e_1\omega \qquad &\text{(adaptive law)} \end{aligned}$$

when the relative degree $n^* = 1$, and the augmented error equation and the adaptive law

$$\begin{aligned} \epsilon_1 &= \phi^T\zeta \qquad &\text{(augmented error)} \\ \dot{\phi} &= -\frac{\epsilon_1\zeta}{1+\xi^T\xi} \qquad &\text{(adaptive law)} \end{aligned}$$

when $n^* \geq 2$. When the plant parameters vary with time, the derivation of the error equations proves more difficult because of the noncommutative property of time-varying operators, and perturbation terms are invariably present. The adaptive laws are also modified since the time derivative of the desired control parameter vector is present as an additive disturbance. In view of the reasons above, the robust adaptive control problem in the presence of time-varying parameters can be stated in terms of the error models discussed in Chapter 7 with input perturbations. Precisely how these perturbation terms arise must be understood if a solution to the problem is to be attempted.

This is best illustrated by considering a second-order plant with time-varying parameters [35] described by

$$R_p(s,t)y_p = u$$

$$\text{where} \qquad R_p(s,t) = s^2 + a_1(t)s + a_2(t)$$

and s denotes the differential operator d/dt. With a controller structure as in the ideal case given by the equations

$$\dot{\omega}_1 = -\omega_1 + u \qquad \dot{\omega}_2 = -\omega_2 + y_p \qquad u = r + \theta_1^*\omega_1 + \theta_0^* y_p + \theta_2^*\omega_2,$$

where θ_0^*, θ_1^*, and θ_2^* are the desired parameter values, we have

$$\begin{aligned} r(t) &= R_c(s,t)(s+1)^{-1}y_p(t) \\ R_c(s,t) &= (s+1-\theta_1^*)(s+1)^{-1}(s^2+a_1(t)s+a_2(t))(s+1) \\ &\quad - \left(\theta_2^* + \theta_0^*(s+1)\right). \end{aligned} \tag{8.50}$$

This shows that $R_c(s,t)$ cannot be chosen arbitrarily. For example, parameters $\theta_i^*(t)$, $i = 0, 1, 2$, cannot be readily determined in Eq. (8.50), to make $R_c(s,t) = s^3+3s^2+3s+1$, since $a_1(t)$ and $a_2(t)$ are functions of time. At best, they can be chosen so that $R_c(s,t)$ is equivalent to

$$R_c(s,t) = s^3 + 3s^2 + 3s + 1 + \widetilde{R}(s,t)$$

where $\widetilde{R}(s,t)$ is a polynomial in s whose time-varying coefficients depend on $a_1(t)$, $a_2(t)$ and their derivatives. This implies that when the parameters θ_i^*, $i = 0, 1, 2$ are replaced by time-varying parameters $\theta_i(t)$, which are adjusted adaptively, the output error e_1 satisfies the equation

$$e_1 = W_m(s)(r + \phi^T \omega) + \eta_1(t)$$

where $\eta_1(t)$ depends on $\widetilde{R}(s,t)$. Additionally, if an augmented error ϵ_1 is generated as in the ideal case, an additional term η_2 is also present in the error equation where

$$\eta_2 = \left[\theta^{*T} W_m(s) - W_m(s)\theta^{*T}\right] \omega,$$

since $\theta^*(t)$ is a function of time. Hence, the augmented error satisfies the equation

$$\epsilon_1 = \phi^T \zeta + \eta \tag{8.51}$$

where $\eta = \eta_1 + \eta_2$. From the definitions of η_1 and η_2, it follows that η is a function of the derivatives of the plant parameters and $\theta^*(t)$.

If the standard adaptive law is used to adjust the parameter vector $\theta(t)$, since θ^* is time-varying, we have

$$\dot{\phi} = -\frac{\epsilon_1 \zeta}{1 + \xi^T \xi} - \dot{\theta}^*. \tag{8.52}$$

The problem then is to determine conditions under which Eqs. (8.51) and (8.52) result in bounded signals within the adaptive system. We briefly outline below the approach taken in [14,20,22,35] to solve this problem.

In [35], Tsakalis and Ioannou suggest a novel controller structure for generating the control input u. In this controller, the location of the time-varying parameters $\theta_i(t)$ is changed. This eliminates the term η_1 in the error equation (8.51). The adaptive law in Eq. (8.52) is modified as

$$\dot{\phi} = -\frac{\epsilon_1 \zeta}{1 + \xi^T \xi} - \sigma \theta f(\theta) - \dot{\theta}^*$$

where

$$f(\theta) = \begin{cases} 0 & \|\theta\| < \theta^*_{\max} \\ \left[\dfrac{\|\theta\|}{\theta^*_{\max}} - 1\right] & \theta^*_{\max} \le \|\theta\| \le 2\theta^*_{\max} \\ 1 & \|\theta\| > 2\theta^*_{\max} \end{cases}$$

and $\theta^*_{\max}$ is a known upper bound on $\|\theta\|$. It is assumed that the derivatives of $\theta^*(t)$ contain a small parameter μ so that $|\eta(t)| \le \mu\|\xi(t)\|$. It is then shown that there exists a $c^* > 0$ such that for all $\mu \in [0, c^*)$, all the signals in the adaptive system are bounded. By this method, slow variations of the plant parameters [and hence of $\theta^*(t)$] need not be assumed to prove the boundedness of ϕ. Such assumptions are needed only to prove that the other signals in the system are bounded.

An indirect approach is adopted by Middleton and Goodwin in [22]. It is shown that the plant output y_p can be expressed as

$$y_p(t) = p^T(t) X(t) + \eta(t)$$

where $p(t)$ is a vector of unknown plant parameters, $X(t)$ is a vector of sensitivity functions that can be measured at each instant t, and $\eta(t)$ is a function of the first n derivatives of $p(t)$. It is assumed that there exist

(a) constants $\mu_1 > 0, k_1 > 0$ such that

$$\int_t^{t+T} \|\dot{p}(\tau)\| d\tau \leq \mu_1 T + k_1 \qquad \text{for all } t, T > 0,$$

(b) known constants $\mu_2, k_2 > 0$ such that

$$|\eta(t)| \leq \mu_2 \sup_{0 \leq \tau \leq t} \{\exp[-k_2(t-\tau)] \, \|X(t)\|\},$$

(c) a known convex region D, where $p(t) \in D$ for all t, a constant $k_3 > 0$ such that $\|p_1 - p_2\| \leq k_3$ for all $p_1, p_2 \in D$, and

(d) the estimates $\widehat{Z}_p(s,t)$ and $\widehat{R}_p(s,t)$ of the polynomial operators of the plant, which are coprime for all $\widehat{p} \in D$.[2]

Under these assumptions, it is shown that the solutions of the closed-loop system are globally bounded for a sufficiently small μ_1 and μ_2.

Under very similar assumptions to (a)-(d), Kreisselmeier establishes the boundedness of all the signals for a discrete time plant with unknown time-varying parameters in [14].

In [20], Lee and Narendra establish a bound on the parameter $\theta^*(t)$ and its time variations for which the standard adaptive law ensures boundedness. Neither modifications in the adaptive law, nor persistent excitation of the reference input are used in the proof. The result applies to discrete-time plants.

8.7 REDUCED-ORDER CONTROLLERS

In all the robust adaptive control problems we have considered thus far, the plant is assumed to satisfy the assumptions (I) given in Chapter 5. Within this framework, we discussed stable methods of controlling it when bounded external disturbances are present or when the plant parameters vary with time. The class of problems discussed in this section represents a marked departure from those considered earlier in that there is uncertainty even in the structure of the plant. All physical systems can be modeled only partially so that, in general, the order of such a plant-model is smaller than that of the plant. This is true in all the applications considered later in Chapter 11. It is to these reduced order plant-models that the design methods developed are applied. Since the characteristics of the plant to be controlled are different from those of the plant-model, there is a need to justify the efficacy of the design method applied. For example, in

[2]This assumption is commonly made in most papers based on the indirect approach. This is elaborated on in Chapter 9.

linear systems theory, after the controller has been designed for a plant-model, it must be shown that the controller will also result in satisfactory performance for structural perturbations on it. Recent work in robust control has shown that a solution can be obtained for this problem when the reduced plant-model transfer function is known and the true plant belongs to a set that is defined in terms of the former by a suitable metric.

In this section, our interest is to determine the class of plants that can be controlled adaptively by a controller that has been designed for a specific class of plant-models. However, since neither the plant-model nor the plant is known, even a precise statement of the corresponding adaptive control problem is difficult. At the present time, this problem of adaptively controlling an unknown plant using a reduced order model is perhaps the most important one in the field. Although a great deal of work is in progress in this area, only partial results are currently available. Some of these results are discussed in this section.

8.7.1 Instability with Reduced-Order Controllers: An Example

When bounded disturbances are present in the plant to be adaptively controlled, it was shown in Section 8.3 that the control parameters can grow in an unbounded fashion. This can occur whether or not a persistently exciting reference input is present (refer to Figs. 8.6 and 8.8). A similar instability can occur in the presence of unmodeled dynamics in the plant as well. This was demonstrated in [33] using simulation studies and is given in the following example:

Example 8.1

The plant to be adaptively controlled has a transfer function

$$W_p(s) = \frac{458}{(s+1)(s^2+30s+229)}$$

and an input-output pair $\{u(\cdot), y_p(\cdot)\}$. Assuming that the reduced plant-model is of first order, the corresponding reference model chosen would also be of first order. Let the transfer function of the latter be $3/(s+3)$ with an input r and an output y_m. If a control input is determined as $u(t) = kr(t) + \theta y_p(t)$, and the control parameters $k(t)$ and $\theta(t)$ are adjusted using the standard adaptive law, it was shown in [33] that with a reference input $r(t) = 0.3 + 1.85 \sin 16.1t$, both the plant output y_p as well as the control parameters will grow in an unbounded fashion.

The results above caused considerable stir in the adaptive control community and generated many discussions especially among researchers working in problems related to robust adaptive control. A number of arguments were put forth to account for the observed phenomena, and methods were suggested to overcome the instability problems. In this section, we will touch upon some of the questions raised by this example.

The example above, although interesting, is not surprising. It simply reinforces the comments that have been made many times earlier in this book that, when dealing with nonlinear systems, the behavior of the solutions in the presence of perturbations cannot be concluded directly from that of the unperturbed system (refer to Section 2.9). A more appropriate theoretical formulation of the problem is to determine conditions under which the behavior of the system remains bounded even when subjected to different perturbations.

8.7.2 Statement of the Problem

Unlike the other problems described in this chapter, even a precise formulation of the reduced-order model problem that would find general acceptance is not a straightforward one. In view of the importance of the problem, the difficulties encountered in its analysis, as well as the fact that only partial results have been derived thus far, efforts are constantly under way to reformulate the problem to be able to draw more general conclusions. Hence, the problem formulation and the method of analysis given here are only tentative. In this spirit the reader should take the problem statement.

An unknown plant $\overline{P}$ (of dimension $\overline{n}$) is assumed to consist of two linear time-invariant parts P (of dimension n) and $\widetilde{P}$ (of dimension $\overline{n}-n$) where $\widetilde{P}$ can appear as an additive and/or multiplicative perturbation of P. P represents the plant-model and $\overline{P}$ the plant. The transfer functions of $\overline{P}$, P, and $\widetilde{P}$ are $\overline{W}_p(s)$, $W_p(s)$, and $\widetilde{W}_p(s)$ respectively. A number of parametrizations of P and $\overline{P}$ have been proposed in the adaptive literature and a typical one due to Tsakalis and Ioannou [11] has the form

$$\overline{W}_p(s) = W_p(s)\left[1 + \mu_1\Delta_1(s)\right] + \mu_2\Delta_2(s)$$

where $\Delta_1(s)$ and $\Delta_2(s)$ are asymptotically stable transfer functions constrained in some fashion and μ_1 and μ_2 are small positive constants. The design of a controller for the plant is carried out assuming that $W_p(s)$ represents the true transfer function of the plant and that it satisfies assumptions (I). This involves the choice of a reference model with a transfer function $W_m(s)$ whose relative degree is the same as that of $W_p(s)$ and a controller with a $2n$-dimensional parameter vector $\theta(t)$. The aim then is to determine adaptive laws for adjusting $\theta(t)$ so that the error between plant and model outputs remain bounded for all $t \geq t_0$ and satisfies some desired performance criterion. Since the dimension of the control parameter vector $\theta(t)$ is less than that needed theoretically to control the plant in the ideal case, we refer to the problem as the reduced-order controller problem. Other authors have also referred to it as the adaptive control problem in the presence of unmodeled dynamics, or the reduced-order model problem.

Robust Control and Adaptive Control. The difficulty in the statement of the adaptive control problem arises from the fact that neither P nor $\widetilde{P}$ is known and the objective is to control $\overline{P}$, which is the composite system. It was stated in Chapter 1 that parametrically adaptive systems would be addressed in this book. The problem under consideration, however, includes a structural uncertainty $\widetilde{P}$ which makes a realistic statement of the problem nontrivial.

The inclusion of both uncertainties in the plant brings the problem closer in spirit to that of robust control. In the latter, it is often assumed that the plant-model P is completely known, and the problem is to determine necessary and sufficient conditions on the structural uncertainty $\widetilde{P}$ under which a stable controller can be designed. This similarity in the problem statements has brought together researchers working in the areas of robust control and adaptive control.

8.7.3 Small Parameter μ in $\widetilde{P}$

We first consider the case where $\widetilde{P}$ is an additive perturbation of P so that

$$\overline{W}_p(s) = W_p(s) + \widetilde{W}_p(s).$$

A reference model with a transfer function $W_m(s)$ of order n and an adaptive controller identical to that described in Section 8.4 are chosen. If the output of $\widetilde{P}$ is $\widetilde{y}_p(t)$, the error equation for the overall system can be derived as

$$e_1 = W_m(s)\phi^T\omega + \eta(t)$$

where η is given by $\eta(t) = \left[1 - [\theta_0^* + \theta_2^{*^T}(sI - \Lambda)^{-1}l\,]W_m(s)\right]\widetilde{y}_p(t)$. If an augmented error ϵ_1 is generated as described in Chapter 5, we obtain the error equation

$$\phi^T\zeta + \eta = \epsilon_1. \tag{8.53}$$

The error equation (8.53) is identical to Eq. (8.51) obtained in the case of time-varying plant parameters and Eq. (8.24) obtained with bounded disturbances. Other parametrizations of $\widetilde{P}$ discussed in the literature are also found to yield the same error equation.

The importance of the error equation (8.53) has already been stressed several times in this chapter. When η is known a priori to be bounded as in Section 8.4, the adaptive schemes (a)-(d), which involve various modifications of the adaptive law, were shown to result in the boundedness of all the signals in the system. In Section 8.6, where time-varying plant parameters were considered, the same adaptive laws resulted in bounded signals when the plant parameters vary sufficiently slowly. In the present context, the term η is derived from $\widetilde{y}_p$, which can be considered as the output of the unmodeled part and, hence, cannot be assumed to be bounded. However, the schemes (a)-(d) with minor modifications can once again be shown to result in bounded solutions. Although several authors have considered this problem and have proposed somewhat different solutions [11,17,24,32], we prefer to consider the problem in a unified fashion and indicate merely the changes needed in the adaptive laws (a)-(d) in Section 8.4. The following comments are relevant to understand the nature of these changes.

Dead Zone. In Section 8.4, the disturbance present was known to be bounded and a dead zone was included in the adaptive law. The magnitude of the dead zone was finite and depended on the magnitude of the disturbance. In the present case, $\eta(t)$ is not known to be bounded. Hence, if the adaptive law in Eq. (8.25) is to be used, it should include a dead zone that depends on an upper bound of $|\eta(t)|$ at every instant of time.

Bound on $\|\theta\|$. When a bound on $\|\theta\|$ was given, all signals were shown to be bounded without any knowledge of the magnitude of the disturbance. This is found to be the case in the present problem also.

σ-modification and e_1-modification. In these modifications, additional terms of the form $-\sigma\theta/(1+\xi^T\xi)$ and $-|\epsilon_1|\theta/(1+\xi^T\xi)$ are included in the adaptive laws to assure the boundedness of the parameters. Once again, in Section 8.4, these were chosen for situations where the disturbances are bounded. If they are to be applicable to the problem where $\eta(t)$ depends on the unmodeled part, then a bound on $|\eta(t)|$ has to feature in the modifications.

We state the modified adaptive laws (a)-(d), which include an upper bound $\eta_0(t)$ on $|\eta(t)|$ and result in bounded parameter and state errors.

$$\begin{aligned} \dot{\phi} &= -\frac{\epsilon_1\zeta}{1+\xi^T\xi} \quad \text{if } |\epsilon_1(t)| \geq (1+\delta)\eta_0(t) \\ &= 0 \quad\quad\quad\quad \text{if } |\epsilon_1(t)| < (1+\delta)\eta_0(t) \end{aligned} \tag{A}$$

$$\begin{aligned} \dot{\phi} &= -\frac{\epsilon_1\zeta}{1+\xi^T\xi} - \theta f(\theta) \\ f(\theta) &= \begin{cases} \left(1 - \dfrac{\|\theta\|}{\theta^*_{\max}}\right)^2 & \text{if } \|\theta\| \geq \|\theta^*\|_{\max} \\ 0 & \text{elsewhere} \end{cases} \end{aligned} \tag{B}$$

$$\dot{\phi} = -\frac{\epsilon_1\zeta}{1+\xi^T\xi} - \frac{\sigma\eta_0^2(t)}{1+\xi^T\xi}\theta \qquad \sigma > 0 \tag{C}$$

$$\dot{\phi} = -\frac{\epsilon_1\zeta}{1+\xi^T\xi} - \gamma\frac{|\epsilon_1|\eta_0(t)}{1+\xi^T\xi}\theta \qquad \gamma > 0. \tag{D}$$

In (A), adaptation ceases when the error ϵ_1 becomes smaller than the upper bound on the disturbance due to the unmodeled dynamics. In (B), it is assumed that a bound on $\|\theta^*\|$ is available, where θ^* is the control parameter for which the overall system, in the presence of $\widetilde{P}$, is stable. In (C) and (D), the constants σ and γ are replaced by $\sigma\eta_0^2(t)$ and $\gamma\eta_0(t)$ respectively. When any one of the adaptive laws (A)-(D) is used, boundedness of all the signals in the system can be established. This is stated in Theorem 8.3.

Theorem 8.3 There exists a constant c^* such that if $\mu \in [0, c^*)$, all the signals in the adaptive system are globally bounded if the adaptive law (A), (B), (C), or (D) is used, provided

$$|\eta(t)| \leq \eta_0(t) \leq \mu\left[1 + \xi^T(t)\xi(t)\right]^{\frac{1}{2}} \qquad t \geq t_0. \tag{8.54}$$

Since $\xi(t)$ is the equivalent state of the system (see comments in Appendix C), Eq. (8.54) implies that the output of the unmodeled part of the plant is small compared to the state of the overall system. The proof of Theorem 8.3 follows along the same lines as that given in Appendix D for bounded disturbances [25]. The boundedness of the parameter error vector $\phi(t)$ follows directly from the adaptive law used; the assumption concerning $\eta(t)$ given by Eq. (8.54) leads to a contradiction, as in Appendix D, if the state of the system is assumed to grow in an unbounded fashion.

From the discussion above it is seen that $\eta_0(t)$, which is used in the adaptive laws, plays a central role in the proof of stability. The question of generating $\eta_0(t)$ using available information has been addressed by various authors [11,15,32]. We describe below one such procedure [24].

Let $\widetilde{W}_p(s) = \mu\widetilde{W}(s)$ where $\widetilde{W}(s)$ is an asymptotically stable transfer function whose impulse response $w(t)$ is such that $|w(t)| \leq k_1 \exp\{-\beta t\}$, for all $t \geq t_0$, where μ, k_1 and β are known positive constants. Let the eigenvalues of the matrix Λ and the poles of $W_m(s)$ have real parts less than $-f$ and $-a_m$ respectively, where f and a_m are positive constants and let $\alpha = \min(\beta, f, a_m)$. Then, if $\eta_0(t)$ and $\xi(t)$ are chosen as

$$\begin{aligned} \dot{m}(t) &= -\alpha m(t) + k_2|u(t)| \qquad m(t_0) \geq k_2|\eta(t_0)| \\ \eta_0(t) &= \mu m(t) \\ \xi(t) &= \left[\omega^T(t), \zeta^T(t), m(t)\right]^T \end{aligned}$$

where k_2 is a constant which can be computed from the known parameters of the system, it can be shown that the condition in Eq. (8.54) holds.

8.7.4 Transfer Function $\overline{W}_m(s)$

In Section 8.7.3, it was assumed that in the absence of $\widetilde{P}$, a constant control parameter vector θ^* exists such that the plant transfer function together with that of the controller matches $W_m(s)$, the transfer function of the reference model. For tractability, we make a further assumption that, in the presence of $\widetilde{P}$, the same controller also stabilizes the overall system with a closed-loop transfer function $\overline{W}_m(s)$. The reduced-order model problem can then be restated as the determination of stable adaptive laws for adjusting the parameter $\theta(t)$ so that the adaptive system has bounded signals.

In terms of $\overline{W}_m(s)$, the output $y_p(s)$ can be expressed as

$$y_p = \overline{W}_m(s)[r + \phi^T\omega].$$

Since the output $y_m(t)$ of the reference model is given by $y_m(t) = W_m(s)r(t)$, it follows that the output error satisfies the error equation

$$e_1 = \overline{W}_m(s)[\phi^T\omega + \overline{\nu}(t)] \tag{8.55}$$

where $\overline{\nu}(t) \stackrel{\Delta}{=} [1 - \overline{W}_m^{-1}(s)W_m(s)]r(t)$. In the following paragraphs we discuss the problem in two stages. In the first stage $\overline{W}_m(s)$ is assumed to be SPR and a global result is given. In the second stage, $\overline{W}_m(s)$ is not assumed to be SPR and local stability results are presented.

$\overline{W}_m(s)$ **SPR: Global Stability.** Since $\overline{\nu}(t)$ is bounded, Eq. (8.55) can be analyzed using the methods in Section 8.4 for bounded disturbances. Any one of the four methods discussed can be used to determine the adaptive law, provided the corresponding information is available. In the absence of prior information concerning $\overline{\nu}(t)$, only the σ-modification and the e_1-modification methods can be used.

An alternate method of assuring the boundedness of all the signals is to use a persistently exciting input. The results of Section 8.4 can be applied directly to the present problem to obtain sufficient conditions for global boundedness. These are stated in Theorem 8.4.

Theorem 8.4 Let $\omega(t) \equiv \omega^*(t)$ when $\theta(t) \equiv \theta^*$ in the controller, with a degree of PE ϵ_0. If $|\overline{\nu}(t)| \leq \overline{\nu}_0$, then all the signals in the adaptive system are bounded if

$$\epsilon_0 \geq \gamma\overline{\nu}_0 + \delta$$

where γ depends on system parameters and δ is an arbitrary positive constant.

Since both ϵ_0 and $\overline{\nu}_0$ are proportional to the amplitude of the reference input, the condition has to be satisfied based on the frequency response of the transfer functions $\overline{W}_m(s)$ and $W_m(s)$ and the spectrum of r. Hence the choice of the reference input, the transfer function of the reference model, and the state variable filters of the controller are critical in assuring the boundedness of all the signals in the overall system. The result is particularly appealing since it indicates that the adaptive system will be well behaved if the frequency response $\overline{W}_m(i\omega)$ is large compared with $[\overline{W}_m(i\omega) - W_m(i\omega)]$ in the frequency range of interest.

In the analysis above, we have assumed that $\overline{W}_m(s)$ is SPR so that analytical conditions can be derived for the adaptive system to have globally bounded solutions. Although such conditions provide interesting insights regarding the adaptive process, the assumption of SPR of $\overline{W}_m(s)$ is quite unrealistic. In fact, it is not clear when and if such a condition would ever be satisfied. Hence, the problem when $\overline{W}_m(s)$ is not SPR is significantly more important and is considered below.

$\overline{W}_m(s)$ not SPR: Local stability. In this book, our emphasis has been on global methods of adaptive control, that is, stable adaptive control of a system when the initial conditions are arbitrary. A global analysis of the adaptive system posed in the previous section when $\overline{W}_m(s)$ is not SPR in Eq. (8.55) is a truly formidable task. The problem has been attempted by a number of researchers and it is the opinion of some of them that the development of a theory of robust adaptive control in its most general form must be undertaken in a local context [1]. Although we reserve judgment on this matter, we readily admit that while global stability results are not available for the problem above, several novel results have been derived using local concepts. Hence, we deviate temporarily from our original objective and present one local result in this section for the sake of completeness. This result is conceptually similar to those obtained earlier for a strictly positive real system but at the same time exemplifies the weaker and, hence, more realistic conditions under which local stability results can be derived. Since our treatment has to be cursory, we refer the reader for a thorough treatment of this subject to the book [1].

Our starting point for the problem is

$$e_1 = \overline{W}_m(s)(\phi^T\omega + \overline{\nu}) \tag{8.55}$$

where $\overline{W}_m(s)$ is asymptotically stable. As a first step, we ignore the perturbation $\overline{\nu}$ and study the equation

$$e_1 = \overline{W}_m(s)\phi^T\omega \tag{8.56}$$

where the vector ϕ is adjusted using the standard adaptive law

$$\dot{\phi} = -\mu e_1\omega \qquad \mu > 0. \tag{8.57}$$

If the origin in the (e_1, ϕ) space of Eqs. (8.56) and (8.57) is exponentially stable, then by Malkin's theorem (Theorem 2.23) on total stability, the solutions of Eq. (8.55) are bounded for sufficiently small perturbations represented by initial conditions and $\overline{\nu}(t)$.

The method adopted here differs from those used in earlier chapters in the choice of the parameter μ. If μ is sufficiently small, ϕ and, hence, the parameter vector θ can be assumed to be almost constant over an interval of time. The method of averaging is then used to relate the behavior of the nonlinear system in Eqs. (8.56) and (8.57) to that of an averaged system that is well behaved.

Method of Averaging. This is a standard method which has been used extensively in the study of nonlinear oscillations. Originally suggested by Krylov and Bogoliuboff [18] it was later developed by Bogoliuboff and Mitropolsky and more recently by Hale [9] and Arnold [2]. In this method, assuming that the right-hand side of a differential equation contains a small parameter μ, an approximate solution can be obtained. Thus, if

$$\dot{x} = \mu f(x, t, \mu) \qquad x(0) = x_0 \tag{8.58}$$

then $\dot{x}$ is small and x varies slowly. Hence, it is logical to expect that the rapidly varying terms in f do not effect the slow variation of x in the long term. By a process of averaging, the nonautonomous differential equation (8.58) can be approximated by an autonomous differential equation in the variable x_{av}, which represents the average value of x where the latter is easier to analyze. This method has been applied to the problem under consideration in [1,6,12,13]. In what follows, we present the result due to Kokotovic *et al.* in [12].

If $\overline{W}_m(s) = h^T(sI - A)^{-1}b$, then Eqs. (8.56) and (8.57) can be rewritten as

$$\begin{bmatrix} \dot{e} \\ \dot{\phi} \end{bmatrix} = \begin{bmatrix} A & b\omega^T \\ -\mu\omega h^T & 0 \end{bmatrix} \begin{bmatrix} e \\ \phi \end{bmatrix}. \tag{8.59}$$

Theorem 8.5 Let $\omega(t)$ be bounded, almost periodic, and persistently exciting. Then there exists a $c^* > 0$ such that for all $\mu \in (0, c^*]$, the origin of Eq. (8.59) is exponentially stable if

$$\text{Re}\left[\lambda_i\left(\int_0^T \omega(t)\overline{W}_m(s)\omega^T(t)dt\right)\right] > 0, \qquad \forall\, i = 1, \ldots, n \tag{8.60}$$

and is unstable if

$$\text{Re}\left[\lambda_j\left(\int_0^T \omega(t)\overline{W}_m(s)\omega^T(t)dt\right)\right] < 0, \qquad \text{for some } j = 1, \ldots, n \tag{8.61}$$

In [12], it is further shown that $\omega(t)$ can be expressed as $\omega(t) = \sum_{k=-\infty}^{\infty} \Omega(i\nu_k) \exp(i\nu_k t)$ and the inequality in Eq. (8.60) can be satisfied if the condition

$$\sum_{k=-\infty}^{\infty} Re[\overline{W}_m(i\nu_k)]\ Re[\Omega(i\nu_k)\overline{\Omega}^T(i\nu_k)] > 0 \tag{8.62}$$

is satisfied, where $\overline{\Omega}(i\nu_k)$ is the complex-conjugate of $\Omega(i\nu_k)$. The condition in Eq. (8.62) implies that the stability of the origin of Eq. (8.59) is critically dependent on the spectrum of the excitation of ω in relation to the frequency response $\overline{W}_m(i\nu)$. Given a general transfer function $\overline{W}_m(s)$, there exists a large class of functions ω that satisfies the conditions of the theorem, even when $\overline{W}_m(s)$ is not SPR.

ω in Theorem 8.5 is not an independent variable but rather an internal variable of the nonlinear system in Eq. (8.59). Hence, it cannot be shown to be bounded or persistently exciting. If ω_* represents the signal corresponding to ω when $\widetilde{P} = 0$ (that is, the tuned solution), it can be made to satisfy Eq. (8.60) by the proper choice of the reference input. Expressing $\omega = \omega_* + \omega_e$, ω will also be bounded, persistently exciting, and satisfy Eq. (8.60) if ω_e is small. This can be achieved by choosing the initial conditions $e(t_0)$ and $\phi(t_0)$ in Eq. (8.59) to be sufficiently small. The conditions of Theorem 8.5 are then verified, and for a sufficiently small μ, exponential stability of the origin of Eq. (8.59) follows.

The theorem provides conditions for exponential stability and instability when the solutions of the adaptive system are sufficiently close to the tuned solutions. These are very valuable in understanding the stability and instability mechanisms peculiar to adaptive control in the presence of different types of perturbations. Many of these results have been summarized and presented in a unified fashion in [1].

8.8 SUMMARY

The behavior of adaptive systems in the presence of exogenous and endogenous disturbances is considered in this chapter. The latter includes variations of plant parameters with time as well as signals caused by the imperfect modeling of the plant. In all cases, the aim is to ensure that the signals in the overall system are bounded and that the performance of the adaptive system is satisfactory.

The effect of bounded external disturbances on the adaptation process is first dealt with in detail. Since a bounded disturbance in a nonlinear system which is u.a.s.l. does not assure a bounded response, it becomes evident that some precautionary measures have to be taken if the signals of the system are to remain bounded. Methods for achieving stable adaptation by modifying the adaptive laws are discussed in this context. Following this discussion, it is shown that robust adaptation can also be assured by increasing the degree of PE of the reference input. Qualitatively, it is found that if the latter is larger than some measure of the disturbance, the overall system has bounded solutions. Methods for extending these concepts to time-varying parameters and partially modeled plants are discussed in the following sections.

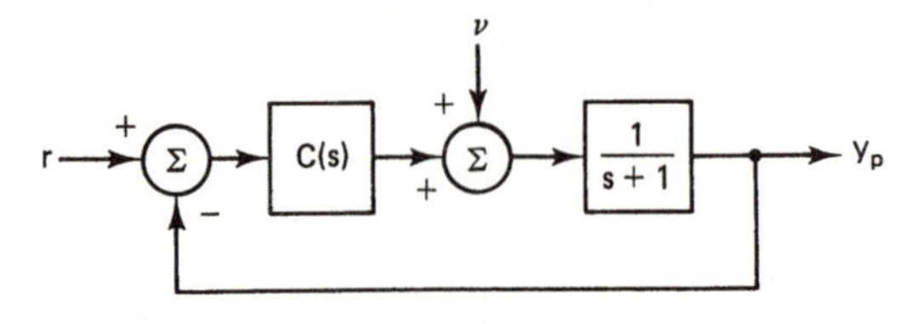

Figure 8.23 Disturbance rejection using integral control.

Hybrid adaptive control, in which the controller parameters are adjusted at discrete instants of time, is found to be more robust than adaptive control where such adjustments are made continuously. Information collected over an interval of time is used to determine the change in parameter values at the end of the interval. Increasing the degree of PE of the reference input results in faster convergence of the parameters and smaller parameter and output errors, in all cases.

PROBLEMS

Part I

1. **(a)** In the system shown in Fig. 8.24, $C(s)$ denotes the transfer function of a linear time-invariant controller. Show that if a constant additive disturbance $\nu(t) \equiv \nu_0$ is present as shown in the figure, its effect at the output tends to zero if $C(s)$ contains an integrator.

(b) If $\nu(t) = \sin\, \omega t$, show that if $C(s)$ has poles at $\pm\, i\omega$, the effect of the disturbance tends to zero.

(c) If the disturbance ν is such that $D(s)\nu(t) = 0$, where $D(s)$ is a polynomial in s, determine the structure of the controller such that the effect of the disturbance at the output tends to zero.

2. $\nu_1(t)$ and $\nu_2(t)$ are disturbances at the input and output of a plant described by the differential equation

$$\begin{aligned} \dot{x} &= Ax + bu + d\nu_1 \\ y &= h^T x + \nu_2. \end{aligned}$$

If ν_1 is a bounded piecewise-continuous function and ν_2 is a bounded differentiable function of time, and the control input u is given by

$$u = r + W_1(s)u + W_2(s)y$$

show that the output of the plant can be expressed as

$$y = W_c(s)r + \nu,$$

where ν is an equivalent bounded disturbance. Also show that the overall system can be expressed as

$$\begin{aligned} \dot{z} &= \overline{A}z + \overline{b}r + \overline{\nu} \\ y &= \overline{h}^T z \end{aligned}$$

where $\overline{\nu}$ is an equivalent bounded vector disturbance at the input.

Part II

3. **(a)** If, in Problem 1, the disturbance is known to be a sinusoid of unknown frequency, how would you design an adaptive controller to cancel its effect at the output?

   (b) If, in addition, the plant transfer function in Problem 1 is assumed to be of the form $1/(s+a)$ where a is unknown, determine how you would design a controller so that the output $y_p(t)$ of the plant follows the output $y_m(t)$ of a reference model with a transfer function $1/(s+1)$ asymptotically.

4. In each of the adaptive laws (b)-(d) in Section 8.3, which include a bound on θ, σ-modification, and e_1-modification, show that the function $V(\phi) = \phi^T\phi/2$ is a Lyapunov function outside a compact set that contains the origin in the (e_1, ϕ) space.

5. **(a)** Let the error equations resulting from a σ-modification scheme be given by

$$\begin{array}{rcl} \dot{e}_1 & = & -e_1 + \phi(e_1 + 5) \\ \dot{\phi} & = & -e_1(e_1 + 5) - \sigma(\phi + 5) \end{array}, \qquad \sigma = 1.$$

   Determine the equilibrium states of these equations.

   (b) Repeat Problem (a) for the error equations resulting from an e_1-modification scheme described by

$$\begin{array}{rcl} \dot{e}_1 & = & -e_1 + \phi(e_1 + 5) \\ \dot{\phi} & = & -e_1(e_1 + 5) - \gamma|e_1|(\phi + 5) \end{array}, \qquad \gamma = 1.$$

6. **(a)** The error differential equations of an adaptive system are

$$\begin{array}{rcl} \dot{e}_1 & = & -e_1 + \phi(e_1 + y_m) \\ \dot{\phi} & = & -e_1(e_1 + y_m). \end{array}$$

   Show that $\lim_{t\to\infty} e_1(t) = 0$. If $y_m(t)$ is persistently exciting, show that $\lim_{t\to\infty} \phi(t) = 0$.

   (b) When an additive disturbance $\nu(t)$ is present, the error equations are modified as

$$\begin{array}{rcl} \dot{e}_1 & = & -e_1 + \phi(e_1 + y_m) + \nu \\ \dot{\phi} & = & -e_1(e_1 + y_m). \end{array}$$

   Use the method described in Section 8.3.4 to show that the solutions of the equations above are bounded if the degree of persistent excitation ϵ_0 of y_m is greater than ν_0 where $|\nu(t)| \leq \nu_0$.

7. Derive the three relations (i)-(iii) for the schemes (b), (c), and (d) in Section 8.4.1.

Part III

8. The error equations of an adaptive system for a first-order plant with time-varying coefficients were derived as

$$\begin{array}{rcl} \dot{e}_1 & = & -e_1 + \phi(e_1 + y_m) \\ \dot{\phi} & = & -e_1(e_1 + y_m) - \dot{\theta}^* \end{array}$$

in Section 8.6.1. Can you determine a $\theta^*(t)$

(i) whose time derivative is bounded, or

(ii) which is piecewise-constant,

for which the system above has unbounded solutions for some initial conditions?

9. The error equations for an adaptive system containing perturbations due to (1) bounded disturbances, (2) time-varying parameters, and (3) unmodeled dynamics were shown in Sections 8.4, 8.6, and 8.7 respectively, to have the form

$$\phi^T \zeta + \eta = \epsilon_1$$

where η is a term due to the perturbations. Discuss the similarities and differences in the proof of boundedness of solutions in the three cases.

REFERENCES

1. Anderson, B.D.O., Bitmead, R., Johnson, C.R., Kokotovic, P.V., Kosut, R.L., Mareels, I., Praly, L., and Riedle, B. *Stability of adaptive systems.* Cambridge, MA:M.I.T Press, 1986.
2. Arnold, V.I. *Geometric methods in the theory of differential equations.* New York:Springer-Verlag, 1982.
3. Bunich, A.L. "Rapidly converging algorithm for the identification of a linear system with limited noise." *Autom. and Remote Control*, 44:1047–1054, 1983.
4. M. de La Sen. "A method for improving the adaptation transient using adaptive sampling." *International Journal of Control*, 40:639–665, 1984.
5. Egardt, B. *Stability of adaptive controllers.* Berlin:Springer-Verlag, 1979.
6. Fu, L.-C., Bodson, M., and Sastry, S.S. "New stability theorems for averaging and their applications to the convergence analysis of adaptive identification and control schemes." In *Proceedings of the IEEE CDC*, Ft. Lauderdale, FL, 1985.
7. Gawthrop, P.J. "Hybrid self-tuning control." In *IEE Proceedings*, 127:229-336, Part D, 1980.
8. G.C. Goodwin and D. Mayne. "A continuous-time adaptive controller with integral action." *IEEE Transactions on Automatic Control*, 30:793-795, 1985.
9. Hale, J.K. *Ordinary differential equations.* New York, NY:Wiley-Interscience, 1969.
10. Ioannou, P.A. and Kokotovic, P.V. *"Adaptive systems with reduced models."* New York:Springer-Verlag, 1983.
11. Ioannou, P.A., and Tsakalis, K. "A robust direct adaptive controller." *IEEE Transactions on Automatic Control*, 31:1033–1043, Nov. 1986.
12. Kokotovic, P., Riedle, B., and Praly, L. "On a stability criterion for continuous slow adaptation." *Systems & Control Letters* 6:7–14, June 1985.
13. Kosut, R.L., Anderson, B.D.O., and Mareels, I.M.Y. "Stability theory for adaptive systems: Methods of averaging and persistent excitation." In *Proceedings of the IEEE CDC*, Ft. Lauderdale, FL, 1985.

14. Kreisselmeier, G. "Adaptive control of a class of slowly time-varying plants." *Systems & Control Letters* 8:97–103, 1986.

15. Kreisselmeier, G., and Anderson, B.D.O. "Robust model reference adaptive control." *IEEE Transactions on Automatic Control* 31:127–133, 1986.

16. Kreisselmeier, G., and Narendra, K.S. "Stable model reference adaptive control in the presence of bounded disturbances." *IEEE Transactions on Automatic Control* 27:1169–1175, Dec. 1982.

17. Kreisselmeier, G., and Smith, M.C. "Stable adaptive regulation of arbitrary n^{th}-order plants." *IEEE Transactions on Automatic Control* 31:299–305, 1986.

18. Krylov, A.N., and Bogoliuboff, N.N. *Introduction to nonlinear mechanics.* Princeton:Princeton University Press, 1943.

19. Lee, T.H., and Narendra, K.S. "Robust adaptive control of discrete-time systems using persistent excitation and hybrid adaptation." *Automatica* Nov. 1988.

20. Lee, T.H., and Narendra, K.S. "Stable direct adaptive control of time-varying discrete-time systems." Technical Report No. 8720, Center for Systems Science, Yale University, New Haven, CT, 1987.

21. Ljung, L. "On positive real transfer functions and the convergence of some recursive schemes." *IEEE Transactions on Automatic Control* 22:539–551, Aug. 1977.

22. Middleton, R., and Goodwin, G.C. "Adaptive control of time-varying linear systems." *IEEE Transactions on Automatic Control* 33:150–155, 1988.

23. Morgan, A.P., and Narendra, K.S. "On the stability of nonautonomous differential equations $\dot{x} = [A+B(t)]x$ with skew-symmetric matrix $B(t)$." *SIAM Journal of Control and Optimization* 15:163–176, Jan. 1977.

24. Narendra, K.S., and Annaswamy, A.M. "A new adaptive law for robust adaptive control without persistent excitation." *IEEE Transactions on Automatic Control* 32:134–145, Feb. 1987.

25. Narendra, K.S., and Annaswamy, A.M. "Robust adaptive control in the presence of bounded disturbances." *IEEE Transactions on Automatic Control* 31:306–315, April 1986.

26. Narendra, K.S., Annaswamy, A.M., and Singh, R.P. "A general approach to the stability analysis of adaptive systems." *International Journal of Control* 41:193–216, 1985.

27. Narendra, K.S., Khalifa, I.H., and Annaswamy, A.M. "Error models for stable hybrid adaptive systems." *IEEE Transactions on Automatic Control*, 30:339–347, April 1985.

28. Narendra, K.S., and Lin, Y.H. "Stable discrete adaptive control." *IEEE Transactions on Automatic Control* 25:456–461, June 1980.

29. Narendra, K.S., and Peterson, B.B. "Bounded error adaptive control - Part I." Technical Report No. 8005, Center for Systems Science, Yale University, New Haven, CT, December 1980.

30. Ortega, R., and Lozano-leal, R. "A note on direct adaptive control of systems with bounded disturbances." *Automatica*, 23:253–254, 1987.

31. Peterson, B.B., and Narendra, K.S. "Bounded error adaptive control." *IEEE Transactions on Automatic Control* 27:1161–1168, Dec. 1982.

32. Praly, L. "Robustness of model reference adaptive control." In *Proceedings of the Third Yale Workshop on Applications of Adaptive Systems Theory*, Yale University, New Haven, CT, 1983.

33. Rohrs, C., Valavani, L., Athans, M., and Stein, G. "Robustness of continuous-time adaptive control algorithms in the presence of unmodeled dynamics." *IEEE Transactions on Automatic Control* 30:881–889, 1985.

34. Samson, C. "Stability analysis of adaptively controlled systems subject to bounded disturbances." *Automatica* 19:81–86, 1983.

35. Tsakalis, K., and Ioannou, P. "Adaptive control of linear time-varying plants: A new model reference controller structure." Technical Report No. 86-10-1, University of Southern California, 1987.

9

The Control Problem: Relaxation of Assumptions

9.1 INTRODUCTION

The preceding chapters have attempted to develop a systematic procedure for the design of stable adaptive controllers for linear time-invariant plants with unknown parameters. The four assumptions (I) made regarding the plant transfer function $W_p(s)$, for the solution of the adaptive control problem in the ideal case, are listed below for easy reference:

(i) the sign of the high frequency gain k_p,

(ii) the relative degree n^*, and

(iii) an upper bound on the order n of $W_p(s)$ (I)

are known, and

(iv) the zeros of $W_p(s)$ lie in $\mathbb{C}^-$.

Although (ii) is used to set up a reference model and (iii) to choose the order of the controller, (i) and (iv) are required to prove global stability using Lyapunov's direct method. In the course of time, with constant use, it came to be generally believed by many researchers that (I) is also necessary for a stable solution of the adaptive control problem to exist. At the same time, it was realized that these assumptions were too restrictive, since in practice they are violated by most plants, even under fairly benign conditions. The fact that many practical adaptive controllers were found to perform satisfactorily under circumstances where (I) is not valid, led others to believe that it may be sufficient but not necessary. It was argued that the assumptions in (I) were dictated entirely by the stability methods used, that they were not intrinsic to the problem, and that new approaches might result in less restrictive and perhaps more realistic assumptions. This, in turn, led to further discussions concerning the minimum amount of information that is needed about the transfer function of an unstable plant to design a stable adaptive controller. The answer to such questions would involve the derivation of necessary and sufficient conditions for the solution of the adaptive control problem.

While a complete answer to this question is not yet available, many partial solutions have been suggested. In the first part of this chapter, we present the progress that has been made in these directions. In Section 9.2, the relaxation of the assumption (i) based on the work of Nussbaum in 1983 [23], is described. This had a major impact on the thinking of the workers in the field and triggered a wave of activity that spawned many new ideas related to assumptions (ii)-(iv).

Since the early 1980s, it was felt by some workers that the requirement (iv) in (I) could be dispensed with. For stable control, the LQG as well as the pole placement methods do not require the minimum phase (zeros in $\mathbb{C}^-$) condition, when the plant parameters are known. Hence, it was felt that this assumption may not be necessary for the solution of the adaptive control problem also, but was required only because of the approach used. In Section 9.4, we discuss several results in which the stability of the overall system is shown without requiring assumption (iv) in (I).

A further simplification in the adaptive control problem can be realized by limiting the class of reference inputs to the system. The analysis in the preceding eight chapters is based on the assumption that the reference input r is bounded and piecewise-continuous. However, in many practical systems, the reference input may be sufficiently smooth so that, over a finite interval, it can be approximated by the homogeneous solution of a linear time-invariant system. Assuming that the reference input belongs to the latter class, the model reference adaptive control problem can be reformulated. In Section 9.5, we discuss the simplifications resulting from such an approach as well as its limitations.

The different directions in which research progressed in the early 1980s following the solution of the ideal adaptive control problem are shown in the form of a tree structure in Fig. 9.1. The left branch of the tree concerns the class of problems in which (I) holds. Much of the research here is related to the analysis of the behavior of the overall system under various perturbations and has already been treated in Chapter 8. The right branch of the tree deals with efforts aimed at relaxing (I) and is the subject of this chapter.

9.2 ASSUMPTION (i): SIGN OF k_p

The need for assumption (i) in the solution of the adaptive control problem using Lyapunov's direct method is best illustrated by considering the simple case of a scalar plant described by the differential equation

$$\dot{y}_p = a_p y_p + k_p u. \tag{9.1}$$

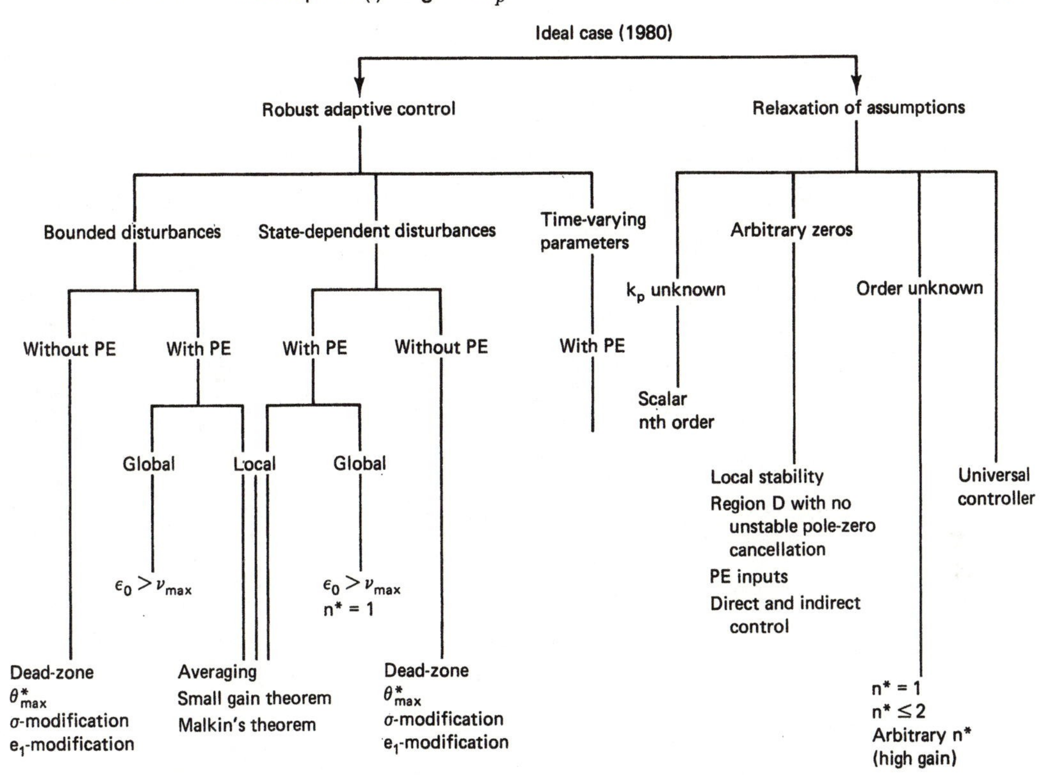

Figure 9.1 Recent developments in adaptive control.

u and y_p are the scalar input and output respectively, a_p and k_p are unknown constants, and the input u has to be determined such that the plant output $y_p(t)$ is bounded for all $t \geq t_0$ and $\lim_{t\to\infty} y_p(t) = 0$. The methods described in Chapter 3 to assure the existence of a quadratic Lyapunov function result in a feedback controller described by the equations

$$\begin{aligned} u &= sgn(k_p)\theta y_p \\ \dot{\theta} &= -y_p^2 \end{aligned}$$

where θ is a scalar time-varying parameter. Hence, the sign of k_p is used explicitly in the design of the controller.

The question then is to determine whether the same objective can also be realized without such explicit knowledge of $sgn(k_p)$. In attempting to answer this question, Morse [18] reformulated the problem as follows:

> Do there exist integers $m \geq 1$ and differentiable functions $f : \mathbb{R}^{m+1} \to \mathbb{R}$ and $g : \mathbb{R}^{m+1} \to \mathbb{R}^m$ such that for every $k_p \neq 0$, and for every initial condition $y_p(t_0) = y_0 \in \mathbb{R}$ and $\theta(t_0) = \theta_0 \in \mathbb{R}^m$, the solution of

$$\dot{y}_p = a_p y_p + k_p u$$

$$
\begin{aligned}
u &= f(\theta, y_p) \\
\dot{\theta} &= g(\theta, y_p)
\end{aligned} \tag{9.2}
$$

is bounded and $\lim_{t\to\infty} y_p(t) = 0$?

Qualitatively, the question above addresses the existence of a general nonlinear controller for stabilizing the system, independent of the sign of k_p.

Morse further conjectured that such functions do not exist, implying that the sign of k_p is necessary to solve the stabilization problem. In 1983, Nussbaum, attempting the problem above with $m = 1$, conclusively showed that, although Morse's conjecture was valid for rational functions $f(\theta, y_p)$ and $g(\theta, y_p)$, it was incorrect if f and g were transcendental functions. Hence, valid solutions could be found for the adaptive control problem without explicit knowledge of the sign of k_p. In view of the impact that Nussbaum's paper had on the field, we deal with it in some detail in this section.

Theorem 9.1 Consider the system in Eq. (9.2) with initial conditions (y_0, θ_0). Let $f, g : \mathbb{R}^2 \to \mathbb{R}$ and satisfy the following conditions:

1. There exist $c > 0$ and θ_1 such that

$$\text{either } g(\theta_1, y_p) > 0 \;\forall\, y_p \geq c \qquad \text{or} \qquad g(\theta_1, y_p) < 0 \;\forall\, y_p \geq c.$$

2. There exists $y_1 \geq c$ such that $f(\theta, y_1)$ is bounded above or bounded below for $\theta \geq \theta_1$ and $f(\theta, y_1)$ is bounded above or bounded below for $\theta \leq \theta_1$.

Then a domain Q exists such that if $(y_0, \theta_0) \in Q$, then $\lim_{t\to\infty} y_p(t) \neq 0$.

Proof. Let $a_p > 0$. Let constants θ_1 and $c > 0$ exist such that $g(\theta_1, y_p) > 0 \;\forall\, y_p \geq c$. If $f(\theta, y_1)$ is bounded below for $\theta \geq \theta_1$, $y_1 \geq c$, then there exists a positive k_p so that

$$a_p y_1 + k_p f(\theta, y_1) > 0 \qquad \forall \theta \geq \theta_1.$$

If

$$Q \triangleq \{(\theta, y_p) \mid \theta \geq \theta_1,\; y_p \geq y_1\}$$

then the vector field $(a_p y_p + k_p f(\theta, y_p), g(\theta, y_p))$ points into Q for any (θ, y_p) on the boundary of Q (see Fig. 9.2). Hence, any trajectory that starts in Q remains in Q. Similarly, if $f(\theta, y_1)$ is bounded above for $\theta \geq \theta_1$, then for a negative k_p, $a_p y_1 + k_p f(\theta, y_1) > 0$ for $\theta \geq \theta_1$. Once again, any trajectory that starts in Q remains in Q and, hence, $\lim_{t\to\infty} y_p(t) \neq 0$. A similar proof can be given for the case when $g(\theta_1, y_p) < 0$ $\forall y_p \geq c$.

From the discussions above, it follows that the sign of k_p can be chosen in such a fashion that the output $y_p(t)$ does not tend to zero asymptotically with time for some initial conditions. Following is a typical example where f and g are rational functions.

Example 9.1

Choose f and g to be of the form

$$f(\theta, y_p) = \theta^3 + y_p^3, \qquad g(\theta, y_p) = \frac{1}{y_p + \theta}$$

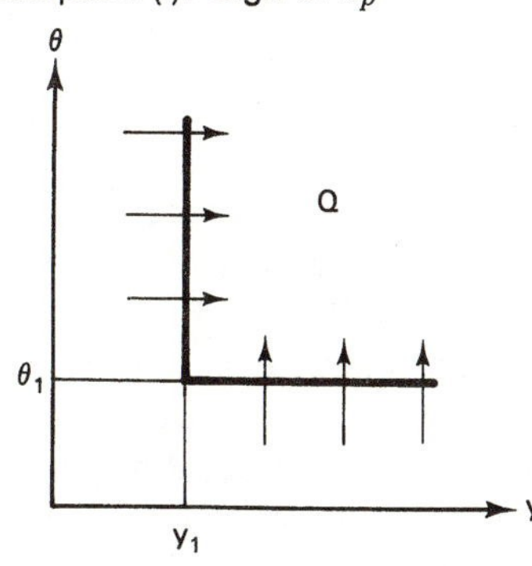

Figure 9.2 Proof of Theorem 9.1.

For a positive value θ_1 of θ, $g(\theta_1, y_p)$ is of constant sign for all $y_p > 0$. Similarly, for $y_p = y_1$, where y_1 is a positive constant, $f(\theta, y_1)$ is bounded below for all $\theta \geq 0$ and is bounded above for all $\theta \leq 0$. Hence f and g satisfy the conditions of Theorem 9.1 and $\lim_{t\to\infty} y_p(t) \neq 0$.

From Example 9.1, it follows that functions f and g, which are not rational, must be chosen if conditions 1 and 2 of Theorem 9.1 are to be violated. Transcendental functions, whose sign changes an infinite number of times as their arguments tend to infinity, satisfy this requirement. The following controller structure was suggested in [17] to generate the plant input u:

$$\begin{aligned} u &= N(\theta) y_p \\ \dot{\theta} &= y_p^2 \end{aligned} \tag{9.3}$$

where

$$N(\theta) \triangleq \theta^2 \cos\theta. \tag{9.4}$$

In Eq. (9.3), $N(\theta)$ plays the role of a gain and is referred to as the *Nussbaum gain*. As θ increases in magnitude and tends to ∞, it is clear that $N(\theta)$ changes its sign an infinite number of times. Further $\sup_\theta[N(\theta)] = +\infty$ and $\inf_\theta[N(\theta)] = -\infty$. With such an input function, the following theorem can be stated:

Theorem 9.2 [23] The adaptive system described by Eqs. (9.1), (9.3), and (9.4) has globally bounded solutions and $\lim_{t\to\infty} y_p(t) = 0$.

Proof. From Eqs. (9.1), (9.3), and (9.4), we derive the following expression:

$$\frac{d}{d\theta}\left[y_p^2\right] = 2\left[a_p + k_p\theta^2 \cos\theta\right].$$

Hence, integrating the two sides of the equation above between the limits $\theta(t_0)$ and $\theta(t)$, we obtain

$$\begin{aligned} y_p^2(t) = {} & y_p^2(t_0) + 2a_p\theta(t) + 2k_p\left[\theta^2(t)\sin\theta(t) + 2\theta\cos\theta(t) - 2\sin\theta(t)\right] \\ & - 2a_p\theta(t_0) - 2k_p\left[\theta^2(t_0)\sin\theta(t_0) + 2\theta(t_0)\cos\theta(t_0) - 2\sin\theta(t_0)\right] \end{aligned} \tag{9.5}$$

From Eq. (9.3), it follows that $\theta(t)$ is a monotonic nondecreasing function. We assume that $\theta(t)$ grows without bound. The term $2k_p\theta^2(t)\sin\,\theta(t)$ can assume a large negative value independent of the sign of k_p, so that the right hand side of Eq. (9.5) is negative. But this is a contradiction since the left-hand side of Eq. (9.5) is $y_p^2(t)$. This assures the boundedness of $\theta(t)$ and, hence, of $y_p(t)$ for all $t \geq t_0$. From Eq. (9.3), we have $y_p \in \mathcal{L}^2$ and, since $\dot{y}_p(t)$ is bounded, it follows that $\lim_{t\to\infty} y_p(t) = 0$.

The importance of the result above lies in the fact that it has been derived without using arguments based on Lyapunov theory. Many of the subsequent results also share this feature and are based on similar arguments. Using the approach of Nussbaum, Mudgett and Morse [22] showed that the general adaptive control problem treated in Chapter 5 can be solved without explicitly knowing the sign of the high frequency gain k_p.

Extension to n^{th} order plants. Let the plant to be adaptively controlled have a transfer function $W_p(s) = k_p Z_p(s)/R_p(s)$. Let the relative degree and an upper bound on the order of $W_p(s)$ be known and let its zeros lie in $\mathbb{C}^-$. It is then desired to determine an adaptive controller such that the output of the plant y_p approaches asymptotically the output y_m of a reference model whose transfer function is $W_m(s)$.

The adaptive controller has an identical structure to that described in Chapter 5 [refer to Eq. (5.16)]. Since the sign of k_p is unknown, an augmented error different from that given in Eq. (5.50) has to be used and is described below.

$$\text{Augmented error: } \epsilon_1 = e_1 + N(x)k_1(t)e_2 \tag{9.6}$$

where e_1 is the true output error, $e_2 = \theta^T W_m(s)I\omega - W_m(s)\theta^T\omega$, and the variable x has to be chosen appropriately. The main result in [22] can be stated in the form of the following theorem:

Theorem 9.3 If the adaptive laws for adjusting the control parameters $\theta(t)$ and $k_1(t)$ are given by

$$\begin{aligned} \dot{\theta} &= N(x)\frac{\epsilon_1\zeta}{1+\zeta^T\zeta} \\ \dot{k}_1 &= -N(x)\frac{\epsilon_1 e_2}{1+\zeta^T\zeta} \end{aligned} \tag{9.7}$$

where

$$x = z + \frac{k_1^2}{2}; \quad \dot{z} = \frac{\epsilon_1^2}{1+\zeta^T\zeta}; \quad N(x) = x^2\cos x$$

then all signals are bounded and $\lim_{t\to\infty} |y_p(t) - y_m(t)| = 0$.

Before proving Theorem 9.3, we consider the following error model which provides the motivation for the choice of the adaptive laws in Eq. (9.7).

Consider the error equation

$$\epsilon_1 = k_p \phi^T \zeta$$

where the sign of k_p is unknown. We recall from Chapter 5 that the adaptive law for adjusting ϕ is given by $\dot{\phi} = -sgn(k_p)\epsilon_1\zeta/(1+\zeta^T\zeta)$. The discussion based on the Nussbaum gain in the earlier section then suggests the adaptive law

$$\dot{\phi} = N(x)\frac{\epsilon_1 \zeta}{1+\zeta^T\zeta} \tag{9.8}$$

where the variable x is yet to be defined. The time derivative of $\phi^T\phi/2$ is

$$\frac{d}{dt}\left[\frac{1}{2}\phi^T\phi\right] = \frac{1}{k_p}N(x)\frac{\epsilon_1^2}{1+\zeta^T\zeta}. \tag{9.9}$$

If x is defined by the differential equation

$$\dot{x} = \frac{\epsilon_1^2}{1+\zeta^T\zeta}$$

then Eq. (9.9) becomes

$$\frac{d}{dt}\left[\frac{1}{2}\phi^T\phi\right] = \frac{1}{k_p}N(x)\dot{x}. \tag{9.10}$$

Integrating Eq. (9.10) on both sides we obtain

$$\frac{1}{2}\|\phi(t)\|^2 - \frac{1}{2}\|\phi(t_0)\|^2 = \frac{1}{k_p}\int_{t_0}^{t} N(\sigma)d\sigma\,.$$

Choosing $N(x) = x^2 \cos x$, it can then be shown that x is bounded, ϕ is bounded, and that $\dot{\phi} \in \mathcal{L}^2$.

Proof of Theorem 9.3. The underlying error equation in the adaptive control problem that describes the relation between the output error e_1 and the signals in the closed-loop system is given by

$$e_1 = \frac{k_p}{k_m}W_m(s)\phi^T\omega \tag{5.35}$$

where ω is the state of the controller, ϕ is the parameter error, and k_p and k_m are the high frequency gains of the plant and the model. From Eq. (5.35), Eq. (9.6) can be expressed as

$$\epsilon_1 = \frac{k_p}{k_m}\phi^T\zeta + N(x)k_1e_2 - \frac{k_p}{k_m}e_2.$$

If the adaptive laws for adjusting ϕ and k_1 are chosen as in Eq. (9.7), the time derivative of $\|\phi\|^2$ can be computed as

$$
\begin{aligned}
\frac{d}{dt}\left[\frac{1}{2}\phi^T\phi\right] &= \frac{k_m}{k_p}N(x)\frac{\epsilon_1}{1+\zeta^T\zeta}\left[\epsilon_1 - N(x)k_1e_2 + \frac{k_p}{k_m}e_2\right] \\
&= \frac{k_m}{k_p}N(x)\left[\frac{\epsilon_1^2}{1+\zeta^T\zeta} - \frac{N(x)\epsilon_1e_2}{1+\zeta^T\zeta}\cdot k_1\right] + \frac{N(x)\epsilon_1e_2}{1+\zeta^T\zeta} \\
&= \frac{k_m}{k_p}N(x)\left[\dot{z} + \dot{k}_1k_1\right] - \dot{k}_1 \qquad (9.11) \\
&= \frac{k_m}{k_p}N(x)\dot{x} - \dot{k}_1 .
\end{aligned}
$$

Integrating Eq. (9.11) on both sides, we have

$$
\frac{1}{2}\|\phi(t)\|^2 - \frac{1}{2}\|\phi(t_0)\|^2 = \frac{k_m}{k_p}\int_{x(t_0)}^{x(t)} N(\sigma)d\sigma - [k_1(t) - k_1(t_0)] .
$$

Since $k_1^2(t) \leq 2x(t)$ for all $t \geq t_0$, a contradiction results if $N(x) = x^2\cos x$. This implies that x and therefore k_1 and ϕ are bounded and that $\dot{\phi} \in \mathcal{L}^2$.

The proof of stability of adaptive systems in Chapter 5 depends on (i) the boundedness of ϕ and (ii) the fact that $\dot{\phi} \in \mathcal{L}^2$. Since both (i) and (ii) are satisfied in this problem, and since a controller structure identical to that discussed in Chapter 5 is used here, it follows that all the signals in the adaptive system are globally bounded and $e_1(t) \to 0$ as $t \to \infty$.

The discussions in this section demonstrate conclusively that the assumption regarding the knowledge of the sign of the high frequency gain k_p is not necessary for stable adaptive control.

9.3 ASSUMPTION (iii): UPPER BOUND ON THE ORDER OF THE PLANT

In Chapter 5, the knowledge of the order n of the plant was needed to determine a controller structure with enough freedom to generate the desired control input u to the plant. In particular, if an upper bound on the order of the plant is specified as $\overline{n}$, the controller has to be of dimension $2\overline{n}$. Following the relaxation of assumption (i) using the methods described in the previous section, similar efforts were also made at relaxing the other assumptions in (I). The examples following indicate that explicit knowledge of $\overline{n}$ may not be necessary for the adaptive control problem to have a positive solution.

9.3.1 Regulation of Plant with Relative Degree $n^* = 1$

Consider a plant with a transfer function $W_p(s)$ of relative degree unity satisfying assumptions (i) and (iv) in (I). From root locus arguments [27], it is evident that the system

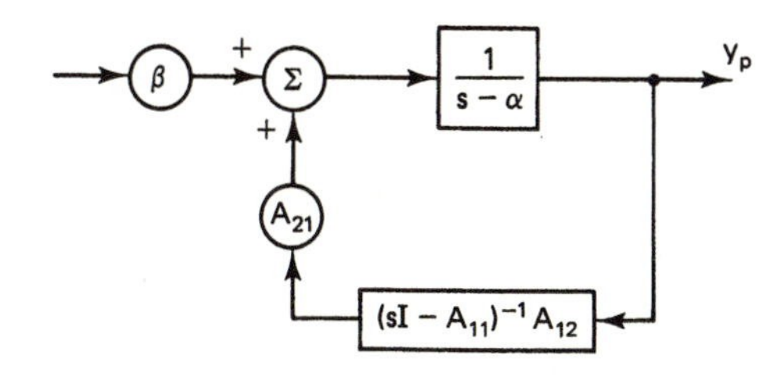

Figure 9.3 Minimal compensator: $n^* = 1$.

will be stable for a sufficiently large feedback gain. Since the sign of k_p is known, the proper sign of the feedback can also be chosen. The adaptive controller is given by

$$u = -sgn(k_p)\theta y_p, \qquad \dot{\theta} = y_p^2 \tag{9.12}$$

and can be shown to result in globally bounded solutions. In this simple problem, the order of the system is not explicitly used and a high gain output feedback is necessary.

A more general situation prevails when the sign of the high frequency gain k_p is also not known in the problem above and, hence, the controller in Eq. (9.12) cannot be used. Since the method outlined in the previous section was developed precisely to deal with this situation, attempts were made to incorporate a function $N(\theta)$ in the adaptive law in place of $sgn(k_p)$ in Eq. (9.12). The adaptive controller

$$u = N(\theta)\theta y_p, \qquad \dot{\theta} = y_p^2 \tag{9.13}$$

can be shown to result in global stability[28]. This is stated in Theorem 9.4 below.

Theorem 9.4 Assume that the plant described by the differential equation

$$\dot{x}_p = A_p x_p + b_p u; \qquad y_p = h_p^T x_p \tag{9.14}$$

is of order n, has relative degree unity and has zeros in $\mathbb{C}^-$. Then an adaptive controller of the form in Eq. (9.13) ensures that all the signals in the closed loop are bounded and $\lim_{t\to\infty} y_p(t) = 0$.

Proof. The plant in Eq. (9.14) can be described equivalently, as shown in Fig. 9.3, by a feedback system with a first-order transfer function in the forward path and an $(n-1)^{\text{th}}$ order transfer function in the feedback path. The corresponding state equations are given by

$$\begin{bmatrix} \dot{x}_1 \\ \dot{y}_p \end{bmatrix} = \begin{bmatrix} A_{11} & A_{12} \\ A_{21} & \alpha \end{bmatrix} \begin{bmatrix} x_1 \\ y_p \end{bmatrix} + \begin{bmatrix} 0 \\ \beta \end{bmatrix} u$$

where $x_1 : \mathbb{R}^+ \to \mathbb{R}^{n-1}$, and A_{11}, A_{12}, and A_{21} have appropriate dimensions. It can be shown that $\beta \neq 0$, and that A_{11} has all its eigenvalues in $\mathbb{C}^-$. Since $u = N(\theta)\theta y_p$, where $N(\theta) = \theta \cos\theta$, we have

$$\begin{aligned} \frac{d}{dt}\left[y_p^2\right] &= 2\left[\alpha + \beta\theta^2 \cos\,\theta\right] y_p^2 + 2A_{21}x_1 y_p \\ &= 2\left[\alpha + \beta\theta^2 \cos\,\theta\right] \dot{\theta} + 2A_{21}x_1 y_p. \end{aligned}$$

Hence,

$$\begin{aligned} y_p^2(t) - y_p^2(t_0) &= 2\alpha[\theta(t) - \theta(t_0)] + 2\beta\left[\theta^2(t)\sin\,\theta(t) + 2\theta(t)\cos\,\theta(t)\right. \\ &\quad \left. -2\sin\,\theta(t)\right] + 2\int_{t_0}^{t} A_{21}x_1(\tau)y_p(\tau)d\tau \\ &\quad - 2\beta\left[\theta^2(t_0)\sin\,\theta(t_0) + 2\theta(t_0)\cos\,\theta(t_0) - 2\sin\,\theta(t_0)\right]. \end{aligned}$$

Since A_{11} is asymptotically stable, it can be shown that

$$\int_{t_0}^{t} x_1(\tau)y_p(\tau)d\tau \;\leq\; M_1\|x_1(t_0)\|^2 + M_2\int_{t_0}^{t} |y_p(\tau)|^2 d\tau$$

where $M_1, M_2 \in \mathbb{R}^+$. Hence,

$$\begin{aligned} y_p^2(t) - y_p^2(t_0) &\leq 2\alpha[\theta(t) - \theta(t_0)] + 2\beta\left[\theta^2(t)\sin\,\theta(t) + 2\theta(t)\cos\,\theta(t)\right. \\ &\quad \left. -2\sin\,\theta(t)\right] + 2N_1\|x(t_0)\|^2 + 2N_2\theta(t) \\ &\quad + 2\beta\left[\theta^2(t_0)\sin\,\theta(t_0) + 2\theta(t_0)\cos\,\theta(t_0) - 2\sin\,\theta(t_0)\right] - 2N_2\theta(t_0) \end{aligned}$$

where N_1 and $N_2 \in \mathbb{R}^+$. Using the same arguments as in the proof of Theorem 9.2, we can conclude that θ and, therefore, y_p and $\dot{y}_p$ are bounded. Also $y_p \in \mathcal{L}^2$ and, hence, $\lim_{t\to\infty} y_p(t) = 0$.

9.3.2 $n^* = 2$

The specific case discussed in Section 9.3.1 indicates that condition (iii) in (I) is not necessary and that stable controllers that are substantially simpler than that given in Chapter 5 exist. A similar solution was provided for the case $n^* = 2$ in [21] under additional assumptions on the parameters of the system. This is based on the fact that any plant with relative degree two can be expressed as

$$\begin{aligned} y_p(t) &= \frac{1}{s^2 + \alpha_1 s + \alpha_2}(A_{21}x_1(t) + k_p u(t)) \\ x_1(t) &= [(sI - A_{11})^{-1}A_{12}]y_p(t) \end{aligned} \tag{9.15}$$

or equivalently by replacing the first-order transfer function in Fig. 9.3 by the second-order transfer function $k_p/(s^2 + \alpha_1 s + \alpha_2)$. If $\alpha_1 > 0$, a control input

$$u(t) = -sgn(k_p)\theta y_p(t)$$

where θ is a sufficiently large positive constant stabilizes the system so that $\lim_{t\to\infty} y_p(t) = 0$. In [21] it is shown that in the adaptive control problem, the adaptive law

$$\dot{\theta}(t) = y_p^2(t)$$

will also result in bounded solutions with $\lim_{t\to\infty} y_p(t) = 0$.

For an LTI plant with a transfer function $W_p(s)$ with zeros in $\mathbb{C}^-$ and whose parameters as well as the order, and sign of high frequency gain are unknown, and whose relative degree n^* is less than or equal to two, a stable controller was shown to exist in [16]. The same problem has been extended to the case when $n^* \leq 3$ in [19]. In the latter case, a nonlinear function of a single parameter $\theta \in \mathbb{R}$ is used in the controller which is shown to be stable $\forall\, \theta \geq \theta_0$, $\theta_0 \in \mathbb{R}$. In the following sections, we discuss a result by Morse [20], where once again, use is made of a high gain but the parametrization of the controller is chosen so as to assure the existence of a stabilizing controller. This is contained in the following section.

9.3.3 Arbitrary n^*

Let a plant satisfy assumptions (i), (ii), and (iv), that is, the order of the plant transfer function is unknown but the sign of the high frequency gain, and the relative degree n^* are known, and zeros of the plant transfer function lie in $\mathbb{C}^-$. The theorem following assures the existence of a high gain controller for such a plant, provided an upper bound on the absolute value of the high frequency gain is known.

Theorem 9.5 Let a linear time-invariant finite dimensional plant be described by the differential equation

$$\dot{x}_p = A_p x_p + b_p u; \qquad y_p = h_p^T x_p \tag{9.16}$$

where the triple $\{h_p, A_p, b_p\}$ is controllable and observable. Let assumptions (i), (ii), and (iv) in (I) be satisfied and the high frequency gain k_p be such that $|k_p| \in (0, \overline{k})$. Then a controller of the form

$$\begin{aligned}
\dot{\omega} &= A_c\omega + b_c u && (9.17\text{a})\\
u &= f_1(\theta)y_p + f_2^T(\theta)\omega && (9.17\text{b})\\
\dot{\theta} &= y_p^2 && (9.17\text{c})
\end{aligned}$$

exists such that all the solutions of the overall system are bounded and $\lim_{t\to\infty} y_p(t) = 0$, where

$$A_c = \begin{bmatrix} -\lambda & 1 & & & \\ 0 & -\lambda & 1 & & \\ & & \ddots & & \\ & & & -\lambda & 1 \\ & & & & -\lambda \end{bmatrix} \in \mathbb{R}^{(n^*-1)\times(n^*-1)}, \quad b_c = \begin{bmatrix} 0 \\ 0 \\ \vdots \\ 0 \\ 1 \end{bmatrix},$$

$$f_1(\theta) = -sgn(k_p)a_1\theta^{n^*}, \qquad f_2(\theta) = \begin{bmatrix} -a_2\theta^{n^*-1} \\ -a_3\theta^{n^*-2} \\ \vdots \\ -a_{n^*-1}\theta^2 \\ -a_{n^*}\theta \end{bmatrix}, \qquad \lambda > 0$$

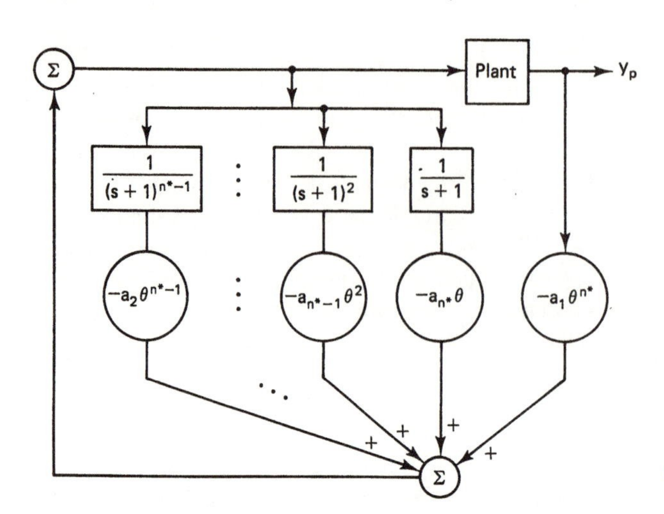

Figure 9.4 High gain stabilization.

and the polynomial

$$s^{n^*} + a_{n^*}s^{n^*-1} + \cdots + a_2 s + a_1\overline{k} \text{ is Hurwitz.} \tag{9.18}$$

Figure 9.4 contains a block diagram of the plant together with the controller with $\lambda = 1$ and $sgn(k_p) = 1$. A few comments regarding the structure of the controller are in order. The vector $\omega(t)$ is the output of a linear dynamical system with an input u and the latter is a linear combination of y_p and ω. The coefficients are nonlinear functions of a single parameter θ with $f^T(\theta) = [f_1(\theta), f_2^T(\theta)]$ having elements with decreasing powers of θ. The algebraic part of the problem is to assure that the overall system is asymptotically stable for all $\theta \geq \theta_0$ and some constant θ_0. The analytic part consists in showing that the adaptive law of Eq. (9.17c) results in $\theta(t)$ converging to some finite value asymptotically.

The motivation for the choice of the controller structure can be illustrated by considering a plant with relative degree two for which $\alpha_1 < 0$ in Eq. (9.15). The difficulty in assuring stability arises in this case because the poles of the overall system will lie in the right half plane for a high gain. However, the root locus can be shifted to the left half plane using a cascade compensator of the form $(s+1)/(s+1+\theta)$ for a large positive value of θ. A corresponding feedback gain of $O(\theta^2)$[1] will move the closed loop poles into $\mathbb{C}^-$. This results in the controller structure shown in Fig. 9.5.

The following lemmas related to Hurwitz polynomials are found to be useful in the proof of Theorem 9.5.

Lemma 9.1 If $p(s)$ and $q(s)$ are polynomials in the variable s of degrees $n-1$ and n respectively with $p(s)$ Hurwitz and $q(s)$ monic, then the polynomial $q(s) + \theta p(s)$ is Hurwitz for all $\theta \geq \theta_0$ for some constant $\theta_0 > 0$.

The proof of Lemma 9.1 follows directly from root locus arguments.

[1]Here $O(\theta^2)$ implies that $\lim_{\theta\to\infty} \frac{g(\theta)}{\theta^2} = k_1$, where $g(\theta) = O(\theta^2)$ and $k_1 \in \mathbb{R}$.

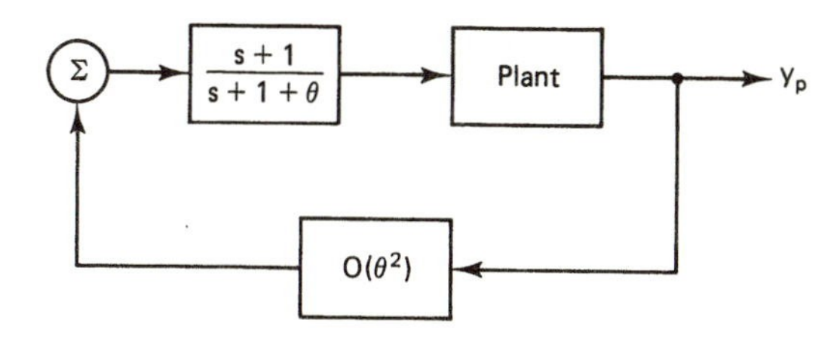

Figure 9.5 High gain stabilization: $n^* = 2$.

Lemma 9.2 The polynomial

$$s^n + f_1(\theta)a_n s^{n-1} + f_2(\theta)a_{n-1}s^{n-2} + \cdots + f_n(\theta)a_1$$

is Hurwitz for all $\theta \geq \theta_0 > 0$ if $f_i(\theta)$ is a monic polynomial in θ of degree i and

$$s^n + a_n s^{n-1} + a_{n-1}s^{n-2} + \cdots + a_1 \text{ is a Hurwitz polynomial in } s. \tag{9.19}$$

Proof. This follows from the fact that if the condition in Eq. (9.19) is satisfied, then

$$s^n + \theta a_n s^{n-1} + \theta^2 a_{n-1}s^{n-2} + \cdots + \theta^n a_1 \text{ is a Hurwitz polynomial } \forall \theta > 0 \tag{9.20}$$

since the roots of Eq. (9.20) are at $-\lambda_i\theta$, where $-\lambda_i$ are the roots of (9.19).

Lemma 9.3 If

$$q_c(s) = s^m + a_m s^{m-1} + \cdots + a_2 s + a_1$$

is Hurwitz, $q_n(s)$ and $q_{n-1}(s)$ are monic polynomials in s of degree n and $n-1$ respectively, and $q_{n-1}(s)$ is Hurwitz, then the polynomial

$$\begin{aligned} q_m(s) &= q_n(s)\left[(s+\lambda)^{m-1} + \theta a_m(s+\lambda)^{m-2} + \cdots + \theta^{m-1}a_2\right] \\ &\quad + \theta^m a_1 q_{n-1}(s) \text{ is Hurwitz } \forall \theta \geq \theta_0, \lambda > 0 \end{aligned}$$

for some positive constant θ_0.

Proof. Since $q_c(s)$ is Hurwitz, we obtain from Lemma 9.2 that

$$p(s) = s^m + \theta a_m s^{m-1} + \cdots + \theta^{m-1}a_2 s + \theta^m a_1 \text{ is Hurwitz } \forall\, \theta > 0$$

and that $p(s)q_{n-1}(s)$ is Hurwitz. If $q_{n-1}(s) = s^{n-1} + q_{n-2}(s)$ where $q_{n-2}(s)$ is a polynomial of degree $n-2$,

$$\begin{aligned} p(s)q_{n-1}(s) &= s^n\left[s^{m-1} + \theta a_m s^{m-2} + \cdots + \theta^{m-1}a_2\right] + \theta^m a_1 q_{n-1}(s) \\ &\quad + \left[s^m + \theta a_m s^{m-1} + \cdots + \theta^{m-1}a_2 s\right] q_{n-2}(s). \end{aligned}$$

Hence, there exists a constant θ_1 such that the polynomial

$$s^n\left[s^{m-1} + \theta a_m s^{m-2} + \cdots + \theta^{m-1}a_2\right] + \theta^m a_1 q_{n-1}(s) \text{ is Hurwitz for } \theta \geq \theta_1.$$

Therefore $q_m(s)$ is Hurwitz for all $\theta \geq \theta_0$ for some $\theta_0 \geq \theta_1$.

Proof of Theorem 9.5. From Lemma 9.3 and Eq. (9.18), it follows that the polynomial $R^*(s)$ given by

$$\begin{aligned} R^*(s) &= R_p(s)\left[(s+\lambda)^{n^*-1} + \theta a_{n^*}(s+\lambda)^{n^*-2} + \cdots + \theta^{n^*-1}a_2\right] \\ &\quad + \theta^{n^*} a_1 k_p Z_p(s)(s+\lambda)^{n^*-1} \end{aligned}$$

has its roots in $\mathbb{C}^-$ for all $\theta \geq \theta^*$, where θ^* is a positive constant. Since $R^*(s)$ is the characteristic polynomial of the closed-loop system for constant values of θ, the system is asymptotically stable for every constant value $\theta \geq \theta^*$.

The closed-loop equations describing the plant together with the controller are obtained from Eqs. (9.16) and (9.17) as

$$\begin{aligned} \dot{x} \triangleq \begin{bmatrix} \dot{x}_p \\ \dot{\omega} \end{bmatrix} &= \begin{bmatrix} A_p + b_p f_1(\theta) h_p^T & b_p f_2^T(\theta) \\ b_c f_1(\theta) h_p^T & A_c + b_c f_2^T(\theta) \end{bmatrix} \begin{bmatrix} x_p \\ \omega \end{bmatrix} \triangleq A(\theta)x \\ y_p &= \begin{bmatrix} h_p^T & 0 \end{bmatrix} \begin{bmatrix} x_p \\ \omega \end{bmatrix} \triangleq C(\theta)x. \end{aligned} \tag{9.21}$$

Since the plant is controllable and the zeros of the plant lie in the left-half plane, $(C(\theta), A(\theta))$ is detectable for all θ. Further, the parametrization of the controller is such that the closed-loop system is asymptotically stable for all $\theta \geq \theta_0$, where θ_0 is some constant, that is, $A(\theta)$ is asymptotically stable for $\theta \geq \theta_0$.

From Eq. (9.17c) it follows that $\theta(t)$ is a nondecreasing function of time and, hence, can either converge to a finite limit $\overline{\theta}$ or be unbounded. We show that the latter leads to a contradiction and, hence, $\lim_{t\to\infty} \theta(t) = \overline{\theta}$.

Let $\lim_{t\to\infty} \theta(t) = \infty$. Then there exists a time t_1 such that $\theta(t) \geq \theta^*$ for all $t \geq t_1$ and, hence, $A(\theta)$ is an asymptotically stable matrix for every constant value θ for $t \geq t_1$. Since $(C(\theta), A(\theta))$ is detectable for every value of θ, there exists a symmetric, proper, rational matrix $Q(\theta)$ such that [20]

$$Q(\theta)A(\theta) + A^T(\theta)Q(\theta) = -C^T(\theta)C(\theta)$$

and $Q(\theta)$ is positive-semidefinite for all $\theta \geq \theta^*$. Hence, if $R(\theta)$ is defined as $R(\theta) = Q(\theta)/\theta$, $R(\theta)$ is symmetric, strictly proper, and positive-semidefinite for $\theta \geq \theta^*$. Also it can be shown that [20]

$$\frac{dR}{d\theta} \leq 0 \qquad \forall\, \theta \geq \theta^*. \tag{9.22}$$

Defining an indicator function $V(x, \theta)$ as

$$V(x,\theta) = x^T R(\theta) x,$$

the time derivative of V along the trajectory of Eqs. (9.21) and (9.17c) can be computed as

$$\begin{aligned} \dot{V}(x,\theta) &= x^T\left[R(\theta)A(\theta)+A^T(\theta)R(\theta)\right]x+2x^T\frac{dR}{d\theta}\dot{\theta}x \\ &= -\frac{y_p^2}{\theta}+2x^T\frac{dR}{d\theta}x\dot{\theta} \\ &\leq -\frac{y_p^2}{\theta} = -\frac{\dot{\theta}}{\theta} \qquad \forall\theta\geq\theta^* \end{aligned} \tag{9.23}$$

since $\dot{\theta}\geq 0$ and $dR/d\theta\leq 0$ for all $\theta\geq\theta^*$. By integrating both sides of the inequality in Eq. (9.23), it follows that

$$V(t) \leq -\log(\theta(t))+c_1$$

where c_1 is a constant. If $\theta(t)$ were to grow in an unbounded fashion, $V(t)$ will become negative at some time t_1, which results in a contradiction. Hence, $\theta(t)$ is bounded and $\lim_{t\to\infty}\theta(t)=\overline{\theta}$.

From the boundedness of $\theta(t)$ it readily follows that $\lim_{t\to\infty}y_p(t)=0$ as shown below: Since $\lim_{t\to\infty}\theta(t)=\overline{\theta}$, $\lim_{t\to\infty}A(\theta)=\overline{A}$. Since $(C,\overline{A})$ is detectable, there exists a constant vector h such that $\overline{A}+hC$ is asymptotically stable. Hence, the differential equation describing the overall system has the form

$$\dot{x} = \left[\overline{A}+hC\right]x+\left[A(t)-\overline{A}\right]x-hy_p.$$

Since $[\overline{A}+hC]$ is asymptotically stable, $\lim_{t\to\infty}|A(t)-\overline{A}|=0$ and $y_p\in\mathcal{L}^2$, it follows that x is bounded,

$$\lim_{t\to\infty}x(t) = 0 \qquad \text{and} \qquad \lim_{t\to\infty}y_p(t) = 0.$$

Comment 9.1 By Theorem 9.5, a stabilizing controller exists whose order depends on the relative degree rather than the order of the system. From a practical point of view, a high gain approach may not be attractive but it nevertheless solves the regulation problem without explicit knowledge of the order of the plant. The difficulty lies in extending the result to model-following.

9.3.4 Universal Controller

Another class of adaptive controllers that requires less prior information than that considered in the previous chapters has been termed by its supporters, somewhat exaggeratedly, as *universal controllers* [8,15]. Some of the controllers discussed in the previous section also naturally fall into this category. The authors of [8,15] readily admit that the practicability of such controllers is quite dubious and that their merit lies primarily in the demonstration that stable adaptive controllers can be designed, in theory, with very little prior information. Since we anticipate more research effort in this direction in the future, we comment briefly on two representative papers.

If $\mathcal{P}$ denotes the class of plants to be stabilized, Martensson makes the rather weak assumption in [15] that each element of the set $\mathcal{P}$ admits an lth order stabilizing controller, $l \geq 0$. The adaptive controller carries out an automated dense search throughout the set $\mathcal{C}$ of all possible linear controllers, which includes a subset $\mathcal{C}_s$ of stabilizing controllers. The search scheme is such that the controller passes through the set C_s sufficiently slowly, which ensures asymptotic regulation and the cessation of the search. For a first-order plant, it is shown using simulation studies in [8] that the controller, as described above, results in a peak overshoot of 300,000!

In [8], two more assumptions are made besides those mentioned in [15] regarding the existence of a stabilizing controller. The first is that an upper bound on the order of the plant is known and the second is that the set of all admissible plants is compact. It is then shown that a switching control law results in the exponential stability of the system. The gain matrix of the controller can assume one of a finite number of values and the control law determines the instants at which the controller switches from one matrix to another. Hence, the controller is piecewise linear and time-invariant. It is shown that after a finite number of switches, the controller remains fixed with a constant parameter matrix. Although this method is also far from being practicable, its theoretical merits warrant a brief discussion.

Since $\mathcal{P}$ is assumed to be compact, there exists a finite number of sets $\mathcal{P}_1, \ldots, \mathcal{P}_r$ such that (i) $\cup_i \mathcal{P}_i = \mathcal{P}$, and (ii) a constant gain matrix K_i exists such that any plant in $\mathcal{P}_i$ together with a matrix K_i in the controller is asymptotically stable, $\forall i$. The switching compensator switches from K_1 to K_2 to K_3 and so on and the critical point is to determine when the switching is to take place. If the given unstable plant belongs to $\mathcal{P}_\ell$, according to the rule suggested, no further switching takes place when the gain matrix assumes the value K_ℓ. The decision regarding the switching instants is made by observing the growth of an indicator function over finite windows of time.

9.4 ASSUMPTION (iv): ZEROS OF $W_p(s)$ IN $\mathbb{C}^-$

In Section 9.2 we considered cases where assumptions (i) and (iii) could be relaxed simultaneously. However, most of the methods assumed that the relative degree n^* of $W_p(s)$ is known and that the zeros of $W_p(s)$ lie in $\mathbb{C}^-$. The former is needed to set up a reference model, while the latter assures that the state of the overall system does not grow in an unbounded fashion even while the output error tends to zero.

As mentioned in the introduction, after the adaptive control problem in the ideal case was solved, it was felt that assumption (iv) was particularly restrictive since nonminimum phase plants are ubiquitous. Further, when a continuous-time system is discretized, the resultant discrete-transfer function almost invariably has zeros outside the unit circle in the complex plane. In view of these facts, many efforts were made to determine conditions under which stable adaptive control is possible even when the zeros of the plant transfer function lie in the right-half of the complex plane. In this context, a natural distinction can be made between direct control and indirect control as well as model-following and regulation problems.

9.4.1 Direct Control

The direct control approach can be used for both regulation and tracking and can be based on either model-following or adaptive pole placement. In the former, the poles of the overall system are shifted by state feedback while its zeros are located by feedforward compensation. This implies the cancellation of the zeros of the plant transfer function by the poles of the adaptive feedforward compensator. Hence, with the methods currently in use for model-following, assumption (iv) may be necessary if the adaptive system is to be globally stable. However, no rigorous proof of this is available at the present time.

For system regulation, on the other hand, assumption (iv) may not be necessary if the direct approach is used, since the objective can be achieved merely by locating the poles of the overall system at desired values in the open left-half plane. However, while this condition is related to the algebraic part of the solution, the analytic part also has to be taken into account. For example, assume that the feedback parameter $\theta(t)$ tends to some limit $\overline{\theta}$, which is different from the desired value θ^*. The limiting value may be such that one or more poles of the resultant overall system transfer function coincide with the unstable zeros of $W_p(s)$. This can cause the state of the system to grow in an unbounded fashion even though $y_p(t)$ is bounded and $\lim_{t\to\infty} y_p(t) = 0$.

The theoretical questions that arise while dealing with the problems above can be conveniently discussed using the concept of tunability of an adaptive system proposed by Morse [20]. If the overall adaptive system is described by the differential equation

$$\dot{x} = A(\theta)x + b(\theta)r \qquad y = C(\theta)x \tag{9.24}$$

where $\theta : \mathbb{R}^+ \to \mathbb{R}^m$, the parameter θ can generally be shown to be bounded and hence belongs to some compact set $\subset \mathbb{R}^m$. In such a case, the following definition of tunability is found to be relevant.

Definition 9.1 [20] The system of Eq. (9.24) is said to be *tunable* with respect to $S \subset \mathbb{R}^m$ if for each fixed $\theta \in S$ and each bounded input r, every possible solution $x(t)$ of Eq. (9.24) for which $y(t) \equiv 0$ is bounded for all $t \geq t_0$.

In Eq. (9.24), the output signal y together with some of the state variables determines how the parameter is to be adjusted. Hence, for stability, we have to assure that when $y(t)$ is identically zero, (that is, when θ is a constant) the state $x(t)$ of the system is bounded. This is expressed in the lemma following.

Lemma 9.4 The system in Eq. (9.24) is tunable with respect to S if and only if $(C(\theta), A(\theta))$ is detectable for each $\theta \in S$.

From the definitions above, it follows that if tunability in $\mathbb{R}^m$ is desired, then assumption (iv) is necessary for the class of systems described by Eq. (9.24). Hence, any attempt made to relax this assumption must imply either

(a) the space S in which θ lies is limited in some fashion to be a proper subset of $\mathbb{R}^m$ and satisfies the conditions of Lemma 9.4, (9.25a)

or

(b) the form of the equations describing the adaptive system are different from that given in Eq. (9.24). (9.25b)

The comments above apply to both direct and indirect methods.

Pole placement using the direct method. Although different ways of using the direct method for adaptive pole placement have been suggested, the approach due to Elliott [6] is particularly noteworthy. Using an interesting parametrization of the plant, this approach leads to stable adaptive control of plants with arbitrary zeros. Since this parametrization of the plant appears to have wider applications in adaptive control theory, we present its essential features below.

Let the plant to be controlled adaptively be described by the differential equation

$$\dot{x}_p = A_p x_p + b_p u; \qquad y_p = h_p^T x_p$$

and let its transfer function be $W_p(s) = Z_p(s)/R_p(s)$. It is assumed that $R_p(s)$ is a monic polynomial of degree n, $Z_p(s)$ is a polynomial of degree $\leq n-1$, and $Z_p(s)$ and $R_p(s)$ are relatively prime. The objective is to determine an adaptive feedback controller for generating the control input u such that the transfer function of the overall system tends to that of a linear time-invariant system whose poles are at desired locations in the open left-half of the complex plane. To achieve this, the following standard controller structure is used:

$$\begin{aligned} \dot{\omega}_1 &= F\omega_1 + gu \\ \dot{\omega}_2 &= F\omega_2 + gy_p \\ u &= \theta_1^T(t)\omega_1 + \theta_2^T(t)\omega_2 + r \end{aligned} \tag{9.26}$$

where $F \in \mathbb{R}^{n\times n}$ is asymptotically stable, (F, g) is controllable, and $\theta_1, \theta_2 : \mathbb{R}^+ \to \mathbb{R}^n$. For constant values of θ_1, and θ_2, the control input can be expressed as

$$u = \frac{C(s)}{\lambda(s)}u + \frac{D(s)}{\lambda(s)}y_p + r$$

where $\lambda(s)$ is the characteristic polynomial of F and is Hurwitz, $C(s)$ and $D(s)$ are polynomials of degree $n-1$ and r is an external reference input which is bounded.

The method in [6] is based on the following two facts, which can be deduced from the coprimeness of $Z_p(s)$ and $R_p(s)$.

Fact 1. Constant parameters θ_1^* and θ_2^* and hence polynomials $C^*(s)$ and $D^*(s)$ exist such that

$$R_p(s)[\lambda(s) - C^*(s)] - Z_p(s)D^*(s) = R_m(s)\lambda(s).$$

Fact 2. Polynomials $A^*(s)$ and $B^*(s)$ of degree $\leq n-1$ exist such that

$$A^*(s)R_p(s) + B^*(s)Z_p(s) = 1.$$

As mentioned in Chapter 5, both fact 1 and fact 2 follow directly from the Bezout identity.

From fact 1, we have

$$\lambda(s) = C^*(s) + D^*(s) \cdot \frac{Z_p(s)}{R_p(s)} + R_m(s)\lambda(s) \cdot \frac{1}{R_p(s)} \tag{9.27}$$

and from fact 2,

$$\frac{1}{R_p(s)} = A^*(s) + B^*(s) \cdot \frac{Z_p(s)}{R_p(s)}. \tag{9.28}$$

Combining Eqs. (9.27) and (9.28), we obtain

$$\lambda(s) = C^*(s) + D^*(s) \cdot \frac{Z_p(s)}{R_p(s)} + R_m(s)\lambda(s)\left(A^*(s) + B^*(s) \cdot \frac{Z_p(s)}{R_p(s)}\right). \tag{9.29}$$

Let $f(s)$ be a monic Hurwitz polynomial of degree $3n$. Dividing both sides of Eq. (9.29) by $f(s)$, we obtain

$$\begin{aligned}\frac{\lambda(s)}{f(s)} &= \frac{C^*(s)}{f(s)} + \frac{D^*(s)}{f(s)} \cdot \frac{Z_p(s)}{R_p(s)} + \frac{R_m(s)\lambda(s)A^*(s)}{f(s)} \\ &\quad + \frac{R_m(s)\lambda(s)B^*(s)}{f(s)} \cdot \frac{Z_p(s)}{R_p(s)}.\end{aligned} \tag{9.30}$$

Every term in Eq. (9.30) is now a proper rational transfer function. Defining $\overline{u}(t) = [1/f(s)]u(t)$ and $\overline{y}_p(t) = [1/f(s)]y_p(t)$, and assuming zero initial conditions, we have

$$\overline{y}_p(t) = W_p(s)\overline{u}(t).$$

From Eq. (9.30), using the input u to generate $\overline{u}$, we obtain

$$\begin{aligned}\lambda(s)\overline{u}(t) &= C^*(s)\overline{u}(t) + D^*(s)\overline{y}_p(t) + R_m(s)\lambda(s)\left(A^*(s)\overline{u}(t) + B^*(s)\overline{y}_p(t)\right) \\ &= \overline{\theta}^{*T}\overline{\omega}(t)\end{aligned}$$

where

$$\begin{aligned}\overline{\theta}^* &= [c_1, \ldots, c_n, d_1, \ldots, d_n, a_1, \ldots, a_n, b_1, \ldots, b_n,]^T \\ \overline{\omega} &= [\overline{u}, s\overline{u}, \ldots, s^{n-1}\overline{u}, \quad \overline{y}_p, s\overline{y}_p, \ldots, s^{n-1}\overline{y}_p, \\ &\qquad R_m(s)\lambda(s)\overline{u}, sR_m(s)\lambda(s)\overline{u}, \ldots, s^{n-1}R_m(s)\lambda(s)\overline{u}, \\ &\qquad R_m(s)\lambda(s)\overline{y}_p, sR_m(s)\lambda(s)\overline{y}_p, \ldots, s^{n-1}R_m(s)\lambda(s)\overline{y}_p]^T, \\ C^*(s) &= c_1 + c_2 s + \cdots + c_n s^{n-1}, \qquad D^*(s) = d_1 + d_2 s + \cdots + d_n s^{n-1} \\ A^*(s) &= a_1 + a_2 s + \cdots + a_n s^{n-1}, \qquad B^*(s) = b_1 + b_2 s + \cdots + b_n s^{n-1}.\end{aligned}$$

If

$$\lambda(s)\overline{u}(t) \stackrel{\triangle}{=} z(t)$$

then $z(t)$ and $\overline{\omega}(t)$ are signals that can be measured for all $t \geq t_0$, while $\overline{\theta}^*$ represents the set of of unknown coefficients of the polynomials $C^*(s)$, $D^*(s)$, $A^*(s)$, and $B^*(s)$. The importance of this parametrization lies in the fact that the first $2n$ parameters of $\overline{\theta}^*$ are also precisely the control parameters needed to locate the closed-loop poles of the adaptive system at the roots of $R_m(s)\lambda(s)$. Hence, the problem is to estimate $\overline{\theta}^*$ in the equation

$$\overline{\theta}^{*T}\overline{\omega}(t) = z(t).$$

If a time-varying parameter $\overline{\theta}(t)$ defined by the equation

$$\overline{\theta}^T(t)\overline{\omega}(t) = \widehat{z}(t)$$

is used, the error equation

$$e_z(t) \stackrel{\triangle}{=} \widehat{z}(t) - z(t) = \overline{\phi}^T(t)\overline{\omega}(t)$$

is obtained where $\overline{\phi} = \overline{\theta} - \overline{\theta}^*$. The adaptive procedure must be such that $\overline{\phi}(t) \to 0$ as $t \to \infty$ and all the signals in the closed-loop system remain bounded. The structure of this controller is as shown in Fig. 9.6.

In [6], $\overline{\theta}(t)$ is held constant, but is estimated N times using a least-squares estimation procedure over an interval. At the end of the interval, $\overline{\theta}$ is changed using the final update obtained over this interval and the covariance matrix containing the adaptive gains is reset to a predetermined constant matrix (refer to Chapter 6, Page 265). A persistently exciting reference input with $2n$ distinct frequencies is used. It is shown that $\overline{\phi}$ tends to zero and therefore all the signals are bounded. Unstable pole-zero cancellations are avoided in this problem by using a persistently exciting reference input. Therefore the adaptive law ensures that the parameter asymptotically enters the region where $(C(\overline{\theta}), A(\overline{\theta}))$ is detectable and remains there.

Several other approaches have been suggested for the direct control of nonminimum phase systems. In [1,25], a conceptual approach is presented where a bilinear estimation problem needs to be solved, but the practical implementation of such an algorithm is not discussed. In [9], an adaptive law based on input-matching is shown to result in global stability, but is restricted to plants that are known a priori to be stable.

9.4.2 Indirect Control

The indirect approach to adaptive control was defined in Chapter 1 as that in which the plant parameters are estimated continuously and these estimates, in turn, are used to find the control parameters. In Chapter 3, using such an approach, the adaptive control of first-order plants and, more generally, plants whose state variables are accessible was discussed. Later, in Chapter 5, when the plant transfer function satisfies the assumptions

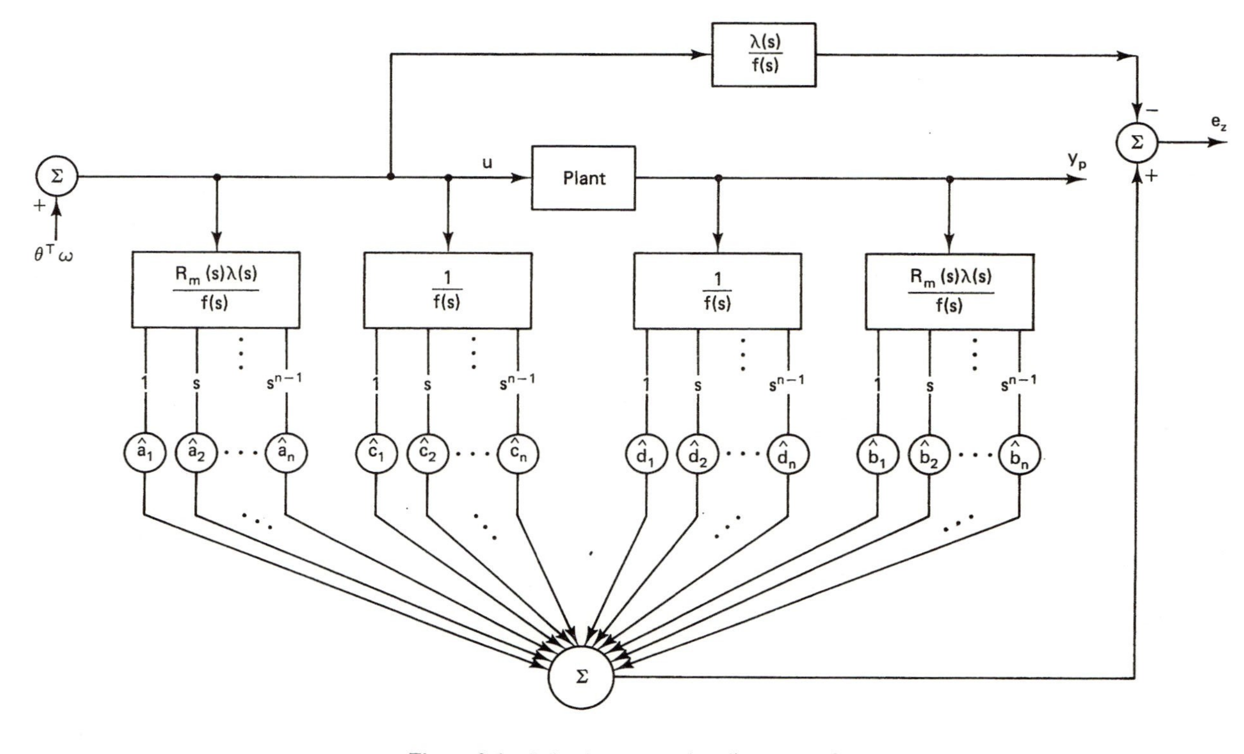

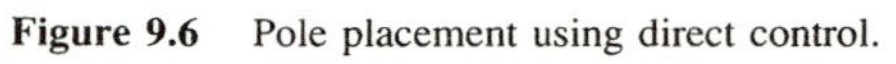

Figure 9.6 Pole placement using direct control.

in (I), the method was shown to be equivalent to the direct control method when used in the context of MRAC.

This implies that in the ideal case, there is little basis to choose between direct and indirect control methods. The researchers who used the latter have maintained that its advantages are best realized precisely when the plants do not satisfy the assumptions (I). Since the plant parameters can be estimated with no more prior information than the order of the plant transfer function (as shown in Chapter 4), it is argued that, if successful indirect control control can be achieved with these estimates, it would follow that indirect control needs less information than the direct one. This argument motivated attempts to adaptively control plants with arbitrarily located zeros. We discuss some of these attempts below.

Statement of the problem. Let the plant to be controlled have the transfer function $Z_p(s)/R_p(s)$ where the roots of both polynomials $Z_p(s)$ and $R_p(s)$ can lie anywhere in $\mathbb{C}$. Let a feedback controller identical to that in Eq. (9.26) be used. For a constant value of the vector $\theta = [\theta_1^T, \theta_2^T]^T$, the transfer function of the feedforward and feedback controllers are given by $(\lambda(s) - C(s))/\lambda(s)$ and $D(s)/\lambda(s)$ where the elements of the vectors θ_1 and θ_2 are the coefficients of the polynomials $C(s)$ and $D(s)$ respectively, and $\lambda(s)$ is a known monic Hurwitz polynomial of degree n. Let

$$\begin{aligned} Z_p(s) &= z_0 + z_1 s + \cdots + z_m s^m, \quad R_p(s) = r_0 + r_1 s + \cdots + r_{n-1} s^{n-1} + s^n \\ z &= [z_0, z_1, z_2, \ldots z_m]^T, \quad r = [r_0, r_1, \ldots, r_{n-1}]^T \text{ and } p = [z^T, r^T]^T. \end{aligned}$$

The plant parameter vector p is estimated at every instant as $\widehat{p}$, and the control parameter estimate $\widehat{\theta}$ is determined using $\widehat{p}$ as

$$\widehat{R}_p(s)[\lambda(s) - \widehat{C}(s)] - \widehat{Z}_p(s)\widehat{D}(s) = R_m(s)\lambda(s) \tag{9.31}$$

where

$$\begin{aligned} P_n &= [1, s, \ldots s^n]^T, \quad \widehat{Z}_p(s) = \widehat{z}^T P_{n-1}, \quad \widehat{R}_p(s) = \widehat{r}^T P_{n-1} + s^n, \\ \widehat{p} &= [\widehat{z}^T, \widehat{r}^T]^T, \qquad \widehat{\theta} = [\widehat{\theta}_1^T, \widehat{\theta}_2^T]^T, \\ \widehat{C}(s) &= \widehat{\theta}_1^T P_{n-1}, \qquad \widehat{D}(s) = \widehat{\theta}_2^T P_{n-1}, \end{aligned}$$

and $R_m(s)$ is a Hurwitz polynomial of degree n whose roots are the desired poles of the overall system. The transfer function of the overall system for constant values of $\widehat{\theta}$ is given by $\widehat{W}_c(s)$ where

$$\widehat{W}_c(s) = \frac{Z_p(s)\lambda(s)}{R_p(s)\left[\lambda(s) - \widehat{C}(s)\right] - Z_p(s)\widehat{D}(s)}. \tag{9.32}$$

Two difficulties are encountered in the procedure above. The first occurs when $\widehat{Z}_p(s)$ and $\widehat{R}_p(s)$ have a common root. In such a case, solutions for $\widehat{C}(s)$ and $\widehat{D}(s)$ cannot be found for Eq. (9.31). The second difficulty arises when the zeros of $\lambda(s) - \widehat{C}(s)$ and $Z_p(s)$ have a common factor. This implies that the overall transfer function in Eq. (9.32) has hidden modes that are unstable. Hence, the output of the plant may tend to zero, while the input becomes unbounded. The methods suggested for the control of plants with arbitrarily located zeros should therefore circumvent both these difficulties. As mentioned in Section 9.4.1, relaxation of assumption (iv) is possible when either θ is limited to a proper subset of $\mathbb{R}^m$, or the form of the differential equations describing the adaptive system is different from that given in Eq. (9.24). The methods can be divided into the following four categories and can be considered as instances where a stable adaptive controller is designed by using one of the two methods above. Only the principal ideas contained in each category are described here. For a detailed description, as well as the proof of stability, the reader is referred to the source papers. Also, while the original results are stated in discrete time, the discussions are presented using their continuous-time counterparts.

(i) Local stability. It is well known that the coprimeness of two polynomials $Z(s)$ and $R(s)$ can be established by demonstrating the nonsingularity of the related Sylvester matrix. If the two polynomials are given by

$$\begin{aligned} Z(s) &= z_0 + z_1 s + z_2 s^2 + \cdots + z_n s^n \\ R(s) &= r_0 + r_1 s + r_2 s^2 + \cdots + r_n s^n \end{aligned}$$

the Sylvester matrix has the form [29]

$$S(Z,R) = \begin{bmatrix} z_0 & z_1 & \cdots & \cdots & z_n & 0 & 0 & \cdots & 0 \\ 0 & z_0 & \cdots & \cdots & z_{n-1} & z_n & 0 & \cdots & 0 \\ \vdots & & & & \vdots & & & & \vdots \\ 0 & 0 & \cdots & z_0 & z_1 & z_2 & \cdots & & z_n \\ r_0 & r_1 & \cdots & \cdots & r_n & 0 & 0 & \cdots & 0 \\ 0 & r_0 & \cdots & & r_{n-1} & r_n & 0 & \cdots & 0 \\ \vdots & & & & \vdots & & & & \vdots \\ 0 & 0 & \cdots & r_0 & r_1 & r_2 & \cdots & & r_n \end{bmatrix}.$$

Eq. (9.31) can be expressed using the Sylvester matrix $S[\widehat{Z}_p, \widehat{R}_p]$ of $\widehat{Z}_p$ and $\widehat{R}_p$ as

$$S[\widehat{Z}_p, \widehat{R}_p]\widehat{\theta} = q \tag{9.33}$$

where $R_m(s)\lambda(s) = q^T P_{2n-1} + s^{2n}$. Since $Z_p(s)$ and $R_p(s)$ are coprime, it follows that $S[Z_p, R_p]$ is nonsingular [29]. Since the determinant of a matrix is a continuous function of the matrix elements, it follows that there exists an $\epsilon > 0$ such that if

$$D_\epsilon \stackrel{\Delta}{=} \{\widehat{p} \mid \|\widehat{p} - p\| < \epsilon\}$$

where, as defined earlier, $p = [z^T ; r^T]$, then $S[\widehat{Z}_p, \widehat{R}_p]$ is nonsingular for all $\widehat{p} \in D_\epsilon$. Therefore Eq. (9.33) has a solution for all $\widehat{p} \in D_\epsilon$. This implies that the overall system is detectable for all $\widehat{p} \in D_\epsilon$. Therefore the closed loop system is tunable and it can be shown that an adaptive law for adjusting $\widehat{p}$ can be determined and together with Eq. (9.33), it leads to bounded solutions. This idea is contained in [10,14].

(ii) Region D where no unstable pole-zero cancellation takes place. Let a set D in parameter space be defined by

$$D \stackrel{\Delta}{=} \{p | \, p_{\min,i} \leq p_i \leq p_{\max,i}\}$$

where p_i is the ith element of p, and $p_{\min,i}$ and $p_{\max,i}$ are specified scalars, $i = 1, \ldots, 2n$. Let it be further assumed that $\widehat{Z}_p(s)$ and $\widehat{R}_p(s)$ are coprime for all $\widehat{p}$ in D. Then it is possible to design an adaptive law for adjusting $\widehat{p}$ which ensures that $\widehat{p}$ converges to D asymptotically. This, in turn, assures the boundedness of all the signals in the system [12]. The difficulty encountered here is in determining the region D that is practically meaningful.

(iii) Persistently exciting inputs and a bound on the Sylvester determinant. Another method of dealing with plant zeros in the right half plane has been used in [2,3,6,11]. This is based on evaluating the Sylvester matrix of the estimated polynomials $\widehat{Z}_p(s)$ and $\widehat{R}_p(s)$ of the plant transfer function on-line.

One measure of the coprimeness of $Z_p(s)$ and $R_p(s)$ is the magnitude of the determinant of S, denoted by $|S|$. The closer $|S|$ is to zero, the closer the polynomials come to having a common factor. In the procedure suggested in [2], it is assumed that a lower bound on the Sylvester determinant is known, that is, $|S| \geq \epsilon$ and the Sylvester determinant of the estimates $\widehat{Z}_p(s)$ and $\widehat{R}_p(s)$ is computed at each instant. Denoting this by $|\widehat{S}|$, the adaptive procedure is continued if $|\widehat{S}| \geq \epsilon$. If $|\widehat{S}| < \epsilon$, the adaptation is stopped and a fixed controller is used. To assure convergence, it is necessary to choose any one of n fixed controllers at random every time the constraint is violated. If the reference input is persistently exciting, it is shown in [2] that this procedure will be globally stable. Similar methods have been proposed in [3,6,11] [2].

(iv) Use of a nonlinear feedback signal. A method of combining both direct and indirect approaches was proposed in [13] for the regulation of a plant for which only the order is known. The input u to the plant consists of two parts and can be expressed as two signals u_1 and u_2, that is, $u = u_1 + u_2$. The feedback signal u_1 is obtained exactly in the same manner as in Section 9.4.1 by estimating the control parameter vector as $\widehat{\theta}(t)$ at every instant. An estimate $\widehat{p}(t)$ of the plant parameter vector is simultaneously obtained using a representation similar to that used in Chapter 4. When the plant parameter vector

[2] In [6,11], the reference input is restricted to be a sum of sinusoids at distinct frequencies. In [3], the problem is treated for a general class of persistently exciting reference inputs.

p and the desired control parameter vector θ^* are known, the closed-loop system has a characteristic polynomial $R_m(s)\lambda(s)$. Since only the estimates $\widehat{p}$ and $\widehat{\theta}$ of p and θ^* are available, the polynomial

$$\widetilde{R}(s) = \left[\widehat{R}_p(s)(\lambda(s) - \widehat{C}(s)) - \widehat{Z}_p(s)\widehat{D}(s)\right] - R_m(s)\lambda(s) \tag{9.34}$$

can be used as a measure of the deviation of the estimates from their true values. In Eq. (9.34), $\widehat{R}_p$ and $\widehat{Z}_p$ are determined by $\widehat{p}$ while $\widehat{C}$ and $\widehat{D}$ are determined by $\widehat{\theta}$. The signal u_2 is nonlinearly related to the coefficients of the polynomial $\widetilde{R}$ as well as signals in the system. In [13] it is shown that with an input $u = u_1 + u_2$ the state of the overall system is bounded and that the output of the plant tends to zero.

Comment 9.2 Methods (i) and (ii) simply require that the region S in $\mathbb{R}^m$ in which θ lies, be restricted so that no unstable pole-zero cancellations take place. Hence, the condition in Eq. (9.25a) is satisfied. The method described in (iii) uses, in addition to the condition in Eq. (9.25a), the persistent excitation of the reference input. The Sylvester determinant is simply used as an indicator function to avoid undesirable regions in the θ space. Finally, in the method described in (iv), the auxiliary input is not directly a function of the tuning signal but depends on the error between the desired and estimated closed-loop characteristic polynomial. The latter can be present even when the tuning signal is identically zero. Hence, the equations describing the overall system do not have the same form as those given in Eq. (9.24). In this case, the condition in Eq. (9.25b) is satisfied.

9.5 COMMAND GENERATOR TRACKING METHOD

In Sections 9.1-9.4, the attempts made in recent years to relax the assumptions (I) while designing stable adaptive controllers were discussed. An independent direction, in which work has been in progress for many years and which is claimed by its proponents to be practically viable in many applications, is the command generator tracking (CGT) method [4,26]. In this method, the class of reference trajectories that the output of the plant has to follow is restricted. It was mentioned in Chapter 1 that in model reference control [7], the desired output $y_m(t)$ was originally modeled as the homogeneous solution of a linear time-invariant differential equation. This class of reference trajectories is considered in the CGT method. Our aim in this section is to delineate the assumptions that have to be made regarding the plant transfer function to achieve perfect model-following, and how such assumptions fit into the general framework of adaptive controller design discussed in the previous chapters. As before, we shall try to indicate the role of the various assumptions in the algebraic and the analytic parts.

9.5.1 Statement of the Problem

Let the plant to be controlled be described by the equations

$$
\begin{aligned}
\dot{x}_p &= A_p x_p + B_p u \\
y_p &= C_p x_p
\end{aligned}
\tag{9.35}
$$

where $A_p \in \mathbb{R}^{n \times n}$, $B_p \in \mathbb{R}^{n \times m}$, and $C_p \in \mathbb{R}^{m \times n}$ are constant unknown matrices. The output of the plant is required to follow the signal y_m which is specified as the output of a reference model described by

$$
\begin{aligned}
\dot{x}_m &= A_m x_m \\
y_m &= C_m x_m
\end{aligned}
\tag{9.36}
$$

where $A_m \in \mathbb{R}^{n_m \times n_m}$ and $C_m \in \mathbb{R}^{m \times n_m}$. The claim made by the researchers in this area is that this represents a sufficiently large class of signals and, hence, adequate for many practical applications. The problem is to determine a control input u to the plant using a differentiator-free controller so that $\lim_{t \to \infty}[y_p(t) - y_m(t)] = 0$ with all the signals in the system remaining bounded.

Comment 9.3 Note that A_m in Eq. (9.36) is not assumed to be an asymptotically stable matrix. Hence, the jth component, $y_{mj}(t)$, of the output $y_m(t)$ can be expressed in the general form

$$
y_{mj}(t) = \sum_i \beta_i t^{k_i} \exp(\alpha_i t) \sin(\omega_i t + \phi_i) \qquad \beta_i, k_i, \alpha_i, \omega_i, \phi_i \in \mathbb{R}
$$

where the summation is over all the modes of the system, for $j = 1, \ldots, m$.

Comment 9.4 The order of the model is n_m, while that of the plant is n, where n_m and n are arbitrary. Hence, low order plants with a reference trajectory with a large number of modes as well as high order plants with y_m containing only a few modes can be considered.

The Algebraic Part

Assumption 1. It is assumed that a constant feedback parameter matrix Θ_1^* exists such that the matrix

$$
\overline{A}_p \triangleq A_p + B_p \Theta_1^* C_p \qquad \text{is asymptotically stable.}
$$

Comment 9.5 Assumption 1 implies that the plant should be output-stabilizable. Necessary and sufficient conditions on the plant transfer matrix under which this is possible are currently not known.

The following assumptions are also needed for the solution of the algebraic part:

Assumption 2. The matrix

$$A \triangleq \begin{bmatrix} A_p & B_p \\ C_p & 0 \end{bmatrix} \quad \text{is nonsingular.}$$

Assumption 3. No transmission zero of the plant is equal to any eigenvalue of A_m.

Theorem 9.6 Let $e_1(t) = y_p(t) - y_m(t)$. Under Assumptions 1-3, matrices $\Theta_1^* \in \mathbb{R}^{m\times m}$, and $\Theta_2^* \in \mathbb{R}^{m\times n_m}$ exist such that if

$$u(t) \quad = \quad \Theta_2^* x_m(t) + \Theta_1^* e_1(t) \tag{9.37}$$

then $\lim_{t\to\infty} e_1(t) = 0$.

Proof. Let $y_p(t_0) = y_m(t_0)$. Define $u^*(t) = \Theta_2^* x_m(t)$ and $x_p^*(t) = \Theta_3^* x_m(t)$, where $\Theta_2^* \in \mathbb{R}^{m\times n_m}$ and $\Theta_3^* \in \mathbb{R}^{n\times n_m}$ and satisfy the equation

$$\begin{bmatrix} A_p & B_p \\ C_p & 0 \end{bmatrix} \begin{bmatrix} \Theta_3^* \\ \Theta_2^* \end{bmatrix} = \begin{bmatrix} \Theta_3^* A_m \\ C_m \end{bmatrix}. \tag{9.38}$$

It can be shown that under Assumptions 2 and 3, such a Θ_2^* and Θ_3^* exist[5,24]. It then follows that

$$\dot{x}_p^*(t) \quad = \quad A_p x_p^*(t) + B_p u^*(t); \qquad C_p x_p^*(t) \quad = \quad C_m x_m(t). \tag{9.39}$$

Hence, if the input $u(t)$ is chosen as $u(t) = \Theta_2^* x_m(t)$, where Θ_2^* is defined by Eq. (9.38), it follows that $y_p(t) = y_m(t)$ for all $t \geq t_0$.

Let $y_p(t_0) \neq y_m(t_0)$. Then, if $u(t) = \Theta_2^* x_m(t) + \Theta_1^* e_1(t)$, the plant Eq. (9.35) becomes

$$\dot{x}_p \quad = \quad \overline{A}_p x_p + B_p \Theta_2^* x_m - B_p \Theta_1^* C_m x_m; \qquad y_p \quad = \quad C_p x_p \tag{9.40}$$

where $\overline{A}_p = A_p + B_p \Theta_1^* C_p$. If $e(t) = x_p(t) - x_p^*(t)$, from Eqs. (9.39) and (9.40), it follows that $\dot{e} = \overline{A}_p e$. From Assumption 1, $\lim_{t\to\infty}[y_p(t) - y_m(t)] = \lim_{t\to\infty}[C_p x_p(t) - C_p x_p^*(t)] = 0$.

Figure 9.7 shows the structure of the fixed controller given by Eq. (9.37).

The Analytical Part. The fixed controller described in Eq. (9.37) consists of the matrices Θ_1^* and Θ_2^* which, in turn, depend on the plant matrices A_p, B_p, and C_p. Hence when the plant parameters are unknown, the fixed controller is replaced by an adaptive controller of the form

$$u(t) \quad = \quad \Theta_1(t) e_1(t) + \Theta_2(t) x_m(t). \tag{9.41}$$

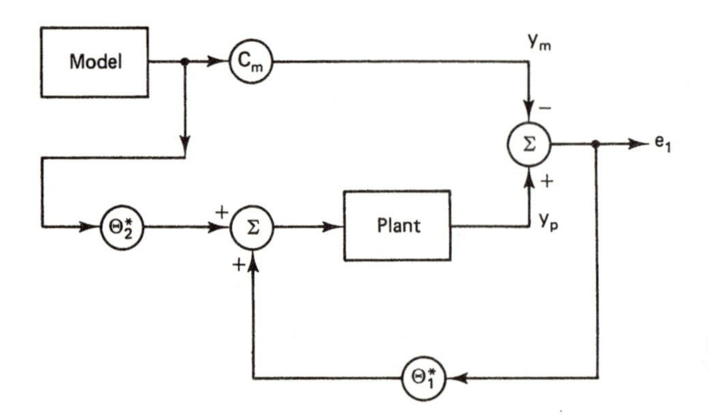

Figure 9.7 Command generator tracking: The algebraic part.

The objective of the adaptive control problem is then to determine adaptive laws for adjusting $\Theta_1(t)$ and $\Theta_2(t)$ so that the error $e(t) \to 0$ as $t \to \infty$.

Defining $\Theta(t) = [\Theta_1(t), \Theta_2(t)]$, and $\omega(t) = [e_1^T(t), x_m^T(t)]^T$, Eqs. (9.35) and (9.41) can be rewritten as

$$\dot{x}_p(t) = A_p x_p(t) + B_p \Theta(t) \omega(t). \tag{9.42}$$

If $\Theta(t) = \Theta^* + \Phi(t)$ where $\Theta^* = [\Theta_1^*, \Theta_2^*]$, and $\Phi(t) = [\Phi_1(t), \Phi_2(t)]$, Eq. (9.42) can be rewritten as

$$\dot{x}_p(t) = A_p x_p(t) + B_p \Theta_1^* C_p e(t) + B_p \Phi_1(t) e_1(t) + B_p \Theta_2(t) x_m(t). \tag{9.43}$$

From Eqs. (9.43) and (9.39), we have

$$\dot{e} = \left(A_p + B_p \Theta_1^* C_p\right) e + B_p \Phi \omega; \qquad e_1 = C_p e. \tag{9.44}$$

Equation (9.44) can also be expressed as

$$e_1(t) = \overline{W}_m(s) \left[\Phi(t) \omega(t)\right]$$

where

$$\overline{W}_m(s) = C_p \left(sI - \left(A_p + B_p \Theta_1^* C_p\right)\right)^{-1} B_p. \tag{9.45}$$

The Adaptive Law

Assumption 4. The transfer matrix $\overline{W}_m(s)$ in Eq. (9.45) is SPR.

With the assumption above, the following adaptive laws (refer to third error model, Chapter 10)

$$\dot{\Phi}_1 = -e_1 e_1^T$$
$$\dot{\Phi}_2 = -e_1 x_m^T$$

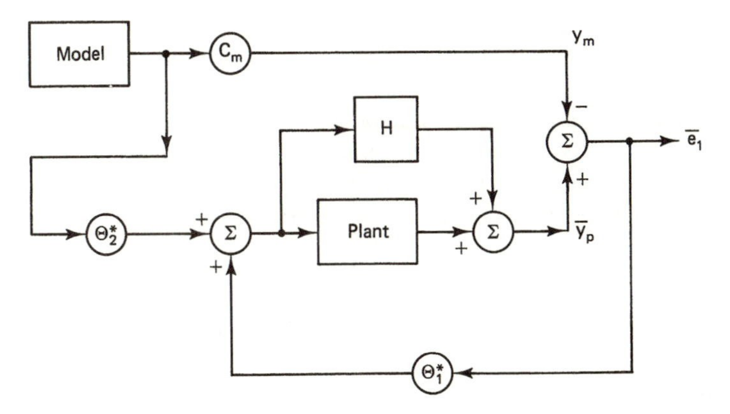

Figure 9.8 Augmented plant: CGT method.

can be written by inspection. Using the results of Section 10.2, it can be shown that the parameter $\Theta(t)$ and the signals in the closed-loop system are bounded and $\lim_{t\to\infty} e_1(t) = 0$.

Comment 9.6 The assumption made here that $\overline{W}_m(s)$ is SPR subsumes Assumption 1 where the unknown plant was required to be output-stabilizable. Conditions on the plant under which $W_m(s)$ can be made SPR using output feedback are not available at present. However, if this assumption can be justified in a practical problem, stable adaptive laws can be derived using the methods in Chapter 5 (or Chapter 10, for multivariable plants).

Comment 9.7 By Assumption 4, $\overline{W}_m(s)$ exists; however, it is unknown. Hence, an augmented error cannot be determined.

Since the requirement that $\overline{W}_m(s)$ is SPR was generally recognized to be a difficult one to satisfy, efforts were made to obtain weaker conditions on the plant transfer function. This included adding a constant or a transfer matrix H in the feedforward path as shown in Fig. 9.8 so that the augmented plant satisfies Assumptions 1-4. Such modifications have also been suggested for the adaptive control problems discussed in the preceding chapters, but were not included in earlier discussions. This is because, in all such cases, the prior information regarding the plant parameters is generally unrealistic. Assuming that the plant augmentation suggested is possible, $\lim_{t\to\infty} \overline{e}_1(t) = 0$. This implies that the true error of the system would be the output of the system H. Work is currently in progress to determine conditions under which this output can be made small enough to satisfy specifications.

9.6 SUMMARY

The efforts made during the past few years to relax the assumptions on the plant while achieving stable adaptive control are treated in this chapter. Since this is an active area of research, it is not possible to deal in a coherent fashion with all the different efforts that are in progress. In Sections 9.2, 9.3, and 9.4, efforts made to relax assumptions (i), (iii), and (iv) in (I) are discussed. In some cases, the solutions suggested include the relaxation of more than one assumption.

The last section deals with situations in which the desired output of the plant is restricted to be the homogeneous solution of a linear time-invariant differential equation. The assumptions that the plant has to satisfy for perfect tracking are indicated.

PROBLEMS

Part I

1. Show that if the equilibrium state of the differential equation

$$x^{(4)} + ax^{(3)} + b\ddot{x} + c\dot{x} + dx = 0$$

is asymptotically stable, where $x^{(n)}$ denotes the nth derivative of x with respect to time, then the equilibrium state of the differential equation

$$x^{(4)} + a\theta x^{(3)} + b\theta^2\ddot{x} + c\theta^3\dot{x} + d\theta^4 x = 0$$

is also asymptotically stable for $\theta \geq \theta_0 > 0$.

2. **(a)** If $a(\theta)$ and $p(\theta)$ are scalar functions of a parameter θ that belongs to a compact set S such that $a(\theta)p(\theta) = 1$, and $a(\theta) > 0$ for all $\theta \in S$, and the derivative of $a(\theta)$ with respect to θ is continuous for all $\theta \in S$, show that the derivative of $p(\theta)$ with respect to θ exists and is bounded for all $\theta \in S$.

(b) If $A : S \to \mathbb{R}^{m\times m}$ and $P : S \to \mathbb{R}^{m\times m}$, show that if

$$A^T(\theta)P(\theta) + P(\theta)A(\theta) = -Q < 0$$

and the partial derivative of A with respect to θ is denoted by A_θ is continuous with respect to θ for all $\theta \in S$, then $\partial/\partial\theta(P(\theta))$ exists and is bounded.

Part II

3. A system is described by the differential equations

$$\begin{aligned} \ddot{x} + (1+\theta)\dot{x} + \theta^2 x &= 0 \qquad x(t_0) = x_0 \\ \dot{\theta} &= x^2 \qquad \theta(t_0) = \theta_0 \end{aligned}$$

Are the solutions of the system bounded?

4. Show that the system of differential equations

$$\begin{aligned} \dot{x} &= x + \theta^2 \cos \theta x \qquad x(t_0) = x_0 \\ \dot{\theta} &= x^2 \qquad \theta(t_0) = \theta_0 \end{aligned}$$

has bounded solutions for all initial conditions (x_0, θ_0).

5. If, in Problem 4, the first equation has the form

$$\dot{x} = (1 + f(\theta))\, x$$

where $f(\theta)$ is a rational function of θ, why doesn't the same result hold?

6. A plant has a transfer function $W_p(s)$ whose zeros lie in $\mathbb{C}^-$. The standard adaptive controller structure and adaptive laws are used to control it adaptively. Show that in such a system the state can grow in an unbounded fashion while the output is identically zero. Give a specific example to illustrate your answer.

Part III

7. **(a)** A plant transfer function $W_p(s)$ can have one of two values, $W_1(s)$ and $W_2(s)$, where

$$W_1(s) = \frac{1}{s^2 + s - 4}, \qquad W_2(s) = \frac{1}{s^3 + s^2 + 3s - 1}.$$

To stabilize the plant, a single gain k is used in a negative feedback loop. The gain can assume one of two values, 2 and 6, each of which stabilizes only one of the two transfer functions, $W_1(s)$ or $W_2(s)$. If you want to stabilize the plant by controlling the adaptive gain k without prior knowledge as to whether the plant transfer function is $W_1(s)$ or $W_2(s)$, indicate what strategy you would adopt.

(b) Can the method be extended to the case where $W_p(s)$ is one of a large number N of systems, assuming that a single value K_i of the feedback gain will only stabilize one of the N systems?

REFERENCES

1. Åström, K.J. "Direct methods for nonminimum phase systems." In *Proceedings of the IEEE CDC*, Albuquerque, New Mexico, 1980.
2. Anderson, B.D.O. "Adaptive systems, lack of persistency of excitation and bursting phenomena." *Automatica* 21:247–258, 1985.
3. Bai, E.W., and Sastry, S.S. "Global stability proofs for continuous-time indirect adaptive control schemes." *IEEE Transactions on Automatic Control* 32:537–543, June 1987.
4. Barkana, I. "Adaptive control – a simplified approach." In *Advances in Control and Dynamic Systems*, edited by C.T. Leondes, New York:Academic Press, 1987.
5. Davison, E.J. "The feedforward control of linear multivariable time-invariant systems." *Automatica* 9:561–573, 1973.
6. Elliott, H., Cristi, R., and Das, M. "Global stability of adaptive pole placement algorithms." *IEEE Transactions on Automatic Control* AC-30:348–356, April 1985.
7. Erzberger, H. "Analysis and design of model following control systems by state space techniques." *Proceedings of the JACC*, Ann Arbor, Michigan, 1968.

8. Fu, M., and Barmish, B.R. "Adaptive stabilization of linear systems via switching control." *IEEE Transactions on Automatic Control* 31:1097–1103, Dec. 1986.

9. Goodwin, G.C., Johnson, C.R., and Sin, K.S. "Global convergence for adaptive one-step ahead optimal controllers based on input matching." *IEEE Transactions on Automatic Control* 26:1269–1273, Dec. 1981.

10. Goodwin, G.C., and Sin, K.S. "Adaptive control of nonminimum phase systems." *IEEE Transactions on Automatic Control* 26:478–483, April 1981.

11. Goodwin, G.C., and Teoh, E.K. "Persistency of excitation in the presence of possibly unbounded signals." *IEEE Transactions on Automatic Control* AC-30:595–597, June 1985.

12. Kreisselmeier, G. "An approach to stable indirect adaptive control." *Automatica* 21:425–431, 1985.

13. Kreisselmeier, G., and Smith, M.C. "Stable adaptive regulation of arbitrary n^{th}-order plants." *IEEE Transactions on Automatic Control* 31:299–305, 1986.

14. De Larminat, Ph. "On the stabilizability condition in indirect adaptive control." *Automatica* 20:793–795, 1984.

15. Martensson, B. "The order of any stabilizing regulator is sufficient information for adaptive stabilization." *Systems & Control Letters* 6:87:91, July 1985.

16. Morse, A.S. "A three-dimensional universal controller for the adaptive stabilization of any strictly proper minimum phase system with relative degree not exceeding two." *IEEE Transactions on Automatic Control* 30:1188–1191, 1985.

17. Morse, A.S. "An adaptive control for globally stabilizing linear systems with unknown high-frequency gains." *Proceedings of the Sixth International Conference on Analysis and Optimization of Systems*, Nice, France, 1984.

18. Morse, A.S. "Recent problems in parameter adaptive control." In *Outils et modèles mathēmatiques pour l'automatique l'analyse de Systèmes et le traitment du signal*, edited by I.D. Landau, Paris, CNRS, 1983.

19. Morse, A.S. "High gain feedback algorithms for adaptive stabilization." In *Proceedings of the Fifth Yale Workshop on Applications of Adaptive Systems Theory*, New Haven, CT, 1987.

20. Morse, A.S. "High-gain adaptive stabilization." In *Proceedings of the Carl Kranz Course*, Munich, W. Germany, 1987.

21. Morse, A.S. "A model reference controller for the adaptive stabilization of any strictly proper, minimum phase linear system with relative degree not exceeding two." In *Proceedings of the 1985 MTNS Conference*, Stockholm, Sweden, 1985.

22. Mudgett, D. R., and Morse, A. S. "Adaptive stabilization of linear systems with unknown high frequency gains." *IEEE Transactions on Automatic Control* 30:549–554, June 1985.

23. Nussbaum, R.D. "Some remarks on a conjecture in parameter adaptive control." *Systems and Control Letters* 3:243–246, 1983.

24. O'Brien, M.J., and Broussard, J.R. "Feedforward control to track the output of a forced model." In *Proceedings of the IEEE CDC*, Phoenix, Arizona, 1978.

25. Praly, L. "Towards a globally stable direct adaptive control scheme for not necessarily minimum phase systems." *IEEE Transactions on Automatic Control* 29:946–949, 1984.

26. Sobel, K., and Kaufman, H. "Direct model reference adaptive control for a class of MIMO systems." In *Advances in Control and Dynamic Systems*, edited by C.T. Leondes, New York:Academic Press, 1987.

27. Truxal, J.G. *Automatic feedback control system synthesis*. New York:McGraw-Hill Book Company, 1955.
28. Willems, J.C., and Byrnes, C.I. "Global adaptive stabilization in the absence of information on the sign of the high frequency gain." *Proceedings of INRIA Conference on Analysis and Optimization of Systems* 62:49–57, June 1984.
29. Wolovich, W.A. *Linear multivariable systems*. Berlin:Springer Verlag, 1974.

10

Multivariable Adaptive Systems

10.1 INTRODUCTION

From the start, the emphasis of control theory has been on single-input single-output (SISO) systems. It is therefore not surprising that many of the early design approaches that were developed using both frequency-domain and time-domain methods were primarily applicable only to such systems. It was only in the 1970s that a well developed theory of linear multivariable systems began to emerge. Since most practical systems are multivariable in character, the newly developed theory immediately found application in a wide variety of problems. Further, because large uncertainty is a concomitant property of complex systems, the need for adaptive control is also greater in such cases. Hence, adaptive control methodologies naturally find more use in multivariable systems.

The contents of Chapters 3-9 dealt almost exclusively with the adaptive control of SISO systems. As mentioned earlier, soon after this problem was successfully resolved in 1980, research efforts were made toward extending the results in many directions. Problems of robust adaptation in the presence of disturbances were addressed in Chapter 8 and efforts made toward reducing the prior information needed for adaptive control were discussed in Chapter 9. Yet another important direction in which an extension of the theory was attempted concerned the stable adaptive control of multivariable systems with unknown parameters. This is the subject matter of this chapter.

The adaptive control problem in the multi-input multi-output (MIMO) case leads to the same types of analytical questions we have already discussed for the SISO case in previous chapters. However, structural problems arise in parametrizing the controller, which makes the algebraic part of the adaptive control problem considerably more involved in the multivariable case. It is well known that the most important consideration that distinguishes the design of linear multivariable control systems from their single variable counterparts is the coupling that exists between inputs and outputs. This is implicit in the representation of plant dynamics by a transfer matrix rather than a scalar transfer function as in the SISO case. Whether a right- or left-matrix fraction description of the plant is to be used for control purposes, how a reference model is to be chosen, what the multivariable counterparts of k_p, n^*, n, and the zeros of the plant transfer function in the SISO case are, and whether a fixed controller structure exists so the outputs of the plant can follow the outputs of the model asymptotically, involve concepts of multivariable control theory. Many of the basic algebraic notions needed for their discussion are presented in Appendix E.

Multivariable adaptive observers are treated briefly in Section 10.3 and the importance of plant parametrization needed in such problems is emphasized. The extension of the adaptive observers to the MIMO problem was carried out by Anderson [1], [9] and [10]. The bulk of the chapter, however, is devoted to the control problem, and emphasis is placed on obtaining conditions on the multivariable plant for stable adaptive control, corresponding to those given in (I), in Chapter 5, for single variable plants. Early attempts to extend SISO results to the MIMO case were made by Monopoli and Hsing [7] for continuous time systems and by Borison [2] and Goodwin *et al.* [6] for discrete time systems. All of them tacitly assumed that the plant transfer matrix can be diagonalized. Later Monopoli and Subbarao [8] considered a special class of such 2×2 systems for practical applications. In [4] Elliott and Wolovich introduced the concept of the interactor matrix and later Goodwin and Long used this concept in [5] to generalize the results. The material presented in Sections 10.4 to 10.6 related to this problem is based on the doctoral dissertation of Singh [11] as well as the results in the paper by Singh and Narendra [12].

The results in Section 10.4 indicate that substantial amount of prior information is needed for the stable adaptive control of multivariable systems and that decoupled reference models may be prerequisites for practical model-following. The concepts developed in Section 10.4 are elucidated by considering the adaptive control of MIMO systems with two inputs and two outputs. The last section deals with general MIMO systems and the practical application of the theory to such systems is relegated to Chapter 11.

10.2 MATHEMATICAL PRELIMINARIES

Let a multivariable plant with m inputs, m outputs and n state variables be described by the triple $[\overline{C}, \overline{A}, \overline{B}]$, which is controllable and observable. The transfer matrix of the system is then given by

$$W_p(s) \stackrel{\triangle}{=} \overline{C}\left(sI - \overline{A}\right)^{-1}\overline{B} \in \mathbb{R}_p^{m\times m}(s).$$

If $W_p(s)$ is strictly proper, it can be factored using the right coprime factorization as[1]

[1]For pertinent details concerning the notation used in multivariable systems, the reader is referred to Appendix E.

$$W_p(s) = N_r(s)D_r^{-1}(s)$$

where $N_r(s)$ and $D_r(s)$ are $m \times m$ polynomial matrices, right coprime, with $D_r(s)$ column proper and column degrees such that $\partial_{ci}(N_r(s)) < \partial_{ci}(D_r(s)) = \mu_i$, $i = 1, \ldots, p$. The sum of all the column degrees is $\sum_{i=1}^{p} \mu_i = n$. The roots of the polynomials $\det[N_r(s)]$ and $\det[D_r(s)]$ are respectively the zeros and poles of $W_p(s)$. By a suitable nonsingular transformation, it can be shown that an equivalent representation of the plant is given by

$$\dot{x} = Ax + Bu \qquad y = Cx$$

where $\{C, A, B\}$ are in control canonical form with

$$A = \begin{bmatrix} \begin{matrix} 0 & 1 & 0 & \cdots & 0 \\ 0 & 0 & 1 & \cdots & 0 \\ \vdots & \vdots & \vdots & & \vdots \\ 0 & 0 & 0 & \cdots & 1 \\ \times & \times & \times & \cdots & \times \end{matrix} & \begin{matrix} & & 0 & & \\ & & & & \\ & & & & \\ & & & & \\ \times & \cdots & \times & \times & \times \end{matrix} & & \begin{matrix} & & 0 & & \\ & & & & \\ & & & & \\ & & & & \\ \times & \times & \times & \cdots & \times \end{matrix} \\ \begin{matrix} & & & & \\ & & 0 & & \\ & & & & \\ \times & \times & \times & \cdots & \times \end{matrix} & \begin{matrix} 0 & 1 & 0 & \cdots & 0 \\ 0 & 0 & 1 & \cdots & 0 \\ \vdots & \vdots & \vdots & & \vdots \\ 0 & 0 & 0 & \cdots & 1 \\ \times & \cdots & \times & \times & \times \end{matrix} & \cdots & \begin{matrix} & & & & \\ & & 0 & & \\ & & & & \\ \times & \times & \times & \cdots & \times \end{matrix} \\ \vdots & & & \vdots \\ \begin{matrix} & & & & \\ & & & & \\ & & 0 & & \\ & & & & \\ \times & \times & \times & \cdots & \times \end{matrix} & \begin{matrix} & & & & \\ & & & & \\ & & 0 & & \\ & & & & \\ \times & \cdots & \times & \times & \times \end{matrix} & & \begin{matrix} 0 & 1 & 0 & \cdots & 0 \\ 0 & 0 & 1 & \cdots & 0 \\ \vdots & \vdots & \vdots & & \vdots \\ 0 & 0 & 0 & \cdots & 1 \\ \times & \times & \times & \cdots & \times \end{matrix} \end{bmatrix},$$

$$B = \left[0,\ 0,\ \ldots,\ 0,\ b_{1\mu_1},\ 0,\ 0,\ \ldots,\ 0,\ b_{2\mu_2},\ \ldots,\ 0,\ 0,\ \ldots,\ 0,\ b_{m\mu_m}\right]^T,$$

$\times$ denote arbitrary scalars, $b_{i\mu_i} \in \mathbb{R}^m$, $i = 1, \ldots, m$, and C is a constant matrix [3]. Similarly if $W_p(s)$ is expressed using the left coprime factorization so that $W_p(s) = D_\ell^{-1}(s)N_\ell(s)$, we can obtain the observable canonical form

$$
\dot{x} = \left[\begin{array}{cccccccccccccccc}
0 & 0 & \cdots & 0 & \times & & & & & \times & & & & & & \times \\
1 & 0 & \cdots & 0 & \times & & & & & \times & & & & & & \times \\
0 & 1 & \cdots & 0 & \times & & 0 & & & \times & & & & & & \times \\
\vdots & \vdots & & \vdots & \vdots & & & & & \vdots & & & & & & \vdots \\
0 & 0 & \cdots & 1 & \times & & & & & \times & & & & & & \times \\
 & & & & \times & 0 & 0 & \cdots & 0 & \times & & & & & & \times \\
 & & & & \times & 1 & 0 & \cdots & 0 & \times & & & & & & \times \\
 & & 0 & & \times & 0 & 1 & \cdots & 0 & \times & & & & & & \times \\
 & & & & \vdots & \vdots & \vdots & & \vdots & \vdots & & & & & & \vdots \\
 & & & & \times & 0 & 0 & \cdots & 1 & \times & & & & & & \times \\
 & & \vdots & & & & & & & & & & & \vdots & & \\
 & & & & \times & & & & & \times & & 0 & 0 & \cdots & 0 & \times \\
 & & & & \times & & & & & \times & & 1 & 0 & \cdots & 0 & \times \\
 & & 0 & & \times & & & 0 & & \times & & 0 & 1 & \cdots & 0 & \times \\
 & & & & \vdots & \vdots & \vdots & & \vdots & \vdots & & & & & & \vdots \\
 & & & & \times & & & & & \times & & 0 & 0 & \cdots & 1 & \times
\end{array}\right] x + Bu
$$

$$
y = \begin{bmatrix} 0\,0 \cdots 0\; c_{1\nu_1}\; 0\,0 \cdots 0\; c_{2\nu_2} \cdots 0\,0 \cdots 0\; c_{m\nu_m} \end{bmatrix} x
$$

where $\times$ denote arbitrary scalars, $c_{i\nu_i} \in \mathbb{R}^m$, $i = 1, \ldots, m$, and B is a constant matrix. We also note that one of the irreducible realizations above can be obtained from the other by applying an equivalent transformation.

Hermite Normal Form. By performing elementary column operations, a nonsingular matrix $T(s)$ in $\mathbb{R}_p^{m \times m}(s)$ can be transformed into its special triangular structure $H_p(s)$, known as the right Hermite normal form of $T(s)$. This operation is equivalent to multiplying the matrix $T(s)$ on the right by a unimodular matrix $U(s)$, that is, $T(s)U(s) = H_p(s)$. A similar definition can be given for the left Hermite normal form. In the paragraphs that follow, we use only the right Hermite normal form for the control problem. The matrix $H_p(s)$ is lower triangular and has the form

$$
H_p(s) = \begin{bmatrix}
\dfrac{1}{\pi^{n_1}(s)} & & 0 \\
h_{21}(s) & \ddots & \\
\vdots & & \ddots \\
h_{m1}(s) & & \dfrac{1}{\pi^{n_m}(s)}
\end{bmatrix}
$$

where $h_{ij}(s) = \delta_{ij}(s)/\pi^{n_{ij}}(s)$ and is proper, $n_{ij} < n_i$, n_i and n_{ij} are positive integers, and $\pi(s)$ is any monic polynomial of degree 1. When $h_{ij}(s) = 0, i \neq j$ and the matrix is diagonal, it is obvious that $H_p(s)$ is uniquely determined by $\pi(s)$ and the integers $n_1, n_2, \ldots, n_m$, where n_i corresponds to the minimum relative degree of the elements of $W_p(s)$ in the ith row. However, when $h_{ij}(s) \neq 0$, the coefficients of the polynomial $\delta_{ij}(s)$ are dependent on the parameters of the plant. The implications of a nondiagonal Hermite form to the adaptive control of multivariable systems are discussed in Section 10.5. If $\pi(s)$ is asymptotically stable, then $H_p(s)$ corresponds to a stable Hermite form. Two Hermite forms $H_1(s)$ and $H_2(s)$ of a given plant, determined by $\pi_1(s)$ and $\pi_2(s)$ respectively, are dynamically equivalent, where $\pi_1(s)$ and $\pi_2(s)$ are monic polynomials of degree one. For simple examples of transfer matrices whose Hermite normal forms are not diagonal, the reader is referred to Section 10.5.

Decoupling of Linear Multivariable Systems. Dynamic decoupling of a MIMO plant, in which one output is affected by one and only one input, is obviously a desirable feature in multivariable control. If, by a suitable choice of a controller, a MIMO plant can be decoupled, then SISO control methods can be used for each loop to obtain the desired closed-loop response. It is clear that, given a transfer matrix $T(s)$, whether or not it can be decoupled is closely related to the Hermite normal form of $T(s)$. A diagonal Hermite form implies that the MIMO plant can be dynamically decoupled with a causal feedforward controller, or more realistically (to avoid unstable pole-zero cancellations), using an equivalent feedback controller. As seen in the following sections, in the adaptive control problem, only plants with diagonal Hermite normal forms can be realistically adaptively controlled in a stable fashion. Hence, it is desirable to know the amount of prior information that is needed to conclude whether or not a given transfer matrix has a diagonal Hermite form. The following lemma, derived from the decoupling problem, provides a necessary and sufficient condition for this.

Let $G(s)$ be a nonsingular strictly proper matrix in $\mathbb{R}_p^{m\times m}(s)$. Let $r_i(G(s))$ denote the minimum relative degree in the ith row of $G(s)$. Let a constant row vector E_i of dimension m be defined as

$$E_i \stackrel{\triangle}{=} \lim_{s\to\infty} s^{r_i} G(s).$$

It is known that r_i and E_i are invariant under a linear state feedback [9].

Lemma 10.1 Let $G(s)$ and $E_i, i = 1, 2, \ldots, m$ be defined as above. Then $G(s)$ can be decoupled by a linear state feedback controller if, and only if, the constant matrix

$$E(G(s)) = \begin{bmatrix} E_1 \\ E_2 \\ \vdots \\ E_m \end{bmatrix} \tag{10.1}$$

is nonsingular.

The reader is referred to [3] for the proof.

Corollary 10.1 Given a nonsingular transfer matrix $W_p(s)$, its right Hermite form $H_p(s)$ is diagonal in structure if, and only if, the constant matrix $E(W_p(s))$ is nonsingular, where the matrix $E(\cdot)$ is defined as in Eq. (10.1).

Proof. The proof follows directly from the definition of $H_p(s)$, Lemma 10.1 and the fact that the unimodular matrix which transforms $W_p(s)$ to $H_p(s)$ is equivalent to a state feedback controller.

From Corollary 10.1 it is clear that the only information that is needed to conclude whether or not a matrix $W_p(s)$ has (generically) a diagonal Hermite form is the relative degree of each of its entries.

Bezout Identity. The design of the adaptive controller for a MIMO system is similar to that described in Chapter 5 for an SISO system and involves two parts. In the first, referred to as the algebraic part, the existence of a feedback controller structure that can achieve the desired objective is established. In the second, the controller parameters are adjusted using an adaptive law so that the output error tends to zero asymptotically. The following lemma, which is a matrix version of the Bezout identity stated in Chapter 5, plays an important role in the first part of the design problem.

Lemma 10.2 Let $Q(s), T(s) \in \mathbb{R}^{m\times m}[s]$ be right coprime with their column degrees such that $\partial_{cj}[T(s)] \leq \partial_{cj}[Q(s)] = d_j > 0$ for all $j = 1,\ldots,m$. Let ν be the observability index of the minimal transfer matrix $T(s)Q^{-1}(s)$. Then polynomial matrices $P(s), R(s) \in \mathbb{R}^{m\times m}[s]$ each having the highest degree $\nu - 1$ exist such that

$$P(s)Q(s) + R(s)T(s) \;=\; M(s) \tag{10.2}$$

where $M(s)$ is any arbitrary polynomial matrix in $\mathbb{R}^{m\times m}[s]$ with $\partial_{cj}[M(s)] \leq d_j + \nu - 1$ for $j = 1,\ldots,m$.

Proof. Since $Q(s)$ and $T(s)$ are right coprime polynomial matrices, there exist polynomial matrices $A(s)$ and $B(s)$ such that

$$A(s)Q(s) + B(s)T(s) \;=\; I.$$

Let $M(s)$ be an $m \times m$ arbitrary polynomial matrix with column degree $\leq d_j + \nu - 1$. Then

$$M(s)A(s)Q(s) + M(s)B(s)T(s) \;=\; M(s). \tag{10.3}$$

The right coprime factorization $T(s)Q^{-1}(s)$ can also be expressed by a left coprime factorization $C^{-1}(s)D(s)$ with $C(s)$ row proper, and row degrees such that $\delta_{ri}(D(s)) \leq \delta_{ri}(C(s))$, $i = 1,\ldots,m$. The highest degree of $C(s)$ is ν, the observability index of the minimal transfer matrix [13]. Hence,

$$T(s)Q^{-1}(s) \;=\; C^{-1}(s)D(s). \tag{10.4}$$

Equations (10.3) and (10.4) can be represented as

$$\begin{bmatrix} M(s)A(s) & M(s)B(s) \\ D(s) & -C(s) \end{bmatrix} \begin{bmatrix} Q(s) \\ T(s) \end{bmatrix} = \begin{bmatrix} M(s) \\ 0 \end{bmatrix}.$$

By elementary column operations, the matrix equation above can be reduced to

$$\begin{bmatrix} P(s) & R(s) \\ D(s) & -C(s) \end{bmatrix} \begin{bmatrix} Q(s) \\ T(s) \end{bmatrix} = \begin{bmatrix} M(s) \\ 0 \end{bmatrix}$$

such that every column degree of $R(s)$ is strictly less than the corresponding column degree of $C(s)$. Since the highest degree of $C(s)$ is ν, the highest degree of $R(s)$ can be at most $\nu - 1$. Further, since $Q(s)$ is column proper and each column of $M(s)$ has degree less than or equal to $d_j + \nu - 1$, the polynomial matrix $P(s)$ can have degree at most $\nu - 1$.

Multivariable Error Models. In Chapter 7, three error models were developed for the SISO case. We present below the multivariable counterparts of the error models 1 and 3.

Error Model 1. Let $\omega : \mathbb{R}^+ \to \mathbb{R}^n$, $\Phi : \mathbb{R}^+ \to \mathbb{R}^{m\times n}$, $K_p \in \mathbb{R}^{m\times m}$, and $e_1 : \mathbb{R}^+ \to \mathbb{R}^m$ and let K_p satisfy the matrix equation

$$\Gamma K_p + K_p^T \Gamma = Q_0$$

where Γ is a symmetric positive-definite matrix, and Q_0 is a symmetric sign-definite matrix. In the following paragraphs, we assume that Q_0 is positive definite. Appropriate changes in the adaptive law can be made when $Q_0 < 0$. If e_1 and Φ are related by the equation

$$e_1 = K_p \Phi \omega \tag{10.5}$$

and Φ is adjusted as

$$\dot{\Phi} = -e_1 \omega^T \tag{10.6}$$

then Φ is bounded for all $t \geq t_0$. This follows by choosing a quadratic function

$$V = Tr(\Phi^T \Gamma \Phi).$$

The time derivative $\dot{V}$ along the trajectories of Eqs. (10.5) and (10.6) is given by

$$\begin{aligned} \dot{V} &= -Tr\left(\Phi^T \Gamma e_1 \omega^T\right) - Tr\left(\omega e_1^T \Gamma \Phi\right) \\ &= -\omega^T \Phi^T \left(\Gamma K_p\right) \Phi \omega - \omega^T \Phi^T \left(K_p^T \Gamma\right) \Phi \omega \\ &= -\omega^T \Phi^T Q_0 \Phi \omega \leq 0. \end{aligned}$$

Error Model 3. Let $\omega : \mathbb{R}^+ \to \mathbb{R}^n$, $\Phi : \mathbb{R}^+ \to \mathbb{R}^{m\times n}$, $K_p \in \mathbb{R}^{m\times m}$, $H(s) \in \mathbb{R}^{m\times m}(s)$ is SPR, and $e_1 : \mathbb{R}^+ \to \mathbb{R}^m$. Let K_p satisfy the matrix equation

$$\Gamma K_p + K_p^T \Gamma = Q_0 > 0$$

where $\Gamma = \Gamma^T > 0$. If e_1 and Φ are related by the equation

$$\dot{e} = Ae + BK_p\Phi\omega; \qquad e_1 = Ce \tag{10.7}$$

where $C(sI - A)^{-1}B = H(s)$ and Φ is adjusted as

$$\dot{\Phi} = -P_0 e_1 \omega^T \qquad P_0 = \Gamma^{-1} \tag{10.8}$$

then $e_1(t)$ and $\Phi(t)$ are bounded for all $t \geq t_0$.

This can be shown as follows: Since $H(s)$ is SPR, a positive-definite matrix P exists so that

$$\begin{aligned} A^T P + PA &= -Q \\ PB &= C^T. \end{aligned}$$

If a Lyapunov function candidate V is chosen as

$$V = e^T Pe + Tr\left(\Phi^T \left(K_p^T \Gamma\right) \Phi\right)$$

its time derivative along the solutions of Eqs. (10.7) and (10.8) is given by

$$\begin{aligned} \dot{V} &= e^T(A^T P + PA)e + e^T PBK_p\Phi\omega + \omega^T \Phi^T K_p^T B^T Pe \\ &\quad -Tr\left(\Phi^T K_p^T e_1 \omega^T\right) - Tr\left(\omega e_1^T K_p \Phi\right) \\ &= -e^T Qe \leq 0. \end{aligned}$$

10.3 MULTIVARIABLE ADAPTIVE IDENTIFICATION

The extension of the adaptive procedures discussed in Chapter 4 to the multivariable case is difficult, since the canonical forms for such systems, as shown in Section 10.2, have a much more complicated structure. In this section, we consider three methods that have been suggested for the identification of multivariable systems. We first briefly describe some of the efforts made to extend the ideas presented in Chapter 4.

10.3.1 Statement of the Problem

Let a multivariable plant with m inputs u and m outputs y be described by the differential equation

$$\dot{x}_p = A_p x_p + B_p u; \qquad y = C_p x_p$$

where A_p, B_p, and C_p are respectively constant matrices of dimensions $(n \times n)$, $(n \times m)$, and $(m \times n)$, whose elements are unknown. The identification problem is to determine the matrices $\{C_p, A_p, B_p\}$ from input-output data and to generate an estimated $\hat{x}_p$, which tends to x_p asymptotically.

10.3.2 The Adaptive Observer: Minimal Realization

As in the SISO case discussed in Chapter 4, the multivariable plant can also be represented as [3]

$$\dot{x}_p = [A + GC]\, x_p + Bu(t); \qquad y = Cx_p \tag{10.9}$$

where A and C are in observable canonical form and contain elements that are known. The identification problem then reduces to the determination of the unknown matrices G and B in Eq. (10.9). Since (C, A) is observable, an $n \times m$ matrix G_0 exists such that $A + G_0C$ is asymptotically stable. Hence, Eq. (10.9) can also be expressed as

$$\dot{x}_p = [A + G_0C]\, x_p + [G - G_0]\, Cx_p + Bu(t) \tag{10.10}$$

where $A + G_0C$ is an asymptotically stable matrix. An adaptive observer can then be constructed so that G, B, and the state x_p are estimated asymptotically. This has the form

$$\dot{\widehat{x}}_p(t) = [A + G_0C]\, \widehat{x}_p(t) + \left[\widehat{G}(t) - G_0\right] y(t) + \widehat{B}(t)u(t) + v(t) \tag{10.11}$$

where $\widehat{x}_p(t)$ is the estimate of the state of the plant, $\widehat{G}(t)$ and $\widehat{B}(t)$ are the estimates of the parameters G and B respectively, and $v(t)$ is a signal that must be chosen to ensure stable identification.

From Eqs. (10.10) and (10.11), the state and parameters are related by the equation

$$\begin{aligned} \dot{e}(t) &= [A + G_0C]\, e(t) + \left[\widehat{G}(t) - G\right] y(t) + \left[\widehat{B}(t) - B\right] u(t) + v(t) \\ &= [A + G_0C]\, e(t) + \Phi(t)y(t) + \Psi(t)u(t) + v(t) \end{aligned} \tag{10.12}$$

where $e(t) = \widehat{x}_p(t) - x_p(t)$, $\Phi(t) = \widehat{G}(t) - G$, and $\Psi(t) = \widehat{B}(t) - B$. We note that error equation (10.12) is of the same form as Eq. (4.11). The problem is to determine the signal $v(t)$ and the adaptive laws for adjusting $\Phi(t)$, and $\Psi(t)$ so that the error $e(t)$ tends to zero asymptotically.

Define an $[n \times 2mn]$ matrix X as

$$X \triangleq [X_1 | X_2 | \cdots | X_{2m}]$$

where X_i is an $n \times n$ matrix for $i = 1, 2, \ldots, 2m$. Let

$$\left.\begin{aligned} \dot{X}_i &= [A + G_0C]\, X_i + Iu_i \\ \dot{X}_{m+i} &= [A + G_0C]\, X_{m+i} + Iy_i \end{aligned}\right\} \qquad i = 1, \ldots, m$$

where I is an $n \times n$ identity matrix. The matrices X_i can be considered as sensitivity matrices that correspond to the auxiliary state variables ω_i in the single variable case. The input vector $v(t)$ as well as the adaptive laws for updating $\dot{\Phi}$ and $\dot{\Psi}$ can be determined in terms of the sensitivity matrices as

$$\begin{aligned} \dot{L}(t) &= X^T(t)C^T\left[\widehat{y}(t) - y(t)\right] \\ v(t) &= -X(t)X^T(t)C^T\left[\widehat{y}(t) - y(t)\right] \end{aligned}$$

where $L = [\widehat{G}^T,\ \widehat{B}^T]^T$. Proceeding in the same manner as in Chapter 4, it can be shown that $\lim_{t\to\infty} e(t) = 0$.

10.3.3 The Adaptive Observer: Nonminimal Realization

The identification procedure discussed in Section 4.3.3 using a nonminimal realization of the plant can also be extended directly to multivariable systems. This was first shown in [1]. As in Section 4.3.3, the aim of the approach is to represent the plant in such a fashion that all the unknown parameters of the plant appear as the elements of a single matrix Θ. This enables an observer to be set up to include an adjustable parameter matrix $\widehat{\Theta}$ so that adaptive laws can be derived. The properties of error model 1 or error model 3 can then be used to establish its stability.

In Chapter 4, it was shown that given a scalar input-output pair $\{u(\cdot), y_p(\cdot)\}$ with a transfer function $Z_p(s)/R_p(s)$, the output can be expressed as

$$y_p(t) = c^T(sI - F)^{-1}gu(t) + d^T(sI - F)^{-1}gy_p(t) \tag{10.13}$$

where the pair (F, g) is known, controllable, F is asymptotically stable and c and d are n-dimensional unknown vectors. Equation (10.13) can also be rewritten as

$$\begin{aligned} y_p(t) &= \Theta\left[W_1(s)u(t) + W_2(s)y_p(t)\right] \qquad (10.14) \\ &= [I - \Theta W_2(s)]^{-1}\Theta W_1(s)u(t) \ \text{ or } \ \Theta[I - W_2(s)\Theta]^{-1}W_1(s)u(t) \end{aligned}$$

where

$$\Theta = [c^T, d^T], \quad W_1(s) = [g^T(sI - F)^{-T}, 0]^T, \quad \text{and} \quad W_2(s) = [0, g^T(sI - F)^{-T}]^T.$$

The representation in Eq. (10.14) can be extended to the multivariable case where $W_p(s) \in \mathbb{R}_p^{m\times m}(s)$, and is given by (Fig. 10.1)

$$y_p(t) = \Theta\,[I - W_2(s)\Theta]^{-1}\,W_1(s)u(t)$$

where $\Theta \in \mathbb{R}^{m\times 2nm}$, $W_1(s) \in \mathbb{R}_p^{2nm\times p}(s)$, and $W_2(s) \in \mathbb{R}_p^{2nm\times m}(s)$. To determine the unknown parameter Θ, an observer is constructed as

$$y_m(t) = \widehat{\Theta}(t)\omega(t)$$

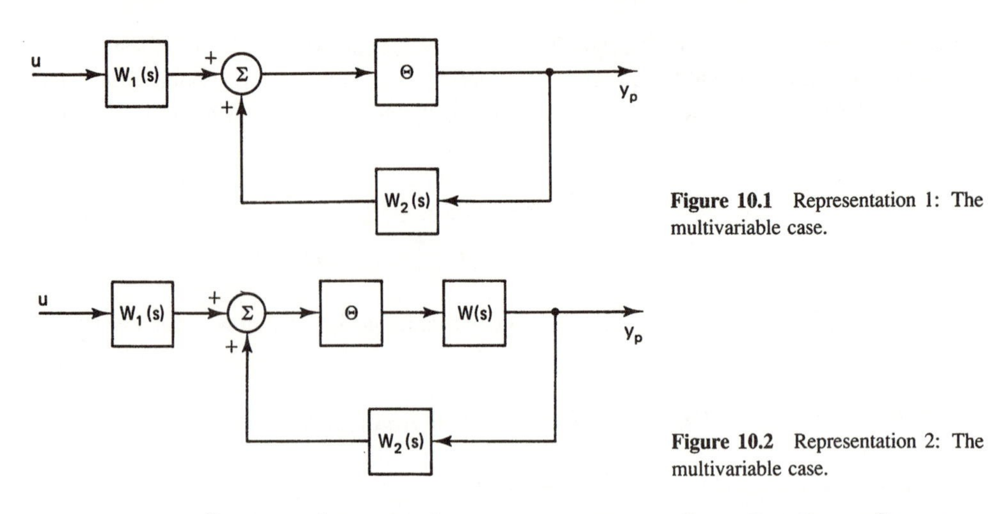

Figure 10.1 Representation 1: The multivariable case.

Figure 10.2 Representation 2: The multivariable case.

where $\widehat{\Theta} \in \mathbb{R}^{m \times 2nm}$ and $\omega(t) = [W_1(s), W_2(s)][u^T(t), y_p^T(t)]^T \in \mathbb{R}^{2nm}$. If $e_1(t) = y_p(t) - y_m(t)$, then we obtain the first error model $e_1(t) = \Phi(t)\omega(t)$. Hence, from Section 10.2, an adaptive law of the form

$$\dot{\widehat{\Theta}}(t) = -e_1(t)\omega^T(t)$$

assures that all signals are bounded, and that $\lim_{t\to\infty} e_1(t) = 0$.

In the second representation, which is a multivariable generalization of representation 2 in Section 4.3.4,

$$W_p(s) = W(s)\Theta\,[I - W_2(s)W(s)\Theta]^{-1}\,W_1(s)$$

where a strictly positive real transfer matrix $W(s)$ is included in the forward path of the feedback loop along with the gain matrix Θ (Fig. 10.2). As in Representation 2, an observer with adjustable parameters $\widehat{\Theta}(t)$ can be designed to estimate Θ. In all the cases above, if the reference input $u(t)$ is persistently exciting with sufficient number of spectral lines, it can be shown that the parameter estimates converge to their true values.

10.4 ADAPTIVE CONTROL PROBLEM

An $m \times m$ MIMO linear time-invariant finite-dimensional plant P is completely represented by the m-input m-output vector pairs $\{u(\cdot), y_p(\cdot)\}$. It is assumed that P can be modeled by a rational transfer matrix

$$W_p(s) = Z_p(s)R_p^{-1}(s)$$

where $Z_p(s)$ and $R_p(s)$ are right coprime polynomial matrices of dimension $m \times m$, $R_p(s)$ is column-proper, and $W_p(s)$ is of full rank and strictly proper. The output of a reference model M represents the behavior expected from the plant, when the latter is augmented with a differentiator-free controller. M is linear, time-invariant, and has a piecewise-continuous and uniformly bounded reference input vector r and an output vector y_m. The transfer matrix, denoted by $W_m(s)$ is strictly proper and asymptotically stable. A control vector u is to be determined so that the error e_1 between plant and model outputs tends to zero asymptotically. As in the SISO case this implies that, for a solution to exist, some assumptions have to be made regarding the plant transfer matrix $W_p(s)$ and that the reference model $W_m(s)$ must be chosen appropriately.

The four assumptions corresponding to (I) in Chapter 5 are listed below:

(i) The high frequency gain matrix K_p satisfies the matrix equation

$$\Gamma K_p + K_p^T \Gamma = Q_0, \qquad \Gamma = \Gamma^T > 0$$

where Q_0 is sign-definite, for some positive-definite symmetric matrix Γ.

(ii) The right Hermite normal form $H_p(s)$ of $W_p(s)$ is known. (II)

(iii) An upper bound ν on the observability index of $W_p(s)$ is known.

(iv) The zeros of $W_p(s)$ lie in $\mathbb{C}^-$.

As in the scalar case, conditions (ii) and (iii) are needed for the algebraic part and (i) and (iv) are required for the analytic part. The relevance of each assumption and its relation to that made for SISO systems is treated here. The role played by these assumptions is clarified in Section 10.6 where the adaptive control problem is discussed in detail.

The Reference Model. The choice of the reference model $W_m(s)$ in the single variable case is determined entirely by the relative degree of the plant transfer function $W_p(s)$. Assuming that a precompensator $G_c(s)$ exists such that $W_p(s)G_c(s) = W_m(s)$, if $G_c(s)$ is to be a proper rational transfer function, the relative degree n_m^* of the model must be greater than or equal to the relative degree n^* of the plant. The reference model is generally chosen to have the same relative degree as the plant so that $G_c(s)$ is an invertible transfer function; a higher relative degree would involve the inclusion of a known strictly proper transfer function in series with the plant. Hence, the class of reference models can be said to consist of all transfer functions with the same relative degree as the plant. As seen in this section, similar concepts also arise in the MIMO case.

The Hermite Normal Form. Denoting the class of reference models that can be used in the MIMO adaptive control problem as $\mathcal{W} \in \mathbb{R}_p^{m\times m}(s)$, let $W_m(s)$ be a typical element of $\mathcal{W}$. Given the plant transfer matrix $W_p(s)$, this implies that a proper transfer matrix $G_c(s)$ exists such that

$$W_p(s)G_c(s) = W_m(s). \tag{10.15}$$

If $H_p(s)$ and $H_m(s)$ are, respectively, the right Hermite forms of $W_p(s)$ and $W_m(s)$, it follows that unimodular matrices $Q_p(s)$ and $Q_m(s)$ exist such that

$$\begin{aligned} W_p(s)Q_p(s) &= H_p(s) \\ W_m(s)Q_m(s) &= H_m(s) \end{aligned}$$

or

$$H_p(s)Q_p^{-1}(s)G_c(s)Q_m(s) = H_m(s).$$

Since $Q_p^{-1}(s)$ and $Q_m(s)$ are unimodular and $W_m(s)$ is proper, it follows that a proper matrix $T(s) = Q_p^{-1}(s)G_c(s)Q_m(s)$ exists such that

$$H_p(s)T(s) = H_m(s)$$

or $H_p(s)$ is a left divisor for $H_m(s)$. For the particular case where $G_c(s)$ is also unimodular, $T(s)$ in turn is unimodular and, hence, $H_p(s)$ and $H_m(s)$ are equivalent. In other words the class of reference models consists essentially of those asymptotically stable transfer matrices that are generated by the Hermite normal form of the plant. This is summarized in the following theorem.

Theorem 10.1 The controller with a transfer matrix $G_c(s)$ given by Eq. (10.15) is causal if, and only if, the class of asymptotically stable reference models $\mathcal{W}$ is parametrized by $H_p(s)$, that is,

$$\mathcal{W} \triangleq \{W_m(s) |\; W_m(s) = H_p(s)V(s)\}$$

where $V(s) \in \mathbb{R}_p^{m\times m}(s)$, $H_p(s)$ is the Hermite normal form of $W_p(s)$, and $H_p(s)$ and $V(s)$ are asymptotically stable.

The left divisibility property of $H_p(s)$ and $H_m(s)$ is the multivariable generalization of the condition that the relative degree of the reference model must be greater than or equal to that of the plant in the SISO problem. A unimodular matrix $G_c(s)$ corresponds to the case where the relative degrees of the plant and the model are equal.

Since the plant that is being adaptively controlled may be unstable, the transfer matrix $G_c(s)$ specified in Eq. (10.15) must be realized by a combination of feedback and feedforward controllers. As in the SISO case, the poles of the plant transfer matrix can be changed by feedback, but the zeros of the plant transfer matrix can be modified only by a feedforward compensator. Hence, to achieve stable adaptive control, the zeros of the plant transfer matrix must lie in the open left-half plane, and the right Hermite form $H_p(s)$ of $W_p(s)$ must be asymptotically stable. The structure of the controller needed to realize $G_c(s)$ is treated in Section 10.6.1.

For a SISO plant whose transfer function is given by $k_pZ_p(s)/R_p(s)$, the Hermite form is given by $h(s) = 1/\pi^{n^*}(s)$ where $\pi(s)$ is a monic Hurwitz polynomial of degree unity and n^* is the relative degree of the plant transfer function. The unimodular form is a unit in $\mathbb{R}_p(s)$, that is, a transfer function of relative degree zero. The entire class of models is then seen to have the form $h(s)v(s)$ where $v(s)$ is an asymptotically stable transfer function in $\mathbb{R}_p(s)$. Just as the relative degree of a system is invariant with respect to feedback, the structure $H_p(s)$ is also preserved with feedback and therefore this information must be contained in the model transfer matrix.

The High Frequency Gain Matrix K_p. The high frequency gain matrix K_p in the multivariable case is similar to the high frequency gain k_p of the SISO plant. However the structure of K_p is not transparent but depends on the structure of the right Hermite form $H_p(s)$ of $W_p(s)$. For example, in the single variable case, if n^* is the relative degree of the plant, the high frequency gain k_p is defined as

$$\lim_{s\to\infty} s^{n^*} W_p(s) \triangleq k_p.$$

Since the Hermite normal form in the multivariable case corresponds to n^* of the single variable system, the matrix K_p may be defined as

$$\lim_{s\to\infty} H_p^{-1}(s)W_p(s) \triangleq K_p.$$

As shown in the sections following, the form of K_p is simple only when $H_p(s)$ is diagonal and depends on plant parameters when $H_p(s)$ is triangular. In the former case $H_p(s)$ has the form

$$H_p(s) = \begin{bmatrix} \frac{1}{\pi^{n_1}(s)} & & & 0 \\ & \frac{1}{\pi^{n_2}(s)} & & \\ 0 & & \ddots & \\ & & & \frac{1}{\pi^{n_m}(s)} \end{bmatrix}$$

so that $H_p^{-1}(s)W_p(s)$ can be written as

$$H_p^{-1}(s)W_p(s) = \begin{bmatrix} \pi^{n_1}(s) & & 0 \\ & \pi^{n_2}(s) & \\ & & \ddots & \\ 0 & & & \pi^{n_m}(s) \end{bmatrix} \begin{bmatrix} k_{11}W_{11}(s) & \cdots & k_{1m}W_{1m}(s) \\ \vdots & & \\ k_{m1}W_{m1}(s) & \cdots & k_{mm}W_{mm}(s) \end{bmatrix}$$

and by the definition of $n_i (i = 1, 2, \ldots, m)$ it follows that

$$\lim_{s\to\infty} H_p^{-1}(s)W_p(s) = K_p .$$

Hence, the elements of K_p are identical to the high frequency gains of the scalar transfer functions corresponding to the minimum relative degrees in each row and zero elsewhere, that is, $K_p = E[W_p(s)]$. When $H_p(s)$ has a triangular structure, such a direct relation between K_p and high frequency gains does not exist as can be seen in the next section.

Observability Index. An upper bound ν on the observability index can be obtained by knowing an upper bound $n_{ij}(i,j = 1,2,\ldots,m)$ on the order of the ij^{th} scalar entry in $W_p(s)$. Assuming that no one particular output is excessively favored over others, ν is given by

$$\nu = \left\lceil \frac{1}{m} \sum_{i.j} n_{ij} \right\rceil$$

where $\lceil \cdot \rceil$ denotes the smallest integer larger than the argument. This is, however, quite a conservative estimate.

In the next section, we illustrate the difficulties encountered in satisfying the assumptions (II) by considering 2×2 systems.

10.5 ADAPTIVE CONTROL OF TWO INPUT - TWO OUTPUT PLANTS

The concepts developed in the earlier sections are best illustrated by considering in detail their application to multivariable plants with two inputs and two outputs. We shall consider several simple examples and in each case indicate the prior information needed to determine the Hermite normal form and the high frequency gain matrix K_p of the plant.

Example 10.1

Let

$$W_p(s) = \begin{bmatrix} \dfrac{k_1}{s+\alpha_1} & \dfrac{k_2}{s+\alpha_2} \\ \dfrac{k_3}{s+\alpha_3} & \dfrac{k_4}{s+\alpha_4} \end{bmatrix}$$

represent the plant transfer matrix where $k_i, \alpha_i (i = 1,2,3,4)$ are unknown parameters. We shall further assume that the zeros of $W_p(s)$ lie in the open left-half of the complex plane.

Case (i). Let $k_1 k_4 \neq k_2 k_3$. In this case, the Hermite normal form $H_p(s)$ is diagonal and the high frequency gain matrix K_p is the same as E, that is,

$$H_p(s) = \begin{bmatrix} \dfrac{1}{s+a} & 0 \\ 0 & \dfrac{1}{s+a} \end{bmatrix} \quad a > 0; \qquad K_p = \begin{bmatrix} k_1 & k_2 \\ k_3 & k_4 \end{bmatrix}.$$

Case (ii). Let $k_1 k_4 = k_2 k_3$ and $\alpha_1 + \alpha_4 \neq \alpha_2 + \alpha_3$. In this case $E[W_p(s)]$ is singular and, hence, the Hermite normal form is triangular. Using the methods described in Section 10.2, $H_p(s)$ may be computed as

$$H_p(s) = \begin{bmatrix} \dfrac{1}{s+a} & 0 \\ \dfrac{k_3/k_1}{s+a} & \dfrac{1}{(s+a)^2} \end{bmatrix} \quad a > 0$$

and the high frequency gain matrix K_p can be computed as

$$\lim_{s\to\infty} H_p^{-1}(s)W_p(s) = K_p = \begin{bmatrix} k_1 & k_2 \\ k_3(\alpha_1 - \alpha_3) & k_4(\alpha_2 - \alpha_4) \end{bmatrix}.$$

The plant parameters k_1 and k_3 are needed to determine the Hermite normal form of $W_p(s)$ while the high frequency gain matrix K_p depends on all the plant parameters α_i and $k_i (i = 1, 2, 3, 4)$. For the case where $k_1 k_4 = k_2 k_3$ and $\alpha_1 + \alpha_4 = \alpha_2 + \alpha_3$, the prior information needed is even greater and both $H_p(s)$ and K_p are found to depend on all the parameters of the plant.

Example 10.2

Let the plant transfer matrix have the form

$$W_p(s) = \begin{bmatrix} \dfrac{1}{s+\alpha_1} & \dfrac{1}{(s+\alpha_2)^2} \\ \dfrac{k_3}{s+\alpha_3} & \dfrac{1}{(s+\alpha_4)^2} \end{bmatrix} \qquad k_3 \neq 1 \tag{10.16}$$

Once again, by inspection, we see that the matrix E is singular since the transfer functions of minimum relative degree occur in the first column. Hence, the Hermite normal form is triangular and is of the form

$$H_p(s) = \begin{bmatrix} \dfrac{1}{s+a} & 0 \\ \dfrac{k_3}{s+a} & \dfrac{1}{(s+a)^2} \end{bmatrix} \qquad a > 0$$

and the high frequency gain matrix K_p can be computed as

$$K_p = \begin{bmatrix} 1 & 0 \\ k_3(\alpha_1 - \alpha_3) & 1 - k_3 \end{bmatrix}.$$

As in the previous example, prior information regarding plant parameters is needed in this case also if the plant is to be controlled adaptively.

The two examples above indicate that the prior information needed to determine $H_p(s)$ and K_p is substantially reduced when the Hermite normal form of the plant transfer function is known a priori to be diagonal. In that case the matrix K_p can be written by inspection. It also follows that, generically, the transfer matrix $W_p(s)$ of a multivariable plant will not have a diagonal Hermite normal form. The following example shows that by using a known precompensator, the transfer matrix considered in Example 10.2 can be modified so that its Hermite normal form is diagonal.

Example 10.3

Consider a plant with the transfer matrix $W_p(s)$ given in Eq. (10.16), whose Hermite form $H_p(s)$ is triangular. If such a plant is augmented by a matrix $W_c(s)$ where

$$W_c(s) = \begin{bmatrix} \dfrac{1}{s+\beta} & 0 \\ 0 & 1 \end{bmatrix},$$

then the transfer matrix of the plant together with the compensator is

$$W_p^1(s) \triangleq W_p(s)W_c(s) = \begin{bmatrix} \frac{1}{(s+\alpha_1)(s+\beta)} & \frac{1}{(s+\alpha_2)^2} \\ \frac{k_3}{(s+\alpha_3)(s+\beta)} & \frac{1}{(s+\alpha_4)^2} \end{bmatrix},$$

which has a diagonal Hermite form $H_p^1(s)$ if $k_3 \neq 1$. $H_p^1(s)$ and the corresponding high frequency gain matrix K_p can then be written by inspection as

$$H_p^1(s) = \begin{bmatrix} \frac{1}{(s+a)^2} & 0 \\ 0 & \frac{1}{(s+a)^2} \end{bmatrix}; \qquad K_p = \begin{bmatrix} 1 & 1 \\ k_3 & 1 \end{bmatrix} \quad a > 0, \;\; k_3 \neq 1$$

It was stated in Section 10.4 that to control a plant with a transfer matrix $W_p(s)$ adaptively, its Hermite normal form should be known and its high frequency gain matrix K_p should satisfy condition (II)-(i). It was also stated that whether the Hermite normal form of $W_p(s)$ will be diagonal or not can be concluded (generically) from a knowledge of the relative degrees of the individual elements of $W_p(s)$. Examples 10.1 and 10.2 reveal that determination of $H_p(s)$ and K_p may be relatively straightforward when the former is known to be diagonal, but may require substantially more information regarding the plant parameters, when it is not. Finally, from Example 10.3 it follows that even in those cases where $H_p(s)$ is triangular in form, the Hermite normal form of the system $W_p(s)W_c(s)$ can be made diagonal (generically) by the proper choice of the compensator $W_c(s)$. Hence, practical adaptive control can be considered to be feasible only when $H_p(s)$ [or $H(s)$ of the compensated system] is diagonal. A complete classification of plants with (2×2) transfer matrices is given by Singh [11] establishing the prior information needed to control them adaptively on the basis of the high frequency scalar gains k_i of the individual transfer functions as well as their relative degrees.

10.6 DYNAMIC PRECOMPENSATOR FOR A GENERAL MULTIVARIABLE PLANT

The simple examples of 2×2 systems considered in the previous section showed that an inordinate amount of prior information regarding the plant parameters may be needed in the choice of the reference model when $W_p(s)$ has a nondiagonal Hermite normal form. On this basis we concluded that practical adaptive control is feasible only when $H_p(s)$ is diagonal. In view of this, it becomes quite important to know whether or not a precompensator $W_c(s)$ can be designed, with limited prior information regarding the plant, such that $W_p(s)W_c(s)$ has (generically) a diagonal Hermite normal form. A positive answer to this was given by Singh and Narendra [12] and is stated in the following theorem.

Theorem 10.2 Let $W_p(s)$ be an $m \times m$ rational transfer matrix with unknown parameters, but with elements whose relative degrees are known exactly. There exists an asymptotically stable, diagonal, and nonsingular dynamic precompensator $W_c(s) \in \mathbb{R}_p^{m\times m}(s)$ such that $E(W_p(s)W_c(s))$ is generically nonsingular if, and only if, $W_p(s)$ is nonsingular. The Hermite form of $W_p(s)W_c(s)$ is a nonsingular diagonal matrix in $\mathbb{R}_p^{m\times m}(s)$.

Proof: Necessity. The necessity of nonsingularity of $W_p(s)$ follows directly from the fact that the nonsingularity of $E(T(s))$ implies the nonsingularity of $T(s)$ for any $m \times m$ matrix $T(s)$. If

$$E(T(s)) = \lim_{s\to\infty} \begin{bmatrix} s^{r_1} & & 0 \\ & \ddots & \\ 0 & & s^{r_m} \end{bmatrix} T(s) \triangleq \lim_{s\to\infty} Q(s)T(s),$$

equating determinants on both sides and noting that it is a continuous operation, it follows that $T(s)$ is nonsingular.

Sufficiency. A detailed proof of sufficiency is beyond the scope of this book and the reader is referred to [12]. We merely indicate here some principal arguments used in that note.

Because only the relative degrees of the elements of $W_p(s)$ are needed in the proof, a matrix $R(s)$ is defined whose ij^{th} elements are of the form $k_{ij}/s^{n_{ij}}$, where k_{ij} is the high frequency gain and n_{ij} is the relative degree of the ij^{th} element of $W_p(s)$. Let $W_c(s)$ be a diagonal transfer matrix in $\mathbb{R}_p^{m\times m}(s)$ and $D(s)$ be a diagonal matrix whose elements are of the form $1/s^{n_i}$ where n_i is the relative degree of the ith diagonal entry of $W_c(s)$. From this, it follows that the Hermite normal form of $R(s)D(s)$ is a diagonal matrix if, and only if, the same holds true for the Hermite normal form of $W_p(s)W_c(s)$.

Since $W_p(s)$ is nonsingular, it also follows that $R(s)$ is nonsingular and can be expressed as

$$R(s) = G(s)\left[K_0 + K_1 s^{-1} + \cdots + K_\ell s^{-\ell}\right]$$

where $\ell = \max_{i,j}(n_{ij}^*)$, $G(s)$ is a diagonal matrix with elements of the form $1/s^n$, and $K_i (0 \leq i \leq \ell)$ are constant matrices in $\mathbb{R}^{m\times m}$. Since it follows that if K_0 is nonsingular, the Hermite normal form is diagonal, we assume that K_0 is singular. The proof of the sufficiency part of Theorem 10.2 consists in demonstrating that a matrix $D(s)$ exists such that $E[R(s)D(s)]$ is generically nonsingular. In [12], it is shown that if the rows and columns of $R(s)$ are suitably permuted a diagonal matrix $D_1(s)$ can be determined so that $E[R(s)D_1(s)]$ has a rank that is greater than or equal to that of $E[R(s)]$. This, in turn, is used to demonstrate that a finite sequence $\{D_i\}$ of matrices exists so that if the permutation p_i of $R(s)D_{i-1}(s)$ is multiplied by $D_i(s)$, after a finite number of iterations,

$$E\left(p_t\left(\left(p_{t-1}\left((\cdots((p_1((p_0R(s))D_1(s)))D_2(s))\cdots D_{t-1}(s)\right)\right)\right)$$

is generically nonsingular.

The following example of a 5×5 system is given in [12] to illustrate the method described above.

Example 10.4

Let a 5×5 matrix $R(s)$ be given by

$$R(s) = \begin{bmatrix} k_1 s^{-2} & k_2 s^{-4} & k_3 s^{-1} & k_4 s^{-4} & k_5 s^{-7} \\ k_6 s^{-1} & k_7 s^{-3} & k_8 s^{-2} & k_9 s^{-4} & k_{10} s^{-6} \\ k_{11} s^{-1} & k_{12} s^{-8} & k_{13} s^{-9} & k_{14} s^{-7} & k_{15} s^{-5} \\ k_{16} s^{-1} & k_{17} s^{-2} & k_{18} s^{-2} & k_{19} s^{-2} & k_{20} s^{-3} \\ k_{21} s^{-1} & k_{22} s^{-3} & k_{23} s^{-3} & k_{24} s^{-4} & k_{25} s^{-6} \end{bmatrix}.$$

Then $E[R(s)]$ is given by

$$E[R(s)] = \begin{bmatrix} 0 & 0 & k_3 & 0 & 0 \\ k_6 & 0 & 0 & 0 & 0 \\ k_{11} & 0 & 0 & 0 & 0 \\ k_{16} & 0 & 0 & 0 & 0 \\ k_{21} & 0 & 0 & 0 & 0 \end{bmatrix}$$

and has rank 2. Using the procedure outlined in [12] the diagonal matrix $D(s)$ can be constructed as

$$D(s) = \begin{bmatrix} \frac{1}{s^4} & & & & 0 \\ & \frac{1}{s^2} & & & \\ & & \frac{1}{s^3} & & \\ & & & \frac{1}{s} & \\ 0 & & & & 1 \end{bmatrix}$$

so that $E[R(s)D(s)]$ is nonsingular. Hence the Hermite form of $R(s)D(s)$ or equivalently that of $W_p(s)W_c(s)$ is also diagonal.

10.6.1 Design of the Adaptive Controller

The design of an adaptive controller for a multivariable plant proceeds along the same lines as those for the single variable case discussed in Chapter 5. This includes the choice of a reference model, the selection of a controller structure, the demonstration that a constant controller parameter matrix exists so that the transfer matrix of the plant together with the controller matches that of the model, and finally the derivation of adaptive laws for adjusting the controller parameter vector $\theta(t)$ so that the output error $e(t)$ tends to zero asymptotically. In this section we briefly describe each of these steps.

Choice of the reference model. In the multivariable adaptive control problem described in Section 10.4, if the relative degree n_{ij}^* of the $(i,j)^{\text{th}}$ element of $W_p(s)$ for all i,j is known, we can conclude whether its Hermite normal form is diagonal or not. In the latter case, a precompensator with a transfer matrix $W_c(s)$ can be designed so that $W_p(s)W_c(s)$ has a diagonal Hermite form. Hence, we shall assume in this section, with no loss of generality, that $H_p(s)$, the right Hermite normal form of the plant is diagonal and stable. As stated in Section 10.4, the reference model can then be chosen to have

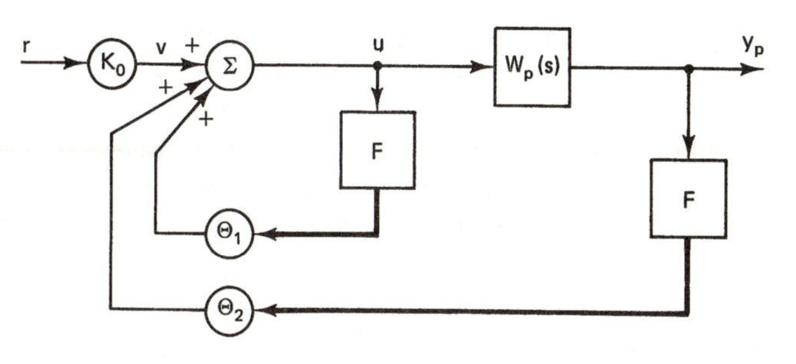

Figure 10.3 The adaptive controller structure.

the form $W_m(s) = H_p(s)Q_m(s)$ where $Q_m(s)$ is an asymptotically stable unimodular matrix. For purposes of simplicity, and with no loss of generality, we shall assume that $Q_m(s) = I$, the identity matrix.

The Controller Structure. The controller consists of a feedforward matrix $K_0 \in \mathbb{R}^{m \times m}$ and filters F_1 and F_2 in the feedback path (Fig. 10.3) with transfer matrices $R_q^{-1}(s)Z_c(s)$ and $R_q^{-1}(s)Z_d(s)$ respectively, for constant values of Θ_1 and Θ_2. The $m \times m$ matrix $R_q^{-1}(s)$ is chosen to be row proper and to commute with $Z_c(s)$ and $Z_d(s)$. One such choice is a diagonal matrix whose entries are all equal to $1/r_q(s)$ where $r_q(s)$ is a Hurwitz monic polynomial of degree $\nu - 1$. The matrices $Z_c(s)$ and $Z_d(s)$ are determined by the input-output relation

$$\omega_i(t) = \frac{s^{i-1}}{r_q(s)}u(t), \quad \omega_j(t) = \frac{s^{j-\nu}}{r_q(s)}y_p(t), i = 1, \ldots, \nu - 1, \; j = \nu, \ldots, 2\nu - 1.$$

The $2m\nu$ dimensional vector ω and the $m \times 2m\nu$ matrix Θ are defined as

$$\omega = \left[r, \omega_1^T, \ldots, \omega_{\nu-1}^T, \omega_\nu^T, \ldots \omega_{2\nu-1}^T\right]^T, \Theta = \left[K_0, C_1, \ldots, C_{\nu-1}, D_0, \ldots, D_{\nu-1}\right]$$

where K_0, C_i, D_j are $m \times m$ matrices for $i = 1, \ldots, \nu - 1$, and $j = 0, \ldots, \nu - 1$. The control input to the plant can then be expressed as

$$u(t) = \Theta(t)\omega(t).$$

For constant values of the parameter matrix Θ, the closed loop transfer function is given by $W_o(s)$ where

$$W_o(s) = Z_p(s)\left[\left(R_q(s) - Z_c(s)\right)R_p(s) - Z_d(s)Z_p(s)\right]^{-1}R_q(s)K_0. \quad (10.17)$$

In Eq. (10.17), $R_p(s)$ and $Z_p(s)$ are right coprime and, hence, by the Bezout identity, the matrices $R_q(s) - Z_c(s)$ and $Z_d(s)$ of degree $\nu - 1$ exist such that the transfer matrix within the brackets can be made equal to any arbitrary polynomial of column degree $d_j + \nu - 1$ where d_j is the column degree of $R_p(s)$. Choosing this matrix to be $R_q(s)K_0H_p^{-1}(s)Z_p(s)$, it follows that $W_o(s) = H_p(s)$ as desired. Therefore the existence of $Z_c(s)$ and $Z_d(s)$ implies the existence of a matrix Θ^* in the controller to assure that the transfer matrix of the plant together with the controller is equal to $W_m(s)$.

The Adaptive Law. The next step in the design of the adaptive controller is the determination of adaptive laws for adjusting the parameter $\Theta(t)$ so that the output error between plant and model tends to zero. Since the transfer function of the plant feedback loop when $\Theta(t) \equiv \Theta^*$ is $H_p(s)$, it follows that the transfer function from $v(t)$ in Fig. 10.3 to the output is $H_p(s)K_0^{*-1} = H_p(s)K_p$. Further, if $\Phi(t) = \Theta(t) - \Theta^*$ and $e_1(t) = y_p(t) - y_m(t)$, the output of the model, the output of the plant and the error $e_1(t)$ are described by the equations

$$\begin{aligned} y_m(t) &= H_p(s)r(t) \\ y_p(t) &= H_p(s)\left[r(t) + K_p\Phi(t)\omega(t)\right] \\ e_1(t) &= H_p(s)K_p\Phi(t)\omega(t). \end{aligned}$$

As in the SISO case, the adaptive law depends on the information available concerning $H_p(s)$ and K_p. As shown below, when K_p is known and $H_p(s)$ is a strictly positive real transfer matrix, the generation of adaptive laws is simplified considerably. When K_p is unknown, some prior information is needed to assure that the overall system is stable. Finally, when $H_p(s)$ is not strictly positive real, an augmented error must be generated as in the single variable case. This makes the adaptive controller substantially more complex. The adaptive laws corresponding to these cases are given below.

Case (i)* K_p *known. In this case, K_0 is chosen as the fixed gain matrix K_p^{-1} and the modified plant has a unity high frequency gain. The adaptive controller contains $m \times m(2\nu - 1)$ parameters, which are the elements of a parameter error matrix $\overline{\Phi}(t)$, and the corresponding $m(2\nu - 1)$ dimensional vector signal is denoted by $\overline{\omega}(t)$.

(a) $H_p(s)$ is SPR. When $H_p(s)$ is SPR, the parameter error matrix is updated according to the law

$$\dot{\overline{\Theta}} = \dot{\overline{\Phi}} = -e_1\overline{\omega}^T. \tag{10.18}$$

The discussions of the multivariable error models in Section 10.2 indicate that a scalar function

$$V(e, \overline{\Phi}) = e^T Pe + Tr(\overline{\Phi}^T\overline{\Phi})$$

leads to a time derivative $\dot{V}$, which when evaluated along the trajectories of Eq. (10.18), given by

$$\dot{V} = -e^T Qe$$

from the matrix version of the Kalman-Yakubovich lemma. Therefore $e_1(t)$ and $\overline{\Phi}(t)$ are bounded. Since the signals in the model are bounded, it implies that $\overline{\omega}(t)$ is bounded. As in the SISO case, we conclude that $e_1(t)$ as well as $\dot{\overline{\Phi}}(t)$ tend to zero as $t \to \infty$.

(b) $H_p(s)$ is not SPR. Let $\overline{\theta}_i^T(t)$ and $\overline{\phi}_i^T(t)$ denote the ith row in $\overline{\Theta}(t)$ and $\overline{\Phi}(t)$ respectively, and $h_i(s)$ be the ith diagonal entry in $H_p(s)$, for $i = 1, \ldots, m$. Then an auxiliary signal $e_2(t)$ is defined as

$$e_2(t) = \begin{bmatrix} \overline{\theta}_1^T(t)h_1(s)\overline{\omega}(t) \\ \vdots \\ \overline{\theta}_m^T(t)h_m(s)\overline{\omega}(t) \end{bmatrix} - H_p(s)\overline{\Theta}(t)\overline{\omega}(t).$$

The augmented error ϵ_1 is then realized as

$$\begin{aligned} \epsilon_1(t) &= e_1(t) + e_2(t) \\ &= \begin{bmatrix} \epsilon_1(t) \\ \vdots \\ \epsilon_m(t) \end{bmatrix} = \begin{bmatrix} \overline{\phi}_1^T(t)\overline{\zeta}_1(t) \\ \vdots \\ \overline{\phi}_m^T(t)\overline{\zeta}_m(t) \end{bmatrix} \end{aligned}$$

where $\overline{\zeta}_i(t) = h_i(s)\overline{\omega}(t)$, $i = 1, \ldots, m$. The adaptive laws are therefore chosen as

$$\begin{aligned} \dot{\overline{\theta}}_1(t) = \dot{\overline{\phi}}_1(t) &= -\frac{\epsilon_1(t)\overline{\zeta}_1(t)}{1 + \overline{\zeta}_1^T(t)\overline{\zeta}_1(t)} \\ \vdots &= \vdots \\ \dot{\overline{\theta}}_m(t) = \dot{\overline{\phi}}_m(t) &= -\frac{\epsilon_m(t)\overline{\zeta}_m(t)}{1 + \overline{\zeta}_m^T(t)\overline{\zeta}_m(t)}. \end{aligned}$$

With a Lyapunov function candidate $V(\overline{\Phi}) = Tr(\overline{\Phi}^T\overline{\Phi})$, as in the SISO case, we can show that $\overline{\Phi}$ is bounded, $\dot{\overline{\Phi}} \in \mathcal{L}^2$, and $\epsilon_i/[1 + \overline{\zeta}_i^T\overline{\zeta}_i]^{1/2} \in \mathcal{L}^2$.

Case (ii) K_p ***unknown.*** (a) $H_p(s)$ is SPR. As in case (i), the generation of stable adaptive laws is simpler when the Hermite form is strictly positive real. However a sign definiteness condition on K_p is required since an additional parameter matrix K_0 needs to be adjusted in the forward path (Fig. 10.3) so that it tends to K_p^{-1}. The error equation in this case is given by $e_1(t) = H_p(s)K_p\Phi(t)\omega(t)$. If K_p satisfies the matrix equation $\Gamma K_p + K_p^T\Gamma = Q_0 > 0$, then a scalar function $V = e^TPe + Tr(\Phi^TQ_0^T\Phi)$ can be shown to be a Lyapunov function if $\dot{\Phi} = -P_0e_1\omega^T$, $P_0 = \Gamma^{-1}$. Hence, all the signals are bounded.

(b) $H_p(s)$ is not SPR. In Chapter 5, it was seen that even for a SISO system, when $H_p(s)$ is not SPR and the high frequency gain is not known, the problem is considerably more complex. This is only amplified in the multivariable case. A detailed description of a general adaptive system here would add very little to our understanding of multivariable adaptive control. Instead, we present below the augmented error and the adaptive laws when the plant has two inputs and two outputs [11]. As before, it can be shown that the parameter error matrix Φ is bounded and that $\dot{\Phi} \in \mathcal{L}^2$. Using the arguments based on the growth rates of signals, it can once again be shown that all the signals of the system remain bounded while $\epsilon_1(t)$ and $e_1(t)$ tend to zero as $t \to \infty$.

$$\epsilon(t) = \begin{bmatrix} \epsilon_1(t) \\ \epsilon_2(t) \end{bmatrix} = \begin{bmatrix} \left(k_1\phi_1^T(t) + k_2\phi_2^T(t)\right)\zeta_1(t) + k_1g_1(t)\xi_1(t) + k_2g_2(t)\xi_2(t) \\ \left(k_3\phi_1^T(t) + k_4\phi_2^T(t)\right)\zeta_2(t) + k_3g_3(t)\xi_3(t) + k_4g_4(t)\xi_4(t) \end{bmatrix}$$

where $\zeta_i(t) = h_i(s)\omega(t), i = 1,2,\ g_i(t),\ i = 1,\ldots,4$ are the multivariable counterparts of the parameter error $\psi_1(t)$ in the single variable case [refer to Eq. (5.52)], and $h_i(s)$, ω, and ϕ_i are as defined earlier.

$$\begin{bmatrix} \xi_1(t) \\ \xi_2(t) \\ \xi_3(t) \\ \xi_4(t) \end{bmatrix} = \left[\begin{bmatrix} \phi_1^T(t) & 0 \\ \phi_2^T(t) & 0 \\ 0 & \phi_1^T(t) \\ 0 & \phi_2^T(t) \end{bmatrix} \begin{bmatrix} h_1(s)I_{4\nu} \\ h_2(s)I_{4\nu} \end{bmatrix} - \begin{bmatrix} h_1(s)I_2 \\ h_2(s)I_2 \end{bmatrix} \Phi(t) \right] \omega(t)$$

and

$$K_p = \begin{bmatrix} k_1 & k_2 \\ k_3 & k_4 \end{bmatrix}.$$

The corresponding adaptive laws are

$$\dot{\phi}_1(t) = -\Gamma \frac{\epsilon_1(t)\zeta_1(t)}{1 + x^T(t)x(t)}$$

$$\dot{\phi}_2(t) = -\Gamma \frac{\epsilon_2(t)\zeta_2(t)}{1 + x^T(t)x(t)}$$

$$\dot{g}_1(t) = -\gamma_1 \frac{\epsilon_1(t)\xi_1(t)}{1 + x^T(t)x(t)} \qquad \dot{g}_2(t) = -\gamma_2 \frac{\epsilon_1(t)\xi_2(t)}{1 + x^T(t)x(t)}$$

$$\dot{g}_3(t) = -\gamma_3 \frac{\epsilon_2(t)\xi_3(t)}{1 + x^T(t)x(t)} \qquad \dot{g}_4(t) = -\gamma_4 \frac{\epsilon_2(t)\xi_4(t)}{1 + x^T(t)x(t)}$$

where $\gamma_i > 0, i = 1,\ldots,4$, and $x(t) = [\zeta_1^T(t), \zeta_2^T(t), \xi_1(t), \xi_2(t), \xi_3(t), \xi_4(t)]^T$.

In the example that follows, a plant with two inputs and two outputs is considered, whose Hermite form is diagonal and strictly positive real. The controller structure and the adaptive laws for adjusting the control parameters are presented.

Example 10.5

Let the plant to be controlled adaptively have two inputs and two outputs, with a transfer matrix

$$W_p(s) = \begin{bmatrix} \dfrac{1}{s-1} & \dfrac{2}{(s-2)^2} \\ \dfrac{1}{(s-1)^2} & \dfrac{1}{s+3} \end{bmatrix}.$$

We note there exists a unimodular matrix $Q_p(s)$ given by

$$Q_p(s) = \begin{bmatrix} \dfrac{s-1}{s+3} & \dfrac{2}{s-2} \\ \dfrac{-1}{s-1} & -\dfrac{s-2}{s-1} \end{bmatrix} \times$$

$$\begin{bmatrix} \dfrac{(s+3)(s-2)^2(s-1)}{(s+1)\left[(s-2)^2(s-1) - 2(s+3)\right]} & 0 \\ 0 & \dfrac{(s-1)^2(s-2)(s+3)}{(s+1)\left[2(s+3) - (s-1)(s-2)^2\right]} \end{bmatrix}$$

so that $W_p(s)Q_p(s) = H_p(s)$, which is diagonal and SPR, and given by

$$H_p(s) = \begin{bmatrix} \frac{1}{s+1} & 0 \\ 0 & \frac{1}{s+1} \end{bmatrix}.$$

The right coprime factorization of $W_p(s)$ is given by

$$W_p(s) = \begin{bmatrix} s-1 & 2(s+3) \\ 1 & (s-2)^2 \end{bmatrix} \begin{bmatrix} (s-1)^2 & 0 \\ 0 & (s+3)(s-2)^2 \end{bmatrix}^{-1}$$

and the left coprime factorization is

$$W_p(s) = \begin{bmatrix} (s-1)(s-2)^2 & 0 \\ 0 & (s+3)(s-1)^2 \end{bmatrix}^{-1} \begin{bmatrix} (s-2)^2 & 2(s-1) \\ s+3 & (s-1)^2 \end{bmatrix}.$$

The state variable description of the plant is given by

$$\dot{x} = Ax + Bu; \qquad y_p = Cx$$

where

$$A = \begin{bmatrix} 0 & 1 & 0 & 0 & 0 \\ -1 & 2 & 0 & 0 & 0 \\ 0 & 0 & 0 & 1 & 0 \\ 0 & 0 & 0 & 0 & 1 \\ 0 & 0 & 12 & -8 & -1 \end{bmatrix}, \quad B = \begin{bmatrix} 0 & 0 \\ 1 & 0 \\ 0 & 0 \\ 0 & 0 \\ 0 & 1 \end{bmatrix}, \quad C^T = \begin{bmatrix} -1 & 1 \\ 1 & 0 \\ 6 & 4 \\ 2 & -4 \\ 0 & 1 \end{bmatrix}$$

Since the rank of $[c_1, A^T c_1, A^{2T} c_1, c_2, A^T c_2]$ is 5, where c_1 and c_2 are the two columns of C^T, it follows that the observability index $\nu = 3$. The latter can also be determined as the highest degree of the denominator in the left coprime factorization of $W_p(s)$.

The plant above has to be controlled adaptively, and the information available to the designer includes the relative degree of each element of the transfer matrix, the sign of the high frequency gains of the diagonal elements, and the observability index of $W_p(s)$. From this information, the designer can determine the Hermite normal form of $W_p(s)$ as $H_p(s)$, which is given above, and conclude that all the conditions (II) needed for adaptive control are satisfied. The reference model can have a transfer matrix $W_m(s) = H_p(s)Q_m(s)$ where $Q_m(s)$ is unimodular, asymptotically stable, and minimum phase. The adaptive controller can now be determined and is given by (Fig. 10.4)

$$\begin{aligned} \omega_1(t) &= R_q(s)u(t) & \omega_3(t) &= R_q(s)y_p(t) \\ \omega_2(t) &= s\omega_1(t) & \omega_4(t) &= s\omega_3(t) \\ & & \omega_5(t) &= s\omega_4(t) \end{aligned}$$

where $R_q(s) = \text{diag}[1/r_q(s), 1/r_q(s)]$, and $r_q(s) = s^2 + 3s + 2$. The control input is then described as

$$u(t) = K_0(t)r(t) + C_1(t)\omega_1(t) + C_2(t)\omega_2(t) + D_0(t)\omega_3(t) + D_1(t)\omega_4(t) + D_2(t)\omega_5(t).$$

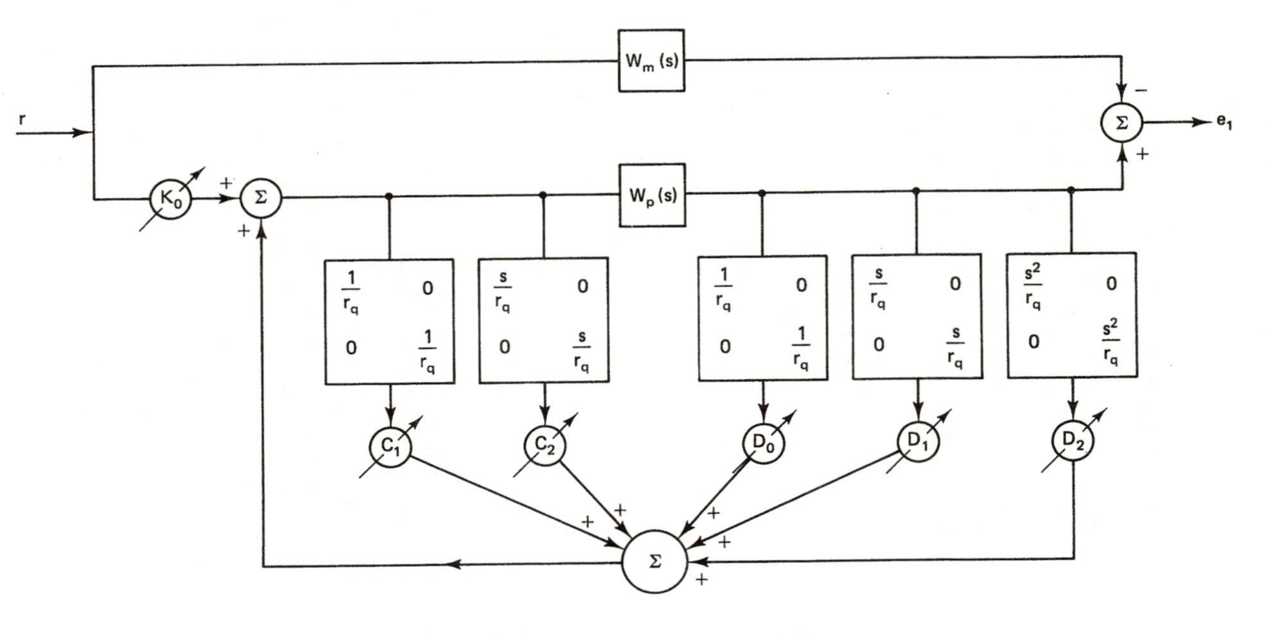

Figure 10.4 The adaptive system for the 2×2 case.

Since $H_p(s)$ is strictly positive real, and $K_p = I$ is symmetric and positive-definite where I is a 2×2 identity matrix, from the discussions in this section, the adaptive laws are given by

$$\begin{array}{rclcrcl} \dot{K}_0 &=& -e_1 r^T & \qquad & \dot{D}_0 &=& -e_1 \omega_3^T \\ \dot{C}_1 &=& -e_1 \omega_1^T & & \dot{D}_1 &=& -e_1 \omega_4^T \\ \dot{C}_2 &=& -e_1 \omega_2^T & & \dot{D}_2 &=& -e_1 \omega_5^T . \end{array}$$

It follows that all the signals are bounded and $\lim_{t\to\infty} e_1(t) = 0$.

10.7 SUMMARY

In this chapter, many of the concepts developed for SISO systems in Chapters 3-7 are extended to multivariable systems. In view of the coupling that exists in such systems, additional prior information is needed to control them adaptively. However, there are no major surprises in the results obtained. The assumptions (II) corresponding to (I) for single variable systems are given in Section 10.4. It is found that knowledge of the right Hermite normal form of the plant transfer matrix $W_p(s)$ (which corresponds to the relative degree in the scalar case), must be known to determine a reference model and the high frequency gain matrix K_p of $W_p(s)$ has to satisfy a matrix equation if stable adaptive laws are to be generated. When the right Hermite normal form of the plant is not diagonal, substantial prior information regarding the plant parameters may be needed to determine it. Hence, for practical adaptive control, only plants with diagonal Hermite normal forms are considered. Theorem 10.2 assures that with no more information concerning the elements of $W_p(s)$ than that needed in the scalar case, a precompensator $W_c(s)$ can be designed so that $W_p(s)W_c(s)$ has a diagonal Hermite normal form. For such systems, if assumption (II) holds, stable adaptive laws can be determined so that the outputs of the plant track the corresponding outputs of the reference model exactly as $t \to \infty$.

We have not discussed robustness of multivariable adaptive systems in this chapter. However, almost all the methods discussed in Chapter 8 for SISO systems can be applied to multivariable systems with minor modifications.

PROBLEMS

Part I

1. Determine the poles and zeros of the following transfer matrices:

$$\text{(a)} \begin{bmatrix} \dfrac{1}{s-1} & \dfrac{s-2}{(s+3)^2} \\ \dfrac{1}{(s+1)(s+3)} & \dfrac{s+2}{s-1} \end{bmatrix} \qquad \text{(b)} \begin{bmatrix} \dfrac{1}{s-1} & \dfrac{1}{s-2} \\ \dfrac{4}{s-3} & \dfrac{2}{s-4} \end{bmatrix}.$$

2. Determine the right Hermite normal forms of the following transfer matrices:

$$\text{(a)} \begin{bmatrix} \dfrac{k_1}{s-\alpha} & \dfrac{k_2}{s+\beta} \\ \dfrac{k_3}{(s+\gamma)^2} & \dfrac{k_4}{(s+\delta)^2} \end{bmatrix} \qquad \text{(b)} \begin{bmatrix} \dfrac{k_1}{s+\alpha} & \dfrac{k_3}{(s+\gamma)^2} \\ \dfrac{k_2}{s+\beta} & \dfrac{k_4}{(s+\delta)^2} \end{bmatrix}.$$

Part II

3. The error equation for a multivariable system has the form

$$e_1 = K_p \Phi \omega$$

where K_p is an $m \times m$ matrix, $\Phi : \mathbb{R}^+ \to \mathbb{R}^{m \times m}$, $\omega : \mathbb{R}^+ \to \mathbb{R}^m$, and $e_1 : \mathbb{R}^+ \to \mathbb{R}^m$. It is known that a matrix $\Gamma > 0$ exists such that

$$\Gamma K_p + K_p^T \Gamma = Q > 0.$$

Determine an adaptive law for adjusting $\dot{\Phi}$ so that $\lim_{t\to\infty} e_1(t) = 0$.

4. **(a)** The transfer matrix $W_p(s)$ of a dynamical system with three inputs and three outputs has the form

$$W_p(s) = \begin{bmatrix} \dfrac{k_{11}}{(s+\alpha)^3} & \dfrac{k_{12}}{s+\gamma} & \dfrac{k_{13}}{(s+\beta)(s+\gamma)} \\ \dfrac{k_{21}}{s+\eta} & \dfrac{k_{22}}{(s+\delta)^4} & \dfrac{k_{23}}{(s+\alpha)(s+\beta)^2} \\ \dfrac{k_{31}}{(s+\eta)^2} & \dfrac{k_{32}}{s+\beta} & \dfrac{k_{33}}{(s+\gamma)^3} \end{bmatrix}$$

where $k_{ij}, i, j = 1, 2, 3$ are known and α, β, γ, δ, and η are unknown. The relative degree of each element of the matrix is known. Show that the right Hermite normal form of $W_p(s)$ is not diagonal.

(b) Determine an asymptotically stable, diagonal, precompensator with a transfer matrix $W_c(s)$ such that $W_p(s)W_c(s)$ has a diagonal Hermite normal form.

5. A plant has a transfer matrix $W_p(s)$ where

$$W_p(s) = \begin{bmatrix} \dfrac{2}{s+\alpha} & \dfrac{1}{s+\beta} \\ \dfrac{1}{s+\gamma} & \dfrac{3}{s+\delta} \end{bmatrix}.$$

α, β, γ, and δ are unknown but negative. Indicate how you would design an adaptive controller to stabilize the plant above.

Part III

6. In Section 10.4, it was stated that from a practical standpoint, only plants with transfer matrices having a diagonal Hermite normal form can be controlled adaptively. Comment on this statement.

REFERENCES

1. Anderson, B.D.O. *"Multivariable adaptive identification."* Technical Report EE7402, University of Newcastle, Australia, 1974.
2. Borison, U. "Self Tuning Regulators for a class of multivariable systems." *Automatica*, 15:209–215, 1979.
3. Chen, C.T. *Introduction to linear systems design.* New York, NY:Holt, Rinehart and Winston, Inc., 1980.
4. Elliott, H. and Wolovich, W.A. "Parameter adaptive control of linear multivariable systems." *IEEE Transactions on Automatic Control*, 27:340–352, 1982.
5. Goodwin, G.C. and Long, R.S. "Generalization of results on multivariable adaptive control." *IEEE Transactions on Automatic Control*, 25:1241–1245, Dec. 1980.
6. Goodwin G.C., Ramadge P.J., and Caines, P.E. "Discrete time multivariable adaptive control." *IEEE Transactions on Automatic Control*, 25:449–456, 1980.
7. Monopoli, R.V. and Hsing, C.C. "Parameter adaptive control of multivariable systems." *International Journal of Control*, 22:313–327, 1975.
8. Monopoli, R.V. and Subbarao, V.N. "Design of a multivariable model following adaptive control system." In *Proceedings of the 1981 CDC*, San Diego, California, 1981.
9. Morse, A.S. "Representation and parameter identification of multi-output linear system." In *Proceedings of the 1974 CDC*, Phoenix, Arizona, 1974.
10. Narendra, K.S. and Kudva, P. "Stable adaptive schemes for system identification and control - Parts I&II." *IEEE Transactions on Systems, Man and Cybernetics*, SMC-4:542–560, Nov. 1974.
11. Singh, R.P. *Stable multivariable adaptive control systems.* Ph.D. thesis, Yale University, 1985.
12. Singh, R.P., and Narendra, K.S. "Prior information in the design of multivariable adaptive controllers." *IEEE Transactions on Automatic Control* 29:1108–1111, Dec. 1984.
13. Wolovich, W.A. *Linear multivariable systems.* Berlin:Springer Verlag, 1974.

11

Applications of Adaptive Control

11.1 INTRODUCTION

The impetus to the development of practical adaptive control design methodologies during the past decade has come from three principal sources. These include the demands made by a rapidly advancing technology for more sophisticated controllers, the availability of microprocessors, which, in spite of their limitations, have rendered the implementation of adaptive algorithms economically feasible, and the recent advances in theory that have provided a rigorous framework for the practical design of adaptive controllers.

The design of industrial controllers to meet speed, accuracy, and robustness requirements is invariably dictated by economic considerations. This, in turn, implies that generally, in any practical situation, the simplest controller capable of meeting performance specifications is chosen. In the highly competitive industrial world, control tolerances are constantly being revised to achieve higher performance and necessitate taking into account uncertainties heretofore neglected in the design. This calls for the use of more complex and, hence, more versatile controllers. Naturally adaptive controllers are candidates in such cases.

As described in Chapters 8 and 9, in the control of a dynamical system uncertainties may arise due to external disturbances, variations in plant and noise parameters, and poor mathematical models of the process. Changes in raw materials used and changes in the characteristics of the processing units are some typical causes of parameter variations. In the presence of such uncertainties, improved performance can be realized by using adaptive control. In many cases small improvements in efficiency resulting from such control may yield significant economic savings.

Another factor that requires the use of more sophisticated control is the high cost of skilled labor in advanced industrialized societies. For example, in complex chemical processes, it is often necessary to tune hundreds of parameters continuously to keep the system operating close to its optimum point. Economic considerations strongly dictate the use of automatic tuning devices without human intervention.

As emphasized in Chapter 1, revolutionary changes that have taken place in computer technology during the last decade are, to a large extent, responsible for many of the system theoretic methods, described earlier, becoming practically viable. The fact that successful self-tuning regulators have been introduced commercially by various industries such as ASEA, Foxboro, and Leeds and Northrup, also provides a strong motivation to others to attempt adaptive solutions. Although a significant gap exists between the theoretical results that have been reported in the preceding chapters and the practical needs of industry, nevertheless the former provide a basis for design and suggest the types of behavior that can be anticipated in their performance.

In this chapter, we deal with applications of adaptive control theory to specific engineering problems. Since uncertainty can assume myriad forms in practice, it is found that adaptive principles must be applied selectively in different cases. At the same time, solutions developed in one adaptive context may be applicable to similar problems in widely different fields. Our aim here is to discuss in some detail five areas where adaptive control has been applied successfully. In each case, we shall attempt to describe the reasons for the choice of an adaptive controller, the practical limitations of such control, and suggest the type of improvement that has been achieved.

11.2 SHIP DYNAMICS

An area that exemplifies many of the characteristics we have described above and where adaptive control has been applied successfully is the automatic steering of ships. Ship steering, especially on long straight courses and under bad weather conditions, is an arduous and monotonous task. Automatic ship steering devices (autopilots) therefore were needed long before control design methods were well established. The introduction of such devices is attributed to Minorsky [18] and Sperry [28] who suggested a simple proportional feedback controller that supplied a corrective signal to the rudder proportional to the heading error. This was followed through the course of time by the introduction of a PID control, which substantially improved the steering performance compared with proportional controllers. Such controllers contain three fixed gains corresponding to the proportional, integral, and derivative of the input signals. Over the years, however, there has been a growing realization that the PID controller is too simple and that the parameters may not be tuned properly for optimal performance. Hence, several investigations were carried out consisting of full-scale experiments, scaled model tests, and computer simulations to study the feasibility of applying adaptive control techniques. As a consequence of these investigations, it has been realized that such techniques could result in superior performance while being economically viable. Hence, adaptive control is being introduced in the design of autopilots for ships.

The need for a transition from fixed PID controllers to adaptive controllers can be justified by examining the control problem in greater detail. Ship steering must often be performed under varying environmental conditions. Some of these variables directly influence the dynamics of the ship, which include trim, loading, and water depth. Others are external disturbances such as wind, waves, and currents that can be considered additive input signals. The aim of the controller then is to either maintain the course of the ship along a straight path or follow a desired course in the presence of the variations above. When the internal or external disturbances become significant, it is tedious and difficult to properly determine the fixed parameters of the PID controller that results in good performance. Fixed controller settings often result in energy losses due to excessive rudder motion (particularly evident in bad weather) or inadequate maneuvering performance (when the ship has to make large maneuvers). If gain scheduling is contemplated, the number of settings of the parameters increases rapidly as environmental changes or operational demands become large. Finally, in view of the complexity of the environment, the relation between parameter settings and environmental changes is not always evident to permit efficient gain scheduling. In addition to the physical reasons mentioned above, other practical considerations that depend on advances and changes in the shipping industry also dictate the need for improved performance. Fuel costs have been growing since 1973 and it is worthwhile to investigate the fuel-saving potential of an improved control algorithm. The increased traffic at sea requires accurate steering and, therefore, imposes stringent conditions on the controller. The sizes of ships have been growing as well, which makes manual steering difficult. All these reasons clearly indicate that if improved performance is the objective, adaptive controllers, in which the parameters are adjusted continuously, may be needed.

The different approaches that can be used while attempting to design adaptive controllers to an existing system are reflected in the works of several individuals and research groups working around the world. Oldenburg [25], Kanamaru and Sato [13], and Sugimoto and Kojima [30] suggest monitoring the environment prior to adjusting the autopilot. Self-tuning regulators have been proposed by numerous authors in [3, 11,12,15,19,31], and MRAC has been used by the group at Delft [1,10,32,33]. Van Amerongen has worked extensively on this problem and his doctoral dissertation [32] contains many practical insights. Hill-climbing methods for optimizing a performance index have been reported in [1,5,25,26,27] and stochastic methods for identification and control have been attempted in [9,16,17,24]. Efforts have also been expended in suitably modeling the dynamics of the ship [2,7,21,23,29] determining variations in it [8] as well as in establishing criteria for the design of suitable controllers [4,6,12,14,20,22,34]. In the following section, we deal only with the model reference approach; however, we shall also make relevant comments, wherever appropriate, regarding other approaches and their implications.

The Dynamic Model. The mathematical model of a ship can be derived from dynamical laws of motion by considering it as a rigid body. In Fig. 11.1, the translational motions on the plane are in the x and y directions and are referred to as *surge* and *sway*. The translational motion in the z direction, orthogonal to the $x - y$ plane, is called heave. The rotational motions roll, pitch, and yaw are, respectively, about the x, y, and z directions. Generally, in the motion of a ship, surge, sway, and yaw motions are

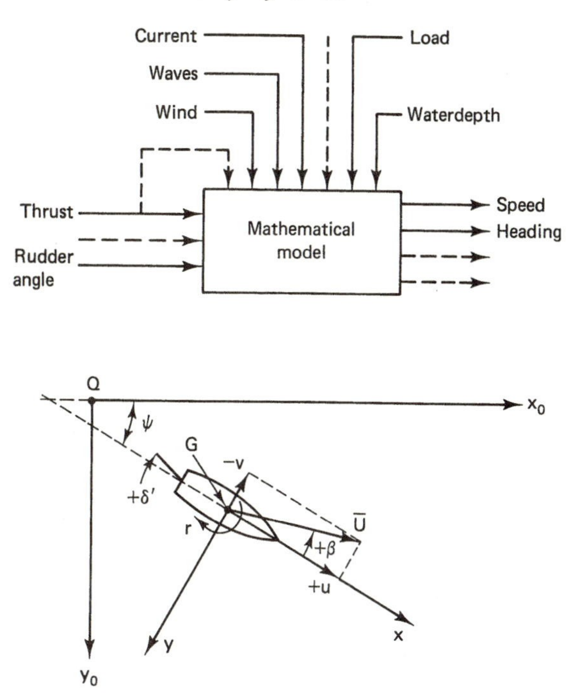

Figure 11.1 The dynamic model. Figure courtesy of van Amerongen, 1982.

considered to be dominant. The corresponding equations of motion are given by [32]

$$
\begin{aligned}
m\dot{u} - v\dot{\psi} &= X \\
m(\dot{v} + u\dot{\psi}) &= Y \\
I_z\ddot{\psi} &= N
\end{aligned}
\tag{11.1}
$$

where ψ is the course heading angle, u is the speed in the forward direction, v is the drift speed, m is the mass of the ship, X and Y are the hydrodynamic forces in the x and y directions, N is the moment with respect to the z-axis, and I_z is the moment of inertia with respect to the z-axis. Different models of ship dynamics have been used by various authors and we present below one general derivation of such a model, described by a third-order differential equation discussed by van Amerongen [32]. Modifications and simplifications of this model will be used in our discussions in this section.

The forces X, Y, and N and the moment I_z are functions of all the variables $u, v, \dot{\psi}$, and δ, the rudder angle, and their first and higher derivatives. Using Taylor series expansion of the forces, we obtain typically for X,

$$
X = X_\delta\delta + X_{\dot{\delta}}\dot{\delta} + X_u u + X_{\dot{u}}\dot{u} + X_v v + X_{\dot{v}}\dot{v} + X_{\dot{\psi}}\dot{\psi} + X_{\ddot{\psi}}\ddot{\psi} + \cdots \tag{11.2}
$$

where X_δ denotes the partial of X with respect to δ. Similar expressions can be derived for Y and N. Assuming that the forward speed is a constant, the first equation in Eq. (11.1) can be omitted and the terms containing u and $\dot{u}$ are zero. Hence, the second and third equations can be written as

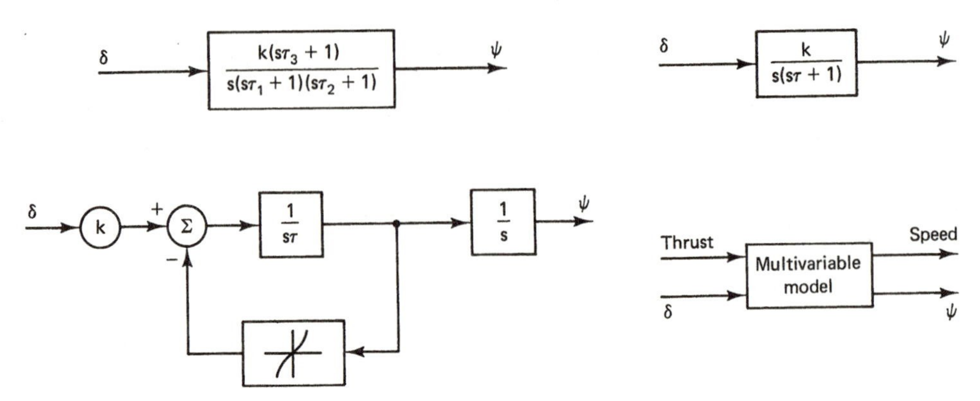

Figure 11.2 Linear and nonlinear models of the ship. Figure courtesy of van Amerongen, 1982.

$$\begin{aligned} m(\dot{v} + u\dot{\psi}) &= Y_\delta \delta + Y_v v + Y_{\dot{v}} \dot{v} + Y_{\dot{\psi}} \dot{\psi} + Y_{\ddot{\psi}} \ddot{\psi} \\ I_z \ddot{\psi} &= N_\delta \delta + N_v v + N_{\dot{v}} \dot{v} + N_{\dot{\psi}} \dot{\psi} + N_{\ddot{\psi}} \ddot{\psi}. \end{aligned} \tag{11.3}$$

Eliminating v and $\dot{v}$ in Eq. (11.3), we can derive the equation

$$\tau_1 \tau_2 \dddot{\psi} + (\tau_1 + \tau_2)\ddot{\psi} + \dot{\psi} = K(\delta + \tau_3 \dot{\delta}). \tag{11.4}$$

Hence, the rudder angle is related to the heading angle by a third order transfer function $G(s)$ where

$$G(s) = \frac{K(\tau_3 s + 1)}{s(\tau_1 s + 1)(\tau_2 s + 1)}. \tag{11.5}$$

The transfer function in Eq. (11.5) is found to be valid for all situations where the rudder angle does not exceed 5°. Also, it does not reflect the phenomenon of course instability. For larger rudder angles, the model suggested by Bech [2] is found to be more appropriate. In this model, the term $\dot{\psi}$ in Eq. (11.4) is replaced by $KH(\dot{\psi})$ where $H(\cdot)$ is a nonlinear function of its argument. In the models suggested by Bech and Norrbin, $H(\cdot)$ has the form

$$\begin{aligned} H(x) &= c_3 x^3 + c_2 x^2 + c_1 x + c_0 \qquad \text{(Bech)} \\ H(x) &= c_4 x^3 \qquad \text{(Norrbin)} \end{aligned} \tag{11.6}$$

respectively. Fig. 11.2(a)-(d) shows the various simplifications that have been made by different authors in the choice of the model, as given in [32]. Assuming that one of these models is chosen to represent the system, the problem is to determine the appropriate control parameters to achieve the desired response in the presence of the disturbances mentioned earlier. Other multivariable models that include the speed characteristics are usually used for simulation and verification purposes rather than control.

Disturbances. The performance required of the adaptive controller naturally depends on the nature and magnitude of the disturbances that can affect the dynamics of the system. As mentioned in the introduction, these can be classified as either changes in the environment that affect the dynamics of the ship, or external disturbances that act as additive inputs. The former category includes factors such as depth of water, load condition, and trim. Wind, waves, and currents are included in the latter category. The adaptive control input must be modified continuously to compensate for the two types of perturbations.

The depth of water, load, and trim conditions affect ship dynamics directly and substantially, but the explicit dependence of the dynamics on these factors is difficult to determine. In particular, the variation of depth of water may even cause instability. The disturbances in the second category mainly appear as forcing terms in the model. Any couplings introduced in the heading to sway and yaw equations are generally small, especially at moderate disturbances, and can be neglected.

A current that is not constant will cause a moment on the ship's hull that can be modeled as a signal added to the rudder input. Similarly, the moment caused by the wind can again be added to the moment of the rudder and appears in the equation describing the heading angle as

$$\begin{aligned} \tau\ddot{\psi} + \dot{\psi} &= k\delta + w(t) \\ \text{where} \qquad w(t) &= w_0 \sin 2\gamma, \qquad \gamma = \text{angle of the wind.} \end{aligned}$$

Waves, on the other hand, are stochastic in nature and different models have been proposed to take their effect into account. In general, waves depend on wind speed and have different amplitudes, frequencies, and phases. They are either approximated by a series of sinusoidal signals or as colored noise outputs of a second-order filter with a white noise input.

Performance Criteria. Course keeping and course changing are two different modes of operation in ship steering. Each mode has a different performance criterion for evaluating the design of an autopilot. In course keeping, precision is the major factor, if steering in confined waters or heavy traffic is required; propulsion economy is of concern in the open sea. This is quantified as

$$J = \frac{1}{T}\int_0^T \left(\epsilon^2 + \lambda\delta^2\right) dt \tag{11.7}$$

where ϵ is the heading error, δ is the rudder angle, λ is a weighting factor, and J is a measure of the percentage loss of speed. $\lambda \sim 0.1$ is used for accurate steering and $\lambda \sim 1$ for economic steering. For larger ships, a smaller value of λ is chosen. Course changing consists of three phases, which include starting, stationary turning (with $\ddot{\psi} = 0$), and the ending of the turn. It is desirable to perform this without any overshoot. This can therefore be specified as a reference trajectory that the ship is required to follow.

The Adaptive Controller. Several control strategies have been proposed in the literature by various authors to realize the desired performance in the presence of disturbances described earlier. The solution proposed by each author, as expected, is different for course keeping and course changing. For the former, Kallstrom *et al.* use a self-tuning regulator to estimate the parameters and a Kalman filter to obtain the state estimate. State feedback using a fixed gain that is obtained from minimizing a quadratic criterion of the form in Eq. (11.7) is used for course keeping. Arie *et al.* [1] use parameter perturbation techniques to minimize Eq. (11.7) by adjusting the proportional and derivative gains of a PID controller. Jia and Song [11] use an extended Kalman filter for state estimation, a least-squares method for parameter estimation, and use both pole placement as well as minimization of a quadratic index for computing the controller gains. Mort and Linkens [19] use a self-tuning controller for course keeping and adjust three parameters in the controller. Van Amerongen [32] estimates the ship parameters using a MRAC approach and uses the LQG method to compute the controller gains.

For course changing, both Arie and van Amerongen use MRAC while Kallstrom uses a nonadaptive high gain control.

In this section we provide further details regarding the controller structure and the adaptive laws used for the adjustment of parameters in the MRAC approach as given by van Amerongen in [32].

The ship is assumed to be modeled by a second-order differential equation

$$\tau_s \ddot{\psi}_p + \dot{\psi}_p = k_s \delta + k_w \tag{11.8}$$

where ψ_p is the course heading angle of the ship, k_s and τ_s are ship parameters that are slowly varying, and k_w is a slowly varying term due to wind disturbances. A reference model of the form

$$\tau_m \ddot{\psi}_m + \dot{\psi}_m + k_{pm}\psi_m = k_{pm}\psi_r \tag{11.9}$$

is chosen where ψ_m is the desired course heading angle, τ_m and k_{pm} are such that a constant rate of turn is achieved without any overshoot. A control input given by

$$\delta = k_i - k_d \dot{\psi}_p + k_p(\psi_r - \psi_p) \tag{11.10}$$

is used where k_i is an integrating action that can be realized by an appropriate adaptive law. This compensates for the wind disturbance. k_p and k_d are the proportional and derivative gains of the PID controller.

Substituting Eq. (11.10) in Eq. (11.8) we have

$$\dot{x}_p = A_p x_p + B_p U \tag{11.11}$$

where

$$x_p = [\psi_p, \dot{\psi}_p]^T, \qquad U = [\psi_r, 1]^T,$$

$$A_p = \begin{bmatrix} 0 & 1 \\ -\dfrac{k_p k_s}{\tau_s} & -\dfrac{(1 + k_d k_s)}{\tau_s} \end{bmatrix}, \quad B_p = \begin{bmatrix} 0 & 0 \\ \dfrac{k_p k_s}{\tau_s} & \dfrac{k_w + k_i k_s}{\tau_s} \end{bmatrix}.$$

Similarly, (11.9) can be rewritten as

$$\dot{x}_m = A_m x_m + B_m U \tag{11.12}$$

where

$$x_m = \begin{bmatrix} \psi_m, \\ \dot{\psi}_m \end{bmatrix}, \quad A_m = \begin{bmatrix} 0 & 1 \\ -\dfrac{k_{pm}}{\tau_m} & -\dfrac{1}{\tau_m} \end{bmatrix}, \quad \text{and } B_m = \begin{bmatrix} 0 & 0 \\ \dfrac{k_{pm}}{\tau_m} & 0 \end{bmatrix}.$$

Subtracting Eq. (11.12) from Eq. (11.11) we have

$$\dot{e} = A_m e + A x_p + BU$$

where $e = x_m - x_p$, $A = A_m - A_p$, and $B = B_m - B_p$. From the results of Chapter 3, we conclude that the following adaptive laws lead to stability:

$$\begin{aligned} \dot{a} &= -\Gamma_1 x_p p_2^T e \\ \dot{b} &= -\Gamma_2 U p_2^T e \end{aligned} \tag{11.13}$$

where a^T and b^T correspond to the second row of A and B respectively,

$$A_m^T P + P A_m = -Q < 0, \qquad P = [p_1, p_2]$$

and Γ_1 and Γ_2 are arbitrary symmetric positive-definite matrices. Assuming that the plant parameters and the disturbance are varying sufficiently slowly, the adaptive laws in Eq. (11.13) can be implemented as

$$\begin{aligned} \dot{k}_p &= -\gamma_1 \left(p_{12}\psi_e + p_{22}\dot{\psi}_e\right)(\psi_p - \psi_r) \\ \dot{k}_d &= -\gamma_2 \left(p_{12}\psi_e + p_{22}\dot{\psi}_e\right)\dot{\psi}_p \\ \dot{k}_i &= \gamma_3 \left(p_{12}\psi_e + p_{22}\dot{\psi}_e\right) \end{aligned} \tag{11.14}$$

where $p_2 = [p_{12}, p_{22}]^T$, $\psi_e = \psi_m - \psi_p$, and $e = [\psi_e, \dot{\psi}_e]^T$.

There are several practical considerations that have not been taken into account in the theory described above. These include variations in course changes, the nonlinearities in the steering machine, in the rudder and rate of turn transfer, and wave and current disturbances. Appropriate changes are made in the reference model to take into account these considerations so that the linear model can still be used for analysis. The nonlinearities and saturators are, however, used to generate the reference input.

Simulations and Experiments. Several experiments have been reported by van Amerongen *et al.* where the fuel-saving potential of the adaptive autopilot described above was explored. These included both full scale trials and scale model tests. Increased speeds of 0.5% for following seas and 1.5% for head seas were observed. Figure 11.3 shows the difference in performance during course keeping between an adaptive controller

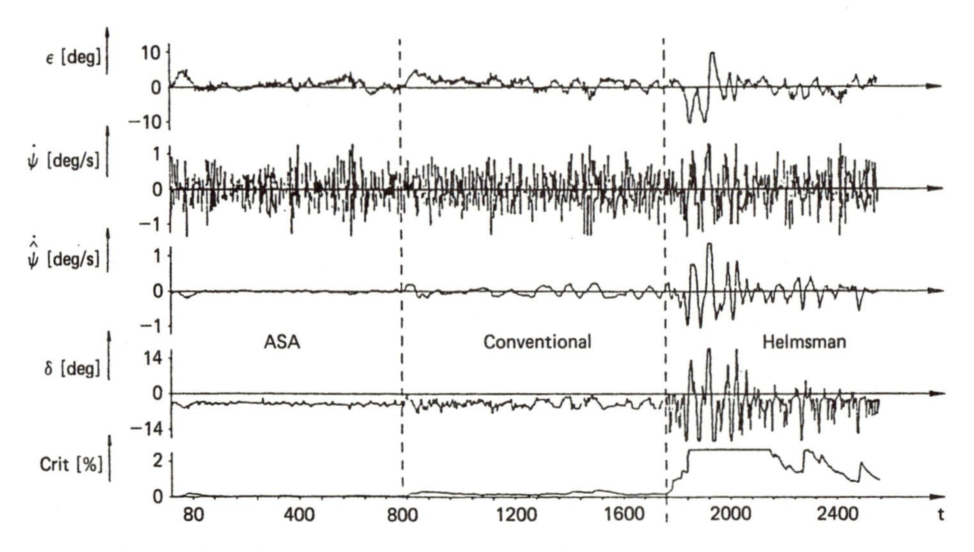

Figure 11.3 Results from van Amerongen [32]. Figure courtesy of van Amerongen, 1982.

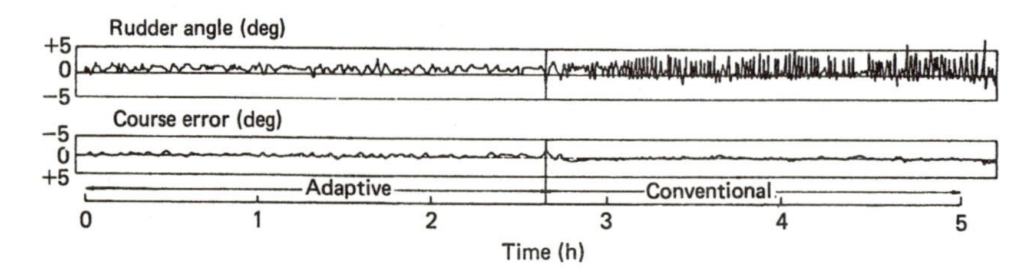

Figure 11.4 Results from Arie *et al.*[1] Courtesy of IEEE Control Systems Magazine, 1986.

(ASA), a conventional controller and a helmsman. During scale model tests, percentage increase in speeds was 0.3-1.5% for following seas and 0.6-5.5% for head seas. A reduction of 1.5% in fuel consumption was also reported; a 1% savings in fuel bills corresponds to roughly $60,000 per year.

Extensive simulation results have also been reported by the other authors referred to earlier and the consensus is overwhelmingly in favor of adaptive control. Kallstrom *et al.* [12] report an improvement of 2% in speed with tankers of 255000tdw (dead-weight-ton) and 355000tdw while Arie *et al.* [1] report a reduction of 1-3.5% in propulsive energy. The rudder angle and course errors as functions of time using adaptive and conventional controllers are shown in Fig. 11.4. Using data from Kallstrom's experiments, Jia and Song have shown that when the wind speed is increased from 5m/sec to 20m/sec, adaptive control is much better than PID control.

Status of Adaptive Control. There appears to be unanimity among the various authors regarding the advantages of adaptive control for ship dynamics. Improvement in performance, smoother control, and safer and better management are claimed in addition to the economic savings reported earlier.

11.3 PROCESS CONTROL

The industrial applications discussed in this chapter contain examples of position control systems (or servomechanisms) and process control systems, which form the two major branches of automatic control engineering. The control of ship dynamics, robots, electrical drives, and the speed control of engines belong to the first category. The control of distillation columns described in this section, as well as the biomedical application discussed in Section 11.5 are typical problems that arise in chemical, petroleum, and food-processing industries. The processes in such industries consist of sequences of operations that are necessary to convert one or more raw materials into finished products. Often, this is accompanied by the simultaneous formation of by-products that must be separated from the end product. Chemical reactors, steel and glass furnaces, paper machines, rotary kilns, and cement mills are some typical examples of industrial processes.

To ensure the efficient operation of a process, control action is essential at every stage and in all but the simplest processes such action must be automatic. Due to the presence of various disturbances, a process, by its very nature, is never in equilibrium for more than a short period of time, and a control action is necessary to constantly drive it back to the desired equilibrium state. Some of the features that distinguish chemical processes from others and consequently call for different control strategies are worth mentioning here. First, good models of such processes are often not available. It is generally agreed that they are enormously complex, characterized by large and time-varying transportation delays, are highly nonlinear, the outputs are strongly interactive, and the key process variables often cannot be measured. On the other hand, the uncontrolled system is generally stable, the dynamics of the process are always well damped so the essential behavior can be retained by models of relatively low order with time constants in the order of several minutes or even hours, and the set points are often constant for extended periods of time.

The perturbations that disturb the processes from their equilibrium states can occur in many ways. These include changes in the properties of the raw materials entering the process, deterioration in the performance of the plants due to aging of catalysts, and the effects of atmospheric and climatic conditions, such as temperature and pressure on the process dynamics. All these changes render accurate modeling very difficult, and when undertaken, invariably expensive. Even when such modeling is successful, the presence of large uncertainties makes the design of optimal controllers an extremely complex task that must be carried out off-line. Adaptive controllers, which can tune themselves on-line without precise information, are consequently found to be particularly attractive in process control systems.

In this section we deal with the adaptive control of distillation columns, which are ubiquitous in chemical processes. They possess several of the features found in complex process control problems. Further, since small improvements can result in substantial savings, they have attracted considerable attention and many results have been reported in the control literature. A good controller for a distillation column, as for any process control system, should ensure long-term steady-state regulation in the presence of disturbances, have good set-point following properties, and ensure plant safety. In the sections that follow, we report some of the principal control methods that have been suggested, the problems that have been addressed, and the insights that have been gained by these studies.

11.3.1 Distillation Columns

Distillation is an efficient process for using sequential evaporation and condensation to separate mixtures of liquids with different boiling points into their original components. Alternately, it is also used to divide them into mixtures having compositions different from those of the original mixture. The history of distillation can be traced back to antiquity when it was used in the separation of essential oils, perfumes, and medicines. Its first recorded description can be traced to the time of Cleopatra [16]. Although distillation was carried out as an art for over two thousand years, it was only in the nineteenth century that Hausbrand and Sorel introduced the first quantitative mathematical discussions for design. For further details regarding the history and development of this technology, the reader is referred to the very readable and informative book by van Winkle [16].

The process of distillation is used to separate compounds using the differences in their vapor pressures. If the original mixture contains two components, the distillation process is said to be binary; if the number of components is greater, it is called a multicomponent process. Distillation processes can also be classified according to the type of separation as equilibrium, differential, or fractionating according to the medium as steam, vacuum, or pressure, and according the mode of operation as batch or continuous. In this section, we describe a typical binary, continuous, steam-based distillation process. We first describe qualitatively the operation of such a process before proceeding to develop a mathematical model that can be used for efficient control.

A typical binary distillation column is shown in Fig. 11.5. Several trays are stacked in the column and numbered $1, 2, \ldots, N$ from top to bottom. A mixture of two components A and B (for example, methanol and water) is fed into the column near the center and onto a feed tray with a flow rate F_f. At the base of the column, the liquid stream is heated using a reboiler as shown in Fig. 11.5. The vapors rising through the column pass through the trays to the top of the column. At the top, the vapors are completely condensed using cooling water and collected in a reflux drum. The liquid from the drum is partly pumped back into the top tray and is partly removed as the distillate product. Similarly, a portion of the bottom product is removed at the bottom. The temperature difference between the top and the bottom of the column gives rise to a concentration gradient throughout the column which causes the two components to separate.

To obtain a mathematical model, we shall denote by x_i the liquid composition, defined as the fraction of A in the mixture in the ith tray. Then x_1, x_N, and x_f refer

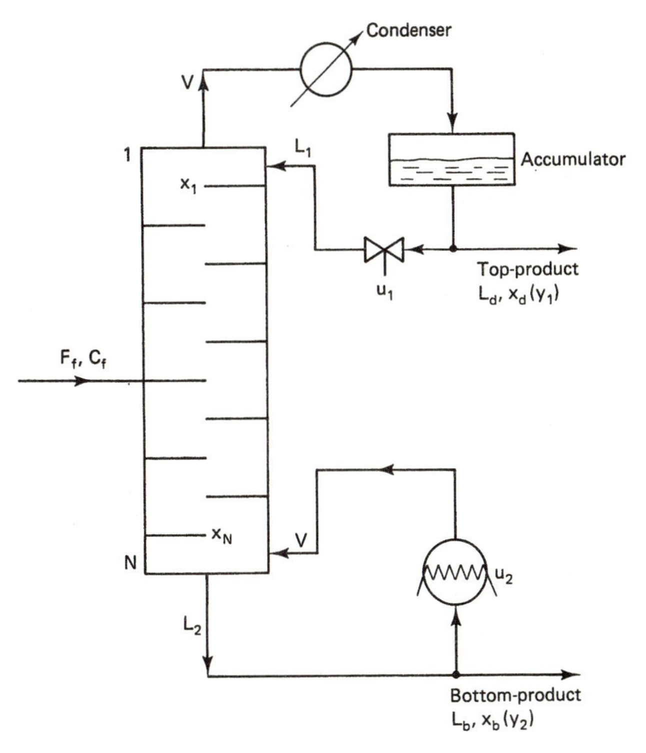

Figure 11.5 A binary distillation column.

to the compositions at the top, bottom, and in the feed tray. Let c_f, x_d, and x_b denote the compositions of the feed, distillate, and bottom product respectively. The various compositions then represent the state variables of the system, and the complexity of the mathematical model depends on both the number of trays in the column as well as the additional elements used to describe the system precisely. In this section, we derive the basic nonlinear state equations of the system that can be used in the control of one such process.

The main control objective in the system above is to maintain product compositions both at the top and bottom so they meet desired specifications. The latter are expressed in terms of product purity, that is, x_d and x_b respectively. The control has to be achieved in the presence of varying flow rate, composition, and temperature of the feed. The liquid flow rate from the reflux drum at the top and the vapor flow rate at the bottom are two typical variables that are used for control purposes. Hence, the valve position u_1 governing the former, and the valve position u_2 determining the steam flow rate in the reboiler, would correspond to the actual control variables used.

The Mathematical Model. The mathematical model of the distillation column is based on mass-energy balances. As mentioned earlier, the number of model state variables depends on both the number of trays as well as any additional elements included in the model. In each case, the mass-energy balance is carried out around each of these elements. This results in a large number of algebraic and differential equations and consequently a high order dynamical system.

For a complete description of the column, the vapor composition y_i of the component A in the mixture in tray i should also be taken into account. It is known that x_i and y_i are related by the relative volatility α_i defined as

$$\alpha_i = \frac{y_i/(1-y_i)}{x_i/(1-x_i)}, \qquad i = 1, \ldots, N. \tag{11.15}$$

If it is assumed that

1. the relative volatility is constant throughout the distillation column (that is, $\alpha_i = \alpha, i = 1, \ldots, N$),

then x_i rather than the pair (x_i, y_i) can be used as a state variable. The dynamics of the column can now be expressed by the typical equation

$$\begin{aligned} \frac{d(H_i x_i)}{dt} &= V_{i+1} y_{i+1} - V_i y_i + L_{i-1} x_{i-1} - L_i x_i \\ \frac{dH_i}{dt} &= V_{i+1} - V_i + L_{i-1} - L_i \end{aligned} \tag{11.16}$$

where H_i is the holdup and L_i and V_i are the liquid flow rate and vapor flow rate respectively, in tray i. Equation (11.16) represents the mass balance in each tray and holds for $i = 1, \ldots, N$, but $i \neq f$, where f is the feed tray. When $i = f$, we have

$$\begin{aligned} \frac{d\left(H_f x_f\right)}{dt} &= V_{f+1} y_{f+1} - V_f y_f + L_{f-1} x_{f-1} - L_f x_f + F_f c_f \\ \frac{dH_f}{dt} &= F_f + L_{f-1} - L_f. \end{aligned} \tag{11.17}$$

Equations for the condenser and reboiler can be similarly derived.

The following assumptions are introduced in [13] before deriving the mathematical model.

2. A single feed stream is fed as a saturated liquid onto the feed tray. The composition of the feed stream is equal to that of the liquid on the feed tray.

3. The overhead vapor is totally condensed in a condenser.

4. The liquid holdups on each tray, condenser, and the reboiler, are constant and perfectly mixed.

5. The holdup of vapor is negligible throughout the system.

6. The molal flow rates of the vapor and liquid through the stripping and rectifying sections are constant, respectively.

7. Any delay time in the vapor line from the top of the column to the condenser and in the reflux line back to the top tray can be neglected.

8. Dynamics of the reflux drum can be neglected.

Under these assumptions, the dynamic model can be expressed by the following equations [3]:

$$
\begin{array}{llcl}
\text{Condenser :} & \dot{x}_d & = & \frac{V}{H_d}(y_1 - x_d) \\
n\text{th tray, } n = 1, \ldots, f-1 : & \dot{x}_n & = & \frac{V}{H}(y_{n+1} - y_n) + \frac{L_1}{H}(x_{n-1} - x_n) \\
\text{Feed tray :} & \dot{x}_f & = & \frac{V}{H}(y_{f+1} - y_f) + \frac{L_1}{H}(x_{f-1} - c_f) \\
m\text{th tray, } m = f+1, \ldots, N : & \dot{x}_m & = & \frac{V}{H}(y_{m+1} - y_m) + \frac{L_2}{H}(x_{m-1} - x_m) \\
\text{Bottom :} & \dot{x}_b & = & \frac{V}{H_b}(x_b - y_b) + \frac{L_2}{H_b}(x_N - x_b)
\end{array}
\qquad (11.18)
$$

where $x_0 = x_d$ and $y_{N+1} = y_b$. We note that V and H are assumed to be constant throughout the column, the liquid flow rate is a constant ($= L_1$) above and ($= L_2$) below the feed tray, and $L_2 = L_1 + F_f$. Also, since a fraction of the top-product is removed as a distillate with a flow rate L_d and a fraction of the bottom-product with a flow rate L_b, we have $L_1 + L_d = V$, and $L_b + V = L_2$.

Despite the simplifying assumptions (1)-(8), Eq. (11.18) representing the distillation column is nonlinear due to the relationship in Eq. (11.15) between x_i and y_i. The outputs of interest are the top distillate and bottom product compositions x_d and x_b respectively. These quantities are desired to be kept within prescribed limits under varying feed rates F_f and different compositions c_f. Various control variables can be chosen to achieve this objective. In the following paragraphs, we consider the liquid flow rate L_1 at the top and the vapor flow rate V in the column to be the control inputs. These, in turn, can be controlled by manipulating the valve position u_1 in the reflux drum and the valve position u_2 in the reboiler, assuming linear valve characteristics.

Even the simplified model of the distillation column above serves to emphasize the enormous complexity of the process. Since Eq. (11.18) is nonlinear, for mathematical tractability, it can be linearized around a nominal operating point as

$$
\dot{z} = \begin{bmatrix} \dot{z}_1 \\ \dot{z}_2 \\ \vdots \\ \vdots \\ \dot{z}_{n-1} \\ \dot{z}_n \end{bmatrix}
= \begin{bmatrix}
a_{11} & a_{12} & 0 & \cdots & & \\
a_{21} & a_{22} & a_{23} & 0 & \cdots & \\
0 & \ddots & & \ddots & & \\
 & & & \ddots & & \\
 & & & a_{n-1,n-2} & a_{n-1,n-1} & a_{n-1,n} \\
 & & & & a_{n,n-1} & a_{n,n}
\end{bmatrix} z
$$

$$
+ \begin{bmatrix}
0 & 0 \\
b_{21} & b_{22} \\
\vdots & \vdots \\
\vdots & \vdots \\
b_{n-1,1} & b_{n-1,2} \\
b_{n,1} & b_{n,2}
\end{bmatrix}
\begin{bmatrix} u_1 \\ u_2 \end{bmatrix}
+ \begin{bmatrix}
0 & 0 \\
\vdots & \vdots \\
d_{f,1} & d_{f,2} \\
0 & d_{f+1,2} \\
\vdots & \vdots \\
0 & d_{n,2}
\end{bmatrix}
\begin{bmatrix} v_1 \\ v_2 \end{bmatrix}
$$

$$y_1 = z_1, \qquad y_2 = z_n \tag{11.19}$$

where the state, input, output, and disturbance variables are defined as

$$z = \begin{bmatrix} z_1 \\ z_2 \\ \vdots \\ z_{n-1} \\ z_n \end{bmatrix} = \begin{bmatrix} \Delta x_d \\ \Delta x_1 \\ \vdots \\ \Delta x_N \\ \Delta x_b \end{bmatrix}, \quad \begin{array}{l} u_1 = \Delta L_1, u_2 = \Delta V, y_1 = \Delta x_d \\ y_2 = \Delta x_b, v_1 = \Delta F_f, v_2 = \Delta c_f. \end{array}$$

Equation (11.19) can be expressed as

$$\dot{z} = Az + Bu + Dw; \qquad y = Cz. \tag{11.20}$$

The model in Eq. (11.20) consists of $N+2$ state variables, two inputs, two outputs and two disturbances. The inputs are perturbations in liquid flow rate and vapor flow rate. The outputs are changes in compositions of top and bottom products. The disturbances are the deviations from the set points of the feed composition and feed flow rate.

One of the first theoretical studies of the control of binary distillation columns was carried out in [10] where a feedforward compensator was used to control a seven-tray column. Subsequently combinations of feedback and feedforward controllers were used to compensate for feed composition and flow rate disturbances [17]. Although the initial efforts were concerned only with the control of top product composition, due to the inherent interaction between control loops, the control of both the top and bottom products were attempted [4,7]. Reviews of multivariable conventional control design methods applied to distillation column control are presented in [2,9].

Since the underlying process is nonlinear, the controller designed by linearizing the nonlinear model is accurate only in a neighborhood of the actual operating point. Hence, if there are large set point disturbances or load changes, a fixed controller may be inadequate and result in poor performance. Further, even if the parameters of the linearized system are known, Eq. (11.19) contains a large number of variables ($\sim N$) as well as parameters ($\sim N^2$), which necessitates the use of a reduced-order model for practical implementation. This may become a difficult task, especially over a wide range of operating points. Hence, the controller may have to be retuned from time to time to provide satisfactory performance under different operating conditions.

While the comments above assume that the plant parameters at any operating point can be computed, in practice, a priori knowledge about the process dynamics is only partially available. Hence, in general, the elements of the matrices A, B, C, and D in Eq. (11.20) can be considered to be unknown. This makes the tuning of control parameters expensive and time-consuming. Therefore, controllers that can compensate for changes in process parameters and operating conditions without sacrificing performance may be desirable. This accounts for the popularity of adaptive control among process engineers for controlling product compositions in distillation columns.

The Adaptive Controller. Equation (11.20) represents a 2-input 2-output multivariable plant. Hence, the design of an adaptive controller can be carried out on the basis of our discussions in Chapter 10. Several approaches have been suggested in the literature for this problem. In [11], a basic self–tuning regulator is applied to the top product composition control. It was shown that the adaptive controller outperformed a PI controller considered in the same study. In [1], a self–tuning regulator is applied to both single and multivariable control of top and bottom product compositions. Multivariable self–tuning control is applied to a pilot scale column in [8]. Multivariable adaptive predictive control of the same distillation column is presented in [12] and the adaptive scheme outperformed the one with a well tuned PID controller. A multivariable MRAC scheme based on the approach from [6] is described in [18] and compared with the one where a conventional PI controller is applied. Simulation results indicated better performance of the adaptive scheme. A comparative study of four multivariable control schemes is given in [14]. These include a conventional PID controller, multivariable linear state feedback (LSF) controller with PI action, generalized multivariable self–tuning minimum variance controller with least-squares parameter estimation (AGMVC), and a multivariable model reference adaptive controller with fixed gain parameter estimation (MRAC). The last two adaptive controllers significantly outperformed PID and LSF controllers. In the paragraphs that follow, we consider the approach presented by Unbehauen and coworkers in [14,15]. Although the analyses in these papers are carried out in discrete time, we present their continuous-time counterparts in the discussion following.

The model of the distillation column in Eq. (11.20) can be expressed in the frequency domain as

$$y_p(t) \quad = \quad G(s)u(t) + D(s)v(t)$$

where $y_p = [y_1, y_2]^T$, $u = [u_1, u_2]^T$, $v = [v_1, v_2]^T$, and $G(s)$ and $D(s)$ are 2×2 transfer matrices. The elements of these matrices in general are transfer functions of order N, where N is the number of trays in the column. Since N is large, the resultant controller will be of high order, and, hence, not practicable. The elements of $G(s)$ are therefore approximated by lower-order transfer functions with a delay. In [14], three elements of $G(s)$ are estimated as second-order transfer functions and the fourth as a first-order transfer function. The adaptive controller used includes an adjustable compensator C_1, a correction network C_2 in parallel with the plant, and a controller C_3 (Fig. 11.6). The correction network C_2 is chosen so that the transfer function $G(s)$ together with C_2 is minimum phase (refer to Section 9.5). The controller C_3 is fixed and incorporates an integral action required to reject any constant disturbances. Twenty-one parameters in the compensator C_1 are adjusted adaptively using the output error e_1 between the plant output and a reference model M.

Simulations and Experiments. When there is no adaptation, in the presence of disturbances, steady-state errors are observed if the integral action is removed (refer to Section 8.1). These errors are large if the control parameters are far from their true values. When these parameters are adjusted adaptively, the performance is considerably improved, that is, much better disturbance rejection is achieved.

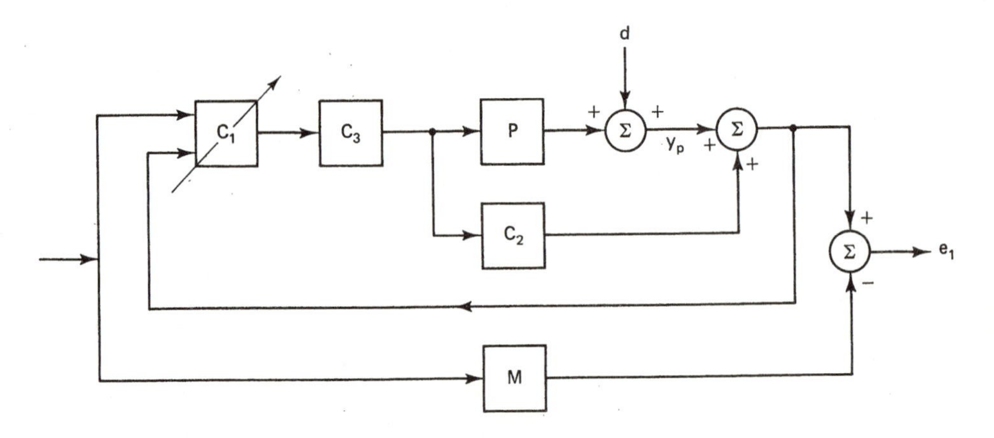

Figure 11.6 Adaptive controller in [14]. From Proceedings of 4th Yale Workshop, 1985.

In Figs. 11.7 and 11.8, the experimental results obtained from a binary pilot distillation column of Bayer AG, Leverkusen are indicated. In both figures, two control variables $\Delta\rho$, Δh and two output variables ΔT_t, ΔT_b are indicated. $\Delta\rho$ corresponds to the deviations in the reflux flow ratio, the ratio of the quantity fed back to the quantity of top product removed. Δh corresponds to the deviations in the heating power. ΔT_t and ΔT_b correspond to the deviations in the top and bottom temperatures in the column. Although the primary interest is in the compositions at the top and the bottom of the column, temperatures, rather than compositions, are used as the output variables. This can be attributed to the diffculty encountered in measuring compositions as well as to the fact that, under isobaric conditions, compositions are proportional to temperatures.

In Fig. 11.7, the transient responses obtained by using an adaptive controller of the form described in Fig. 11.6 are illustrated. Step disturbances due to a 25% change in F_f as well as c_f were introduced. The arrows indicate the direction of the change. The integrator C_3 is removed in (a) and (c) and introduced in (b). Small set-point changes of 1° K were introduced in (c). ΔT_t and ΔT_b are seen to be smaller in (b) as compared with (c), which is due to the integral action. The set-point changes in (c), which introduced persistent excitation, enabled fast convergence of the parameters to their desired values. Low-pass filters of first order were introduced to avoid large control actions in this case. At time Δ, the adaptive gains associated with the parameter in the bottom temperature control loop were decreased.

In Fig. 11.8 the behavior of the control loop, when the control parameters are well adjusted, is shown and indicates good disturbance-rejection. Finally, in Fig. 11.9, the performance of the adaptive controller is compared with that obtained using constant controllers. It is seen that the control inputs generated by the adaptive controller [Fig. 11.9(c) and (d)] have smaller amplitudes implying greater savings in energy. The disturbance rejection properties of the controller are also found to be better.

Status of Adaptive Control. The detailed analysis of distillation columns given earlier, and the results presented for a specific system given in [14,15] indicate that many

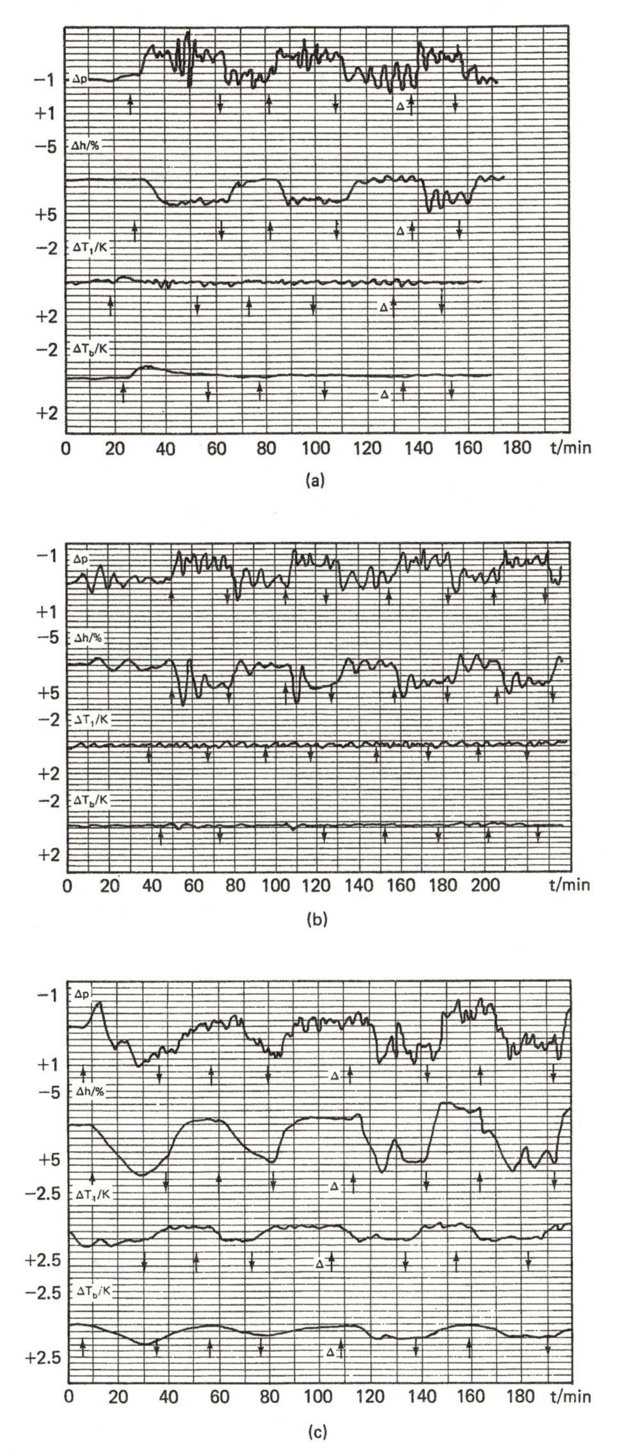

Figure 11.7 Adaptive controller-transient response. From Proceedings of 4th Yale Workshop, 1985.

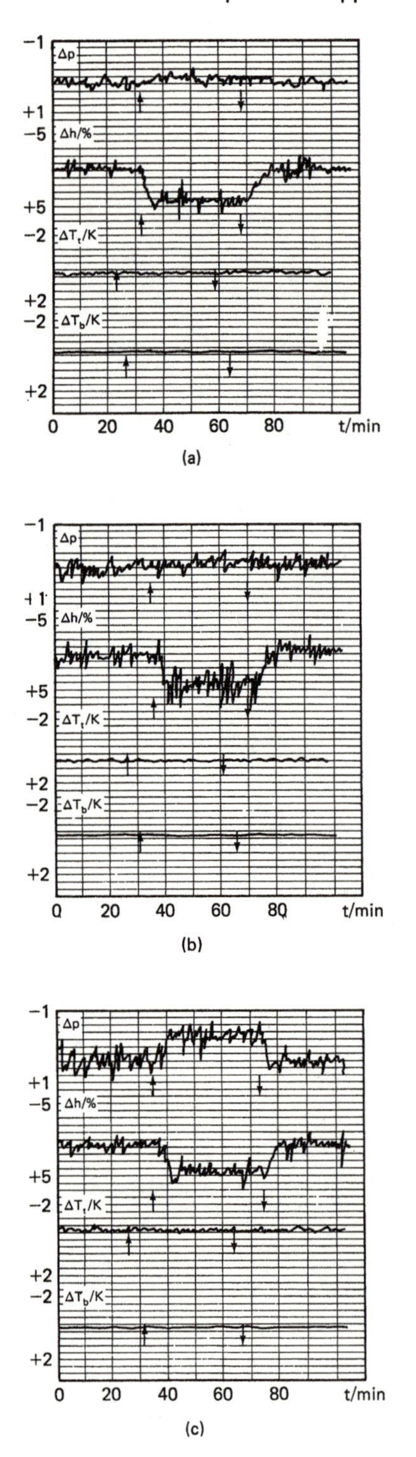

Figure 11.8 Adaptive controller-steady state response. From Proceedings of 4th Yale Workshop, 1985.

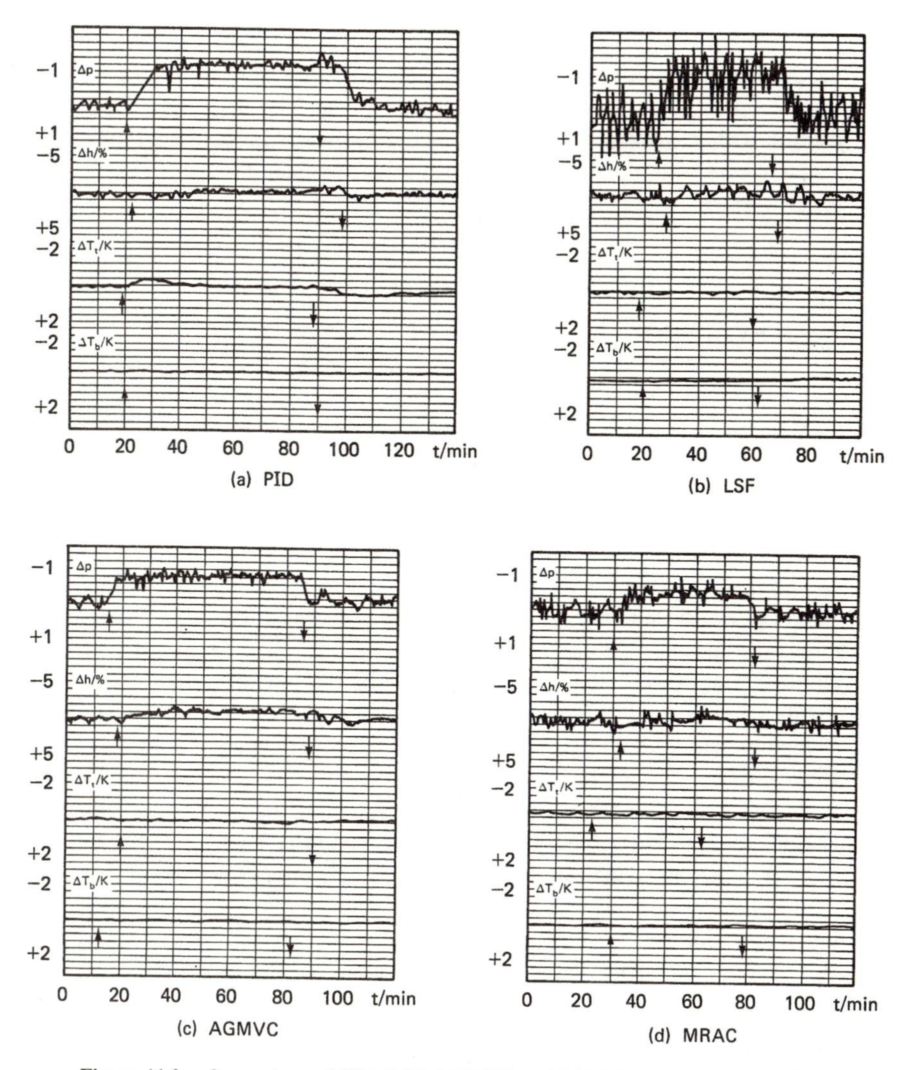

Figure 11.9 Comparison of PID, LSF, AGMVC, and MRAC controllers. From Proceedings of 4th Yale Workshop, 1985.

approximations must be made before adaptive control methods prove viable in such situations. But the results reported here indicate that adaptive methods are sufficiently robust and practically successful even when substantial approximations are made. These simply serve to underscore the fact that theory is still lacking in this area.

Applications of adaptive control methods to large-scale columns are usually trade secrets of each company. The only recently published result is in [5] where the results of [14] have been applied to large scale columns. The adaptive scheme described in [14] indicates that when adaptive schemes were applied to large recycling columns, the energy savings are in the range of 10% compared with conventional control. At the same time, the product continuity is also increased and the purity of the bottom product is improved from 200ppm (parts-per-million) with conventional PID control to 20 ppm with adaptive control. Other workers in the process control area, who have had experience applying adaptive control to distillation columns, appear to share the optimism of Unbehauen and his coworkers.

11.4 POWER SYSTEMS

Power systems are complex multivariable dynamic systems consisting of several electric generators and electrical loads connected in a power grid including transformers, transmission lines, buses, control equipment, and so on. A power system has to maintain a constant frequency and, hence, all the generators connected to it must be in synchronism. The load on the system is dictated by the users and the power must be suitably allocated to the connected generators and controlled to follow the time-varying loads. As the power flow in the system varies, the voltages at different points in the grid tend to change and suitable control action must also be taken to maintain them near predetermined values. Hence, control objectives of the overall power system ensure system stability and maintain frequency, power balance and voltage around desired values with changing load conditions. For this reason, power engineers are finding adaptive control increasingly attractive.

Two areas in adaptive power systems control that have received considerable attention are (i) adaptive load-frequency control and (ii) adaptive generator exciter control.

In a power system, coherent groups of generators (called areas) are connected by tie lines. Each area meets its load changes according to a characteristic that relates area frequency with area load. In (i), the main objective is to bring the frequency back to its normal value whenever changes in frequency occur due to changing load, while maintaining interchange with other areas within prescribed limits. In (ii), the main objective is to keep the voltage at the generator close to a reference value. Although, in the steady state, voltage feedback is sufficient to accomplish this, during transient conditions a feedback signal dependent on speed, frequency, power, or angle is also needed. If constant controller parameters are used, they are, at best, compromises between values that give good damping at light loads and values needed for heavy load conditions. Hence, for improved performance, the parameters of the controller must be varied with time.

For a comprehensive survey of the work in progress in the two areas above, we refer the reader to the article by Pierre [8] where three papers [6,9,10] on (i) and seven papers [1,2,3,4,5,7,11] on (ii) are described. Due to space limitations, we shall confine our attention to only one of these papers, by Irving *et al.* [3], which deals with the problem in (ii) from an MRAC point of view.

The following variables are of importance in the generator-exciter:

P	Active power supplied by the unit	Q	Reactive supplied power
V	Stator voltage	E	Rotor Voltage
V_R	Infinite bus voltage	Ω	Rotation speed of the unit
δ	Angle between E and V_R	V_f	Field voltage
P_m	Mechanical power	ϕ	Power angle
X_q	Quadrature reactance	X_t	Linkage reactance
θ	Internal angle	Z_n	Nominal impedance

The independent variables are V_f and P_m, while the dependent variables are P, V, δ, and Ω and the differential equations that relate the two are nonlinear and time-varying. Linearizing these equations around a nominal operating point, and denoting by Δx the deviation of a variable x from its nominal value, the following linear equations are obtained [3]:

$$\begin{aligned} \begin{bmatrix} \Delta\dot{\delta} \\ \Delta\dot{\Omega} \\ \Delta\dot{P} \end{bmatrix} &= \begin{bmatrix} 0 & 1 & 0 \\ 0 & -D & -H \\ \alpha/\tau & \beta/\tau & -1/\tau \end{bmatrix} \begin{bmatrix} \Delta\delta \\ \Delta\Omega \\ \Delta P \end{bmatrix} + \begin{bmatrix} 0 & 0 \\ 0 & H \\ \gamma/\tau & 0 \end{bmatrix} \begin{bmatrix} \Delta V_f \\ \Delta P_m \end{bmatrix} \\ \Delta V &= h_1 \Delta\delta + h_3 \Delta P \end{aligned} \tag{11.21}$$

The system described in Eq. (11.21) has two inputs ΔV_f, and ΔP_m, three states $\Delta\delta$, $\Delta\Omega$, and ΔP, and an output ΔV. Unlike $\Delta\Omega$ and ΔP, the variable $\Delta\delta$ cannot be measured. However, Eq. (11.21) can be rewritten as

$$\dot{x} = Ax + Bu \tag{11.22}$$

where $x = [\Delta V, \Delta\Omega, \Delta P]^T$, $u = [\Delta V_f, \Delta P_m]^T$, $A \in \mathbb{R}^{3\times 3}$, and $B \in \mathbb{R}^{3\times 2}$, and the state x is completely accessible. The elements of the matrix A and B vary with the linkage reactance X_t, the internal angle θ, the power angle ϕ, and the active power P.

The mechanical power P_m may contain sudden variations, endanger the reliability of the unit when actively changed on-line (for example, in a nuclear unit), and not be possible to control efficiently. For these reasons, P_m is treated as a disturbance that can be measured rather than as a control input. The objectives of the control system are the following:

Static stability: Ensure that Ω and P are sufficiently damped and that V approaches a prespecified constant value V_c despite variations in α, β, γ, h_1, h_3, or τ.

Transient stability: Sudden variations in P_m do not affect Ω.

Disturbance rejection: The controller must work efficiently for different constant disturbances P_m.

Due to the reasons above, ΔV_f is chosen as the control input and is specified as

$$\Delta V_f = k_0(t)\Delta V_c + K^T(t)x(t) - k_4\Delta P_m + k_5\Delta\dot{P}_m$$

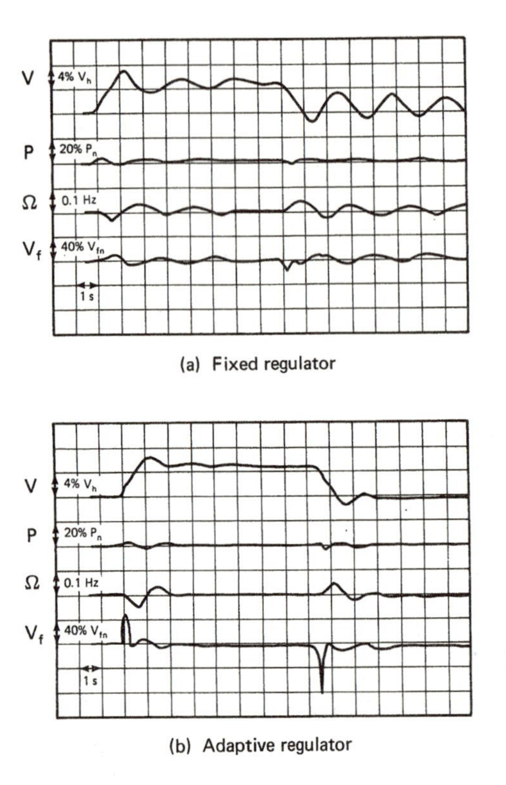

Figure 11.10 Comparison of the fixed regulator and the adaptive regulator for fixed $X_t = 0.9Z_n$. Courtesy Pergamon Press.

where $k_0(t) \in \mathbb{R}$, and $K(t) = [k_1(t), k_2(t), k_3(t)]^T \in \mathbb{R}^3$ are control parameters that are adjusted adaptively. A reference model that has the desired damping and steady-state characteristics is chosen with ΔV_c as the reference input.

In [3] the resultant adaptive controller is implemented on a micro-network with the nominal values V_n=11.15 kilovolts, P_n=1485 megawatts, $Q_n = .48432P_n$, and $X_q = 0.66$ ohm for the generator. The response obtained under different linkage reactance values for X_t is studied, and in each case, it is compared with that obtained using a fixed regulator. In Fig. 11.10(a), and 11.10(b), the responses obtained from the fixed and the adaptive controller are shown respectively. The deviations of the relevant variables from their nominal values are as indicated in Fig. 11.10. In Fig. 11.11, the responses obtained with a varying X_t are shown. The adaptive regulator shows better transient behavior as well as steady-state behavior. According to the authors, the experiments confirm their view that the new nuclear units with delicate stability operating conditions can be operated satisfactorily with adaptive control.

We have discussed a model reference approach to adaptive exciter control in this section. Much of the research on applications of adaptive theory to problems in power systems is, however, based on self-tuning regulators. The adaptive methods developed for exciter control have been tried out on micromachine tests and have also been studied using computer simulations. Based on these tests, the general impression appears to be that adaptive controllers are superior to their fixed-controller counterparts. Applications

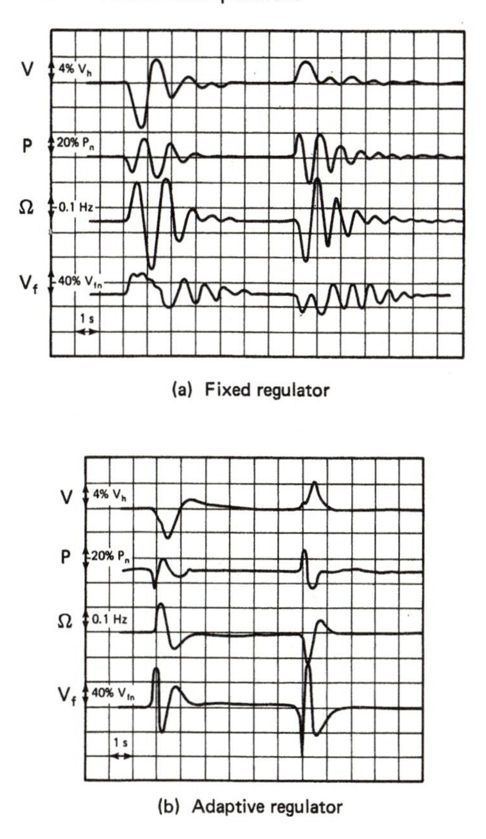

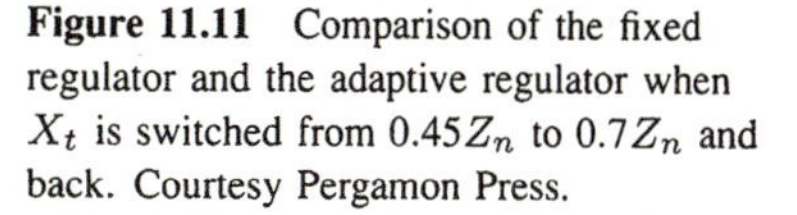

Figure 11.11 Comparison of the fixed regulator and the adaptive regulator when X_t is switched from $0.45Z_n$ to $0.7Z_n$ and back. Courtesy Pergamon Press.

of adaptive methods to load frequency control using self-tuning regulators have shown that they can also perform better under varying operating conditions. For a more detailed account of related adaptive problems in power systems and the numerous questions that must be answered before realistic answers can be provided for the future role of adaptive control in this area, the reader is referred to the excellent survey article by Pierre [8].

11.5 ROBOT MANIPULATORS

The rapidly growing interdisciplinary field of robotics, where systems must operate satisfactorily with different levels of imprecise information, represents yet another area where adaptive control is finding extensive application. Modern industrial robots were born out of a marriage between numerical control and automated manipulation. The analysis and design of robot manipulators is of growing interest in the field of control engineering. The manipulator arm of a robot is basically a sophisticated servomechanism, used in a wide range of applications including welding, materials handling, assembly, paint spraying, and so on. It invariably has a series of links that provide mobility and flexibility to operate in a large workspace. Stability, speed, and accuracy are important considerations when dealing with situations requiring a high level of mobility and dexterity.

Under the constraints above, it is necessary to organize, coordinate, and execute diverse manipulation tasks such as trajectory tracking and obstacle avoidance.

Indirect and Direct Drive Arms. Industrial robot arms can be classified as direct drive arms and indirect drive arms. Most industrial robots are driven indirectly, where the motor torque is amplified using a gear-mechanism. Hence, the effect of inertial variations due to changes in configuration and payload at the motor axis is reduced by $1/r^2$ where r is the gear ratio. It is therefore not surprising that for industrial robots with indirect drive arms, conventional fixed gain controllers provide an acceptable level of performance. Very often, robots are controlled by relatively simple control systems that have proved adequate. There has, however, been a rapid change in the robotic industry where there is interest in designing direct drive robots and lightweight manipulators that can provide better responses. In these robots, high-torque motors are directly coupled to each joint. This leads to a high mechanical stiffness, negligible backlash, and low friction. But it also implies a higher sensitivity to external disturbances, and introduces strong coupling and nonlinear time-varying characteristics. Also, the arms can reach high speeds, which accentuates these characteristics. Hence, some of the fundamental control issues in robot control, such as inertia changes and gravity effects become significant in direct-drive arms. This trend explains the increasing interest in applying adaptive control for direct-drive arms in the robotic industry. In the following paragraphs, we shall be concerned solely with the control of direct drive arms.

Need for Adaptive Control. The problem of robot manipulation can be described as one of dynamic control under kinematic constraints. A manipulator arm can be basically viewed as a series of rigid bodies linked together in a kinematic structure. A kinematic chain consists of a series of links with the base link, which is usually fixed to the ground, and an end-effector attached to the last link. The relative motion of adjacent links is caused by the motion of the joint connecting the two links. The dynamic behavior of the manipulator arms is described in terms of the time rate of change of the arm configuration in relation to the joint torques exerted by the actuators. A typical problem then is to determine the nature of these torques so the end-effector follows a desired trajectory in as fast and accurate a manner as possible.

The use of adaptive controllers requires very little justification in robot manipulators. A variety of factors including parametric and structural uncertainties that are invariably present and nonlinearities that are neglected in the modeling process, make adaptive control mandatory in many cases. In addition, a variety of disturbances are also present, and accurate trajectory following may be called for in the presence of these disturbances. Before proceeding to consider some of the approaches to adaptive control that have been suggested in the literature, we shall briefly outline the sources of these uncertainties and disturbances.

The dynamics of the underlying system is highly coupled and nonlinear. At high speeds, Coriolis and centrifugal forces must be taken into account and introduce additional nonlinearities. The inertial load at each joint varies significantly with the position

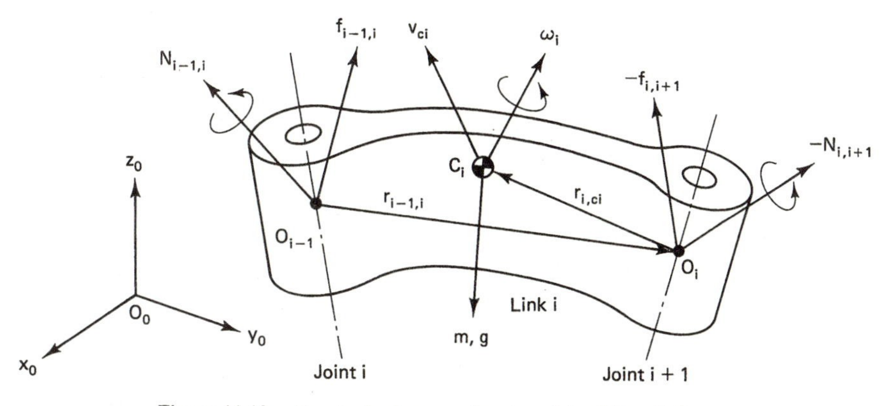

Figure 11.12 Free body diagram. Courtesy John Wiley & Sons.

of the arm. Parametric uncertainties can be present due to imprecise knowledge of the manipulator mass properties, unknown loads, uncertainty in the load position of the end–effector, and inaccuracies in the torque constants of the actuators [4]. Structural uncertainties arise from the presence of high–frequency unmodeled dynamics, resonant modes, actuator dynamics, flexibilities in the links, or finite sampling rates. Additive disturbances are present in the form of Coulomb friction, gravitational forces, measurement noise, stiction, backlash, or actuator saturation [1]. In the presence of such uncertainties, linear feedback controllers (for example, PD or PID) cannot provide consistent performance and, hence, adaptive solutions are required.

The Mathematical Model. A complete derivation of the equations governing the dynamics of a robot manipulator is beyond the scope of this book and the reader is referred to the book by Asada and Slotine [4] for further details. In this section, we merely indicate the manner in which the final equations describing the behavior of the manipulator came about following the methods used in [4] and the features concerning these equations, which are relevant for adaptive control purposes. Although both the Newton-Euler formulation and the Lagrangian formulation can be used to obtain the equations of motion, we use the former in the derivation that follows.

Let $f_{i-1,i}$ be defined as the force exerted by the $(i-1)^{\text{th}}$ link on the ith link, m_i is the mass of link i, v_{C_i} is the linear velocity of the centroid C_i of link i with respect to a fixed coordinate system (see Fig. 11.12). The balance of linear forces then yields

$$f_{i-1,i} - f_{i,i+1} + m_i g - m_i \dot{v}_{C_i} = 0 \qquad i = 1, \ldots, n \tag{11.23}$$

where all vectors are defined with reference to a base coordinate frame and there are n links. Similarly, if $N_{i-1,i}$ is defined as the moment applied to link i by link $i-1$, $r_{i-1,i}$ is the position vector from the point O_{i-1} to O_i in Fig. 11.12, ω_i is the angular velocity vector and I_i is the centroidal inertia of link i, then the balance of moments can be derived as

$$N_{i-1,i} - N_{i,i+1} + r_{i,C_i} \times f_{i,i+1} - r_{i-1,C_i} \times f_{i-1,i} - I_i\dot{\omega}_i - \omega_i \times (I_i\omega_i) = 0, \quad i = 1,\dots,n \tag{11.24}$$

Equations (11.23) and (11.24) can be modified so they are expressed in terms of input-output variables, which can be measured. Choosing the joint torque τ as the input and joint displacements $q = [q_1, \dots, q_n]^T$ as the outputs that locate the whole arm, we derive the dynamic equations, which correspond to a two-degree-of-freedom planar manipulator with two individual links. The equations are in terms of joint torques τ_1 and τ_2 and joint displacements q_1 and q_2 of the two links.

Two Link Arm. Equations (11.23) and (11.24) for link 1 and link 2 are respectively,

$$\begin{aligned} f_{0,1} - f_{1,2} + m_1 g - m_1\dot{v}_{C_1} &= 0 \\ N_{0,1} - N_{1,2} + r_{1,C_1} \times f_{1,2} - r_{0,C_1} \times f_{0,1} - I_1\dot{\omega}_1 &= 0 \end{aligned} \tag{11.25}$$

and

$$\begin{aligned} f_{1,2} + m_2 g - m_2\dot{v}_{C_2} &= 0 \\ N_{1,2} - r_{1,C_2} \times f_{1,2} - I_2\dot{\omega}_2 &= 0. \end{aligned} \tag{11.26}$$

For the planar manipulator, we have $N_{0,1} = \tau_1$ and $N_{1,2} = \tau_2$. Also, the angular velocities ω_1 and ω_2 can be expressed as

$$\omega_1 = \dot{q}_1 \qquad \omega_2 = \dot{q}_1 + \dot{q}_2$$

and the linear velocities v_{C_1} and v_{C_2} can be written as

$$v_{C_1} = \begin{bmatrix} -l_{c_1}\dot{q}_1 \sin q_1 \\ l_{c_1}\dot{q}_1 \cos q_1 \end{bmatrix},$$

$$v_{C_2} = \begin{bmatrix} -\{l_1 \sin q_1 + l_{c_2}\sin(q_1+q_2)\}\,\dot{q}_1 - l_{c_2}\sin(q_1+q_2)\dot{q}_2 \\ \{l_1 \cos q_1 + l_{c_2}\cos(q_1+q_2)\}\,\dot{q}_2 + l_{c_2}\cos(q_1+q_2)\dot{q}_2 \end{bmatrix}$$

where l_1, l_{c_1}, and l_{c_2} are as marked in Fig. 11.13. Using the relationships above, Eqs. (11.25) and (11.26) can be simplified into the equations

$$\begin{aligned} M_{11}\ddot{q}_1 + M_{12}\ddot{q}_2 - h\dot{q}_2^2 - 2C\dot{q}_1\dot{q}_2 + G_1 &= \tau_1 \\ M_{22}\ddot{q}_2 + M_{12}\ddot{q}_1 + h\dot{q}_1^2 + G_2 &= \tau_2 \end{aligned} \tag{11.27}$$

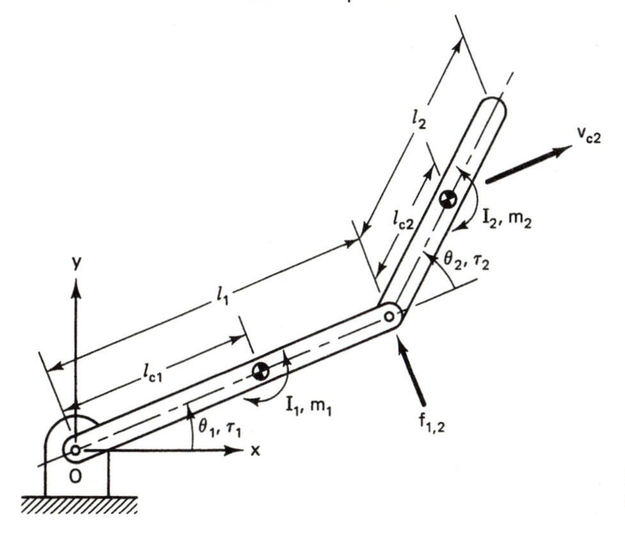

Figure 11.13 Two link robot manipulator. Courtesy John Wiley & Sons.

where

$$
\begin{aligned}
M_{11} &= m_1 l_{c_1}^2 + I_1 + m_2 \left[l_1^2 + l_{c_2}^2 + 2 l_1 l_{c_2} \cos\, q_2 \right] + I_2 \\
M_{22} &= m_2 l_{c_2}^2 + I_2 \\
M_{12} &= m_2 l_1 l_{c_2} \cos\, q_2 + m_2 l_{c_2}^2 + I_2 \\
h &= m_2 l_1 l_{c_2} \sin\, q_2 \\
G_1 &= m_1 l_{c_1} g \cos\, q_1 + m_2 g \left[l_{c_2} \cos(q_1 + q_2) + l_1 \cos\, q_1 \right] \\
G_2 &= m_2 l_{c_2} g \cos(q_1 + q_2).
\end{aligned}
\tag{11.28}
$$

Equation (11.27) can be rewritten as

$$M\ddot{q} + C\dot{q} + G = \tau \tag{11.29}$$

where $q = [q_1, q_2]^T$,

$$M = \begin{bmatrix} M_{11} & M_{12} \\ M_{12} & M_{22} \end{bmatrix}, \; C = \begin{bmatrix} -h\dot{q}_2 & -h\dot{q}_1 - h\dot{q}_2 \\ h\dot{q}_1 & 0 \end{bmatrix}, \text{ and } G = \begin{bmatrix} G_1 \\ G_2 \end{bmatrix}.$$

It is seen that the matrix M is symmetric, positive-definite, and a nonlinear function of q. Further C is a function of both q and $\dot{q}$ and in addition, the matrix $\dot{M} - 2C$ given by

$$\dot{M} - 2C = \begin{bmatrix} -2h\dot{q}_2 & -h\dot{q}_2 \\ -h\dot{q}_2 & 0 \end{bmatrix} + \begin{bmatrix} +2h\dot{q}_2 & 2h\dot{q}_1 + 2h\dot{q}_2 \\ -2h\dot{q}_1 & 0 \end{bmatrix}$$

is skew symmetric. The vector G also depends nonlinearly on q.

The terms G_1 and G_2 are due to gravity. They represent the moment created by the masses m_i about their individual joint axes. Therefore G varies with arm configuration as well as any loads that the arm may carry. M_{ii} corresponds to the total moment of inertia of all links reflected to the ith joint axis. The terms $h\dot{q}_1^2$ and $h\dot{q}_2^2$ are due to centrifugal forces, while $h\dot{q}_1\dot{q}_2$ is due to Coriolis forces that are caused when a mass m moves at a velocity v relative to a moving coordinate frame rotating at an angular velocity ω.

In general, the equations describing an n degree-of-freedom manipulator have the same form as Eq. (11.29), where q and $\dot{q}$ are n–vectors denoting joint displacements and velocities respectively, $M, C,$ and G are $n \times n$ matrices due to inertia, Coriolis and centrifugal forces, and gravitational forces respectively, and $\tau(t) \in \mathbb{R}^n$ is a control input consisting of applied torques or forces. M can be shown to be positive-definite over the entire workspace, as well as bounded from above, since it contains only polynomials involving transcendental functions of q [16]. Matrix C is linear in $\dot{q}$ and bounded in q, since it also involves only polynomials of transcedental functions in the generalized positions. The matrix $\dot{M} - 2C$ is skew-symmetric. Matrix G has a relatively simpler structure than M and C and contains polynomials involving transcedental functions of q.

In the derivation of Eq. (11.27), actuator dynamics, terms due to friction, and effects due to input-saturation were neglected. Even with such simplifications, it is seen that the underlying model of Eq. (11.29) is nonlinear and strongly coupled. This model forms the starting point for all the adaptive control efforts described in the section following.

Adaptive control of robot manipulators. Several adaptive solutions have been suggested in the literature to study the problem above. The reader is referred to [14], which includes a review of some of these solutions.

One approach is to neglect the coupling between the links. In this case, each robot axis is treated as a single-input single-output system that can be modeled as

$$m_i\ddot{q}_i = \tau_i - d_i$$

where q_i is the angular displacement of joint i, τ_i is the actuator torque input and d_i is a disturbance due to Coriolis and centripetal acceleration, gravitational load, viscous friction torque, and Coulomb friction torque. The problem then is to control a single-input single-output second order plant in the presence of input disturbances when the states are accessible. This is suggested in [1,9,12,29,30]. In these schemes, the coupling terms are neglected.

Another approach is to linearize Eq. (11.29) about a nominal trajectory and then discretize it to obtain a model of the form

$$y_k = A_1y_{k-1} + A_2y_{k-2} + B_1u_{k-1} + B_2u_{k-2} + e_k$$

where A_i, B_i $(i = 1,2)$ are 6×6 matrices, and u_k and y_k denote the deviations of the torque inputs and joint displacements, respectively, from their nominal trajectories, and e_k denotes the modeling error. With such a model, a self–tuning controller is designed in [13,20,22,27]. In [22] and [27], only a SISO model rather than an MIMO model is considered.

Computed Torque Method. A third category of adaptive schemes includes algorithms based on what is commonly referred to as the *computed torque method*. The basic idea is to determine M, C, and G in Eq. (11.29) at every instant of time, and cancel the nonlinear terms thereby decoupling the plant. Control is then effected by generating a control torque that results in a linear time-invariant homogeneous error equation whose equilibrium state is asymptotically stable. For example, consider the equation

$$M(q)\ddot{q} + C(q,\dot{q})\dot{q} + G(q) = \tau \tag{11.29}$$

and let the desired trajectory that the end-effector is desired to track be defined by the equation

$$\ddot{q}_d + K_v\dot{q}_d + K_p q_d = r \tag{11.30}$$

where r is a piecewise-continuous bounded reference input and K_v and K_p are symmetric positive-definite matrices so that the equilibrium state of the homogeneous system of Eq. (11.30) is asymptotically stable. The aim is to determine τ in Eq. (11.29) such that $\lim_{t\to\infty}[q(t) - q_d(t)] = 0$.

Comment 11.1 The adaptive model-following problem as posed above is in complete consonance with the problems we have discussed in all the preceding chapters. The output $q(t)$ of the robot follows asymptotically a desired output $q_d(t)$, which is the output of a known dynamical system (the reference model) forced by a known input r. However, we would like to point out that in much of the robotics literature, the reference input $r(t)$ rather than $q_d(t)$ is the signal that $q(t)$ is required to follow. If $\dot{r}(t)$ and $\ddot{r}(t)$ are available, the problem can be recast so the methods outlined result in $[r(t) - q(t)] \to 0$ as $t \to \infty$. If, however, $r(t)$ is known to be twice differentiable but $\dot{r}(t)$ and $\ddot{r}(t)$ are not available, $q(t)$ can follow $r(t)$ only with an error.

In the following paragraphs, we consider three problems of increasing complexity. In all cases, it is assumed that the state of the system represented by the $2n$ vector $x = [q^T, \dot{q}^T]^T$ can be measured.

(i) Parameters of the Manipulator are known. If M, C, and G are completely known, then the desired torque can be computed at every instant as

$$\tau = M\left[r - K_v\dot{q} - K_p q\right] + C\dot{q} + G$$

so that the error equations have the form

$$M\left[\ddot{e} + K_v\dot{e} + K_p e\right] = 0$$

where $e \triangleq q - q_d$. Hence, $\lim_{t\to\infty} e(t) = 0$.

(ii) Parameters of* C *and* G *are unknown. In this case, we assume that the parameters of the matrix M are known and, hence, $M(q)$ can be determined at every instant of time exactly. From a practical standpoint, this is unrealistic since M, C, and G have common elements. Hence, the problem is only of tutorial interest. We consider it here to provide continuity in the analytic reasoning before attempting the more realistic problem in (iii). Assuming that $M^{-1}(q)$ can also be computed at every instant, the following adaptive scheme can be adopted to achieve stable control.

Since the state of the system is accessible, the unknown parameters of C and G can be estimated using the methods outlined in Chapters 3 and 4. Let $\widehat{C}(q,\dot{q})$ and $\widehat{G}(q)$ represent the corresponding operators. If the torque τ is chosen as

$$\tau = M\left[r - K_v\dot{q} - K_p q\right] + \widehat{C}(q,\dot{q})\dot{q} + \widehat{G}(q)$$

the error equation has the form

$$\ddot{e} + K_v\dot{e} + K_p e = M^{-1}\Phi\omega$$

where Φ is a matrix of parameter errors and ω is an m-vector which is nonlinearly related to q and $\dot{q}$ and consequently can be computed. Since this error equation can be cast in the standard form discussed in Chapter 3, the adaptive laws can be readily determined using the errors e and $\dot{e}$ as well as the signal ω. This in turn results in $\lim_{t\to\infty} e(t) = 0$.

(iii) Parameters of* M, C *and* G *are unknown. This is the general adaptive control problem that has been attempted in the field by a number of researchers including [7,16,24,25]. In the following paragraphs, we present two approaches that have been proposed. The first, suggested by many authors, results only in a partial solution. Nevertheless it provides the motivation for the second approach of Slotine and Li [25], which yields a complete solution.

Estimates of the parameters of M, C, and G can be obtained using the identification techniques discussed earlier. However, the procedure for determining the control input is no longer as straightforward since $\widehat{M}$ rather than M is available at every instant of time.

Let $\widehat{M}$, $\widehat{C}$, and $\widehat{G}$ be the operators corresponding to M, C, and G when the unknown parameters of the latter are estimated. Since q_d and $\dot{q}_d$, the desired position and velocity are known, $\ddot{q}_d$ can be computed from Eq. (11.30). The torque to the robot manipulator is chosen as

$$\tau = \widehat{M}(q)\ddot{q}_d + \widehat{C}(q,\dot{q})\dot{q}_d + \widehat{G}(q) - K_P e - K_V\dot{e}. \tag{11.31}$$

This results in the error equation

$$M\ddot{e} + (C + K_V)\dot{e} + K_P e = \Phi_M\ddot{q}_d + \Phi_C\dot{q}_d + \Phi_G \tag{11.32}$$

where Φ_M and Φ_C are error matrices and Φ_G is an error vector defined by

$$\Phi_M(q) = \widehat{M}(q) - M(q),\quad \Phi_C(q,\dot{q}) = \widehat{C}(q,\dot{q}) - C(q,\dot{q}),\ \text{and}\ \Phi_G(q) = \widehat{G}(q) - G(q).$$

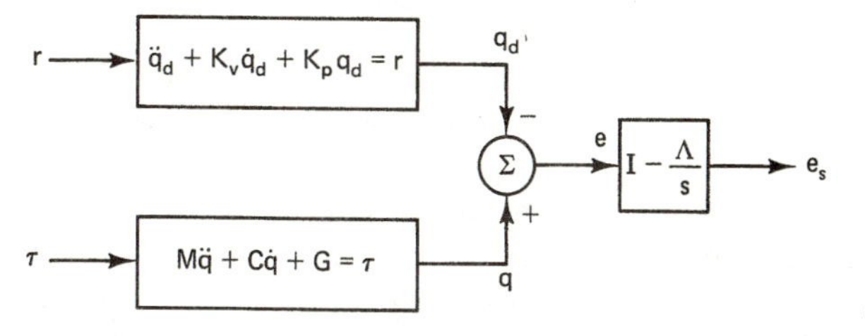

Figure 11.14 Integral action in the computed-torque method [25].

The right-hand side of Eq. (11.32) can be expressed as $\Phi\omega$, where Φ is a matrix of parameter errors and ω is a known vector of nonlinear functions of e, $\dot{e}$, $\dot{q}_d$, and $\ddot{q}_d$. To determine an adaptive law for adjusting the parameter error Φ, we choose the Lyapunov function candidate

$$V = \frac{1}{2}\left[\dot{e}^T M(q)\dot{e} + e^T K_P e + Tr(\Phi^T\Phi)\right], \tag{11.33}$$

which yields a time derivative

$$\dot{V} = -\dot{e}^T K_V \dot{e} + \dot{e}^T \Phi\omega + Tr(\Phi^T\dot{\Phi})$$

along the trajectories of Eq. (11.32). Choosing the adaptive law

$$\dot{\Phi} = -\dot{e}\omega^T \tag{11.34}$$

it follows that $\dot{V} = -\dot{e}^T K_V \dot{e} \le 0$. From the arguments used in Chapters 3-7, it follows that $\dot{e}(t) \to 0$ as $t \to \infty$. However $e(t)$ need not tend to zero.

In the approach above, the error equation of interest is Eq. (11.32), which is expressed in terms of the output error e. In Fig. 11.14, a new error e_s is defined, which is obtained from e using a filter with a transfer function $I - \Lambda/s$, where Λ is an asymptotically stable matrix. If the control input is defined as

$$\tau = \widehat{M}\ddot{q}_s + \widehat{C}\dot{q}_s + \widehat{G} - K_p\dot{e}_s$$

where $q_s = q - e_s$, the error equation can be expressed as

$$M\ddot{e}_s + \left(C + K_p\right)\dot{e}_s = \Phi\omega_s$$

where

$$(\widehat{M}\ddot{q}_s - M\ddot{q}_s) + (\widehat{C}\dot{q}_s - C\dot{q}_s) + \widehat{G} - G = \Phi\omega_s.$$

If the adaptive law

$$\dot{\Phi} = -\dot{e}_s\omega_s^T$$

is used, the time derivative of

$$V = \frac{1}{2}\left[\dot{e}_s^T M\dot{e}_s + Tr\left(\Phi^T\Phi\right)\right]$$

yields

$$\dot{V} = -\dot{e}_s K_p \dot{e}_s \le 0.$$

Using the same arguments as before, it follows that $\dot{e}_s \to 0$. Since e is the output of an asymptotically stable linear time-invariant system with an input $\dot{e}_s$, it follows that $\lim_{t\to\infty} e(t) = 0$. Hence, perfect tracking is achieved as $t \to \infty$. This is the method adopted in [25].

Some researchers argue that since the Lyapunov function V, in the analysis above, is not positive-definite (being only dependent on $\dot{e}_s$ and Φ) the method may be difficult to extend to more complicated problems. As a consequence, a new class of strict Lyapunov functions was developed independently by Arimoto and Miyazaki [3], Wen and Bayard [32], and Koditschek [18,19] for mechanical systems. It was shown that such systems could be made exponentially stable by using proportional and derivative feedback. However, a weakness of this work lies in the dependence of the Lyapunov function on the initial states of the system. Recently, work by Koditschek[17] has shown that strict Lyapunov functions may be constructed free of all a priori information about initial conditions if nonlinear feedback controllers are used.

Other adaptive schemes. In view of the tremendous activity in the field of robotics at the present time and the immense amount of material that is appearing in the published literature, it is impossible to do even partial justice to workers in this area. We present here a few references that the interested reader may find useful.

In [23], the authors apply the hyperstability concept to provide local asymptotic tracking. In [10], a Lyapunov MRAC scheme is implemented under the assumption that elements of the closed–loop system matrix are sufficiently close to that of the reference model. Decentralized adaptive control using high–gain adaptation algorithm is treated in [11]. An adaptive scheme for small perturbations around a known nominal trajectory is considered in [21]. A perturbation method is applied to a linearized model in [28]. Learning theory is applied in [2], and hybrid adaptive control of manipulators is considered in [8,15].

Another interesting approach has been suggested in [6,25,26,31,33] where the theory of variable structure systems is used. This approach takes into account input signal saturation since the input is switched between its minimum and maximum values. Undesirable chattering of the input signal can be overcome by using a continuous rather than a discontinuous control law [4,26]. Such an approach ensures that both position and velocity errors go to zero asymptotically.

Experimental Studies. Despite the great interest in the field and the research in progress, very few adaptive techniques have actually been implemented on industrial robots. However, a number of authors have presented simulation results of their techniques when applied to simple robot arms. To our knowledge, Tomizuka and coworkers are among the few who have performed extensive simulation and experimental studies of adaptive schemes for both laboratory and industrial arms. We briefly present the outcome of some of their studies.

As mentioned in the introduction, earlier versions of robot arms have used indirect drive. It was mentioned that in such cases, adaptive control may not be necessary to meet

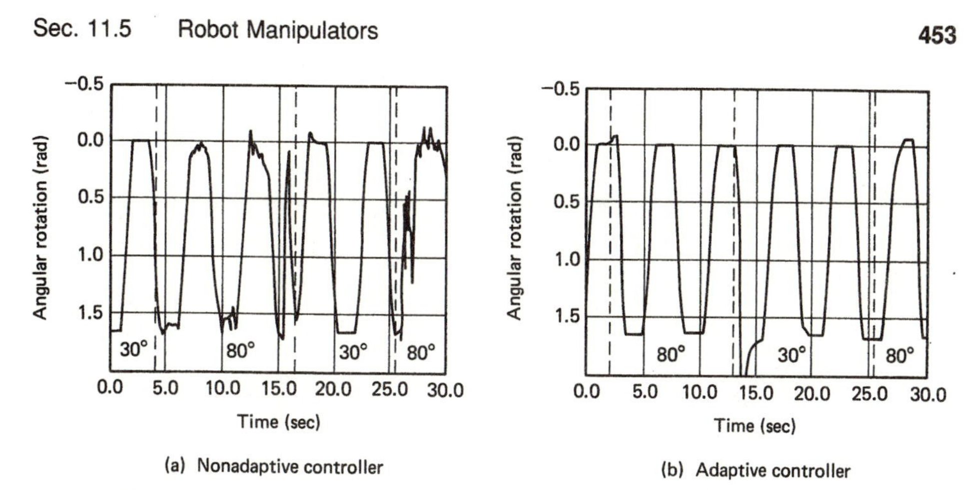

Figure 11.15 Indirect drive arm: comparison of adaptive and nonadaptive controllers. Courtesy ASME.

performance specifications. However the experimental results reported in [30] indicate adaptive control may still improve performance in some situations. These experiments were obtained from a Toshiba TSR-500V six-degree-of-freedom industrial robot arm that employs indirect-drive. The gear reduction ratios of the elbow and the wrist of the TSR are 128:1 and 110:1 respectively. The trajectories obtained using adaptive and nonadaptive controllers are indicated in Fig. 11.15. The two controllers yielded comparable responses in all cases except when their parameters were detuned from their true values. The actual arm motion using the adaptive controller in such a case is much closer to the desired trajectory than that obtained using the nonadaptive controller, as shown in Fig. 11.15.

In [30], further experimental results were reported that were carried out on a test stand with direct drive. It consists of a single arm with a mass at one end and directly attached to a d.c. motor at the other. The total inertia of the system can be varied by changing the angle of the arm with respect to the horizontal plane. Such an apparatus was used to verify that the adaptive controller can stabilize the robot arm under inertial variations. The time responses obtained using an adaptive and nonadaptive controllers are shown in Fig. 11.16. With a periodic trapezoidal velocity profile as the desired trajectory, it is seen that the adaptive controller provided stable and better performance over a wide range of inertial variations compared with the nonadaptive controller.

In both the cases above, the interaction between the axes was neglected in the design of the adaptive controller. Adaptation in the presence of strong coupling among various axes is currently under investigation.

Status of Adaptive Control. The different types of uncertainties enumerated upon earlier make adaptive solutions mandatory in many robot manipulator problems where high performance is required. Experimental results, such as those given in this section, using decoupled models, confirm that adaptive controllers exhibit consistently better performance than their nonadaptive counterparts, particularly when large inertial

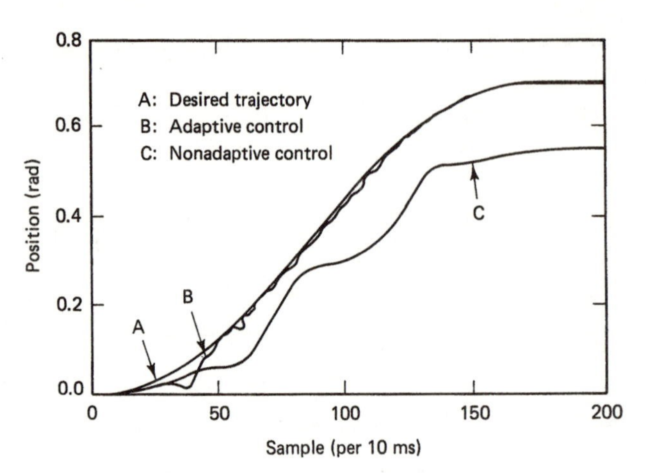

Figure 11.16 Direct drive arm: comparison of adaptive and nonadaptive controllers. Courtesy ASME.

changes are experienced [30]. Adaptive schemes based on the computed torque method, to our knowledge, have not been tested on industrial robot arms with multiple degrees of freedom. Computer simulation results as well as initial experiments with laboratory arms having one or two degrees of freedom, however, suggest the great potential of such methods.

Much of the current work on robot manipulators is related to open kinematic chains where there is no interaction between the manipulator and the environment and the interest is primarily in tracking. However, with rapid advances in the field, interest is shifting to compliant motion controllers where robot arms are in contact with their environments, so that both force control as well as trajectory control are of interest. The general opinion in the community appears to be that adaptive control will also find wide applications in such sophisticated tasks.

11.6 ADAPTIVE CONTROL IN BIOENGINEERING

The various applications we have discussed thus far pertain to engineering systems. In this section, we enter, with trepidation, the sensitive domain of biological systems and discuss some attempts that have been reported to apply adaptive methods for their control. Modeling of biological systems at the cell, organ, and physiological systems level, for a quantitative understanding of their behavior, has been ongoing for over four decades; significant progress has also been made in modeling neuro-muscular control, sensory communication, and dynamic regulation in pulmonary, cardiovascular, and endocrine systems. Despite these efforts, such models are still by and large poorly known. Despite great advances in instrumentation, the collection of input-output data is generally difficult due to the delicate nature of the underlying processes. Hence, while the models have proved useful as means of organizing experimental findings as well as aids in diagnosis, they are not always sufficiently accurate to enable decisions to be made for on-line control. In the applications considered earlier, the adaptive methods suggested were refinements of existing linear controllers. In view of the uncertainties present, as well as the very nature of the system involved, the methods suggested in this section are adaptive in nature from the outset. Further, in the previous sections, approximate models of the plants to be controlled were derived using well established physical laws of mechanics, electrodynamics, and thermodynamics. In contrast to this, the models of the three biological systems described in this section are empirically derived from input and output relations.

11.6.1 Blood Pressure Control

There are several clinical situations where it is necessary to infuse a drug to keep a patient's blood pressure within certain limits, and such drugs often have a strong and short-acting effect on cardiac performance [2]. Often the rate of drug infusion diverges considerably from the desired value and, hence, it must be monitored continuously by attending personnel. Hence, there is a need for a closed loop control system for automatic drug administration that will significantly reduce the burden on attending personnel.

The manner in which blood pressure and blood flow rate vary with drug infusion rate is different for each patient. This is particularly true of patients in critical care. Assuming that the processes above are modeled as dynamical systems, we can say that the parameters of the system vary stochastically with time from patient to patient, as well as for any given patient. Extensive preliminary testing needed to determine the parameters is not practically feasible. All these considerations make adaptive control particularly attractive. The simplicity of the programs required for computer implementation, the elimination of the need for extensive prior studies for the purpose of estimating system parameters, and automatic adjustment to changes in the state of the patient are desirable features of the adaptive control algorithm. [2,5,6] are representative papers in which adaptive control has been applied to such problems. In the paragraphs that follow, we discuss the approach treated in [6].

A linearized dynamical model relating nitroprusside infusion rate as the input u and the mean blood pressure response as the output y derived from correlation analysis with a pseudo–random binary input has the transfer function

$$W_p(s) = \frac{Ke^{-T_i s}(1+\alpha e^{-T_c s})}{1+\tau s} \tag{11.35}$$

where $y(t)$ denotes the deviations in mean arterial blood pressure from a nominal average value, $u(t)$ denotes the infusion rate, T_i and T_c are initial transport and recirculation time delay, respectively, K is the sensitivity to the drug, α denotes recirculating fraction, and τ denotes lag time constant.

The model above can be simplified substantially by assuming that the recirculation term is approximately equal to zero and the initial transport lag is small relative to the system settling time. In such a case, the transfer function in Eq. (11.35) can be approximated by

$$y(t) = \frac{K}{1+\tau s}u(t).$$

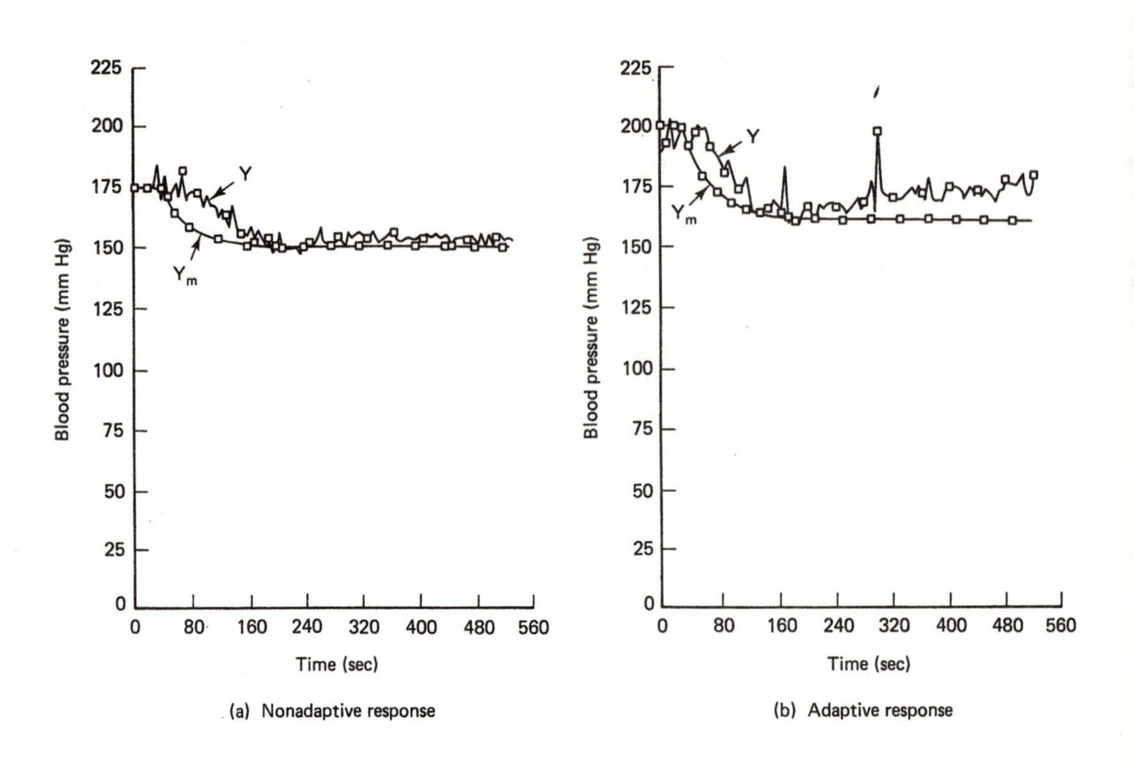

Figure 11.17 Blood pressure response using adaptive and nonadaptive controllers. Courtesy Pergamon Press.

A reference model with an input r and output y_m is chosen to have a transfer function $\exp\{-30s\}/(1+40s)$. The reference input r was designed so that a final blood pressure of 150mm is reached. With the plant and model transfer functions discretized, and adaptive schemes as discussed in Section 9.5, simulation studies were carried out to test if the control scheme assures the desired steady-state value and provides adequate disturbance rejection. Further experimentation was carried out with mongrel dogs as the data source. Infusion rates were regulated using a digitally controlled infusion pump and the arterial pressure was measured using a catheter. Results obtained using an adaptive and nonadaptive controller are shown in Fig. 11.17. In the latter case, the controller supplied a constant infusion rate of 39 ml per hour. The results indicate the effectiveness of the adaptive controller despite the presence of disturbances and parameter variations. As a consequence, the authors of [6] expect adaptive controllers to significantly improve the performance of drug infusion systems in clinical applications.

11.6.2 Adaptive Control for Ventricular Assist Devices

The purpose of the heart is to provide permanent circulation by simultaneous pumping of the deoxygenated blood from the body through the right ventricle and pulmonary artery to the lungs, and the oxygenated blood from lungs through the left ventricle and aorta back into the body [Fig. 11.18(a)]. In the event of heart failure, several devices such as a pacemaker, an intra-aortic balloon pump, a ventricular assist device, or a total artificial heart are currently used to replace many of the heart's functions [4]. Among these, ventricular assist devices are mechanical devices that are capable of maintaining blood circulation and pressure until myocardial functions return. The control systems of ventricular assist devices as well as total artificial hearts currently in use are still manual or semiautomatic with regard to regulation of blood pressures of the recipient. In this section, we describe an attempt by McInnis and coworkers [8] to stabilize the blood pressures of the circulatory system at desired levels by means of a beat–by–beat adjustment of the pump stroke volume. The left ventricular bypass assist device is applied to provide synchronization of pumping with ECG.

A mathematical model of the heart based on the hydraulics of the heart operation is derived in [8] using an electrical analog model [Fig. 11.18(b)]. The model is a network, where resistors and diodes model the inlet and outlet valves of the assist device and the natural ventricles respectively, u_1 and u_2 are voltage sources that represent the strength of the left and right ventricles respectively, and u_3 corresponds to that of the assist device. The state variables x_i, $i = 1, \ldots, 20$ correspond to blood flows and pressures.

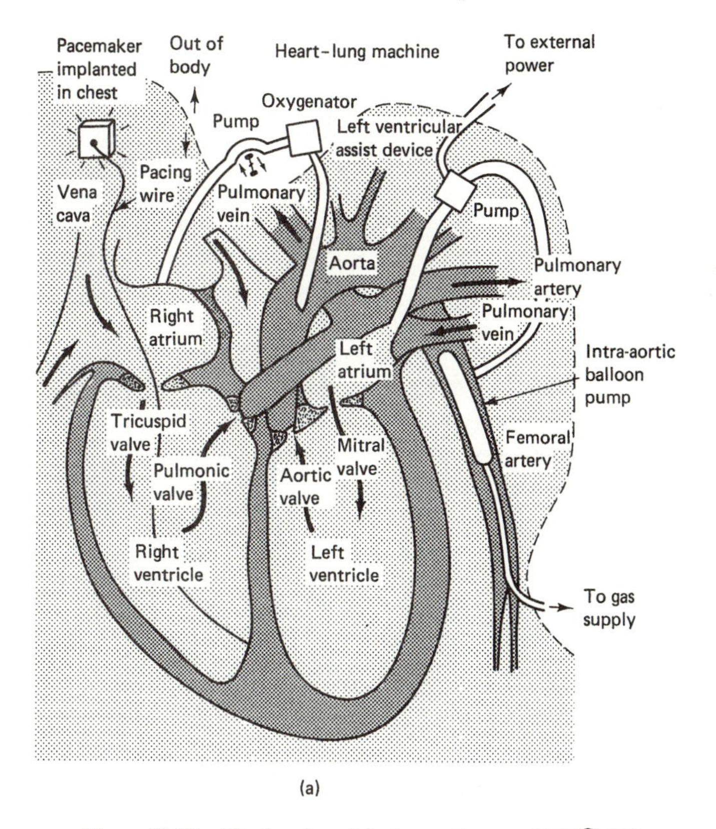

Figure 11.18 The function of the heart. Courtesy IEEE©1983.

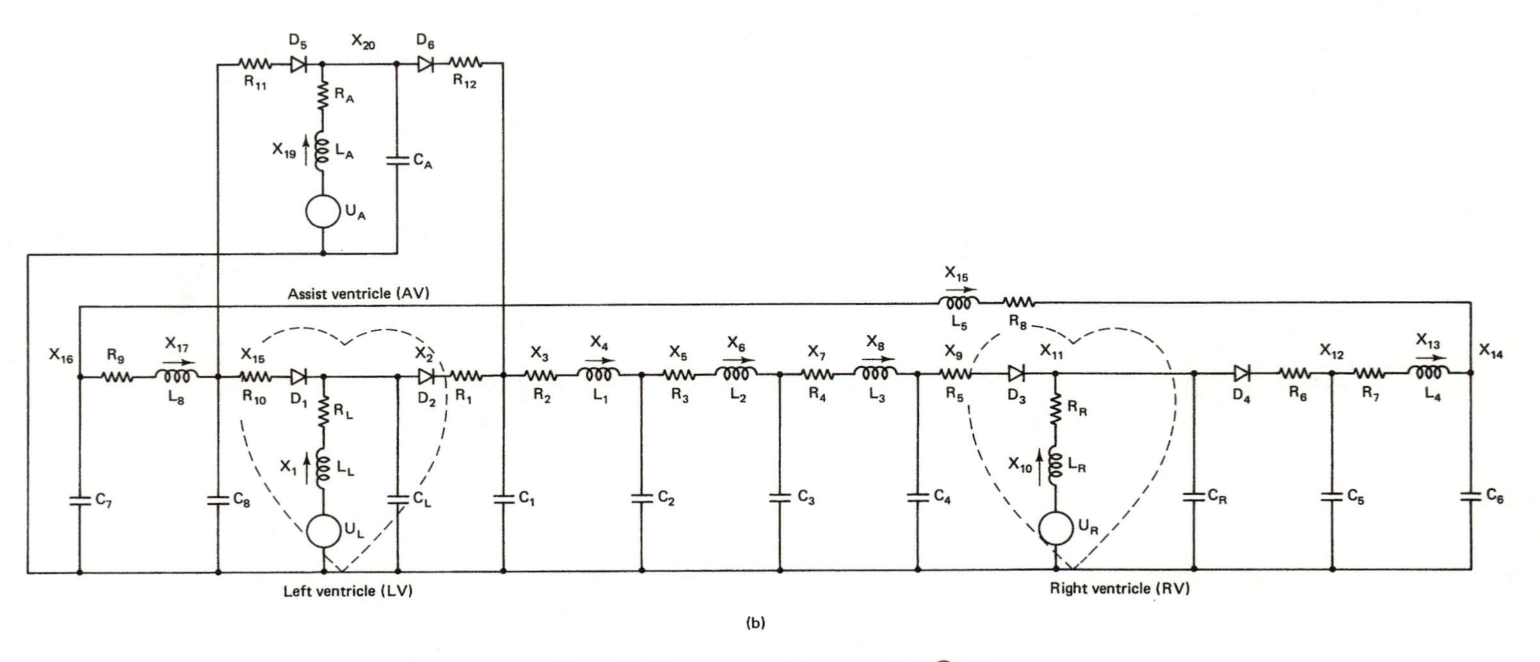

Figure 11.18 (continued) Courtesy IEEE©1985.

The description of the model is given by

$$\dot{x}(t) = A(x)x(t) + Bu(t); \qquad y(t) = x_{18}(t) \tag{11.36}$$

where $x \in \mathbb{R}^{20}, B \in \mathbb{R}^{20\times3}$, and $u^T = [u_1\ u_2\ u_3]$. x_{18} refers to the left atrial pressure. The model in Eq. (11.36) is averaged over the period Δ of a heartbeat (systolic plus diastolic period) and a discrete-time averaged state X_n is defined as

$$X_n \triangleq \frac{1}{\Delta}\int_{t_{n-1}}^{t_n} x(t)\,dt$$

where $t_n = t_{n-1} + \Delta$. This leads to a discrete-time state variable model

$$X_{n+1} = FX_n + GU_n + HU_{n-1}$$

where $F \in \mathbb{R}^{20\times20}$, $G, H \in \mathbb{R}^{20\times6}$, and $U = [u_{1d}, u_{1s},\ u_{2d}, u_{2s},\ u_{3d}, u_{3s}]^T$, u_{id} and u_{is} correspond to the strength of the input u_i during the diastolic and systolic periods, respectively. u_{3d} is treated as a control input since it can be manipulated more efficiently, u_{3s} is chosen so that it provides full emptying of the assist device, while u_1 and u_2 are considered disturbances. If the mean left atrial pressure is the only variable desired to be controlled, an ARMA model of the form

$$A(z^{-1})y(t) = B(z^{-1})u(t) + C(z^{-1})w(t)$$

is obtained where $u = u_{3d}$, $w = [u_{1d}, u_{1s}, u_{2d}, u_{2s}, u_{3s}]^T$, $A(z^{-1}) = 1 + a_1 z^{-1} + a_2 z^{-2}$, and $B(z^{-1}) = z^{-1}(b_0 + b_1 z^{-1})$.

The underlying process as described in Eq. (11.36) is extremely complex and a large number of parameters is involved. In [8], it is reported that in experimental studies with conventional PI and PID controllers, it was found that when major changes in heart operating conditions occur (for example, changes in heart rate, strength of left natural ventricle, system peripheral resistance, and so on), the control system performance was not satisfactory. Determining the gains for conventional controllers using classical tuning procedures is found to be a time consuming task and requires open–loop operation, and therefore not adequate for clinical applications. Hence, an adaptive controller is called for to compensate for nonlinearities in Eq. (11.36), as well as varying operating conditions. In light of the discussions above, a least-squares algorithm with dead-zone and covariance resetting (refer to Chapter 6) was used to obtain parameter estimates and a self–tuning PID controller based on pole placement is used.

A mock circulatory system was used in [8] to implement the resultant controller, where the natural ventricles are represented by artificial ones and a third ventricle was added to simulate the assist device. The afterload of each ventricle is simulated by a series of resist units and compliant chambers. By proper adjustment of the compliance and the peripheral resistors, a wide range of conditions can be produced in the mock circulatory system. A linear differential transformer was used as the position feedback from the pressure cylinder of a hydraulic pump to a PID analog circuit and a hydraulic servovalve. The output of this loop is the control variable that drives the diaphragm of the assist device. In addition to the mean atrial pressure ($\overline{\text{LAP}}$), the mean aortic pressure ($\overline{\text{AOP}}$) and the mean cardiac output flow ($\overline{\text{CO}}$) were measured. The responses were studied in the presence of (a) changes in the heart rate, (b) changes in peripheral resistance, and (c) changes in the simulated left ventricle. These are shown in Figs. 11.19(a)-(c) respectively. The figure indicates that the adaptive controller produces the desired performance.

The result in [8] is extended in [9] for the case of the total artificial heart. A multivariable self–tuning PI controller is applied, and the performances of the adaptive system are demonstrated by *in vitro* experiments on a mock circulatory system and are seen to be satisfactory. A similar problem is considered in [7].

11.6.3 Muscular Control

An important area of research in muscular control is functional neuromuscular stimulation (FNS). One of the major applications of FNS is the restoration of functional movement to paralyzed individuals by applying electrical stimuli to muscular nerves to induce contractions. Over the last two decades, several FNS-based orthoses have also been developed for the improvement of hand function and gait. In this section, we describe the application of adaptive methods suggested by Crago and his coworkers [3] for the control of an electrically stimulated muscle as well as those suggested by Allin and Inbar [1] for the control of an arm.

Adaptive control of electrically stimulated muscle. For many applications, there is a need for accurate regulated and proportional control of electrically elicited contractions. Due to both the nature of techniques of electrical stimulation as well as inherent muscle properties, the relationship between electrical stimulation and muscular contraction is nonlinear and time-varying. By increasing the control pulse width or pulse amplitude, the number of active muscle fibers can be varied. This is referred to as recruitment modulation. The second method of electrical stimulation by which muscular force can be controlled is temporal summation. This is achieved by varying the frequency of the stimuli. The aim then is to determine the electrical stimulation to be used so a desired force is exerted at the output.

In [3], it is proposed that the relation between pulse width and muscle force can be modeled by a second-order deterministic difference equation whose parameters vary with muscle length and frequency. Assuming that U denotes the pulse width and Y is the muscular force, their relation is given by

$$Y_k = a_1 Y_{k-1} + a_2 Y_{k-2} + b_1 U_{k-1}.$$

A direct adaptive controller with three time-varying parameters is designed. Precautionary measures are taken to ensure fast and accurate responses. These include covariance resetting, variable forgetting factor, and pseudo-random-binary-signal (PRBS) perturbations when the parameter estimates are poor. Covariance resetting was discussed in Chapter 6 in the context of least-squares estimation algorithm that ensures fast convergence. Forgetting factor is the discrete counterpart of the weighting term in the integral

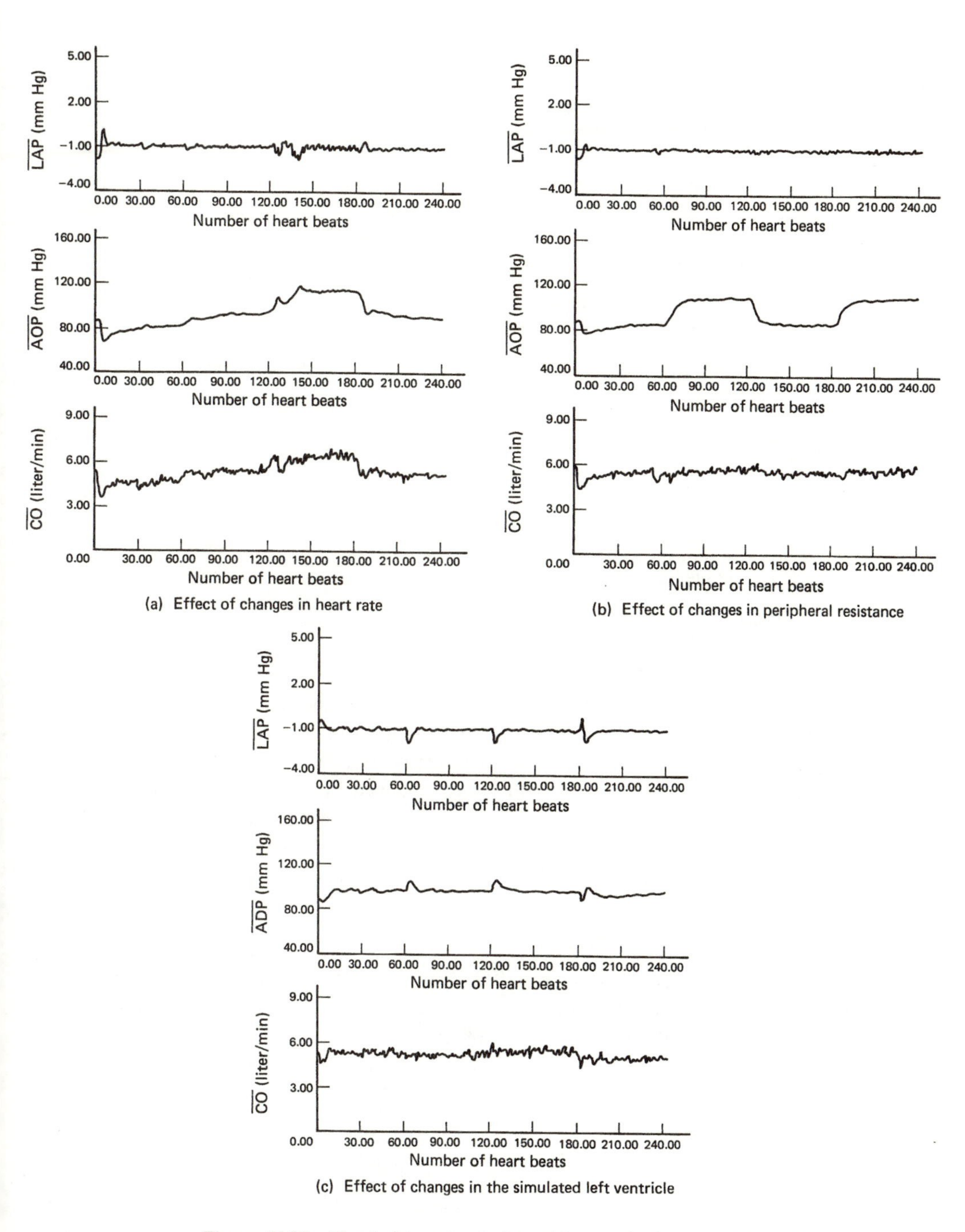

Figure 11.19 Ventricular assist device with an adaptive controller. Courtesy IEEE©1983.

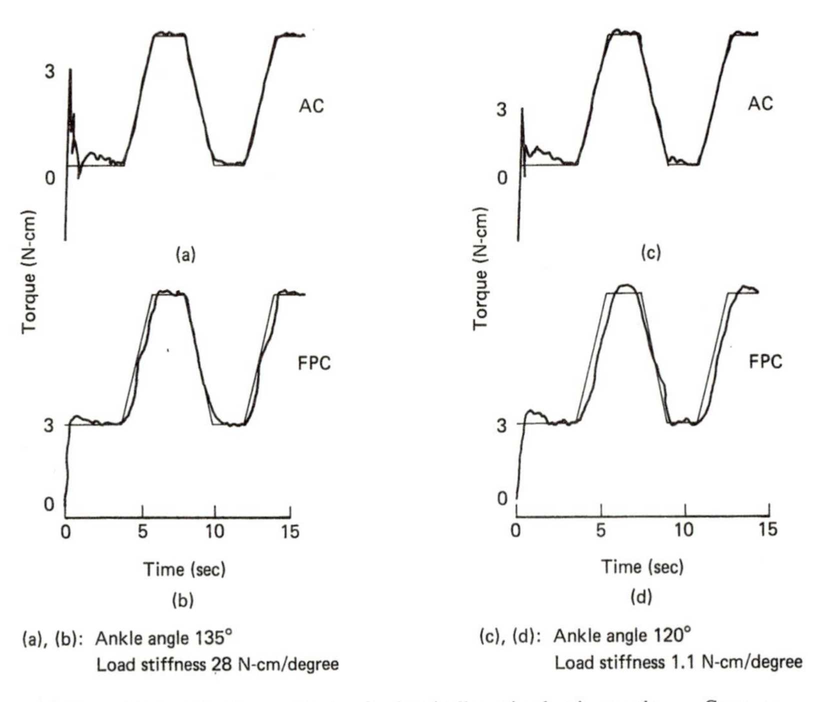

Figure 11.20 Adaptive control of electrically stimulated muscle. Courtesy IEEE©1987.

algorithm discussed in Chapter 6, and PRBS signals introduce persistent excitation in the loop and enable parameter convergence. Such an adaptive controller was used to control the tibialis anterior muscle of a cat under different load conditions, muscle lengths, and command signals, and compared with results obtained by using a fixed–gain controller whose parameters were well tuned. The resultant performance is indicated in Fig. 11.21. The results show that the adaptive controller (AC) yields response comparable to that obtained with a fixed controller (FPC) over the same range of loading conditions and ankle angles.

Adaptive control of an arm. The dynamics of the human forearm can be modeled by a complex nonlinear time-varying system. This is due to the fact that damping of the system depends on the velocity of movement and the level of the active state while the elasticity depends on the joint angle and also the active state. Assuming that the relationship between elbow joint angle and FNS can be described by the transfer function

$$W(s) = \frac{b_0}{s^3 + a_2 s^2 + a_1 s + a_0}$$

the parameters $a_i, i = 0, 1, 2$, and b_0 are found to depend on both the state of the system as well as time for any subject and are also found to vary from subject to subject. Adaptive control is used in [1] to control the elbow and forearm rotation through FNS

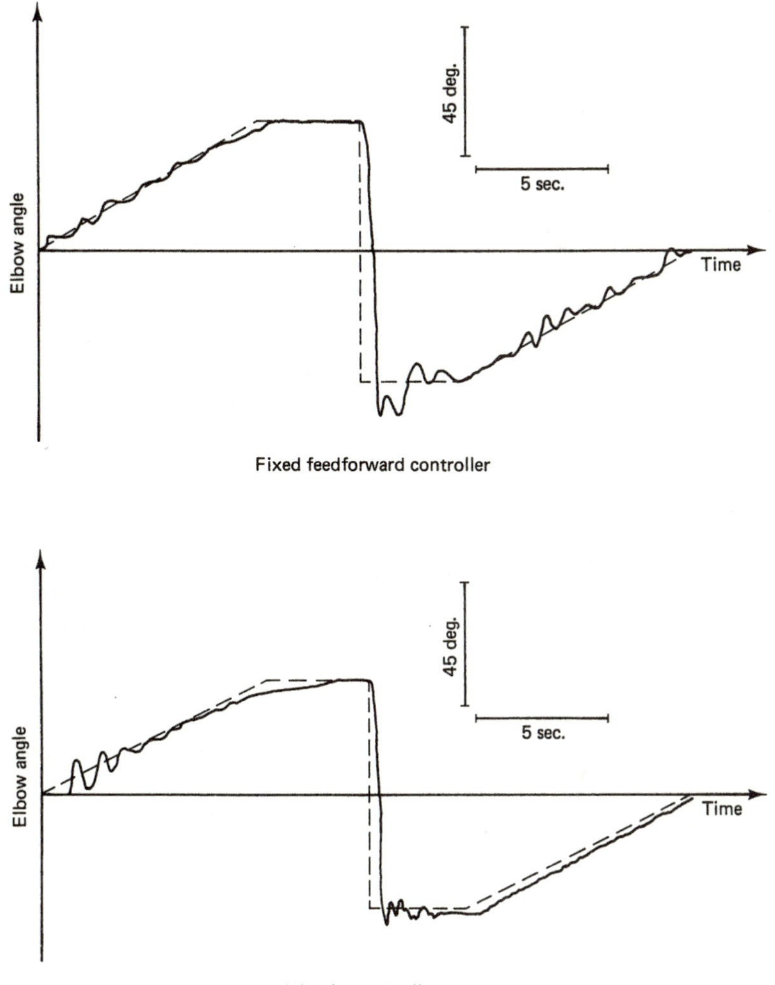

Figure 11.21 Comparison of adaptive and a fixed feedforward control for elbow motion with FNS. Courtesy IEEE©1986.

when large variations in plant parameters are present. The transfer functions of a third-order reference model and the plant are discretized and a discrete adaptive controller with four time-varying parameters is implemented. A step-plus-ramp reference input is used. Human elbow experiments were performed in [1] and the responses obtained using the adaptive controller and a third-order feedforward controller were compared. Both yielded comparable responses, though the latter required lengthy identification procedures prior to each time it was used. The adaptive controller however performs the identification implicitly. The results obtained are indicated in Fig. 11.21.

11.7 COMMENTS AND CONCLUSIONS

In a book whose main aim is to present the principles of adaptive systems theory, we have devoted this fairly long chapter to applications. The present popularity of the field of adaptive control, among theorists and control engineers alike, is due to the fact that it is both a source of interesting theoretical problems on nonlinear systems and a practical methodology for the design of sophisticated control systems. Our aim in this chapter was not to dwell on the practical aspects of the implementation of adaptive techniques but to indicate how the need for adaptive solutions arises in specific applications and to what extent such techniques have been successful. Even a casual reading of the five sections indicates that although the particular reasons given for the choice of adaptive controllers vary from one application to another, the general reasons are remarkably similar.

The applications discussed reflect the gap that currently exists between adaptive theory and adaptive practice. As seen in the preceding chapters, the theoretical analysis of adaptive systems is based on a number of assumptions such as the linearity of the plant characteristics, prior knowledge about the plant, such as its order and relative degree, and the persistent excitation of the reference input. These idealizations are rarely met in practice. In fact, the applications presented are no exceptions since in one way or another they fail to satisfy the assumptions. In all the cases considered in this chapter, the processes to be controlled are approximated by low-order linear systems and the adaptive controllers are designed for these approximate models. In Chapters 8 and 9, only a partial solution was given for the problem of such reduced-order controllers. Hence, the stability of such systems cannot be theoretically assured and various precautions must be taken to assure satisfactory performance when the adaptive algorithms are implemented. Such issues have been discussed at length in [2]. The fact that the adaptive systems are designed to be robust and to achieve a level of performance higher than nonadaptive systems in practical applications provides the theorists with the motivation to reexamine current theories and develop new ones to explain their behavior.

Adaptive controllers are by definition nonlinear and more complex than linear controllers with constant parameters. Simplicity is a virtue in control system design, and this is justified at least from the point of view of the industrialist since simple control systems are economically more viable than complex ones. In fact, the controller chosen in any industrial application must be only of sufficient complexity to meet the performance specifications. This implies that linear time-invariant controllers such as PID and LQG controllers, as well as gain scheduling, should be tried out before resorting to adaptive control.

The application areas we presented in this chapter represent only five of a very large number of areas where adaptive solutions are being attempted. Such applications have been reported in numerous conferences and workshops and have been collected in volumes such as [1,3,4,5,6]. In all cases, the same procedure used in our case studies must be followed and the designer must acquire a fairly detailed knowledge of the process under investigation. Although the broad principles of adaptive control that are applied in these cases are very similar, detailed procedures used will depend on the designer's acquaintance with the process and the ability to make suitable approximations to incorporate prior information judiciously. It is not surprising it is increasingly being realized that close and continuous collaboration between design engineers and adaptive control theorists is essential for successful applications.

References

SHIP DYNAMICS

1. Arie, T., Itoh, M., Senoh, A., Takahashi, N., Fujii, S., and Mizuno, N. "An adaptive steering system for a ship." *IEEE Control Systems Magazine* 6:3–8, Oct. 1986.
2. Bech, M., and Smitt, L.W. "Analog simulation of ship maneuvers." Technical Report Hy-14, Hya Report, 1969.
3. Brink, A.W., Baas, G., Tiano, A., and Volta, E. "Adaptive course keeping of a supertanker and a container ship – A simulation study." In *Proceedings of the 5th Ship Control Systems Symposium*, Annapolis, Maryland, 1978.
4. Broome, D.R., Keane, A.J., and Marshall, L. "The effect of variations in ship operating conditions on an adaptive autopilot optimizing function." In *Proceedings of the International Symposium on Ship Steering Automatic Control*, Genoa, Italy, 1980.
5. Broome, D.R., and Lambert, T.H. "An optimizing function for adaptive ships and autopilots." In *Proceedings of the 5th Ship Control Systems Symposium*, Annapolis, Maryland, 1978.
6. Clarke, D. "Development of a cost function for autopilot optimization." In *Proceedings of the International Symposium on Ship Steering Automatic Control*, Geneva, Italy, 1980.
7. Comstock, J.P. (ed.) *Principles of naval architecture*. The Society of Naval Architects and Marine Engineers, New York, 1967.
8. Fujino, N. "Experimental studies on ship maneuverability in restricted waters–Part I." *International Ship-building progress*, 15:279–301, 1968.
9. Herther, J.C., Warnock, F.E., Howard, K.W., and Vanvelde, W. "Digipilot – A self-adjusting digital autopilot for better maneuvering and improved course and track keeping." In *Proceedings of the International Symposium on Ship Steering Automatic Control*, Genoa, Italy, 1980.

10. Honderd, G., and Winkelman, J.E.W. "An adaptive autopilot for ships." In *Proceedings of the 3rd Ship Control Systems Symposium*, Bath, U.K., 1972.

11. Jia, X.L., and Song, Q. "Self-tuning regulators used for ship course keeping." In *Proceedings of the IFAC Symposium on Adaptive Systems in Control and Signal Processing*, Lund, Sweden, 1987.

12. Källström, C.G., Åström, K.J., Thorell, N.E., Eriksson, J., and Sten, L. "Adaptive autopilots for tankers." *Automatica* 15:241–254, 1979.

13. Kanamaru, H., and Sato, T. "Adaptive autopilot system with minimum energy consumption." In *Proceedings of the International Symposium on Ship Operation Automation*, Tokyo, Japan, 1979.

14. Kasahara, K., and Nishimura, Y. "Overview of optimum navigation systems." In *Proceedings of the International Symposium on Ship Steering Automatic Control*, Genoa, Italy, 1980.

15. Lim, C.C., and Forsythe, W. "Autopilot for ship control." In *Proceedings of the IEE, Part-D*, 130:281–294, 1983.

16. Merlo, P., and Tiano, A. "Experiments about computer controlled ship steering." In *Semana Internacional sobre la Automatica en la Marina*, Barcelona, Spain, 1975.

17. Millers, H.T. "Modern control theory applied to ship steering." In *Proceedings 1st IFAC/IFIP Symposium on Ship Operation Automation*, Oslo, Norway, 1973.

18. Minorsky, N. "Directional stability of automatically steered bodies." *Journal of American Society of Naval Engineers*, 1922.

19. Mort, N., and Linkens, D.A. "Self-tuning controllers for surface ship course-keeping and maneuvering." In *Self-Tuning and Adaptive Control*, The Institution of Electrical Engineers, edited by C.J. Harris and S.A. Billings, London, 1981.

20. Motora, S., and Koyoma, T. "Some aspects of automatic steering of ships." *Japan Shipbuilding and Marine Engineering"*, 18:3–5, 1968.

21. Nomoto, K., Taguchi, T., Honda, K., and Hirano, S. "On the steering qualities of ships." *International Ship-building Progress*, 1957.

22. Norrbin, N.H. "On the added resistance due to steering on a straight course." In *13th ITTC*, Berlin, Hamburg, 1972.

23. Norrbin, N.H. "Theory and observations on the use of a mathematical model for ship maneuvring in deep and confined waters." In *Proceedings of the 8th Symposium on Naval Hydrodynamics*, Pasadena, CA, 1970.

24. Ohtsu, K., Horigome, M., and Kitagawa, G. "A new ship's autopilot design through a stochastic model." *Automatica* 15:255–268, 1979.

25. Oldenburg, J. "Experiment with a new adaptive autopilot intended for controlled turns as well as for straight course keeping." In *Proceedings of the 4th Ship Control Systems Symposium*, The Hague, The Netherlands, 1975.

26. Reid, R.E., and Williams, V.E. "A new ship control design criterion for improved heavy weather steering." In *Proceedings of the 5th Ship Control Systems Symposium*, Annapolis, Maryland, 1978.

27. Schilling, A.C. "Economics of autopilot steering using an IBM system/7 computer." In *Proceedings of the International Symposium on Ship Operation Automation*, Washington D.C., 1976.

28. Sperry, E. "Automatic steering." *Trans. SNAME*, 1922.

29. Strom-Tejsen, J. *A digital computer technique for prediction of standard maneuvres of surface ships."* Technical Report No. 2130, David Taylor Model Basin, 1965.

30. Sugimoto, A., and Kojima, T. "A new autopilot system with condition adaptivity." In *Proceedings of the 5th Ship Control Systems Symposium*, Annapolis, Maryland, 1978.

31. Tiano, A., Volta, E., Brink, A.W., and Verbruggen, T.W. "Adaptive control of large ships in nonstationary conditions – A simulation study." In *Proceedings of the International Symposium on Ship Steering Automatic Control*, Genoa, Italy, 1980.

32. van Amerongen, J. *Adaptive steering of ships – A model reference approach to improved maneuvering and economical course keeping.* Huisdrukkerij, Delft University of Technology, Delft, The Netherlands, 1982.

33. van Amerongen, J., and Udink Ten Cate, A.J. "Model reference adaptive autopilots for ships." *Automatica* 11:441–449, 1975.

34. van Amerongen, J., and van Nauta Lemke, H.R. "Criteria for optimum steering of ships." In *Proceedings of the International Symposium on Ship Steering Automatic Control*, Genoa, Italy, 1980.

PROCESS CONTROL

1. Dahlquist, S. A. "Application of self–tuning regulators to control of distillation columns." In *Proceedings of the 6th IFAC/IFIP Conference on Digital Computer Applications to Process Control*, Dusseldorf, West Germany, 1980.

2. Edgar, T. F. "Status of design methods for multivariable control." *AIChE Symp. Series*, 72(159), 1976.

3. Franks, R. *Modeling and Simulation in Chemical Engineering*. New York:Wiley–Interscience, 1972.

4. Hu, Y. C., and Ramirez, W. F. "Application of modern control theory to distillation columns." *AIChE Journal* 18:479–486, 1972.

5. Krahl, F., Litz, W. and Riehl, F. "Experiments with self-tuning controllers in chemical plants." In *Proceedings of the IFAC Workshop of Chemical Plants*, Frankfurt, pp. 127–132, 1985.

6. Landau, I. D., and Lozano, R. "Unification of discrete-time explicit model reference adaptive control designs." *Automatica* 17:593–611, 1981.

7. McGinnis, R. G., and Wood, R. K. "Control of a binary distillation column utilizing a simple control law." *The Canadian Journal of Chemical Engineering* 52:806–809, 1974.

8. Morris, A., Nazer, Y., and Wood, R. K. "Single and multivariable applications of self–tuning controllers." In *Self–Tuning and Adaptive Control: Theory and Applications*, The Institution of Electrical Engineers, edited by C.J. Harris and S.A. Billings, pp. 249–281, London, England, 1981.

9. Rijnsdorp, J. E., and Seborg, D. E. "A survey of experimental applications of multivariable control to process control problems." *AIChE Symp. Series*, 72(159), 1976.

10. Rippin, D., and Lamb, D. *"A theoretical study of the dynamics and control of binary distillation."* Presented at AIChE Meeting, Washington, Dec. 1960.

11. Sastry, V. A., Seborg, D. E., and Wood, R. K. "Self–tuning regulator applied to a binary distillation column." *Automatica* 13:417–424, 1977.

12. Martin–Sánchez, J., and Shah, S. L. "Multivariable adaptive predictive control of a binary distillation column." *Automatica* 20:607–620, 1984.
13. Stephanopoulos, G. *Chemical Process Control*. Englewood Cliffs, N.J.: Prentice Hall Inc., 1984.
14. Unbehauen, H., and Wiemer, P. "Application of multivariable adaptive control schemes to distillation columns." In *Proceedings of the 4th Yale Workshop on Applications of Adaptive Systems Theory*, Yale University, New Haven, CT, 1985.
15. Wiemer, P., Hahn, V., Schmid, Chr., and Unbehauen, H. "Application of multivariable model reference adaptive control to a binary distillation column." In *Proceedings of the 1st IFAC Workshop on Adaptive Systems in Control and Signal Processing*, San Francisco, 1983.
16. Van Winkle, M. *Distillation*. New York:McGraw–Hill Book Company, 1967.
17. Wood, R. K., and Pacey, W. C. "Experimental evaluation of feedback, feedforward and combined feedforward–feedback binary distillation column control." *The Canadian Journal of Chemical Engineering* 50:376–384, 1972.
18. Yang, D. R., and Lee, W. K. "Experimental investigation of adaptive control of a distillation column." In *Proceedings of 23rd CDC*, Las Vegas, Nevada, 1984.

POWER SYSTEMS

1. Ghosh, A., Ledwich, G., Hope, G. S., and Malik, O. P. "Power system stabilizer for large disturbances." *IEE Proceedings*, C-132(1):39–44, 1985.
2. Ghosh, A., Ledwich, G., Malik, O. P., and Hope, G. S. "Power system stabilizer based on adaptive control techniques." *IEEE Transactions on Power Apparatus and Systems*, PAS-103:1983–1989, 1984.
3. Irving, E., Barret, J. P., Charcossey, C., and Monville, J. P. "Improving power network stability and unit stress with adaptive generator control." *Automatica* 15:31–46, 1979.
4. Kanniah, J., Malik, O. P., and Hope, G. S. "Excitation control of synchronous generators using adaptive regulators, Part I: Theory and simulation results." *IEEE Transactions on Power Apparatus and Systems*, PAS-103:897–903, 1984.
5. Kanniah, J., Malik, O. P., and Hope, G. S. "Excitation control of synchronous generators using adaptive regulators, Part II: Implementation and test results." *IEEE Transactions on Power Apparatus and Systems* PAS-103:904–910, 1984.
6. Kanniah, J., Tripathy, S. C., Malik, O. P., and Hope, G. S. "Microprocessor–based adaptive load–frequency control." *IEE Proceedings* C-131(4):121–128, 1984.
7. Ledwich, G. "Adaptive excitation control." *Proceedings of the Institution of Electrical Engineers* 126(3):249–253, 1979.
8. Pierre, D. A. "A perspective on adaptive control of power systems." *IEEE Transactions on Power Systems* PWRS-2:387–396, 1987.
9. Sheirah, M. A., and Abd-El-Fatah, M. M. "Improved load–frequency self–tuning regulator." *International Journal of Control* 39:143–158, 1984.
10. Vajk, I., Vajta, M., Keviczky, L., Haber, R., Hetthessy, J., and Kovacs, K. "Adaptive load–frequency control of the Hungarian power system." *Automatica* 21:129–137, 1985.
11. Xia, D., and Heydt, G. T. "Self–tuning controller for generator excitation control." *IEEE Transactions on Power Apparatus and Systems* PAS-102:1877–1885, 1983.

ROBOT MANIPULATORS

1. Anwar, G., Tomizuka, M., Horowitz, R., and Kubo, T. "Experimental study on discrete-time adaptive control of an industrial robot arm." In *Proceedings of the 2nd IFAC Workshop on Adaptive Systems in Control and Signal Processing*, Lund, Sweden, 1986.
2. Arimoto, S. "Mathematical theory of learning with applications to robot control." In *Proceedings of the 4th Yale Workshop on Applications of Adaptive Systems Theory*, Yale University, New Haven, CT, 1985.
3. Arimoto, S., and Miyazaki, F. "Asymptotic stability of feedback controls for robot manipulators." In *Proceedings of the 1st IFAC Symposium on Robot Control*, Barcelona, Spain, 1985.
4. Asada, H., and Slotine, J. J. *Robot Analysis and Control*. New York:John Wiley & Sons, 1986.
5. Asare, H. R., and Wilson, D. G. "Evaluation of three model reference adaptive control algorithms for robotic manipulators." In *Proceedings of the 1987 IEEE Conference on Robotics and Automation*, Raleigh, NC, 1987.
6. Balestrino, A., De Maria, G., and Sciavicco, L. "An adaptive model following control for robotic manipulators." *ASME Journal of Dynamical Systems, Measurement and Control* 105:143–151, 1983.
7. Craig, J., Hsu, P., and Sastry, S. S. "Adaptive control of mechanical manipulators." *The International Journal of Robotics Research* 6:16–28, 1987.
8. Das, M., and Loh, N. K. "Hybrid adaptive model–matching controllers for robotic manipulators." In *Proceedings of the 1987 IEEE Conference on Robotics and Automation*, Raleigh, NC, 1987.
9. Dubrowsky, S., and DesForges, D. T. "The application of model–referenced adaptive control to robotic manipulators." *ASME Journal of Dynamical Systems, Measurement and Control* 101:193–200, 1979.
10. Durrant–Whyte, H. "Practical adaptive control of actuated spatial mechanisms." In *Proceedings of the 1985 IEEE Conference on Robotics and Automation*, St. Louis, MO, 1985.
11. Gavel, D., and Hsia, T. "Decentralized adaptive control of robot manipulators." In *Proceedings of the 1987 IEEE Conference on Robotics and Automation*, Raleigh, NC, 1987.
12. Horowitz, R., and Tomizuka, M. "An adaptive control scheme for mechanical manipulators – Compensation of nonlinearities and decoupling control." *ASME Journal of Dynamical Systems, Measurement and Control* 108:127–135, 1986.
13. Houshangi, N., and Koivo, A. J. "Eigenvalue assignment and performance index based force–position control with self–tuning for robotic manipulators." In *Proceedings of the 1987 IEEE Conference on Robotics and Automation*, Raleigh, NC, 1987.
14. Hsia, T. C. "Adaptive control of robot manipulators–A Review." In *Proceedings of the 1986 IEEE Conference on Robotics and Automation*, San Francisco, CA, 1986.
15. Hsu, P., Bodson, M., Sastry, S., and Paden, B. "Adaptive identification and control for manipulators without using joint accelerations." In *Proceedings of the 1987 IEEE Conference on Robotics and Automation*, Raleigh, NC, 1987.
16. Koditschek, D. *"Robot control systems,"* in *Encyclopedia of Artificial Intelligence*. New York:John Wiley & Sons, 1987.

17. Koditschek, D. "Adaptive techniques for mechanical systems." In *Proceedings of the Fifth Yale Workshop on Applications of Adaptive Systems Theory*, Yale University, New Haven, CT, 1987.

18. Koditschek, D. "High gain feedback and telerobotic tracking." In *Workshop on Space Telerobotics*, Pasadena, CA, 1987.

19. Koditschek, D. "Quadratic Lyapunov functions for mechanical systems." Technical Report No. 8703, Yale University, New Haven, CT, 1987.

20. Koivo, A., and Guo, T. H. "Adaptive linear controller for robotic manipulators." *IEEE Transactions on Automatic Control* AC-28:162–171, 1983.

21. Lim, K., and Eslami, M. "Robust adaptive controller design for robot manipulator systems." *IEEE Journal of Robotics and Automation* RA-3:54–66, 1987.

22. Liu, M., and Lin, W. "Pole assignment self–tuning controller for robotic manipulators." *International Journal of Control* 46:1307–1317, 1987.

23. Nicosia, S., and Tomei, P. "Model reference adaptive control algorithms for industrial robots." *Automatica* 20:635–644, 1984.

24. Sadegh, N., and Horowitz, R. "Stability analysis of an adaptive controller for robotic manipulators." In *Proceedings of the 1987 IEEE Conference on Robotics and Automation* Raleigh, NC, 1987.

25. Slotine, J. J., and Li, W. "On the adaptive control of robot manipulators." *The International Journal of Robotics Research* 6:49–59, 1987.

26. Slotine, J. J., and Sastry, S. S. "Tracking control of nonlinear systems using sliding surfaces, with application to robot manipulators." *International Journal of Control* 38:465–492, 1983.

27. Soeterboek, A., Verbruggen, H., and van den Bosh, P. "An application of a self–tuning controller on a robotic manipulator." In *Proceedings of the 1987 IEEE Conference on Robotics and Automation*, Raleigh, NC, 1987.

28. Takegaki, M., and Arimoto, S. "An adaptive trajectory control of manipulators." *International Journal of Control* 34:219–230, 1981.

29. Tomizuka, M., and Horowitz, R. "Model reference adaptive control of mechanical manipulators." In *Proceedings of the 1st IFAC Workshop on Adaptive Systems in Control and Signal Processing*, San Francisco, CA, 1983.

30. Tomizuka, M., Horowitz, R., Anwar, G., and Jia, Y. L. "Implementation of adaptive techniques for motion control of robotic manipulators." *ASME Journal of Dynamical Systems, Measurement and Control* 110:62–69, 1988.

31. Utkin, V. I. "Variable structure systems with sliding modes." *IEEE Transactions on Automatic Control* AC-22:212–222, 1977.

32. Wen, J.T., and Bayard, D.S. "Robust control for robotic manipulators part I: Non-adaptive case." Technical Report No. 347-87-203, Jet Propulsion Laboratory, Pasadena, CA, 1987.

33. Young, K–K. D. "Controller design for a manipulator using theory of variable structure systems." *IEEE Transactions on Systems, Man and Cybernetics* SMC-8:101–109, 1978.

BIOENGINEERING

1. Allin, J., and Inbar, G. "FNS parameter selection and upper limb characterization." *IEEE Transactions on Biomedical Engineering* BME-33:818–828, 1986.

2. Arnsparger, J., McInnis, B., Glover, J., and Norman, N. "Adaptive control of blood pressure." *IEEE Transactions on Biomedical Engineering* BME-30:168–176, 1983.
3. Bernotas, L., Crago, P., and Chizek, H. "Adaptive control of electrically stimulated muscle." *IEEE Transactions on Biomedical Engineering* BME-34:140–147, 1987.
4. Fisheti, M. "The quest for the ultimate artificial heart." *IEEE Spectrum* 20:39–44, March 1983.
5. He, W., Kaufman, H., and Roy, R. "Multiple model adaptive control procedure for blood pressure control." *IEEE Transactions on Biomedical Engineering* BME-33:10–19, 1986.
6. Kaufman, H., Roy, R., and Xu, X. "Model reference adaptive control of drug infusion rate." *Automatica* 20:205–209, 1984.
7. Kitamura, T., Matsuda, K., and Akashi, H. "Adaptive control technique for artificial hearts." *IEEE Transactions on Biomedical Engineering* BME-33:839–844, 1986.
8. McInnis, B. C., Guo, Z. W., Lu, P. C., and Wang, J. C. "Adaptive control of left ventricular bypass assist devices." *IEEE Transactions on Automatic Control* AC-30:322–329, 1985.
9. Wang, J. C., Lu, P. C., and McInnis, B. C. "A microcomputer–based control system for the total artificial heart ." *Automatica* 23:275–286, 1987.

COMMENTS AND CONCLUSIONS

1. In *Proceedings of the IFAC Symposium on Adaptive Systems and Signal Processing*, San Francisco, CA, 1983.
2. Åström, K.J., and Wittenmark, B. *Adaptive Control.* New York:Addison-Wesley, 1988.
3. Chalam, V.V. *Adaptive control systems: Techniques and applications.* New York:Marcel Dekker, Inc., 1987.
4. Harris, C.J., and Billings, S.A. *Self-tuning and adaptive control.* London:The Institution of Electrical Engineers, 1980.
5. Narendra, K.S. *Adaptive and learning systems: Theory and applications.* New York:Plenum Publishing Corporation, 1986.
6. Narendra, K.S., and Monopoli, R.V. *Applications of adaptive control.* New York:Academic Press, 1980.

A

Kalman-Yakubovich Lemma

Proof of Lemma 2.3 (LKY): Let $m(i\omega) \stackrel{\triangle}{=} (i\omega I - A)^{-1}b$. $m(i\omega)$ is a complex vector function of ω. The strict positive realness of $H(s)$ can be written with this notation as

$$\gamma + h^T m + m^* h \; > \; 0$$

where $*$ denotes complex conjugate transpose. From the identity

$$-(i\omega I + A)^T P + P(i\omega I - A) \;\; = \;\; -(A^T P + PA), \tag{A.1}$$

we obtain by premultiplying by $m^*(i\omega)$ and post multiplying by $m(i\omega)$ Eq. (A.1)

$$m^* Pb + b^T Pm \;\; = \;\; m^* q q^T m + \epsilon m^* L m. \tag{A.2}$$

With these preliminiaries, both necessity and sufficiency of Lemma 2.3 can be shown as follows:

Necessity. If Eq. (2.21) is satisfied, substituting for Pb in Eq. (A.2), we obtain

$$2\text{Re}[h^T m] \;\; = \;\; |q^T m|^2 - 2\sqrt{\gamma}\,\text{Re}[q^T m] + \epsilon m^* L m. \tag{A.3}$$

Since L is positive-definite, A is asymptotically stable and $b \not\equiv 0$, we have $\epsilon m^* L m = \delta > 0$. Hence, if $q^T m = \lambda + i\mu$,

$$\gamma + 2\text{Re}[h^T m] \; = \; \left(\lambda - \sqrt{\gamma}\right)^2 + \mu^2 + \delta \; > \; 0,$$

which proves necessity.

Sufficiency. Let $\mathcal{K}(\omega) \stackrel{\Delta}{=} m^*h + h^Tm$ and $\Pi(\omega) = m^*Lm$. Both $\mathcal{K}(\omega)$ and $\Pi(\omega)$ are real rational functions of ω and tend to zero as $\omega \to \infty$. Furthermore they are continuous for ω finite and, hence, have finite upper and lower bounds. Let $\overline{\mu}$ be the upper bound of $\Pi(\omega)$ and $\overline{\nu}$ the lower bound of $\mathcal{K}(\omega)$. Since $\Pi(\omega) > 0$ we have $\overline{\mu} > 0$. Hence,

$$\gamma + m^*h + h^Tm - \epsilon m^*Lm \geq \gamma + \overline{\nu} - \epsilon\overline{\mu}.$$

From the positive realness of $H(s)$ we have $\gamma + \overline{\nu} > 0$. Hence, by choosing ϵ sufficiently small, we have

$$\gamma + m^*h + h^Tm - \epsilon m^*Lm > 0 \tag{A.4}$$

If the characteristic polynomial of A is $\psi(s)$, $\psi(s) = \det[sI - A]$ is a real monic polynomial. The left-hand side of the inequality in Eq. (A.4) may be written as

$$\begin{aligned} \gamma + m^*(i\omega)h + h^Tm(i\omega) - \epsilon m^*(i\omega)Lm(i\omega) &= \frac{\xi(i\omega)}{\psi(i\omega)} \qquad \text{(A.5)} \\ \frac{\xi(i\omega)\psi(-i\omega)}{\psi(i\omega)\psi(-i\omega)} &= \frac{\eta(i\omega)}{\psi(i\omega)\psi(-i\omega)}. \end{aligned}$$

The interesting point to note is that $\eta(i\omega)$ is a polynomial of degree $2n$ in $i\omega$, real, and greater than zero. Hence, if $\eta(i\omega) = \eta_1(\omega^2)$, where η_1 is a polynomial without real roots, $\eta_1(\omega^2)$ can be expressed as

$$\eta_1(\omega^2) = \theta(i\omega)\theta(-i\omega)$$

where $\theta(s)$ is a real polynomial in s of degree n with leading coefficient $\sqrt{\gamma}$. Hence,

$$\frac{\theta(s)}{\psi(s)} = \frac{\nu(s)}{\psi(s)} + \sqrt{\gamma} \stackrel{\Delta}{=} \left[-q^T(sI - A)^{-1}b + \sqrt{\gamma}\right]$$

where $\nu(s)$ is a polynomial of degree at most $(n-1)$ and having real coefficients $q_1, q_2, \ldots, q_n$. Defining $q^T = [-q_1, \ldots, -q_n]^T$, we obtain the positive-definite matrix P by solving the equation

$$A^TP + PA = -qq^T - \epsilon L.$$

From Eq. (A.5), we have

$$\begin{aligned} \gamma + m^*h + h^Tm - \epsilon m^*Lm &= \left(\sqrt{\gamma} - m^*q\right)\left(\sqrt{\gamma} - q^Tm\right) \\ &= \gamma - \sqrt{\gamma}\left(m^*q + q^Tm\right) + m^*qq^Tm \end{aligned}$$

or

$$m^*h + h^Tm = -\sqrt{\gamma}\left(m^*q + q^Tm\right) + m^*qq^Tm + \epsilon m^*Lm.$$

From Eq. (A.2), this reduces to

$$2\text{Re}\left[Pb - h - \sqrt{\gamma}q\right]^T m = 0.$$

Since this is true for all ω and $Pb - h - \sqrt{\gamma}q$ is a constant vector, it follows that

$$Pb - h = \sqrt{\gamma}q.$$

This proves sufficiency.

Proof of Lemma 2.4 (MKY). This primary extension due to Meyer is the removal of the requirement of controllability assumed in the previous lemma. Choosing a coordinate system so that

$$A = \begin{bmatrix} A_1 & A_2 \\ 0 & A_3 \end{bmatrix}, \quad b = \begin{bmatrix} b_1 \\ 0 \end{bmatrix}, \quad h = \begin{bmatrix} h_1 \\ h_2 \end{bmatrix}$$

where A_1, A_2, A_3 are $(p \times p)$, $(p \times (n-p))$, and $(n - p \times n - p)$ matrices, respectively, (A_1, b_1) is controllable, all the eigenvalues of A lie in $\mathbb{C}^-$. Partitioning P, L, and q in the same way,

$$P = \begin{bmatrix} P_1 & P_2 \\ P_2^T & P_3 \end{bmatrix}, \quad L = \begin{bmatrix} L_1 & 0 \\ 0 & L_3 \end{bmatrix}, \quad q = \begin{bmatrix} q_1 \\ q_2 \end{bmatrix}$$

it is found that the following matrix equations must be solved.

(i) $A_1^T P_1 + P_1 A_1 = -q_1 q_1^T - \epsilon L_1.$

(ii) $A_2^T P_1 + A_3^T P_2^T + P_2^T A_1 = -q_2 q_1^T.$

(iii) $A_2^T P_2 + A_3^T P_3 + P_2^T A_2 + P_3 A_3 = -q_2 q_2^T - \epsilon L_3.$

(iv) $P_1 b_1 - h_1 = \sqrt{\gamma} q_1.$

(v) $P_2^T b_1 - h_2 = \sqrt{\gamma} q_2.$

It is shown[1] that if $H(s)$ is SPR, the solutions to the equations above exists. Since $\gamma + 2h_1^T(sI - A)^{-1}b_1$ is SPR, by the previous lemma, a solution exists for (i) and (iv). Since P_1 and q_1 are known, Eqs. (ii) and (v) have only P_2^T and q_2 as unknowns. It is shown that solutions can be found since $q_1^T(sI - A)^{-1}b - \sqrt{\gamma}$ has zeros in the open left half plane. Finally, since L is positive-definite, it follows that P is also positive-definite.

[1] K.R. Meyer, "On the existence of Lyapunov functions for the problem on Lur'e", *SIAM Journal of Control*, 3:373–383, August 1965.

B

Growth Rates of Unbounded Signals

Three classes of signals are needed for the analysis of the systems discussed in this book. The results presented in Chapters 3-5 are derived assuming that the external inputs are uniformly bounded and piecewise continuous. For parameter convergence, it is assumed in Chapter 6 that the external inputs belong to the class $\mathcal{P}_{[0,\infty)}$ (refer to Section 6.3). Finally, while dealing with the behavior of unbounded signals in the system, the signals are assumed to belong to the class $\mathcal{PC}_{[0,\infty)}$ which is defined as follows:

$\mathcal{PC}_{[0,\infty)} \stackrel{\Delta}{=}$ the set of all real piecewise continuous functions defined on the interval $[0,\infty)$ which have bounded discontinuities.

In this appendix, we qualitatively discuss the behavior of unbounded signals which belong to $\mathcal{PC}_{[0,\infty)}$ and state some definitions pertaining to growth rates of such signals. This provides the framework for the derivation of the lemmas in Section 2.8.

The functions t, t^α, $\log t$, and $\exp \alpha t$ with $\alpha > 0$ and $t \geq 0$ belong to $\mathcal{PC}_{[0,\infty)}$ and increase with t in an unbounded fashion. It is evident that the function t^4 grows more rapidly than t^2, or t^4/t^2 increases without bound as t increases. More generally we can say that a function $g(\cdot) \in \mathcal{PC}_{[0,\infty)}$ *grows more slowly* than a function $f(\cdot) \in \mathcal{PC}_{[0,\infty)}$ if

$$\lim_{t\to\infty} \left| \frac{g(t)}{f(t)} \right| = 0.$$

This provides the motivation for the definition of o (small order) given in the latter part of this section. If

$$\lim_{t\to\infty} \left| \frac{g(t)}{f(t)} \right| \leq c_1 \qquad \text{for all } t \geq t_0,\ t_0, c_1 \in \mathbb{R}^+$$

we merely conclude that $|g(t)|$ *does not grow more rapidly* than $|f(t)|$ and this results in the definition of O (large order). Similarly if

$$c_1 \leq \lim_{t\to\infty}\left|\frac{g(t)}{f(t)}\right| \leq c_2 \qquad \text{for all } t \geq t_0,\ c_1, c_2, t_0, \in \mathbb{R}^+$$

we can say that the two functions *grow at the same rate*. Such a comparison, however, cannot be made of all functions. For example, the functions $e^t \cos t$ and $e^t \sin t$ cannot be ordered using the arguments given earlier. This implies that if comparison of signals in terms of their rates of growth is to be made the basis for stability analysis, we must restrict the class of functions considered in some fashion. As a first step, we restrict our attention to the class of positive-monotonic functions $\mathcal{M} \subset \mathcal{PC}_{[0,\infty)}$ defined by

$$\mathcal{M} = \{f \mid f(t) \geq 0,\ f(t_1) \geq f(t_2) \text{ if } t_1 > t_2,\ \forall\, t, t_1, t_2 \in \mathbb{R}^+\}. \tag{B.1}$$

This is accomplished by associating with each function $f(\cdot) \in \mathcal{PC}_{[0,\infty)}$ a function $f_s \in \mathcal{M}$ where

$$f_s(t) \stackrel{\triangle}{=} \sup_{\tau\leq t} |f(\tau)|, \qquad \tau, t \in \mathbb{R} \tag{B.2}$$

We shall use both $f_s(t)$ and $\sup_{\tau\leq t} |f(\tau)|$ to denote the value of the function at time t, the latter being limited to those cases where there is a possibility of ambiguity. If $f : \mathbb{R}^+ \to \mathbb{R}^n$, $\sup_{\tau\leq t} \|f(\tau)\|$ is denoted as $\|f(t)\|_s$.

Some of the concepts introduced earlier are formalized here by several definitions.

Definition B.1 Let $x, y \in \mathcal{PC}_{[0,\infty)}$. We denote $y(t) = O[x(t)]$ if there exist positive constants M_1, M_2, and $t_0 \in \mathbb{R}^+$ such that $|y(t)| \leq M_1|x(t)| + M_2\ \forall t \geq t_0$.

Definition B.2 Let $x, y \in \mathcal{PC}_{[0,\infty)}$. We denote $y(t) = o[x(t)]$ if there exists a function $\beta(t) \in \mathcal{PC}_{[0,\infty)}$ and $t_0 \in \mathbb{R}^+$ such that $|y(t)| = \beta(t)x(t) \quad \forall t \geq t_0$, and $\lim_{t\to\infty} \beta(t) = 0$.

Definition B.3 Let $x, y \in \mathcal{PC}_{[0,\infty)}$. If $y(t) = O[x(t)]$ and $x(t) = O[y(t)]$ then x and y are said to be equivalent and denoted by $x(t) \sim y(t)$.

The definitions above are found to have a wider application when the class of functions is limited to $\mathcal{M}$.

Definition B.4 Let $x, y \in \mathcal{PC}_{[0,\infty)}$. x and y are said to grow at the same rate if $x_s(t) \sim y_s(t)$.

It follows directly from the definitions that two uniformly bounded signals grow at the same rate. The remarks following illustrate some of the properties of unbounded signals.

Remarks

1. $x(t) \sim y(t) \Rightarrow x_s(t) \sim y_s(t)$ but not conversely. For example,

$$\sup_{\tau\leq t} |\exp(\tau)\sin\,\tau| \sim \sup_{\tau\leq t} |\exp(\tau)\cos\,\tau|$$

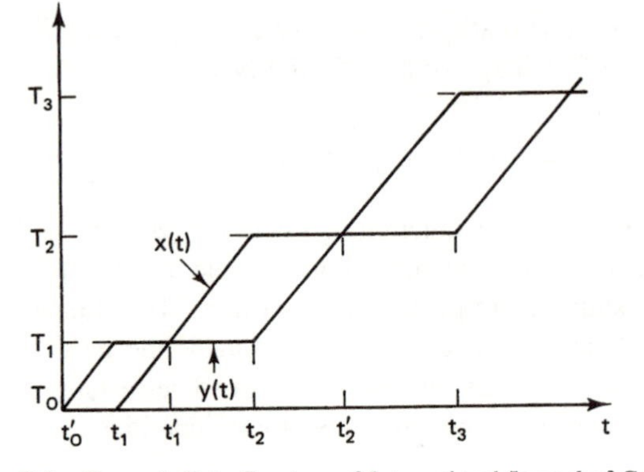

Figure B.1 Example B.1. Courtesy of International Journal of Control.

but $\exp(t)\sin t \not\sim \exp(t)\cos t$.

2. If x, y, and $z \in \mathcal{PC}_{[0,\infty)}$, $x_s(t) = O[z_s(t)]$, $z_s(t) = O[y_s(t)]$, and $x_s(t) n y_s(t)$, then $x_s(t) \sim y_s(t) \sim z_s(t)$.

3. Let x and $y \in \mathcal{M}$. Then $x(t) = O[y(t)] \Rightarrow y(t) \neq o[x(t)]$. However, $x(t)$ cannot be classified as either $x(t) \sim y(t)$ or $x(t) = o[y(t)]$. Example B.1 illustrates this.

Example B.1

Let $\{t_i\}$, $\{t_i'\}$, and $\{T_i\}$ be three unbounded sequences in $\mathbb{R}^+$ such that $t_{i-1}' < t_i < t_i'$, $t_i = [t_{i-1} + t_i']/2$, and $T_{i+1} = T_i + (t_{i+1} - t_i')$ with $T_0 = 0$. Then if

$$\begin{aligned} x(t) &= t - (t_i' - T_i), & y(t) &= T_i & t_i' \le t \le t_{i+1} \\ x(t) &= T_{i+1}, & y(t) &= t - (t_{i+1} - T_i) & t_{i+1} \le t \le t_{i+1}' \end{aligned}$$

by letting $T_{i+1}/T_i \to \infty$ as $i \to \infty$, we have $x(\cdot)/y(\cdot) \to \infty$ on $\{t_i\}$, and $x(\cdot)/y(\cdot) = 1$ on $\{t_i'\}$ so that $y(t)$ is neither equivalent to $x(t)$ nor $o[x(t)]$ [See Fig. B.1].

According to Lemma B.1, order relations derived for x can be used to derive order relations for x_s.

Lemma B.1 Let $x(\cdot) \in \mathcal{PC}_{[0,\infty)}$ and $y(\cdot) \in \mathcal{M}$ where $y(t)$ grows in an unbounded fashion. Then

(i) $x(t) = O[y(t)] \Leftrightarrow x_s(t) = O[y(t)]$.

(ii) $x(t) = o[y(t)] \Leftrightarrow x_s(t) = o[y(t)]$.

In Lemma B.1 (i), but not (ii) holds if $y(\cdot)$ is bounded. In such a case if $x(t) = o[y_s(t)]$, then $\lim_{t\to\infty} x(t) = 0$. Hence, $x_s(t) \neq o[y_s(t)]$ except when $x(t) \equiv 0$. The following corollary can be concluded simply from Lemma B.1 and finds frequent use in the stability analysis of adaptive systems.

Corollary B.1 If $x(t) = o[x_s(t)]$ then $\lim_{t\to\infty} x(t) = 0$. If $x(t) = y(t) + o[x_s(t)]$ where $y(t)$ is uniformly bounded, then $x(t)$ is uniformly bounded.

Thus far all the definitions and results pertain to scalar functions of time. The lemmas following are useful when vector-valued functions are involved and conclusions regarding the growth rates of sub-vectors need to be drawn from that of the entire vector.

Lemma B.2 Let

$$x^T \triangleq [x_1^T,\ x_2^T]$$

where $x : \mathbb{R}^+ \to \mathbb{R}^n$, $x_1 : \mathbb{R}^+ \to \mathbb{R}^{n_1}$, and $x_2 : \mathbb{R}^+ \to \mathbb{R}^{n_2}$, $n_1 + n_2 = n$, and let $x(t)$ grow in an unbounded fashion. Then

1. $\|x_1(t)\|_s \sim \|x_2(t)\|_s \Rightarrow \|x_1(t)\|_s \sim \|x(t)\|_s$
2. $\|x_1(t)\|_s = O[\|x_2(t)\|_s] \Leftrightarrow \|x_2(t)\|_s \sim \|x(t)\|_s$
3. $\|x_1(t)\|_s = o[\|x_2(t)\|_s] \Rightarrow \|x_2(t)\|_s \sim \|x(t)\|_s$

In general, if a vector $x(t)$ grows with time in an unbounded fashion, it cannot be concluded that there exists at least one component $x_i(t)$ such that $\sup_{\tau\leq t}\|x(\tau)\| \sim \sup_{\tau\leq t}|x_i(\tau)|$. The following example illustrates this.

Example B.2

Let $x : \mathbb{R}^+ \to \mathbb{R}^2$ and $x^T = [x_1, x_2]$. Let x_1 and x_2 satisfy the differential equations

$$\begin{aligned}
\dot{x}_1(t) &= \alpha_1(t)x_1(t) \qquad x_1(0) = 1\\
\dot{x}_2(t) &= \alpha_2(t)x_2(t) \qquad x_2(0) = 1.
\end{aligned}$$

Let

$$\begin{aligned}
\alpha_1(t) = 1, \quad \alpha_2(t) &= 0 \qquad t_i' \leq t \leq t_i\\
\alpha_1(t) = 0, \quad \alpha_2(t) &= 1 \qquad t_i \leq t \leq t_{i+1}'
\end{aligned}$$

where $\{t_i\}$, $\{t_i'\}$, and $\{T_i\}$ are unbounded sequences in $\mathbb{R}^+$ such that $t_i' < t_i < t_{i+1}'$, $T_0 = T_1 = 0$,

$$\begin{aligned}
T_{2i} &= (t_i - t_i') + T_{2i-2}\\
T_{2i+1} &= (t_{i+1}' - t_i) + T_{2i-1}
\end{aligned}$$

where $i = 1, 2, \ldots$. Then $x_1(t)/x_2(t) \to \infty$ on $\{t_i\}$ and 0 on $\{t_i'\}$ so that $\|x(t)\|$ is neither equivalent to $x_1(t)$ nor to $x_2(t)$.

C

Properties of the Adaptive System with Bounded Parameters

An important step in the analysis of adaptive systems is the demonstration of the boundedness of the parameter vector $\theta(t)$. This implies that the overall system can be considered as a linear time-varying system with bounded parameters. As a consequence, all the signals in the system belong to $\mathcal{PC}_{[0,\infty)}$.

In the adaptive control problem described in Chapter 5, some conclusions regarding the relations between various signals can be drawn when the parameter vector θ is bounded. We deal with these conclusions in this appendix. These, in turn, are used in Chapter 5 to show that all the signals remain bounded and $e_1(t) \to 0$ as $t \to \infty$ when appropriate adaptive laws are used to adjust $\theta(t)$.

The equations that describe the adaptive system consisting of an nth order plant together with the controller described in Chapter 5 are given below. It is assumed that the relative degree n^* is greater than one and an augmented error is generated using Method 1.

$$\begin{aligned} \dot{x}_p &= A_p x_p + b_p u; \qquad y_p = h_p^T x_p \qquad \text{(Plant)} \\ W_p(s) &= k_p \frac{Z_p(s)}{R_p(s)} = h_p^T (sI - A_p)^{-1} b_p \end{aligned}$$

$$\begin{aligned} \dot{x}_m &= A_m x_m + b_m r; \qquad y_m = h_m^T x_m \qquad \text{(Model)} \\ W_m(s) &= k_m \frac{Z_m(s)}{R_m(s)} = h_m^T (sI - A_m)^{-1} b_m \end{aligned}$$

$$\begin{aligned} \dot{\omega}_1 &= \Lambda \omega_1 + \ell u \\ \dot{\omega}_2 &= \Lambda \omega_2 + \ell y_p \qquad \text{(Controller)} \\ u &= k(t) r + \theta_1^T(t)\omega_1 + \theta_0(t) y_p + \theta_2^T(t)\omega_2 \end{aligned}$$

$$\omega \triangleq \begin{bmatrix} r \\ \omega_1 \\ y_p \\ \omega_2 \end{bmatrix}; \qquad \overline{\omega} \triangleq \begin{bmatrix} \omega_1 \\ y_p \\ \omega_2 \end{bmatrix}; \quad \underline{\omega} \triangleq \begin{bmatrix} \omega_1 \\ \omega_2 \end{bmatrix}$$

$$\theta \triangleq \begin{bmatrix} k \\ \theta_1 \\ \theta_0 \\ \theta_2 \end{bmatrix}; \qquad \overline{\theta} \triangleq \begin{bmatrix} \theta_1 \\ \theta_0 \\ \theta_2 \end{bmatrix}$$

$$\begin{aligned} \epsilon_1 &= y_p - y_m + k_1 e_2 \\ \zeta &= W_m(s) I \omega; \qquad \overline{\zeta} = W_m(s) I \overline{\omega} \qquad \text{(Augmented Error)} \\ e_2 &= \theta^T \zeta - W_m(s) \theta^T \omega \end{aligned}$$

where $A_p, A_m \in \mathbb{R}^{n\times n}$, $b_p, h_p, b_m, h_m \in \mathbb{R}^n$, $\Lambda \in \mathbb{R}^{(n-1)\times(n-1)}, \ell \in \mathbb{R}^{n-1}$. Let the parameter $\vartheta = [\theta^T, k_1]$ be adjusted such that it is bounded for all $t \geq t_0$.

The signals of interest in the stability analysis of the adaptive system are the state of the plant x_p, the state of the model x_m, the reference input r, the state of the controller $\underline{\omega}$, and the signal ζ in the auxiliary network used for generating the augmented error. The state of the overall system (excluding the model) is z, and ξ denotes an equivalent state of the system which is more convenient for purposes of analysis. Theorem C.1 states succinctly the relations that exist between these signals.

Theorem C.1 For the system described above, the following properties hold if the parameter vector θ is bounded:

$$\text{(i)} \qquad x_p(t) = C\overline{\omega}(t) + \delta_1(t) \tag{C.1}$$

$$\|x_p\| \leq \alpha_1 \|\overline{\omega}\| + \alpha_2 \tag{C.2}$$

$$\text{(ii)} \qquad \|\dot{\overline{\omega}}\| \leq \beta_1 \|\overline{\omega}\| + \beta_2 \tag{C.3}$$

$$
\begin{aligned}
\text{(iii)} \qquad \dot{\overline{\zeta}} &= A_1\overline{\zeta} + b_1 y_p + \delta_2(t) && \text{(C.4)}\\
\|\dot{\xi}\| &\leq \gamma_1\|\xi\| + \gamma_2 \qquad \xi = [\overline{\omega}^T, \overline{\zeta}^T] && \text{(C.5)}\\
\text{(iv)} \qquad \|z\| &\leq \rho_1\|\xi\| + \rho_2 \qquad z = [x_p^T, \xi^T]^T && \text{(C.6)}
\end{aligned}
$$

where $C \in \mathbb{R}^{n\times 2n-1}$, $\delta_1(t)$ and $\delta_2(t)$ are decaying exponentials, $\alpha_i, \beta_i, \gamma_i$, and $\rho_i \in \mathbb{R}^+$, $i = 1, 2$, $A_1 \in \mathbb{R}^{(2n-1)\times(2n-1)}$, and $b_1 \in \mathbb{R}^{(2n-1)}$.

By part (i) of the theorem, $\overline{\omega}(t)$ can be considered as the equivalent state of the closed-loop system containing the plant and the controller. In part (ii), it is shown that $\overline{\omega}(t)$ can grow at most exponentially. Parts (iii) and (iv) are used to show that ξ is the equivalent state of an overall system consisting of the plant, the controller, and the auxiliary network.

Proof. (i) Using representation 1 (refer to Chapter 4), the plant can be described by the equations

$$
\begin{aligned}
\dot{x}_1 &= \Lambda x_1 + \ell u\\
\dot{x}_2 &= \Lambda x_2 + \ell y_p && \text{(C.7)}\\
\dot{y} &= -\lambda y - p^T X; \quad y_p = y
\end{aligned}
$$

where $p = [c_0, \overline{c}^T, d_0, \overline{d}^T]^T$, $\lambda > 0$, and $X = [u, x_1^T, y, x_2^T]^T$. Hence,

$$x_i(t) = \omega_i(t) + \exp(\Lambda(t - t_0))\,[x_i(t_0) - \omega_i(t_0)] \qquad i = 1, 2.$$

Let $x_{np} = [x_1^T, y, x_2^T]^T$. Hence,

$$x_{np}(t) = \overline{\omega}(t) + \delta_0(t)$$

where $\delta_0(t)$ is a vector of decaying exponentials. Since x_{np} is a nonminimal state corresponding to the plant, Eqs. (C.1) and (C.2) follow. These imply that $\overline{\omega}$ can be considered as the equivalent state of the plant.

(ii) The controller equations can be rewritten as

$$
\begin{bmatrix} \dot{\omega}_1 \\ \dot{y}_p \\ \dot{\omega}_2 \end{bmatrix} = \begin{bmatrix} \Lambda + \ell\theta_1^T(t) & \ell\theta_0(t) & \ell\theta_2^T(t) \\ h_p^T b_p \theta_1^T(t) & h_p^T b_p \theta_0(t) & h_p^T b_p \theta_2(t)^T \\ 0 & \ell & \Lambda \end{bmatrix} \begin{bmatrix} \omega_1 \\ y_p \\ \omega_2 \end{bmatrix} \qquad \text{(C.8)}
$$

$$
+ \begin{bmatrix} 0 \\ h_p^T A_p \\ 0 \end{bmatrix} x_p + \begin{bmatrix} \ell k(t) \\ h_p^T b_p k(t) \\ 0 \end{bmatrix} r.
$$

From Eq. (C.1), it follows that

$$\dot{\overline{\omega}} = A_0(t)\overline{\omega} + b_0(t) r + \delta_3(t) \qquad \text{(C.9)}$$

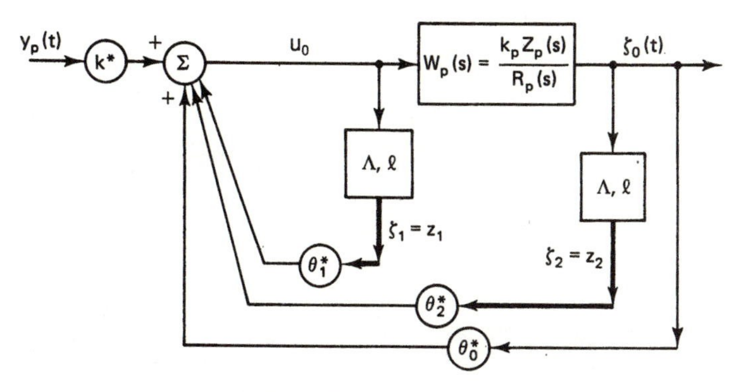

Figure C.1 Eq. (C.10).

where $A_0(t) \in \mathbb{R}^{(2n-1)\times(2n-1)}$, $b_0(t) \in \mathbb{R}^{2n-1}$, and $\delta_3(t)$ is due to decaying exponentials. Since θ is bounded, the elements of $A_0(t)$ and $b_0(t)$ are bounded, and the inequality in Eq. (C.3) follows. This implies that $\|\overline{\omega}(t)\|$ can grow at most exponentially.

(iii) Let $\zeta = [\zeta_r, \zeta_1^T, \zeta_0, \zeta_2^T]^T$ so that $\zeta_r = W_m(s)r$, $\zeta_1 = W_m(s)I\omega_1$, $\zeta_0 = W_m(s)y_p$, and $\zeta_2 = W_m(s)I\omega_2$. Since $W_m(s)$ is equal to the transfer function of the overall adaptive system when $\theta(t) = \theta^*$, ζ_0 can be expressed as the output of a differential equation (Fig. C.1)

$$\begin{aligned}
\zeta_0 &= W_p(s)u_0 \\
\dot{z}_1 &= \Lambda z_1 + \ell u_0 \\
\dot{z}_2 &= \Lambda z_2 + \ell \zeta_0 \\
u_0 &= k^* y_p + \theta_1^{*T} z_1 + \theta_0^* \zeta_0 + \theta_2^{*T} z_2.
\end{aligned} \tag{C.10}$$

It follows that $z_1 = W_p^{-1}(s)z_2$ and $z_2 = (sI - \Lambda)^{-1}\ell\zeta_0$. Therefore, neglecting initial conditions, we have $z_1 = \zeta_1$, and $z_2 = \zeta_2$. Equation (C.4) can be derived along the same lines as Eq. (C.9), and Eq. (C.5) follows from the definition of ξ and Eqs. (C.9), (C.4), and (C.1). Hence ξ is an equivalent state of the complete system consisting of the plant, the controller, and the dynamical system that generates the augmented error.

(iv) z represents the state of the overall system and includes the state x_p besides ξ. However, since $\xi^T = [\overline{\omega}^T, \overline{\zeta}^T]$, and $\overline{\omega}$ is equivalent to x_p from (i), it follows that ξ is equivalent to z. Hence, the inequality in Eq. (C.6) follows from Eq. (C.2), and the definitions of z and ξ.

D

Proof of Boundedness in the Presence of Bounded Disturbances

Given a plant whose input and output are related by the equation

$$\dot{x}_p = A_p x_p + b_p u + d_p \nu_1; \qquad y_p = h_p^T x_p + \nu_2$$

where ν_1 and ν_2 are bounded and the latter is differentiable, a reference model, controller, and augmented error as in Section 8.4, the error equation can be stated as

$$\epsilon_1 = \phi^T \zeta + \nu$$

where ν is an equivalent bounded disturbance, ϕ is the parameter error and ϵ_1 is the augmented error. If the adaptive law used is such that the following conditions (i)-(iii) are satisfied, then all solutions of the adaptive system are bounded.

(i) ϕ and $\dot{\phi}$ are bounded,

(ii) $\displaystyle\int_{t_1}^{t_2} \frac{\epsilon_1^2}{1+\xi^T\xi} dt \leq c_0 + c_1 \int_{t_1}^{t_2} \frac{1}{\sqrt{1+\xi^T\xi}} dt$, and

(iii) $\displaystyle\|\dot{\phi}\| < \frac{c_2|\epsilon_1|}{\sqrt{1+\xi^T\xi}} + \frac{c_3}{1+\xi^T\xi}$

where c_i, $i = 1, 2, 3$, are constants in $\mathbb{R}^+$ and ξ is as defined in Section 8.4 and is the equivalent state of the system.

Since both the σ-modification and e_1-modification schemes satisfy these conditions, boundedness of the solutions follows. Scheme (b), employing the bound on θ^* satisfies condition (i), while (ii) and (iii) are modified as shown below:

(ii′) $$\int_{t_1}^{t_2}\left[\frac{\epsilon_1^2}{1+\xi^T\xi}+\phi^T\theta f(\theta)\right]dt \;\leq\; c_0+c_1\int_{t_1}^{t_2}\frac{1}{\sqrt{1+\xi^T\xi}}dt$$

(iii′) $$\|\dot{\phi}\| < \frac{c_2|\epsilon_1|}{\sqrt{1+\xi^T\xi}}+c_3 f(\theta).$$

However, since $\phi^T\theta f(\theta) > 0$, the same results can be derived.

Proof. The proof is given in three steps. In the first step, the behavior of $|\phi^T\zeta|/(1+\xi^T\xi)^{1/2}$ over an interval $[t_i, t_i+a_i]$ is characterized. In the second step, it is shown that $|\phi^T\omega|/\|z\|$ behaves essentially in the same manner where z is the state of the system. In step 3, by choosing a_i sufficiently large, it is shown by contradiction that conditions (i)-(iii) imply that all solutions of the adaptive system are bounded.

Step 1. Since ϕ is bounded, all the signals in the system can grow at most exponentially. Hence, if the solutions are unbounded, unbounded sequences $\{t_i\}$, and $\{a_i\}$ exist such that

$$\|\xi(t)\| \;\geq\; a_i \qquad \forall t\in[t_i, t_i+a_i]. \tag{D.1}$$

If $\kappa \triangleq |\epsilon_1|/\sqrt{1+\xi^T\xi}$, it follows that

$$|\dot{\kappa}| \;\leq\; c_4{}^1$$

since ϕ, $\dot{\phi}$, and $\dot{\nu}$ are bounded, and from Appendix C, $\|\dot{\xi}\| \leq \gamma_1\|\xi\|+\gamma_2$, $\gamma_1,\gamma_2\in\mathbb{R}^+$.

From (ii) we have

$$\int_{t_i}^{t_i+a_i}\kappa^2(t)\,dt \;\leq\; c_0+c_1\int_{t_i}^{t_i+a_i}\frac{1}{\sqrt{1+\xi^T\xi}}dt.$$

Since, over the interval $[t_i, t_i+a_i]$, $\|\xi(t)\|\geq a_i$, it follows that

$$\int_{t_i}^{t_i+a_i}\kappa^2(t)\,dt \;\leq\; c_5<\infty. \tag{D.2}$$

Let

$$T_{i1} \;\triangleq\; \{t|\;|\kappa|\geq\epsilon,\; t\in[t_i, t_i+a_i]\}$$

$$T_{i2} \;\triangleq\; \{t|\;|\kappa|<\epsilon,\; t\in[t_i, t_i+a_i]\}$$

[1] In what follows, c_i's and k_i's are positive constants.

where ϵ is an arbitrary positive constant. From Eq. (D.2), it follows that

$$\mu\,(T_{i1}) \leq \frac{c_5}{\epsilon^2}$$

where $\mu(S)$ is the Lebesgue measure of a set S. Hence,

$$\mu\,(T_{i2}) \geq a_i - \frac{c_5}{\epsilon^2}.$$

From the definition of the relevant variables, it follows that

$$a_i \geq \frac{\nu_0}{(c_6 - 1)\epsilon} > 0 \text{ and } |\kappa| < \epsilon \Rightarrow \frac{|\phi^T \zeta|}{(1 + \xi^T \xi)^{1/2}} < c_6 \epsilon.$$

Step 2. It is shown below that

$$\frac{|\phi^T \zeta|}{(1 + \xi^T \xi)^{1/2}} < c_6 \epsilon \Rightarrow \frac{|\phi^T \omega|}{\|z\|} < c_7 \epsilon.$$

Without loss of generality, in what follows, the model transfer function can be assumed to be $1/R_m(s)$, where

$$R_m(s) \;=\; [1,\; s, \ldots, s^{n^*}][a_0,\; a_1, \ldots, a_{n^*-1}, 1]^T.$$

Denoting $\zeta^{(i)}$ as the ith derivative of ζ, the vector ω can be written as

$$\omega \;=\; \sum_{i=0}^{n^*} a_i \zeta^{(i)}$$

where $a_{n^*} = 1$. We prove by induction that

$$\frac{|\phi^T \zeta^{(i)}|}{(1 + \xi^T \xi)^{1/2}} \;<\; k_i \epsilon \qquad i = 0, 1, \ldots, n^* \text{ when } \; t \in T_{i2}. \tag{D.3}$$

This is true for $i = 0$. We obtain the following identity by partial integration:

$$\int_t^{t+\Delta t} \frac{\phi^T \zeta^{(i+1)}}{(1 + \xi^T \xi)^{1/2}} dt \;=\; \left.\frac{\phi^T \zeta^{(i)}}{(1 + \xi^T \xi)^{1/2}}\right|_t^{t+\Delta t} - \int_t^{t+\Delta t} \frac{\dot{\phi}^T \zeta^{(i)}}{(1 + \xi^T \xi)^{1/2}} dt$$
$$+ \int_t^{t+\Delta t} \frac{\phi^T \zeta^{(i+1)}}{(1 + \xi^T \xi)} \cdot \frac{\xi^T \dot{\xi}}{(1 + \xi^T \xi)^{1/2}} dt$$

From the inequality in Eq. (C.5) and Condition (iii), it follows that Eq. (D.3) is true for $i + 1$ if it is true for i and $a_i \geq \sqrt{c_3/\epsilon}$. Therefore,

$$\frac{|\phi^T \omega|}{(1 + \xi^T \xi)^{1/2}} \;<\; c_8 \epsilon \qquad \text{when } t \in T_{i2} \text{ and } a_i \geq \max\left(\frac{\nu_0}{(c_6 - 1)\epsilon}, \sqrt{\frac{c_3}{\epsilon}}\right).$$

In addition, since $\|z\| \geq \|\xi\|$, we can also show that

$$\frac{|\phi^T \omega|}{\|z\|} < c_7 \epsilon \qquad \text{when } t \in T_{i2} \text{ and } a_i \geq \max\left(\frac{\nu_0}{(c_6-1)\epsilon}, \sqrt{\frac{c_3}{\epsilon}}, c_9\right).$$

Step 3. The overall adaptive system can be described by the differential equation

$$\dot{z} = A_2 z + b_2\left(\phi^T\omega + r\right) + \overline{\nu}_1.$$

where A_2 is asymptotically stable and $\overline{\nu}_1$ is a bounded vector disturbance. A symmetric positive definite matrix P_2 exists such that $A_2^T P_2 + P_2 A_2 = -Q$ where $Q > 0$. If $W = z^T P_2 z$, the time-derivative of W can be expressed as

$$\begin{aligned}\dot{W} &= -z^T Q z + 2z^T P_2 b_2\left[\phi^T\omega + r\right] + 2z^T P_2 \overline{\nu}_1 \\ &\leq -\|z\|^2\left\{\lambda_2 - 2\lambda_1\|b_2\|\left[\frac{|\phi^T\omega|}{\|z\|} + \frac{|r| + |\overline{\nu}_1|/\|b_2\|}{\|z\|}\right]\right\}.\end{aligned}$$

where λ_1 and λ_2 are the maximum eigenvalue of P_2 and minimum eigenvalue of Q respectively. Hence, if

$$\overline{\epsilon} = \frac{\lambda_2 - k\lambda_1}{4\lambda_1\|b_2\|}, \qquad \frac{|\phi^T\omega|}{\|z\|} \leq \overline{\epsilon} \text{ and } \|z\| \geq \frac{\|b_2\|r_0 + \overline{\nu}_0}{\|b_2\|\overline{\epsilon}}, \tag{D.4}$$

where $|r(t)| \leq r_0$, $|\overline{\nu}_1(t)| \leq \overline{\nu}_0$ and $0 < k < \lambda_1/\lambda_2$, then

$$\dot{W} \leq -kW.$$

If the inequality in Eq. (D.4) is not satisfied, since ϕ is bounded, it is known that the solutions can grow at most exponentially or equivalently

$$\dot{W} \leq \lambda W \qquad \text{for some positive constant } \lambda.$$

Hence, by choosing

$$\epsilon = \frac{\overline{\epsilon}}{c_7}, \qquad \text{and} \qquad a_i > \max\left(\frac{\nu_0}{(c_6-1)\epsilon}, \sqrt{\frac{c_3}{\epsilon}}, c_9, \frac{\|b_2\|r_0 + \overline{\nu}_0}{\|b_2\|\overline{\epsilon}}\right)$$

we obtain that

$$\dot{W} \leq \begin{cases} \lambda W & \text{when } t \in T_{i1} \\ -kW & \text{when } t \in T_{i2} \end{cases}$$

over the interval $[t_i, t_i + a_i]$. Since the Lebesgue measure $\mu(T_{i1})$ is bounded, the measure $\mu(T_{i2})$ of the set over which $W(z)$ decays can be increased by increasing a_i. This leads to the inequality

$$W(t_i + a_i) \leq \exp\left\{-\left(ka_i - \frac{c_5(\lambda + k)}{\epsilon^2}\right)\right\} W(t_i). \tag{D.5}$$

If

$$a_i > \max\left(\frac{\nu_0}{(c_6-1)\epsilon}, \sqrt{\frac{c_3}{\epsilon}}, c_9, \frac{\|b_2\|r_0+\overline{\nu}_0}{\|b_2\|\overline{\epsilon}}, \frac{c_{10}+c_5(\lambda+k)}{\epsilon^2}\right)$$

Eq. (D.5) can be rewritten as

$$W(t_i+a_i) \leq \exp\left\{-\frac{c_{10}}{\epsilon^2}\right\} W(t_i)$$

and hence contradicts Eq. (D.1). Hence, all solutions are bounded.

E

Some Facts About Transfer Matrices

Rings and Fields. A field is a set of elements that is closed under addition, subtraction, multiplication, and division (except by zero) and the elements satisfy associative, distributive, and commutative rules. A commutative ring is a field that is not closed under division, and an integral domain is a commutative ring possessing a multiplicative identity 1 and no zero divisors ($ab = 0$ implies that $a = 0$ or $b = 0$). An element in a ring is a unit u if it divides the identity 1 and it is a prime if it has no factors besides units and itself. If $v : D \to \mathbb{R}^+$ and $a, b \in D$ where D is an integral domain so that $a = bq + r$ where q (quotient) and r (residue) $\in D$, with $v(a) < v(b)$ or $r = 0$, then D is a Euclidean domain. A matrix defined over D whose inverse belongs to D is termed unimodular. The determinant of a unimodular matrix U is a unit in D. Two matrices A and B are called right associates if, and only if, a unimodular matrix U exists such that $A = BU$.

Let $\mathbb{R}[s], \mathbb{R}(s)$, and $\mathbb{R}_p(s)$ denote respectively, the ring of polynomials, the field of rational functions, and the ring of proper rational functions of a single variable with coefficients in $\mathbb{R}$. Matrices and vectors defined over $\mathbb{R}$, $\mathbb{R}[s]$, and $\mathbb{R}(s)$ are denoted by superscripting the set symbol. Thus $\mathbb{R}^{p \times m}[s]$ denotes a polynomial matrix of dimension $p \times m$.

The Ring of Polynomials. The set $\mathbb{R}[s]$ has been studied extensively in the context of multivariable systems theory. A unimodular matrix in $\mathbb{R}^{m \times m}[s]$ is defined to be a matrix whose determinant is a nonzero scalar in $\mathbb{R}$. Hence, the inverse of a unimodular matrix also belongs to $\mathbb{R}^{m \times m}[s]$. The degree of a polynomial matrix $P(s)$

is equal to the degree of the polynomial element of the highest degree in $P(s)$ and is denoted by $\partial[P(s)]$. The degree of the jth column vector of $P(s)$ is then obtained by applying the definition of degree to the jth column of $P(s)$ and is denoted as $\partial_{cj}[P(s)]$. $[P]_c$ denotes a matrix in $\mathbb{R}^{p\times m}$ consisting of the coefficients of the highest degree s terms in each column. $P(s)$ is said to be column proper if, and only if, $[P]_c$ is of full rank. By elementary column operations, any matrix in $\mathbb{R}^{m\times m}[s]$ can be column reduced. Corresponding statements can also be made in terms of the rows of the matrix $P(s)$ instead of its columns. Using similar elementary operations, $P(s)$ can be transformed to a nonsingular matrix with a lower- or upper-triangular structure. A matrix $P(s)$ defined over $\mathbb{R}^{m\times m}[s]$ is nonsingular if $\det[P(s)] \neq 0$ for all s in the complex plane except for a finite set of points.

If $A(s), B(s)$, and $C(s)$ are three matrices in $\mathbb{R}^{m\times m}[s]$, and if $A(s) = B(s)C(s)$, then $C(s)$ is called the right divisor of $A(s)$. The greatest common right divisor (g.c.r.d.) of two matrices $A(s)$ and $B(s)$ is a polynomial matrix that is right divisible by every right divisor of $A(s)$ and $B(s)$. The g.c.r.d. is unique upto multiplication by a unimodular matrix. If the g.c.r.d. of two matrices $A(s)$ and $B(s)$ is unimodular, then $A(s)$ and $B(s)$ are said to be coprime.

If $A(s)$ and $B(s)$ are $m \times m$ polynomial matrices and $B(s)$ is nonsingular, then $A(s)$ and $B(s)$ are right coprime if, and only if, any one of the following conditions holds:

(i) For every $s \in \mathbb{C}$, or for every root of the determinant of $B(s)$, the $2m \times m$ matrix $[B^T(s), A^T(s)]^T$ has rank m in $\mathbb{C}$.

(ii) Polynomial matrices $C(s)$ and $D(s) \in \mathbb{R}^{m\times m}[s]$ exist such that $A(s)C(s) + B(s)D(s) = I$.

(iii) There exist no polynomial matrices $X(s)$ and $Y(s)$ such that $A(s)X(s) = B(s)Y(s)$.

The Ring of Proper Rational Matrices. The set $\mathbb{R}_p(s)$ of proper rational transfer functions is a Euclidean domain. All transfer functions with zero relative degree are units. Any element $r(s) \in \mathbb{R}_p(s)$ can be expressed as

$$r(s) = \frac{1}{[\pi(s)]^{n^*}} \cdot u(s)$$

where $\pi(s)$ is a monic polynomial of degree one, $u(s)$ is a unit in $\mathbb{R}_p(s)$, and n^* is the relative degree of $r(s)$.

A matrix $T(s) \in \mathbb{R}^{p\times m}(s)$ is proper if, and only if, $\lim_{s\to\infty} T(s)$ is a nonzero matrix in $\mathbb{R}^{p\times m}$. If $T(s)$ is a matrix in $\mathbb{R}_p^{m\times m}(s)$, it can be factored as a right polynomial matrix fraction $A(s)B^{-1}(s)$ where $A(s)$ and $B(s)$ are in $\mathbb{R}^{m\times m}[s]$ with $\partial_{cj}[A(s)] \leq \partial_{cj}[B(s)]$. Further $B(s)$ can be chosen to be column reduced. The roots of the polynomials $\det[A(s)]$ and $\det[B(s)]$ are the zeros and poles of $T(s)$ respectively. Similarly, $T(s)$ can also be factored as a left polynomial matrix fraction $C^{-1}(s)D(s)$ with $\partial_{ri}[D(s)] \leq \partial_{ri}[C(s)]$.

Index

Errata List

p. 51, first line, change $f: R^+ \to R^n$ to $f: R^n \times R^+ \to R^n$.
p. 51, def. 2.2, change "... there exists a number" to "there exists a finite number".
p. 53, after eqn (2.10), insert $\forall t \geq t_0$ after "Since $f(o,t) = 0$".
p. 64, def. 2.8 part (b), insert $\lim_{|w| \to \infty} \mathrm{Re} H(jw) > 0$ after $n^* = -1$.
p. 95, bottom of page, change $\|x(t)\| \exp(-at) = 0$ to $\lim_{t \to \infty} \|x(t)\| \exp(-at) = 0$.
p. 117, second paragraph, change "Shackloth" to "Shackcloth" (twice).
p. 135, third line, change Ψ to $\Psi(t)$.
p. 158, fig. 4.8, add "(a)" and "(b)" below figures.
pp. 214 – 215, eqns (5.61) to (5.65), renumber to be (5.61a) to (5.61e).
p. 215, starting at "From Appendix C", replace all text. See revision following this list.
p. 219, add to the list of eqns, $d = \left[1 + \zeta_1^2 + \zeta_2^2 + \zeta_3^2\right]$.
p. 228, align equations.
p. 272, Ref. 18 end of the title of paper, add period: "Persistent ... .".
p. 456, fig. 11.17, interchange the titles of the two figures. The second figure should come first.
p. 461, fig. 11.19 second figure, change $\overline{ADP}$ to $\overline{AOP}$.

Revisions to Page 215

From Appendix C,

$$\left|\frac{d}{dt}\|\bar{w}(t)\|\right| = \|\dot{w}(t)\| \leq c_1\|w(t)\| + c_2, \qquad c_1, c_2 \varepsilon R^+ \tag{5.62}$$

that is, $\bar{w}(t)$ can be considered to be the effective state of the system. Since Λ is a stable matrix and Λ, l is controllable, there exists a positive constant α and a scalar signal $v(t)$ such that

$$v(t) = \frac{1}{s+\alpha}u(t) \tag{5.63}$$

and

$$w_1(t) = O\left[\sup_{\tau\leq t}|v(\tau)|\right]. \tag{5.64}$$

From (5.61a), (5.62), and (5.64) we obtain

$$|\dot{v}(t)| = O\left[\sup_{\tau \varepsilon t}|\bar{v}(t)|\right] \tag{5.65}$$

where $\bar{v} = \left[y_p v\right]$.

Since $y_p(t) = (s+\alpha)W_p(s)v(t)$, and $(s+\alpha)W_p(s)$ is a proper transfer function with zeros in C^-, from Corollary 2.7 and Equation (5.65), we have

$$v(t) = O\left[\sup_{\tau\leq t}|y_p(\tau)|\right]. \tag{5.66}$$

From equations (5.61a), (5.64), and (5.66) it follows that

$$\|\bar{w}(t)\| = O\left[\sup_{\tau\leq t}|y_p(\tau)|\right] \tag{5.67}$$

and from equations (5.61b) and (5.67) we have

$$|\dot{y}_p(t)| = O\left[\sup_{\tau\leq t}|y_p(\tau)|\right]. \tag{5.68}$$

Similarly, since $W_m(s)\bar{w}(t) = \bar{\xi}(\tau)$, from equation (5.62) it follows that

$$\|\bar{w}(t)\| = O\left[\sup_{\tau\leq t}\|\bar{\zeta}(\tau)\|\right]. \tag{5.69}$$

All the results in step 3 can therefore be summarized from equations (5.61a through e), (5.67) and (5.69) as

$$\sup_{\tau\leq t}|y_p(\tau)| \sim \sup_{\tau\leq t}\|w_2(\tau)\| \sim \sup_{\tau\leq t}\|\bar{w}(\tau)\| \sim \sup_{\tau\leq t}\|\bar{\zeta}(\tau)\|. \tag{5.70}$$